T0328617

THERMAL POWER PLANT

THERMAL POWER PLANT
Design and Operation

DIPAK K. SARKAR

AMSTERDAM • BOSTON • HEIDELBERG • LONDON
NEW YORK • OXFORD • PARIS • SAN DIEGO
SAN FRANCISCO • SINGAPORE • SYDNEY • TOKYO

Elsevier
Radarweg 29, PO Box 211, 1000 AE Amsterdam, Netherlands
225 Wyman Street, Waltham, MA 02451, USA
The Boulevard, Langford Lane, Kidlington, Oxford OX5 1GB, UK

ISBN: 978-0-12-801575-9

Library of Congress Cataloging-in-Publication Data
A catalog record for this book is available from the Library of Congress.

British Library Cataloguing-in-Publication Data
A catalogue record for this book is available from the British Library.

For information on all Elsevier publications
visit our website at http://store.elsevier.com/

Publisher: Joe Hayton
Acquisition Editor: Raquel Zanol
Editorial Project Manager: Mariana Kühl Leme
Production Project Manager: Anusha Sambamoorthy
Designer: Greg Harris

Dedicated to my parents

MANORAMA & SISIR KUMAR

CONTENTS

ACKNOWLEDGEMENTS

I am indebted to my colleagues and numerous academicians and friends from the industry for their encouragement for writing a book on Thermal Power Plant that deals with both design and operation aspects. I am particularly grateful to Mr. D. S. Mallick, Executive Director, M/S Development Consultants Private Limited (DCPL), who prompted me to write this book. I am also inspired by Dr. S. Dattagupta, Retired Professor, Power Engineering, Jadavpur University, and Dr. Amitava Datta, Professor, Power Engineering, Jadavpur University. Mr. Samiran Chakraboty, retired Chief Engineer, erstwhile M/S ACC Babcock Limited (ABL) currently M/S Alstom India Limited, impressed upon me the necessity of such book for both institution and industry.

I gracefully acknowledge the support extended by M/S Pinaki Nag and Manabendra Mitra Roy for drafting numerous drawings and figures of this book.

While preparing the manuscript I extensively searched through various electronic information sources that provide latest information and technology developed globally. One such source that needs special mention is Wikipedia, the free encyclopedia.

Myriad collection of information that is frequently visited by me in the course of current pursuit is the Data Center of M/S Development Consultants Private Limited (DCPL). I am indebted to the staff of this Data Center for their guidance.

My uncle Mr. Subir K Sarkar, himself a distinguished engineer and an author, extended all kinds of support.

I also received regular encouragement from my daughter, Purbita, and son-in-law, Sudip, for writing this book.

I am a slow learner in handling modern electronic gadgets, in exploring various software available in the internet and in formatting the manuscript to make it printable. My son, Krishanu, himself a promising mechanical engineer, guided me at every step and helped me in shaping this book in its present form.

My main strength was Anita, my wife, who steered me to write this book and kept the fort under her control all the time. I could engross with my writing without any interference as long as she had been around.

I am also indebted to the editorial team of Elsevier, Inc. for their guidance. Special mention is due to Raquel Zanol, Chelsea Johnston, Mariana Kuhl Leme, and Anusha Sambamoorthy in this regard.

Dipak K. Sarkar
July 03, 2015

PREFACE

This book on **THERMAL POWER PLANT – Design and Operation** deals with various aspects of a thermal power plant starting from fundamentals leading in depth to technical treatment. The book is aimed at providing new dimension to the subject and thrust of the book is focused on technology and design aspect with special treatment on plant operating practices and troubleshooting. Certain chapters also deal with numerical problems along with some worked out examples.

This book is prepared based on author's long association with thermal power plants for more than 40 years in design as well as in field engineering. During this long carrier, author has shared his knowledge and experience with students of various technical institutes as visiting faculty in the under graduate level and found that students are very attentive to his lectures because they found contents of these lectures would be beneficial to their professional carrier. The author also shared his experience with professional engineers under various training schemes, viz. graduate engineers training programme, refreshers training programme, operating personnel training programme. Against the back drop of feedback received during interaction with engineers at various forums, this book aims at sharing author's experience with much wider group of engineers.

The book is so developed as to be used as a core text by Mechanical/Power Engineering students at the undergraduate level and as a special paper on Heat & Work in the Postgraduate level. Diploma engineering students who intend to specialize in Thermal/Power Engineering can use this as a text book. This book can also be used as a reference book in Power Plant Training Institutes and in Graduate Engineers Training Programme on power plants. To Utility Operators and Design Engineers this book would be of immense help as reference book and to execute day-to-day activities. This book on one side addresses basic design aspects of thermal power plants to make it attractive to students pursuing mechanical/power engineering courses, on the other side it discusses how safely to run these plants so that utility operators find it handy as an useful guide book.

Design of a thermal power plant is based on the science of thermodynamics. Chapter 1 deals with treatment on fundamentals of thermodynamics, and comprises vapour cycles, their evolution, merits-demerits and their applications.

Chapter 2 discusses on steam generator covers boiling, circulation, classification, design of heat transfer areas. The intricacies of supercritical boiler are addressed separately.

Fuels and combustion are covered in Chapter 3 elaborating sources, availability, characteristics of fuel, combustion calculation, and design aspects of fuel handling.

From global trend the International Energy Agency (IEA) forecast that coal will remain a dominant fuel worldwide through 2035 for the purpose of power generation. Hence, pulverized coal fired boiler is discussed separately in Chapter 4 in view of large global coal reserves and its acceptance as major power producer in many countries.

Chapter 5 covers fluidized bed boiler that can burn lower grade of coal and other low grade combustible material for the generation of steam.

Steam turbine is the prime mover of steam power plant, and Chapter 6 deals with introduction, type, governing and speed control, losses, performance of steam turbine.

For quick start-up and peak load generation, gas turbine is ideal. Chapter 7 covers introduction, combustion system, and performance of gas turbine. This chapter also covers design, benefits, and use of heat recovery steam generator (HRSG) that facilitates improvement of the efficiency of a gas turbine power plant.

Chapter 8 deals with diesel power plant. Its design, equipment, and associated systems are addressed in this chapter.

A thermal power plant comprises miscellaneous systems comprising electric power supply and distribution systems, as well as solid (coal, ash), liquid (water, oil, acid, alkali) and gaseous (steam, air, flue gas, natural gas, hydrogen) matter supply and distribution systems. Description and purpose of these systems of a steam power plant are covered in Chapter 9.

Operation of modern large power plants is very complex in nature. It requires lot of activities to be executed simultaneously in order to ensure safety of equipment and personnel, as well as stable operation of the unit efficiently. Although discussion on whole gamut of such activities is difficult to be accommodated in this book, an attempt is made to address key aspects of these issues. Thus, Chapter 10 covers automatic control of key parameters of steam generator, steam turbine, and regenerative system. Chapter 11 is developed to address interlock & protection system of steam generator, steam turbine, gas turbine, diesel engine, and generator (alternator). While Chapter 12 covers start-up and shut down of steam generator, steam turbine, gas turbine, and diesel engine; their abnormal operating conditions are discussed in Chapter 13.

Air pollution control is addressed in Chapter 14. This chapter covers emission control of SPM, GHG, SO_x, and NO_x generated from a coal-fired steam generator.

More often than not design engineers get confused and search blindly which code and/or standard are to be followed to design a particular equipment or system. Chapter 15 presents purpose, benefits, and a list of commonly used codes and standards for design and operation of thermal power plants.

Fossil fuels are basically polluting in nature and are major producer of greenhouse gases causing global warming. Hence, to make these fuels environmentally acceptable cost intensive different types of treatment plants are essential to be installed.

To mitigate such complexities and investments, a viable alternative is to adopt renewable energy sources. These sources do not produce greenhouse gases and are free from emitting toxic wastes. So in continuation to aforementioned chapters, a brief discussion on Power from Renewable Energy is addressed under Appendix A. It is more so since renewable energy provides about 16% of global energy consumption.

It is of major concern that on one hand global supply of fossil fuels is depleting; on the other hand world's demand of electricity is rising sharply. As a result, utilities look at nuclear energy as a savior source of bulk power producer. This energy does not produce any air pollution, hence is an attractive alternative in the arena of electricity production even though the reactor area is a potential source of radioactivity and needs special safeguard devices. Appendix B discusses briefly Power from Nuclear Energy.

In accordance with current global practice, SI units have been used all through the book. Nevertheless, for the convenience of readers Conversion Factors from SI units to Metric System of units to Imperial & US System of units are addressed under Appendix C.

Author would earnestly welcome any suggestion for the improvement of the contents of this book both by supplementing with additional information in existing chapters and/or by addressing other areas in consonance with the present intention of this book. These suggestions would be acknowledged gratefully by the author.

Dipak K. Sarkar
March 26, 2015

LIST OF ACRONYMS/ABBREVIATIONS

a, abs	Absolute
A	Ash (content in coal)/Ampere
ABMA	American Boiler Manufacturers Association
A/C	Air/Cloth
AC	Alternating Current
ACF	Activated Carbon Filter
ACW	Auxiliary Cooling Water
ad	Air Dried
AFBC	Atmospheric Fluidized Bed Combustion
AFR	Air-Fuel Ratio
AH	Air Heater
AHS	Ash Handling System
ANSI	American National Standards Institute
APC	Auxiliary Power Consumption
API	American Petroleum Institute
APS	Automatic Plant Start-up & Shutdown System
ar	As Received
AS	Auxiliary Steam
ASME	American Society of Mechanical Engineers
ASTM	American Society for Testing & Materials
atm	Atmosphere
AVR	Automatic Voltage Regulator
AVT	All Volatile Treatment
AWWA	American Water Works Association
b	Bar
B	Billion
BA	Bottom Ash
B&W	The Babcock & Wilcox Company
BDC	Bottom Dead Center
BEI	British Electricity Institute
BF	Base Factor
BFBC	Bubbling Fluidized Bed Combustion
BFP	Boiler Feed Pump
BHRA	British Hydraulic Research Association
BIS	Bureau of Indian Standards
BMCR	Boiler Maximum Continuous Rating

BMS	Burner Management System
BOOS	Burner Out Of Service
BOP	Balance Of Plant
BP	Booster Pump
BPVC	Boiler and Pressure Vessel Code
BSI	British Standards Institution
Btu	British Thermal Unit
BWR	Boiling Water Reactor
C	Carbon/Celsius/Centegrade
Ca	Calcium
CA	Compressed Air
CAA	Clean Air Act, U.S.A.
CAAA	Clean Air Act Amendments
cc	Cubic Centimeter
CC	Combined Cycle
CCCW	Closed Cycle Cooling Water
CCGT	Combined Cycle Gas Turbine
CCPP	Combined Cycle Power Plant
CE	Combustion Engineering Inc./Collecting Electrode
CEA	Central Electricity Authority, India
CEGB	Central Electricity Generating Board
CEN	(Comité Européen de Normalisation)-European Committee for Standardization
CEP	Condensate Extraction Pump
CFBC	Circulating Fluidized Bed Combustion
cfm	Cubic Feet Per Minute
CHF	Critical Heat Flux
CHP	Combined Heat And Power
CHS	Coal Handling System
CI	Combustion Inspection of Gas Turbine
C.I.	Compression Ignition
cm	Centimeter
CO	Carbon Monoxide
CO_2	Carbon Dioxide
cP	Centipoise
CPCB	Central Pollution Control Board, India
CR	Compression Ratio
CRH	Cold Reheat
CSA	Canadian Standards Association
CV	Calorific Value/Control Valve

CW	Circulating (Condenser Cooling) Water
cwt	Hundredweight
D	Drain/Diameter
D, d	Day
DAF	Dry Ash Free
dB	Decibel
DAS	Data Acquisition System
DC	Direct Current
DCA	Drain Cooler Approach
DCS	Distributed Control System
DE	Discharge Electrode
deg	Degree
DIN	Deutsches Institut für Normung
DM	De-mineralized
dmmf	Dry Mineral Matter Free
DMW	De-mineralized Water
DNB	Departure From Nucleate Boiling
DO	Dissolved Oxygen
DSI	Duct Sorbent Injection
EA	Excess Air
ECS	Environmental Control Systems
EDI	Electrical De-ionization Unit
eff/EFF	Efficiency
EHS	Environmental, Health and Safety
EIA	Environmental Impact Assessment
emf	Electromotive Force
EMV	Effective Migration Velocity
EPA	Environmental Protection Agency, U.S.A
EPRI	Electric Power Research Institute, U.S.A
EPRS	Effective Projected Radiant Surface
ESI	Economizer Sorbent Injection
ESP	Electrostatic Precipitator
ESV	Emergency Stop Valve
EU	European Union
EX	Extraction
F	Fahrenheit
FA	Fly Ash
FAC	Flow Accelerated Corrosion

FBC	Fluidized Bed Combustion
FBR	Fast Breeder Reactor
FC	Fixed Carbon (in coal)
FD	Forced Draft
FEGT	Furnace Exit Gas Temperature
FGD	Flue Gas Desulfurization
FGR	Flue Gas Recirculation
FIG	Figure
FFH	Factored Fired Hours
FO	Furnace Oil
FSI	Furnace Sorbent Injection
ft	Foot/Feet
FW	Feed Water
FWH	Feed Water Heater
fpm	Feet Per Minute
g	Gram/Gauge/Acceleration Due to Gravity (1 kg.m/Ns^2)
G	Gallon/Giga
GB	Guojia Biaozhun, China National Standard
GCB	Generator Circuit Breaker
GCR	Gas Cooled Reactor
GCS	Gas Conditioning Skid
GCV	Gross Calorific Value
GE	General Electric Company
GGH	Gas to Gas Heater
GHG	Greenhouse Gas
GJ	Giga Joule
GLR	Generator Lock-out Relay
GOST	Gosudartsvennye Standarty, Russian National Standards
GPHR	Gross Plant Heat Rate
gpm	Gallons Per Minute
gr	Grain
GT	Gas Turbine/Generator Transformer
h	Hour
H	Hydrogen
H, h	Enthalpy
HAP	Hazardous Air Pollutants
HAZ	Heat Affected Zone
HCSD	High Concentration Slurry Disposal System
HEI	Heat Exchange Institute, U.S.A.

HFO	Heavy Fuel Oil
Hg	Mercury
HGI	Hardgrove Grindability Index
HGPI	Hot Gas Path Inspection
HHV	Higher Heating Value
HI	Hydraulic Institute, Inc., U.S.A.
hp	Horse Power
HP	High Pressure
HRSG	Heat Recovery Steam Generator
HR	Heat Rate
HRH	Hot Reheat
HSD	High Speed Diesel
HSI	Hybrid Sorbent Injection
HT	High Tension/Hemispherical Temperature (of ash)
HV	High Volatile/High Voltage
HVAC	Heating-Ventilation & Air Conditioning
HWR	Heavy Water Reactor
Hz	Hertz (Frequency)
IA	Instrument Air
I&C	Instrumentation & Control
IBR	Indian Boiler Regulations
ICE	Internal Combustion Engine
ICS	Integrated Control System
ID	Induced Draft/Inside Diameter
IDT	Initial-Deformation Temperature (of ash)
IEC	International Electrotechnical Commission
IEEE	Institute of Electric and Electronic Engineers
IFC	International Finance Corporation
IGCC	Integrated Gasification Combined Cycle
IM	Inherent Moisture (content in coal)
Imp.	Imperial
in	Inch/Inches
IPB	Isolated Phase Bus
IP	Intermediate Pressure
IPCC	Intergovernmental Panel on Climate Change
IPP	Independent Power Producer
IR	Infrared Radiation
IS	Indian Standards
ISO	International Standards Organization
ISO Condition	Pressure: 1.013 kPa, Temperature: 288 K, Relative Humidity: 65%
IV	Interceptor Valve

J	Joule
JO	Jacking Oil
k	Kilo
K	Kelvin/Potassium
kA	Kilo Ampere
kcal	Kilocalories
kg	Kilogram
kg-mole	Kilogram-mole
kg/s	Kilogram Per Second
kJ	Kilo Joule
km	Kilometer
KOD	Knock Out Drum
kPa	Kilo Pascal
kV	Kilo Volt
kW	Kilo Watt
kWh	Kilo Watt Hour
l	Liter
lb	Pound
LDO	Light Diesel Oil
LEA	Low Excess-Air
LH_2	Liquid Hydrogen
LHV	Lower Heating Value
LMTD	Log Mean Temperature Difference
LNB	Low-NOx Burner
LNG	Liquefied Natural Gas
LO	Lube Oil
LP	Low Pressure
LPG	Liquefied Petroleum Gas
LSHS	Low Sulfur Heavy Stock
LT	Low Tension
LV	Low Volatile
m	Meter
M	Moisture (content in coal)/Million
MAF	Moisture and Ash Free
max	Maximum
MB	Mixed Bed Unit
MCC	Motor Control Center
MCR	Maximum Continuous Rating

M.E.P.	Mean Effective Pressure
MFR	Master Fuel Relay
MFT	Master Fuel Trip
mg	Milligram
Mg	Magnesium
MGD	Million Gallons Per Day
m/h	Miles Per Hour
MI	Major Inspection Of Gas Turbine
min	Minute/Minimum
MJ	Mega Joule
ml	Milliliter
mm	Millimeter
MM	Mineral Matter (content in coal)
MMT	Minimum Metal Temperature
mol	Molecular
MOEF	Ministry of Environment & Forests, India
MPa	Mega Pascal
mph	Miles Per Hour
m/s	Meter/Second
MS	Main Steam
mV	Milli Volt
MW	Mega Watt
μS	Micro-Siemens
N	Newton/Nitrogen
Na	Sodium
NCV	Net Calorific Value
NEMA	National Electrical Manufacturers Association
NFPA	National Fire Protection Association, U.S.A.
Nm^3	Normal Cubic Meter
NO	Nitric Oxide
NO_2	Nitrous Oxide
NO_x	Nitrogen Oxides
NPHR	Net Plant Heat Rate
NPSH	Net Positive Suction Head
NRV	Non-Return Valve
NTP	Normal Temperature & Pressure (273 K & 101.3 kPa)
O	Oxygen
O&M	Operation & Maintenance
OD	Outside Diameter

OEM	Original Equipment Manufacturer
OFA	Over-Fire Air
OH	Operating Hours
OLTC	On-Line Tap Changer
OPEC	Organization of Petroleum Exporting Countries
OSHA	Occupational Safety & Health Administration
OT	Oxygen Treatment
oz	Ounce
P	Power/Poise
P, p	Pressure
Pa	Pascal
PA	Primary Air
PAC	Powdered Activated Carbon
PC	Pulverized Coal
PCC	Power Control Center
PF	Power Factor/Pulverized Fuel
PFBC	Pressurized Fluidized Bed Combustion
PG	Performance Guarantee
pH	Negative Log of Hydrogen Ion Concentration
P&ID	Process & Instrumentation Diagram
PLF	Plant Load Factor
PM	Particulate Matter
ppb	Parts Per Billion (mass)
ppm	Parts Per Million (mass)
ppmv	Parts Per Million by Volume
PRDS	Pressure Reducing And De-superheating
Press	Pressure
PRV	Pressure Reducing Valve
PSF	Pressure Sand Filter
psi	Pounds Per Square Inch
PTC	Performance Test Code
PWR	Pressurized Water Reactor
R	Rankine/Universal Gas Constant
rad	Radian
rev	Revolution
RH	Reheater
RHO	Reheater Outlet
RO	Reverse Osmosis
rpm	Revolution Per Minute

RW	Raw Water
s	Second
S	Sulfur
S, s	Entropy
SA	Secondary Air/Service Air
SAC	Strong Acid Cation Unit/Standardization Administration of the People's Republic of China
SBA	Strong Base Anion Unit
SCA	Specific Collection Area
SCAH or SCAPH	Steam coil air pre-heater
SCC	Submerged Chain Conveyor
scf	Standard Cubic Feet
scfm	Standard Cubic Feet Per Minute
SCR	Selective Catalytic Reduction
SDA	Spray Drier Absorber
sec	Second
SG	Steam Generator
SH	Superheater
SHO	Superheater Outlet
SI	Systeme International D'Unites/International System of Units
S.I.	Spark Ignition
SLD	Single Line Diagram
SM	Surface Moisture (content in coal)
SNCR	Selective Non-Catalytic Reduction
SO_2	Sulfur Dioxide
SO_3	Sulfur Trioxide
SO_x	Sulfur Oxides
sp gr	Specific Gravity
SPM	Suspended Particulate Matter
SS	Stainless Steel
s.s.c.	Specific Steam Consumption
ST	Steam Turbine/Station Transformer/Softening Temperature (of ash)
STP	Standard Temperature & Pressure (288 K & 101.3 kPa)
SWAS	Steam Water Analysis System
t	Ton/Tonne
T	Temperature/Turbine
TAC	Tariff Advisory Committee, India
TDC	Top Dead Center

TDS	Total Dissolved Solids
TEMA	Tubular Exchanger Manufacturer Association
Temp.	Temperature
TG	Turbo-Generator
TLR	Turbine Lock-out Relay
TM	Total Moisture (content in coal)
TMCR	Turbine Maximum Continuous Rating
tph	Tonnes Per Hour
TPSC	Toshiba Power Services Corporation, Japan
T/R	Transformer Rectifier Set
TSP	Total Suspended Particulates
TSS	Total Suspended Solids
TTD	Terminal Temperature Difference
U, u	Internal Energy
UAT	Unit Auxiliary Transformer
UBC	Unburned Carbon
ULR	Unit Lock-out Relay
UPS	Uninterrupted Power Supply System
UV	Ultra Violet
V	Volt
V, v	Volume
VDI	Verlag des Vereins Deutscher Ingenieure (the Association of German Engineers)
VM	Volatile Matter (in coal)
vol	Volume
vs	Versus
VWO	Valve Wide Open condition
W	Watt/Work
WAC	Weak Acid Cation Unit
WB	World Bank
WBA	Weak Base Anion Unit
WFGD	Wet Flue Gas Desulfurization
wg	Water Gauge
WHRB	Waste Heat Recovery Boiler
wt	Weight
yd	Yard/Yards
YGP	Yancy, Geer and Price (Index)
yr	Year

CHAPTER 1

Steam Power Plant Cycles

1.1 INTRODUCTION

The science of *thermodynamics* covers various concepts and laws describing the conversion of one form of energy to another, e.g., conversion of heat energy into mechanical energy as in a steam or gas turbine or conversion of chemical energy into heat energy as observed during the combustion of fuel. The science of thermodynamics also deals with the various systems that are put into service to perform such conversions. A *system* in thermodynamics refers to a definite quantity of matter bounded by a specified region (Figure 1.1), where the transfer and conversion of mass and energy take place. A *boundary* is a surface that separates the quantity of matter under investigation from its surroundings. While the region may not be fixed in either shape or volume, the boundary either may be a physical one, as the walls of a pressure vessel, or it could be an imaginary surface [1,2].

There are two types of thermodynamic systems: open and closed. In an open system, mass enters or leaves through the system boundary (Figure 1.2) as in case of "steam flow through a turbine." In the closed system, mass remains completely within the system boundary throughout the period of thermodynamic study and observation, as in the event of "expansion or compression of steam in a reciprocating steam engine." In this system, there is no interchange of matter between system and surroundings. It is to be noted that in both open and closed systems, heat/work may cross the system boundary [1–4].

Thermodynamics also deals with the relations between properties of a substance and quantities of "work" and "heat," which cause a change of state. *Properties* of a substance that describe its condition or state are characterized as pressure (P/p), temperature (T), volume (V/v), etc. These properties are measurable and depend only on the thermodynamic state and thus do not change over a cycle. From a thermodynamic point of view, the *state* of a system at any given moment is determined by the values of its properties at that moment [1,2,5].

When a system passes through a series of states in a process or series of processes in such a way that the final state of the system becomes identical to its initial state in all respects and is capable of repeating indefinitely then the system has completed a *cycle* [2]. The same principle is also followed in a thermodynamic cycle in which a fluid is returned to its initial state after the transfer of heat/work across the system boundary

Thermal Power Plant
DOI: http://dx.doi.org/10.1016/B978-0-12-801575-9.00001-9

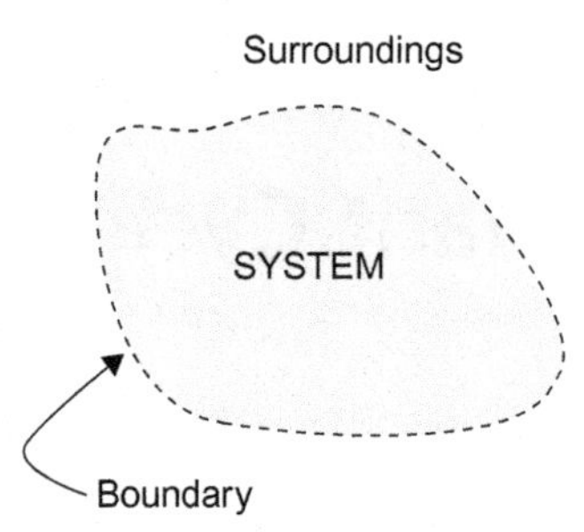

Figure 1.1 Configuration of a system. *Source: (This is Google's cache); http://commons.wikimedia.org/wiki/File:System_boundary.svg.; http://en.wikipedia.org/wiki/Thermodynamic_system.*

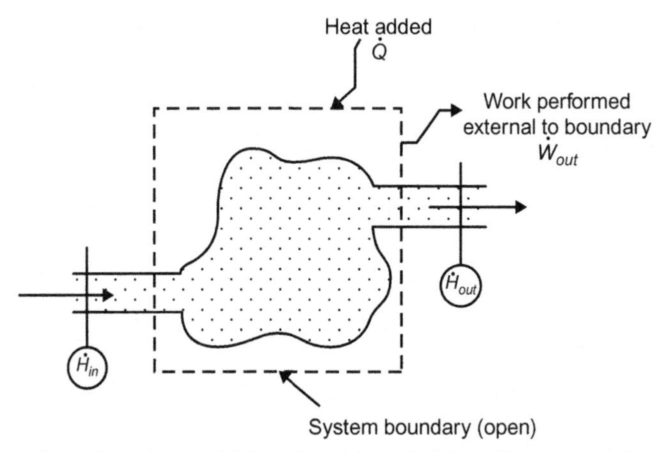

Figure 1.2 System Boundary. *Source: (This is Google's cache) http://commons.wikimedia.org/wiki/File:System_boundary.svg.; http://en.wikipedia.org/wiki/Thermodynamic_system.*

irrespective of whether the system is open or closed. However, heat and work are not zero over a cycle, they are process dependent. If the cyclic process moves clockwise around the loop producing a net quantity of work from a supply of heat, it represents a heat engine, and "work" will be positive. If the cyclic process moves counterclockwise, during which the net work is done on the system while a net amount of heat is rejected, then it represents a heat pump, and "work" will be negative [2].

A *reservoir* is a source of heat or a heat sink so large that no temperature change takes place when heat is added or subtracted from it. When heat from a reservoir is transferred to a working fluid circulating within a thermodynamic cycle mechanical power is produced. Thermodynamic power cycles are the basis of the operation of heat engines, which supply most of the world's electric power and run almost all motor vehicles. The most common power cycles used for internal combustion engines are the *Otto cycle* and the *Diesel cycle*. The cycle used for gas turbines is called the the *Brayton cycle*, and the cycle that supports study of steam turbines is called the *Rankine cycle*.

Table 1.1 Example of thermodynamic cycles

Cycle/Process	Compression	Heat addition	Expansion	Heat rejection
External combustion power cycles:				
Carnot	isentropic	isothermal	isentropic	isothermal
Stirling	isothermal	isochoric	isothermal	isochoric
Ericsson	isothermal	isobaric	isothermal	isobaric
Internal combustion power cycles:				
Otto (Gasoline/Petrol)	adiabatic	isochoric	adiabatic	isochoric
Diesel	adiabatic	isobaric	adiabatic	isochoric
Brayton	adiabatic	isobaric	adiabatic	isobaric

(Note: Details on internal combustion power cycles can be found in Chapters 7 and 8.)

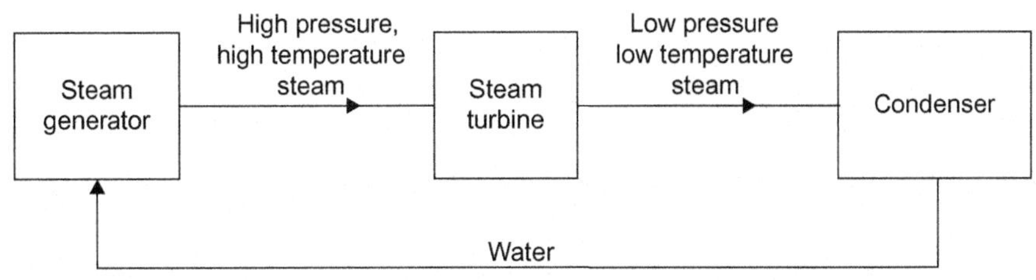

Figure 1.3 Flow diagram of a simple power plant.

A thermodynamic cycle is ideally be made up of any three or more thermodynamic processes as follows:

i. Isothermal (constant temperature) process
ii. Isobaric (constant pressure) process
iii. Isochoric (constant volume) process
iv. Adiabatic (no heat is added or removed from the working fluid) process
v. Isentropic or reversible adiabatic (no heat is added or removed from the working fluid and the entropy is constant) process
vi. Isenthalpic (constant enthalpy) process

Table 1.1 shows some examples of thermodynamic cycles.

In a simple steam power plant as the fluid circulates it passes through a continuous series of cyclic mechanical and thermodynamic states. Water enters a steam generator at a certain pressure and temperature and gets converted to steam, the high-pressure steam then enters a steam turbine and expands to a low pressure while passing through the turbine, the low pressure steam then gets condensed in a condenser, and the condensed water is recycled back to the boiler at original pressure and temperature (Figure 1.3) [6,7].

The performance of combined heat and power (CHP) generation steam power plants is determined by a term *specific steam consumption* (s.s.c.), which is defined as the mass flow of steam required per unit of power output (kg/kWh). The smaller the value of s.s.c., the bigger the plant size and vice-versa, e.g., for a CHP plant of 30 MW or lower capacity the value of s.s.c. is more than 4 kg/kWh, while for a utility plant of 250 MW or higher size the s.s.c. is close to or less than 3 kg/kWh. In utility power plants, however, the *heat rate*, defined as the energy required to be supplied to generate unit of power (kcal/kWh), is the best tool to determine the efficiency of the plant [7].

1.2 LAWS OF THERMODYNAMICS

Before we further discuss thermodynamic cycles it is essential to cover the fundamental laws of thermodynamics on which science of thermodynamics is based. While studying the laws of thermodynamics it should also to be kept in mind that the science of thermodynamics deals with relations between heat and work only.

1.2.1 First law of thermodynamics

In a cyclic process, since the initial and final states are identical, the net quantity of heat delivered to the system is proportional to the net quantity of work done by the system. When heat and work are mutually convertible, we have the first law of thermodynamics. Mathematically speaking,

$$\sum dQ \propto \sum dW \tag{1.1}$$

where

Q = Heat supplied to the system,
W = Work done by the system.

From Eq. 1.1 we can see that the first law of thermodynamics is an expression of *the principle of conservation of energy.* With the help of this law it is possible to calculate the quantities of heat and work that cross the surroundings of a system when given changes in properties occur. In a more generalized way the first law may be defined as *when a system is taken through a cycle, the net work delivered to the surroundings is proportional to the net heat taken from the surroundings* [2].

As a consequence of the first law, in addition to the pressure, temperature, and volume, another property exists in thermodynamics, such that any change in its magnitude is equal to the difference between the heat supplied and the work done during any change of state. This property is known as *internal energy*, denoted by U, and in a thermodynamic cycle change in internal energy (ΔU) is zero. If U_1 represents the energy of the system at the beginning of a process, U_2 its energy on completion of the process, Q the net heat flowing into the system in course of the process, and W the net work done by the system during the process, then the increase in energy of the steady-state open system having one inlet (subscript 1) and one exit (subscript 2), according to the first law of thermodynamics is expressed as [1,2,4]

$$Q - W = U_2 - U_1 + \frac{1}{2}(C_2^2 - C_1^2) + g(Z_2 - Z_1) \quad (1.2)$$

where

C is velocity of the mass at inlet or exit
Z is elevation at inlet or exit
g is acceleration due to gravity

Neglecting the effects of kinetic and potential energies Eq. 1.2 reduces to

$$U_2 - U_1 = Q - W \quad (1.3)$$

Equation 1.3 reveals that when a system passes through a cyclic process $U_2 = U_1$ and $Q = W$. Thus, the net heat flowing to the system equals the net work done by the system, and as a result it is impossible to construct a machine that would operate in a cycle. Consequently, the first law may be stated as *it is impossible to construct a perpetual motion machine of the first kind* [2]. [NOTE: The perpetual motion machine of the first kind is one, which once set in motion would continue to run for ever.]

Equation 1.3 also reveals that $U_2 - U_1$ depends only on the end states and is independent of the process by which the system is taken from one state to the other. The entire science of thermodynamics is in conformity with this conclusion. Hence, in a process whose end states change infinitesimally, the change in the internal energy of the system may be expressed as dU, the change in heat flow as dQ, and the change in net work done as dW. Accordingly Eq. 1.3 changes to:

$$dU = dQ - dW \quad (1.4)$$

In a constant pressure process, as in steam power cycle, $dW = pdV$, where p is pressure and dV represents the change in volume of the system on completion of the process, Eq. 1.4 changes to

$$dU = dQ - pdV$$

or,

$$dQ = dU + pdV \quad (1.5)$$

Since p is constant, $pdV = d(pV)$, therefore, Eq. 1.5 may be written as

$$dQ = dU + d(pV)$$

or,

$$dQ = d(U + pV) \quad (1.6)$$

In thermodynamics the quantity $(U + pV)$ occurs frequently, and is identified with a special property called *enthalpy H*.

Therefore,

$$dQ = dH \tag{1.7}$$

Following the above expression of enthalpy, the specific enthalpy (enthalpy per unit mass), represented by h in thermodynamics, is expressed in a state of equilibrium as

$$h = u + pv \tag{1.8}$$

where

u = Specific internal energy (internal energy per unit mass)
v = Specific volume (volume per unit mass)

Since u and pv are functions of the state of a system only, a change in enthalpy between two states depends only upon the end states, and is independent of the process.

For incompressible fluid or liquid undergoing infinitesimal change in end states the expression in Eq. 1.8 for an isentropic process changes to

$$dh = vdp \tag{1.9}$$

Integrating Eq. 1.9 between state 2 and state 1 it is found that

$$h_2 - h_1 \cong v(p_2 - p_1) \tag{1.10}$$

1.2.2 Second law of thermodynamics

While the first law asserts that net work cannot be produced during a cycle without some supply of heat, the second law expresses that some heat must always be rejected during the course of a cycle. In its simplest form the second law states that "heat cannot, by itself, flow from lower temperature to a higher temperature." One of the classical statements of the second law, as given by Kelvin–Planck, is "*it is impossible to construct a system that will operate in a cycle, extract heat from a reservoir, and do an equivalent amount of work on the surroundings* [2]." Another statement of the second law, as set forth by German physicist Rudolf Clausius, is "*it is impossible to construct a system that will operate in a cycle, remove heat from a reservoir at one temperature, and absorb an equal quantity of heat by a reservoir at a higher temperature* [2,8]."

It is clear from these statements that to have continuous output from a system, it is essential that there be transfer of heat across the system boundary at inlet and exit. Hence, it may be concluded that, "while the first law states that the net work can never be greater than the heat supplied, the second law goes further and says that work must always be less." In other words, all heat input to a system cannot be utilized into work, and a part of this heat must be rejected, thereby involving a term called system efficiency.

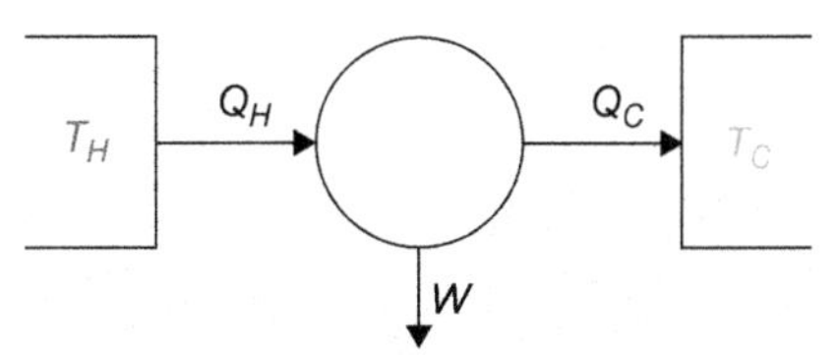

Figure 1.4 A simple cycle. *Source: (This is Google's cache); http://commons.wikimedia.org/wiki/File: System_boundary.svg.; http://en.wikipedia.org/wiki/Thermodynamic_system.*

Suffice it to say that work can be completely and continuously converted to heat, but heat cannot be completely and continuously converted to work, which means within a cycle heat is not entirely available to do work on a continuous basis. This *unavailable energy*, called *anergy*, has to be rejected as low-grade heat once the work has been done. The *available energy*, also known as *exergy*, on the other hand, is the work output obtainable from a certain heat input in a cyclic heat engine [5].

Consider a system that receives heat Q_H from a high-temperature (T_H) heat source, rejects Q_C to a low-temperature (T_C) heat sink, and generates work W while operating in a cycle (Figure 1.4).

Hence, it can be written that

$$Q_H = Q_C + W \tag{1.11}$$

or,

$$W = Q_H - Q_C$$

or,

$$AE = Q_H - UE \tag{1.12}$$

where,

AE = Available energy (exergy)
UE = Unavailable energy (anergy)

Therefore, the cycle efficiency of the system is defined mathematically as η = Work done/Heat supplied:

$$= \frac{W}{Q_H} = \frac{Q_H - Q_C}{Q_H}$$

or

$$\eta = 1 - \frac{Q_C}{Q_H} \tag{1.13}$$

Since both Q_H and Q_C are finite, the value of η must be less than 1, even when all processes involved are ideal and frictionless. Thus, it could be concluded from the second law that the efficiency of any cycle must be less than 100%.

From this discussion it is evident that "a heat cycle which produces work" comprises (i) a high temperature heat source, (ii) a low temperature heat sink, and (iii) a system producing work, e.g., a heat engine. Consequent to this it may be stated that *it is impossible to construct a perpetual motion machine of the second kind.* [NOTE: A perpetual motion machine of the second kind is one, which will produce work continuously, while exchanging heat with only one reservoir.] [2].

1.2.2.1 Concept of entropy

From the first law of thermodynamics it was found that there is a new property called internal energy. Likewise, the second law contains another new property, which is function of a quantity of heat, that shows the possibility of conversion of heat into work. This property is a thermodynamic quantity representing the unavailability of a system's thermal energy for conversion into mechanical work, and is often interpreted as the degree of disorder or randomness in the system. The name of this property was coined in 1865 by Rudolf Clausius as *entropy*. Clausius introduced the concept of entropy to facilitate the study of fluids passing through a reversible process (Section 1.3). Any increase in this new property "entropy" lowers the availability of that energy for doing useful work. In a physical system, entropy provides a measure of the amount of thermal energy that cannot be used to do work [1,8,9].

From the entropy of a fluid we may assess the degree of orderly or disorderly motion of its molecules. While condensing steam the degree of order of its molecular motion increases, which in thermodynamics is interpreted as a decrease in the entropy of the fluid. Likewise, when heat is applied to a fluid, its molecules get more agitated, collisions become more frequent, and the degree of motion of molecules become more chaotic, resulting in an increase in entropy.

Entropy is not a physical phenomenon that would exist in reality; there is no physical instrument to measure entropy. On the contrary, it is calculated from the pressure and temperature of a fluid at a particular state. It is a thermodynamic property that determines the thermal energy that always flows spontaneously from regions of higher temperature to regions of lower temperature in the form of heat. Like internal energy, the entropy of a system is a function of its end states only [7].

In a Carnot cycle (Section 1.3) the total change in entropy is zero, but when heat flows from a hotter substance to a colder one, the hotter substance loses less entropy than the colder substance gains – thus total entropy increases. In thermodynamics, an increase in entropy is small when heat is added at high temperature and is greater when heat is added at lower temperature. Hence, in an isolated system the necessary condition for the equilibrium of a system is that its entropy is at its maximum. Thus, for maximum entropy there is minimum availability of heat for conversion into work and for minimum entropy there is maximum availability of heat for conversion into work. Therefore, the second law of thermodynamics may also be stated as "total entropy of an isolated physical system either increases or remains constant; and can never decrease [2]."

Entropy, S, is not defined directly, but rather by an equation giving the change in entropy of the system to the change in heat of the system or is defined by "the quotient of a quantity of heat divided by its absolute temperature." For constant temperature, the change in entropy, ΔS, is defined by the equation $\Delta S = \Delta Q/T$, where ΔQ is the amount of heat absorbed in a thermodynamic process in which the system goes from one state to another, and T is the absolute temperature at which the process is occurring. Hence, in a process whose end states change infinitesimally, an "infinitesimal change in entropy is equal to an infinitesimal change in heat addition divided by the temperature at which the heat is supplied." Mathematically the above relationship may be expressed as

$$dS = dQ/T \tag{1.14}$$

Integrating Eq. 1.14

$$\int dQ/T = (S_2 - S_1) \tag{1.15}$$

where

S = Entropy of the system
Q = Heat supplied to the system
T = Temperature of the system

2 and 1 stand for final and initial states, respectively.

From Eqs. 1.14 and 1.15 we may note that for a adiabatic process $dQ = 0$, $dS = 0$ and for an isothermal process $T = T_2 = T_1$ and $Q = T(S_2 - S_1)$.

Combining Eq. 1.7 and Eq. 1.14,

$$dS = dH/T$$

or,

$$dH = TdS \tag{1.16}$$

or, per unit mass

$$dh = Tds \tag{1.17}$$

From this discussion it may be concluded that the science of thermodynamics comprises six properties of a fluid that describe its state: pressure (P/p), temperature (T), volume (V/v), specific internal energy (u), specific enthalpy (h), and specific entropy (s) [7].

1.3 CARNOT CYCLE

The Carnot cycle was invented by *Nicholas Le'onard Sadi Carnot* in 1824. It is an ideal cycle in which heat is taken at a constant higher temperature and rejected to a constant lower temperature. This cycle laid the foundation for the second law of thermodynamics and introduced the concept of *reversibility.* A reversible process is an ideal process, where the process traverses the same path during forward and backward travel. Both the fluid and its surroundings in a reversible process can be restored to their original states, and work and heat exchanged in one path is restored in the reverse path. In a reversible process, the entropy of the system remains unchanged throughout various states of the process.

All real processes, however, are irreversible, although the degree of irreversibility varies among processes. The *irreversibility* in a process is developed due to mixing, friction, throttling, heat transfer, etc. One irreversible process in a cycle causes the whole cycle to become irreversible. In an irreversible process, the total entropy of the system increases as the process continues.

In a Carnot cycle a working fluid takes in heat reversibly from a reservoir (heat source) at a constant higher temperature, T_1, expands adiabatically and reversibly to a constant lower temperature, T_2, gives up heat reversibly to a reservoir (heat sink) at T_3 (= T_2), and is then compressed reversibly and adiabatically to its original state T_4 (= T_1)

In P-V (Pressure-Volume) and T-S (Temperature-Entropy) diagrams (Figure 1.5) the cycle is represented by an area bounded by the following two isothermals and two adiabatic processes:

- **i.** 1-2: Reversible adiabatic expansion
- **ii.** 2-3: Reversible isothermal heat rejection
- **iii.** 3-4: Reversible adiabatic compression
- **iv.** 4-1: Reversible isothermal heat addition

State 4 in Figure 1.5 refers to saturated liquid and state 1 refers to saturated vapor.

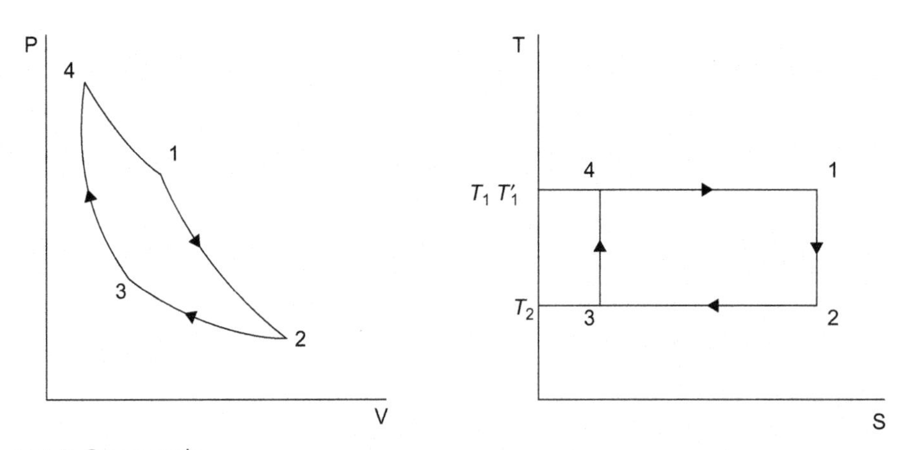

Figure 1.5 Carnot cycle.

From Figure 1.5, it is evident that the changes in entropy during heat addition and heat rejection are equal in magnitude. Therefore, $(s_1 - s_4) = (s_2 - s_3)$. Thus:

Heat added to the system

$$Q_{4-1} = T_1(s_1 - s_4) = h_1 - h_4 \tag{1.18}$$

Heat rejected from the system

$$Q_{2-3} = T_2(s_2 - s_3) = h_2 - h_3 \tag{1.19}$$

The net work output of the system

$$W_o = Q_{4-1} - Q_{2-3} = T_1(s_1 - s_4) - T_2(s_2 - s_3) = (T_1 - T_2)(s_2 - s_3) = (h_1 - h_4) - (h_2 - h_3) \tag{1.20}$$

The efficiency of the Carnot cycle

$$\eta_c = \text{Work Output/Heat Input} = W_o/Q_{4-1},$$

or,

$$\eta_c = \frac{(T_1 - T_2)(s_2 - s_3)}{T_1(s_1 - s_4)} = \frac{(T_1 - T_2)}{T_1} = 1 - \frac{T_2}{T_1} \tag{1.21}$$

$$= 1 - \frac{h_2 - h_3}{h_1 - h_4} \tag{1.22}$$

Equation 1.21 shows that efficiency of the Carnot cycle depends on the temperature of the heat source and heat sink only and is independent of the type of working fluid. The higher the temperature of the heat source the lower the temperature of the heat sink, i.e., the wider the temperature range, the more efficient the cycle. This cycle also establishes that this is the maximum achievable efficiency a heat engine may attain [9]. In a real cycle, however, cycle efficiency will be always lower than the Carnot cycle efficiency.

In practice, T_2 can seldom be reduced below 298–303 K, which corresponds to a condenser pressure of 3.5 kPa. Because the temperature of the sink – e.g., atmosphere, river, ocean – i.e., the temperature of cooling water below 288 K, is rarely available and the temperature difference between the condensing steam and the cooling water, required for the heat transfer with a reasonable size of condenser, must be about 10–15 K. The maximum value of T_1 is limited to the critical point temperature of 647 K. Thus, the maximum efficiency the Carnot cycle may attain is about 54%.

The Carnot cycle is thermodynamically simple, but it is extremely difficult to realize in practice, because the isothermal heat rejection must be stopped at state 3, then subsequent adiabatic compression of a very wet vapor needs to be carried out to restore it to its initial state 4. During compression, the liquid gets separated from the vapor and further compression of the fluid has to work with a non-homogeneous mixture. As such, none of the practical engines operate on this cycle. Nevertheless, the Carnot cycle is an excellent yardstick to compare various thermodynamic cycles on a theoretical basis. All practical cycles, nevertheless, differ significantly from the Carnot cycle.

Example 1.1

Using Figure 1.5, calculate the efficiency and exhaust steam quality of a steam cycle operating between 4 MPa and 10.13 kPa.

Solution: From the steam table the following values have been found corresponding to steam pressures 4 MPa and 10.13 kPa:

Steam pressure	Fluid temperature K	h_f kJ/kg	h_{fg} kJ/kg	h_g kJ/kg	s_f kJ/kgK	s_g kJ/kgK
4.00 MPa	$T_2 = 523.33$	1087.40	1712.90	2800.30	2.7965	6.0685
10.13 kPa	$T_1 = 319.09$	192.89	2392.31	2585.20	0.6526	8.1466

The cycle efficiency is $\eta_c = (T_1 - T_2)/T_1)\}*100 = (204.24 / 523.33) \times 100 = 39.03\ \%$

Using $s_1 = s_2$ (Figure 1.5), it is found that at a condenser pressure of 10.13 kPa, dryness fraction at state 2, exhaust, is

$$x_2 = (6.0685 - 0.6526)/(8.1466 - 0.6526) = 0.723$$

Enthalpy at exhaust is $h_2 = 192.89 + 0.723*2392.31 = 1922.53$ kJ/kg

Example 1.2

Consider the Carnot cycle of Figure 1.5 and calculate the heat transfers, net work output, cycle efficiency, and specific steam consumption (s.s.c), using steam operating between pressures 8 MPa and 9.6 kPa.

Solution: From the steam table the following values have been found corresponding to steam pressures 8 MPa and 9.6 kPa:

Steam pressure	Fluid temperature K	h_f kJ/kg	h_{fg} kJ/kg	h_g kJ/kg	s_f kJ/kgK	s_g kJ/kgK
8.0 MPa	$T_2 = 568$	1316.27	1441.80	2758.07	3.2061	5.7436
9.6 kPa	$T_1 = 318$	188.42	2394.80	2583.22	0.6386	8.1647

Using $s_1 = s_2$ and $s_3 = s_4$ (Figure 1.5), it is found that at a condenser pressure of 9.6 kPa, dryness fraction at states 2 and 3 are as follows:

$$x_2 = (5.7436 - 0.6386)/(8.1647 - 0.6386) = 0.678$$

and

$$x_3 = (3.2061 - 0.6386)/(8.1647 - 0.6386) = 0.341$$

Hence, from $h = h_f + xh_{fg}$, we find

$$h_2 = 188.42 + 0.678 * 2394.80 = 1812.09 \text{ kJ/kg}$$
$$h_3 = 188.42 + 0.341 * 2394.80 = 1005.05 \text{ kJ/kg}$$

The net work output is

$$\begin{aligned} W_o &= (h_1 - h_4) - (h_2 - h_3) \\ &= (2758.07 - 1316.27) - (1812.09 - 1005.05) = 634.76\ \text{kJ/kg} \end{aligned}$$

The heat transfer in the boiler is

$$Q_{4-1} = h_4 - h_1 = 1441.80\ \text{kJ/kg}$$

The heat transfer in the condenser is

$$Q_{2-3} = h_2 - h_3 = 807.04\ \text{kJ/kg}$$

The cycle efficiency is

$$\begin{aligned} \eta_c &= (W_o/Q_{4-1}) * 100 = (634.76/1441.80) * 100 \\ &= 44.03\% \end{aligned}$$

The above value may be verified from Eq. 1.21 as

$$\begin{aligned} \eta_c &= \{(T_1 - T_2)/T_1\} * 100 = (250/568) \times 100 \\ &= 44.01\% \end{aligned}$$

(Theoretically the above two efficiency values should have been the same. A minor difference between these values is the result of the approximation taken in the decimal portion while calculating the former efficiency value.)

Since 1 kWh = 3600 kJ, the specific steam consumption is

$$\text{s.s.c.} = 3600/W_o = 3600/634.76 = 5.67\ \text{kg/kWh}$$

1.4 STIRLING CYCLE

Like the Carnot cycle the Stirling cycle is also a thermodynamic cycle that was invented, developed, and patented before the Carnot cycle in 1816 by *Reverend Dr. Robert Stirling* [9]. The P-V and T-S diagrams of an ideal Stirling cycle (Figure 1.6) include the following four thermodynamic processes:

i. 1-2: Reversible isothermal compression
ii. 2-3: Reversible isochoric heat addition
iii. 3-4: Reversible isothermal expansion
iv. 4-1: Reversible isochoric heat rejection

The working medium in a Stirling cycle is a gaseous matter, such as air, helium, hydrogen, etc., instead of water and steam. Like the Carnot cycle all the processes in an ideal Stirling cycle are reversible in nature, hence when the gas is heated the engine produces work or power and when work is supplied to the cycle it works as the refrigerator or the heat pump. When the processes in the Stirling cycle are reversed they act as cryogenerator and the cycle is used in the field of cryogenics to produce extremely low temperatures or to liquefy gases like helium and hydrogen.

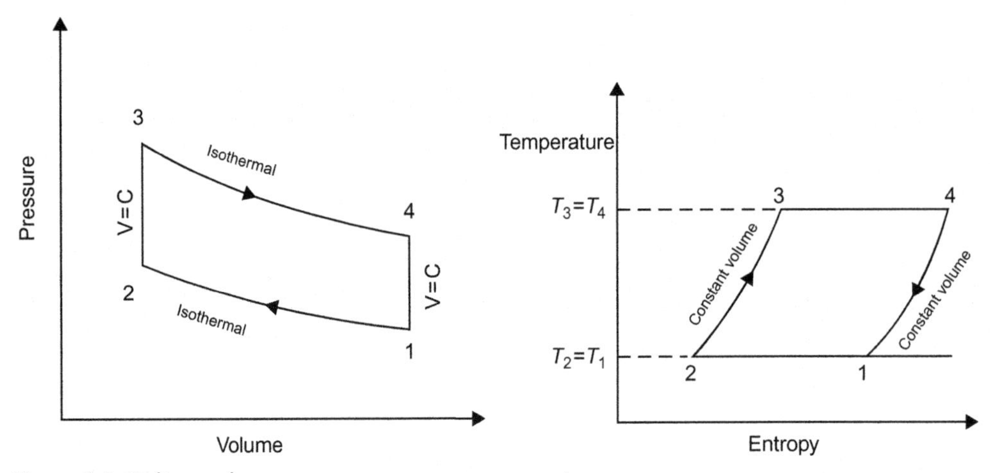

Figure 1.6 Stirling cycle.

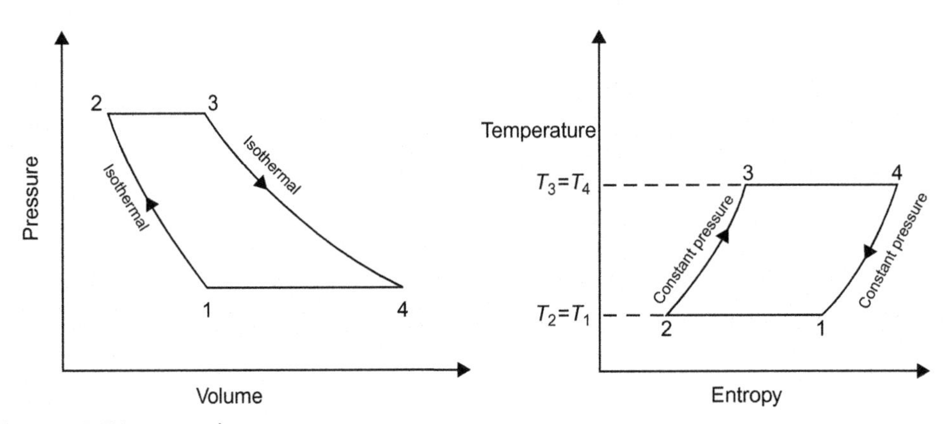

Figure 1.7 Ericsson cycle.

1.5 ERICSSON CYCLE

The Ericsson cycle is another ideal thermodynamic cycle named after inventor *John Ericsson*, who designed and built many unique heat engines based on various thermodynamic cycles. In the P-V and T-S diagrams (Figure 1.7) the ideal Ericsson cycle is represented by an area bounded by following two isothermals and two isobaric processes:

i. 1-2: Reversible isothermal compression
ii. 2-3: Reversible isobaric heat addition
iii. 3-4: Reversible isothermal expansion
iv. 4-1: Reversible isobaric heat rejection

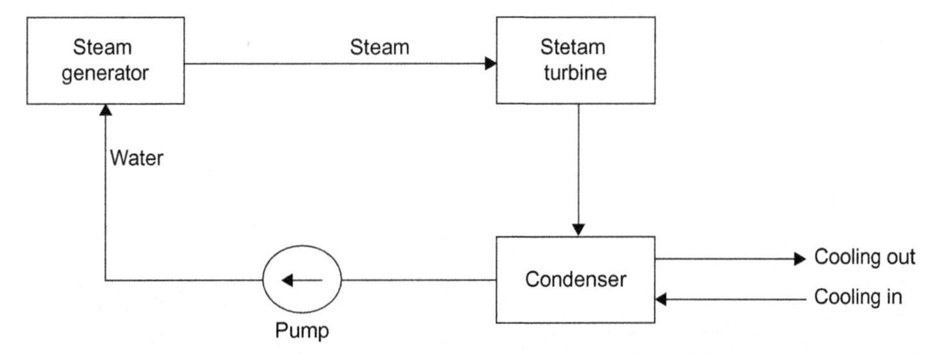

Figure 1.8 Simplest steam cycle.

The Ericsson cycle does not use the vapor-compression cycle, since the vapor never returns back to liquid. Hence, this cycle does not find practical application in piston engines. This cycle, however, may find application in gas turbines.

Thermodynamically, all three cycles, i.e., Carnot, Stirling, and Ericsson cycles are identical in nature, since all of them work between constant temperature at heat source and constant temperature at heat sink. All are theoretically capable of attaining identical efficiency, working between similar hot and cold end temperatures if there are no losses. However, none of these cycles finds practical application in commercial engines.

1.6 RANKINE CYCLE

In a steam power plant supply and rejection of heat is more easily realized at constant pressure than at constant temperature. It was *William John Macquorn Rankine*, after whom the Rankine cycle is named, who first calculated the maximum possible work that could be developed by an engine using dry saturated steam between the pressure limits of the boiler and condenser. The simplest steam cycle using dry saturated steam as the working fluid has the following basic components (Figure 1.8):

a. Steam Generator/Boiler
b. Steam Turbine
c. Condenser
d. Boiler Feed Pump

The Rankine cycle is an ideal thermodynamic cycle involving the following processes:

i. Steam generation in boiler at constant pressure
ii. Isentropic expansion in steam turbine
iii. Condensation in condenser at constant pressure
iv. Pressurizing condensate to boiler pressure by isentropic compression

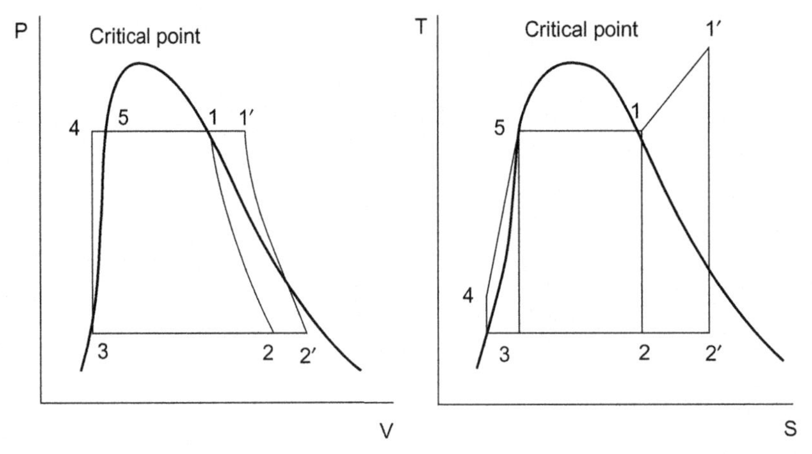

Figure 1.9 Rankine cycle.

Figure 1.9 shows the Rankine cycle in P-V and T-S diagrams. This cycle assumes that all processes are reversible, i.e., all processes take place without any friction and heat transfers take place across infinitesimal temperature drops.

The following states/processes are represented in these diagrams:

- State 1: Condition of saturated vapor at temperature T_1 and pressure p_1.
- Process 1-2: The vapor then expands through the turbine reversibly and adiabatically (isentropicaly) to temperature at T_2 and pressure at p_2.
- The turbine generates power at a magnitude much higher than the power required by the boiler feed pump.
- State 2: The exhaust vapor is usually in the two-phase region at temperature T_2 and pressure p_2.
- Process 2-3: The low pressure wet vapor, at temperature T_2 ($=T_3$) and, being a two-phase mixture process, at constant pressure p_2 ($=p_3$), is then liquefied in the condenser into saturated water and reaches the state 3, thus minimizing the work required by the boiler feed pump.
- State 3: Condition of saturated liquid at temperature T_3 and pressure p_3.
- Process 3-4: The saturated water, at the condenser pressure p_3, is then compressed reversibly and adiabatically (isentropically) by the boiler feed pump to sub-cooled liquid in the steam generator at pressure p_4.
- As the fluid is liquid at this stage the pump requires little input energy.
- State 4: Condition of sub-cooled liquid at temperature T_4 and pressure p_4.
- Process 4-1: The sub-cooled liquid at state 4 is heated at constant pressure p_4, in the "economizer" section of the steam generator, to a saturated liquid at state 5. The saturated liquid is further heated in the "boiler or evaporator" section of the steam generator to saturated vapor at constant temperature and constant pressure (being a two-phase mixture) to its initial state (T_1 and p_1).

Hence, looking at Figure 1.9 following equations can be derived:
Heat added in the steam generator

$$Q_1 = (h_1 - h_4) \tag{1.23}$$

Heat rejected to the condenser

$$Q_2 = (h_2 - h_3) \tag{1.24}$$

Work output of turbine

$$W_T = (h_1 - h_2) \tag{1.25}$$

Work input to feed pump

$$W_P = (h_4 - h_3) \tag{1.26}$$

Thus, net work output

$$W_o = W_T - W_P = (h_1 - h_2) - (h_4 - h_3) \tag{1.27}$$

Therefore the efficiency of the Rankine cycle is

$$\eta_r = \frac{W_o}{Q_1} = \frac{(h_1 - h_2) - (h_4 - h_3)}{(h_1 - h_4)}$$

or,

$$\eta_r = 1 - \frac{(h_2 - h_3)}{(h_1 - h_4)} \tag{1.28}$$

Comparing Eq. 1.22 and Eq. 1.28 it may be noted that while working between identical temperature limits, h_1 and h_2 in both the cycles are the same, but h_4 and h_3 in the Carnot cycle are much higher than h_4 and h_3 in the Rankine cycle. As a result $(h_1 - h_4)$ and $(h_2 - h_3)$ in the Carnot cycle are much smaller than those in the Rankine cycle. Thus, the Rankine cycle offers a lower ideal thermal efficiency for the conversion of heat into work than does the Carnot cycle. This conclusion is also evident from Example 1.2 and Example 1.4.

Example 1.3

Steam at a pressure of 2 MPa has 150 K of superheat. If this steam expands adiabatically until it becomes saturated, what will be the pressure and temperature of the final steam?

Solution: Referring to the steam table we get that the saturated temperature of steam at 2 MPa pressure is 485 K. Therefore, the temperature of the superheated steam is 635 K and entropy of this steam is 7.0019 kJ/kg.K.

For adiabatic expansion, since entropy remains unchanged, the corresponding to above entropy, we find pressure and temperature of dry saturated steam from the steam table as follows:

Pressure: 290 kPa
Temperature: 405 K

Example 1.4

Referring to Figure 1.9, calculate the net work output, cycle efficiency, and specific steam consumption (s.s.c) using steam operating between pressures 8 MPa and 9.6 kPa.

Solution: From the steam table the following values have been found, corresponding to steam pressures 8 MPa and 9.6 kPa:

Steam pressure	Fluid temperature K	h_f kJ/kg	h_{fg} kJ/kg	h_g kJ/kg	s_f kJ/kgK	s_g kJ/kgK	v_f m^3/kg
8 MPa	$T_1 = 568$	1316.27	1441.80	2758.07	3.2061	5.7436	–
9.6 kPa	$T_2 = 318$	188.42	2394.80	2583.22	0.6386	8.1647	0.00101

Using $s_2 = s_1$, it is found that at a condenser pressure of 9.6 kPa and dryness fraction at state 2 is as follows:

$$x_2 = (5.7436 - 0.6386)/(8.1647 - 0.6386) = 0.678$$

Hence, from $h = h_f + xh_{fg}$

$$h_2 = 188.42 + 0.678 * 2394.80 = 1812.09 \text{ kJ/kg}$$

Applying Eq. 1.9,

$$\begin{aligned}(h_4 - h_3) &= v_{f3} * (p_4 - p_3)\\ &= \{0.00101 * (81.600 - 0.098) * 10000\} * (9.8/1000)\\ &= 8.07 \text{ kJ/kg}\end{aligned}$$

(since 1 kPa = 0.0102 kg/cm^2, 1 MPa = 10.2 kg/cm^2, 1 m^2 = 10000 cm^2 and 1000 mkgf = 9.8 kJ). Therefore,

$$h_4 = h_3 + 8.07 = 188.42 + 8.07 = 196.49 \text{ kJ/kg}$$

The heat supplied is

$$Q_1 = (2758.07 - 196.49) = 2561.58 \text{ kJ/kg}$$

The net work output is

$$\begin{aligned}W_o &= (h_1 - h_2) - (h_4 - h_3)\\ &= (2758.07 - 1812.09) - 8.07 = 937.91 \text{ kJ/kg}\end{aligned}$$

The cycle efficiency is

$$\eta_r = (937.91/2561.58) * 100 = 36.61\%$$

The specific steam consumption is

$$\text{s.s.c.} = 3600/W_o = 3600/937.91 = 3.84 \text{ kg/kWh}$$

1.6.1 Real rankine cycle

In an ideal Rankine cycle (Figure 1.9) both the compression in the boiler feed pump (process 3-4) and expansion in the turbine (process 1-2) take place isentropically (reversibly and adiabatically). Hence, the efficiency of the ideal Rankine cycle may be regarded as the highest efficiency achievable in practice with a straight condensing machine. However, in

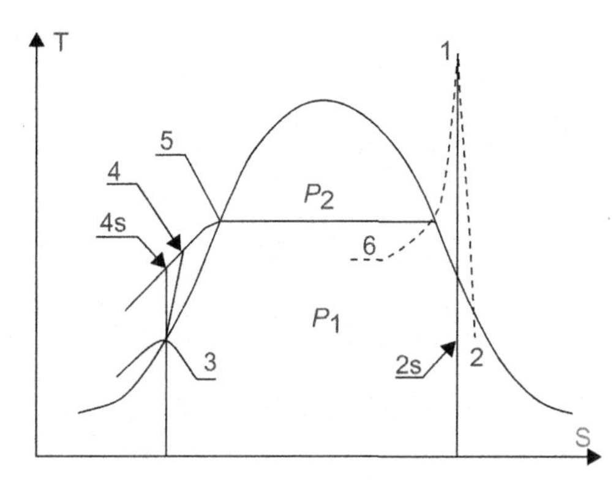

Figure 1.10 Real rankine cycle.

a real Rankine cycle (Figure 1.10) the efficiency that could be achieved is less than the efficiency of the ideal Rankine cycle, since none of the compression and expansion processes are isentropic. This is because these processes on completion are non-reversible causing entropy to increase during compression in boiler feed pump (the isentropic process 3-4s increases to process 3-4) as well as during expansion in turbine (the isentropic process 1-2s increases to process 1-2), which further results in an increase in power required by the boiler feed pump and decrease in power generated by the turbine.

The irreversibility in a real Rankine cycle is the result of irreversibility in the following areas:

i. Friction in bearings, etc.
ii. Pressure drop in steam and water piping, heat exchangers, bends and valves, etc.
iii. Friction in turbine blading and pump resulting in increase in entropy in both
iv. Windage loss in turbine

Further, during the process 1-2, since the vapor is moist it starts condensing and water droplets hit the turbine blades at high speed, causing erosion and pitting. These water droplets are very harsh, particularly at the last few stages of a low-pressure turbine. Thus, the efficiency of the turbine starts reducing gradually. This problem is overcome by superheating the vapor at state 1, which will move to the right to state 1′ (Figure 1.9) and hence produce less wet vapor after expansion (state 2′). However, by superheating the heat addition process in the steam generator is no longer isothermal and the average temperature of heat reception lies somewhere between that corresponding to states 1 and 1′, i.e., between T_1 and T_1'.

In addition to the higher dryness fraction, another advantage of superheating the steam is that the efficiency of the real cycle will also increase, since the cycle receives heat at higher temperature. Improving cycle efficiency almost always involves making a cycle more like a Carnot cycle operating between the same high and low temperature limits.

Nevertheless, the maximum possible temperature of steam in a thermal power plant is restricted by the strength of the available materials for boiler tubes or turbine blades.

The real superheat cycle is slightly different from the cycle shown in Figure 1.10. Although the liquid leaves the boiler feed pump at state 4, steam leaves the steam generator, in lieu of a state 1, at state $1'_{sg}$ (not shown in the figure), and enters the turbine at a different state $1'_{t}$ (not shown in the figure). Further, liquid leaving the boiler feed pump must be at a higher pressure (p_4) than the steam-generator exit steam pressure ($p_{1'sg}$), which also will have to be higher than the steam pressure at turbine inlet ($p_{1't}$) because of friction drops in the heat exchangers, feedwater heaters, pipes connecting boiler feed pump and steam generator, pipes connecting steam generator and turbine, bends, superheater outlet valves, turbine throttle valves, etc. [3].

There would also be heat loss from pipes connecting the steam generator and turbine, causing a drop in entropy from state 1′sg to state 1′t.

Example 1.5

Considering the isentropic efficiencies of the expansion process as 0.80 and the compression process as 0.85 calculate the cycle efficiency and specific steam consumption (s.s.c) of the steam cycle of Example 1.4. If the steam flowing through the turbine is 60 kg/s, find out the power output.

Solution:

Actual expansion work is $W_{EA} = 0.8*(2758.07 - 1812.09) = 756.78$ kJ/kg
and actual compression work is $W_{CA} = 8.07/0.85 = 9.49$ kJ/kg
Therefore, the net work output is $W_o = 756.78 - 9.49 = 747.29$ kJ/kg
The enthalpy at state 4 now is $h_4 = h_3 + W_{CA} = 188.42 + 9.49 = 197.91$ kJ/kg
And the heat supplied is $Q_1 = (2758.07 - 197.91) = 2560.16$ kJ/kg
The actual cycle efficiency is $\eta_r = 747.29/2560.16*100 = 29.19\%$
The specific steam consumption is s.s.c. = 3600/747.29 = 4.82 kg/kWh
The power output is $P = 60*W_o = 60*747.29 = 44837.4$ kJ/s
= 44837.4 kW (since, 1 kW = 1 kJ/s)

(Note: The results of Examples 1.2, 1.4, and 1.5, as presented in the following table, compares the performance of the Carnot, Ideal Rankine, and Real Rankine cycles of steam operating between 8 MPa and 9.6 kPa pressures. It is evident from the table that even though the Carnot cycle provides the highest cycle efficiency, its net work output is the least and the specific steam consumption is the highest.)

Parameters	Carnot cycle	Ideal saturated rankine cycle	Real saturated rankine cycle
Heat Added, kJ/kg	1441.80	2561.58	2560.16
Cycle Efficiency, %	44.01	36.61	29.19
Net Work Output, kJ/kg	634.76	937.91	747.29
Specific Steam Consumption, kg/kWh	5.67	3.84	4.82

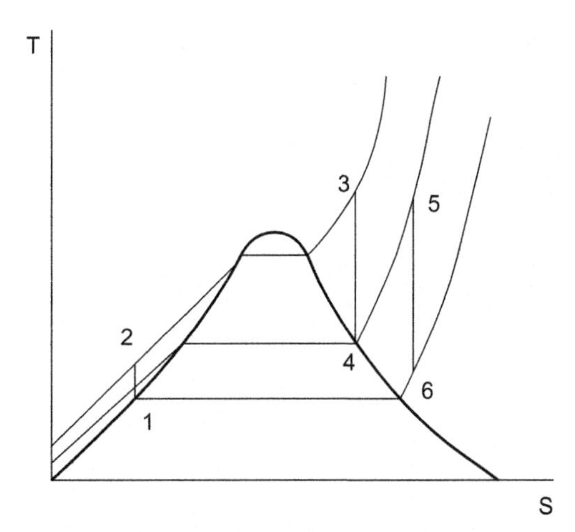

Figure 1.11 Reheat rankine cycle.

1.6.2 Reheat rankine cycle

For large steam power plants it is vital to maximize thermal efficiency and minimize specific steam consumption. Hence, these plants operate on high-pressure and high-temperature steam cycles. However, the problem encountered by raising the turbine inlet pressure is that the steam coming out of the turbine (at state 2, Figure 1.10) is very wet. Water droplets accompanying steam impinge turbine blades and damage the blades severely, impairing the efficiency of the turbine.

For all practical purposes it is unwise to allow steam with dryness fraction less than around 88% to remain in the turbine [7], even though higher-grade low-pressure turbine blade materials may accept a lower dryness fraction.

Accordingly, with superheating alone, the turbine exhaust dryness fraction will not increase enough to restrict erosion of turbine blades. To prevent blade erosion and inefficient turbine operation, wet steam is sometimes extracted from the turbine from an intermediate stage and then passed through reheater, where steam becomes re-superheated and is returned to the next stage in the turbine for further expansion (Figure 1.11). The fundamental purpose of reheat is not to improve cycle efficiency but to reduce the moisture content of the vapor in the exhaust of the turbine. Nevertheless, with reheating the thermal efficiency of the cycle increases significantly as compared with non-reheat cycle.

In practice steam is extracted from high-pressure (HP) turbine exhaust, circulated to the boiler for re-superheating or reheating, and returned to the intermediate-pressure (IP) or low-pressure (LP) turbine for further work. Exhaust steam from the HP turbine is called "cold reheat" steam, while reheated steam coming out of the steam generator is known as "hot reheat" steam.

Reheating is best suited in high-pressure units since it provides low specific volume of steam, thereby reducing the overall surface area of the reheater along with a consequent reduction in capital expenditure.

From Figure 1.11, neglecting the work input to the boiler feed pump, the work done in reheat cycle is

$$W_o = (h_3 - h_4) + (h_5 - h_6) \tag{1.29}$$

The heat added in the steam generator

$$Q_1 = (h_3 - h_2) + (h_5 - h_4) \tag{1.30}$$

Therefore, the efficiency of the Reheat Rankine cycle is

$$\eta_{rh} = \frac{(h_3 - h_4) + (h_5 - h_6)}{(h_3 - h_2) + (h_5 - h_4)} \tag{1.31}$$

(Note: It should be kept in mind that by increasing the turbine inlet steam temperature and accordingly the steam pressure the cycle efficiency will always increase.)

Example 1.6

Using Figure 1.11, calculate the ideal cycle efficiency and specific steam consumption (s.s.c.) of a reheat cycle operating between pressures 8 MPa and 9.6 kPa with a superheat temperature and reheat temperature of 773 K and 773 K, respectively. (Assume steam condition after first expansion as dry saturated and neglect work input to boiler feed pump.)

Solution: From the steam table we find $h_2 = h_1 = 188.42$ kJ/kg (since the given condition is to neglect the work input to the boiler feed pump.)

$$h_3 = 3398.80 \text{ kJ/kg}, s_3 = 6.7262 \text{ kJ/kgK}$$

In order to find the reheat pressure p_4/p_5, we refer to the steam table to find the saturation pressure corresponding to $s_3 = s_4$. It is found that

$$p_4 = p_5 = 0.665 \text{ MPa, so } h_4 = 2761.2 \text{ kJ/kg}$$

Further,

$$h_5 = 3482.0 \text{ kJ/kg and } s_5 = 7.9545 \text{ kJ/kgK}$$

In contrast to Figure 1.11 the expansion from state 5 to 9.6 kPa, i.e., state 6 of this example, is in the wet region. Hence at the turbine exhaust

$$h_6 = 2516.3 \text{ kJ/kg and } x_6 = 0.972$$

The total heat added to the boiler is

$$Q_1 = Q_{31} + Q_{54} = (h_3 - h_1) + (h_5 - h_4) = (3398.80 - 188.42) + (3482.0 - 2761.2)$$
$$= 3931.18 \text{ kJ/kg}$$

and the total work output from turbine is

$$W_T = W_{34} + W_{56} = (h_3 - h_4) + (h_5 - h_6) = (3398.80 - 2761.20) + (3482.0 - 2516.3)$$
$$= 1603.30 \text{ kJ/kg}$$

Therefore, the cycle efficiency is

$$\eta_r = 1606.30/3931.18 \times 100 = 40.86\%$$

The specific steam consumption is

$$\text{s.s.c.} = 3600/1606.3 = 2.24 \text{ kg/kWh},$$
$$(\text{since1 kWs} = 1 \text{ kJ}, 1 \text{ kWh} = 3600 \text{ kJ}).$$

1.6.3 Regenerative rankine cycle

Note from Figure 1.9 that the efficiency of the un-superheated cycle is less than the efficiency of the Carnot cycle, i.e., $(T_1-T_2)/T_1$, because a certain portion of the heat supplied to the boiler is transferred to raise the feedwater temperature from T_4 to T_1 [10]. If this additional heat to the feedwater could be transferred reversibly from the steam part of the cycle, then the heat supplied to the boiler would be transferred to the feedwater at a temperature somewhere between T_4 and T_1. This cycle, where efficiency is raised as explained, is known as the regenerative cycle. In this cycle, a specified quantity of energy remains circulating within the cycle, and as a result, irreversibility in the process of mixing relatively cold water with hot steam gets reduced.

The purpose of regeneration or "stage bleeding" is to improve cycle efficiency, which in turn minimizes steam consumption and results in plant efficiency. This is realized by adding regenerative feedheating to the superheat-reheat cycle. Feedheating involves extracting a fraction of the steam flowing through the turbine from one or more positions along the turbine expansion and using heat in the steam to preheat the water in feedwater heaters prior to entering the boiler (Figure 1.12) [3,11,12].

Each unit of extracted steam does a certain amount of work in the turbine from throttle conditions to point of extraction and transfers the remainder of its heat to the feedwater, thus conserving the total heat, instead of losing part of the heat to the circulating water passing through the main condenser.

Figure 1.13 shows T-s diagram of a multistage feedheating system, which shows that the heat addition to the cycle is reduced from the area bounded by 4′-4-4″-5-1-1′-2-2′-4′ to the area bounded by 3′-3-4″-5-1-1′-2-2′-3' by adopting regenerative

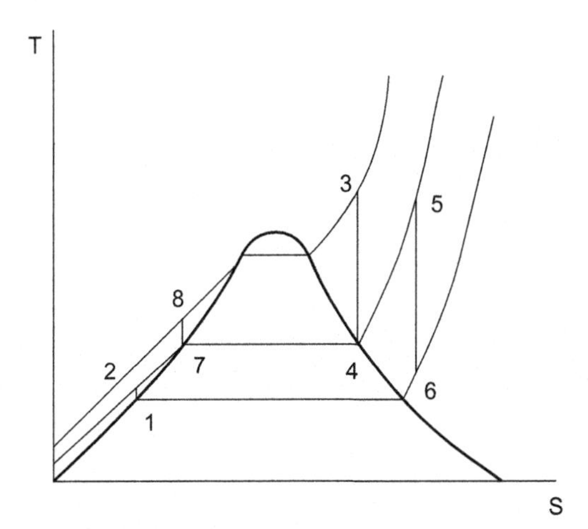

Figure 1.12 Regenerative rankine cycle.

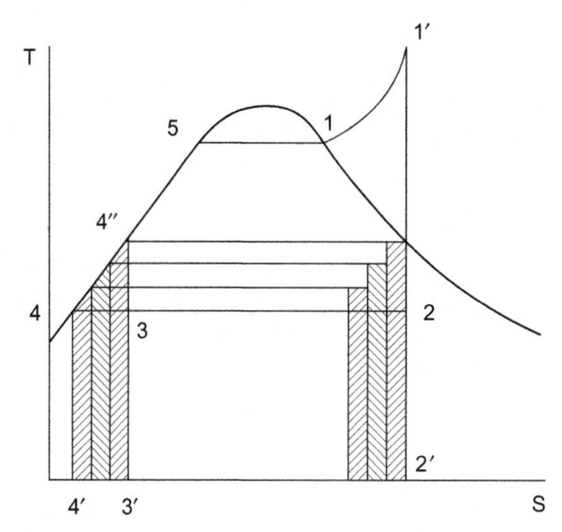

Figure 1.13 Multistage feedheating system.

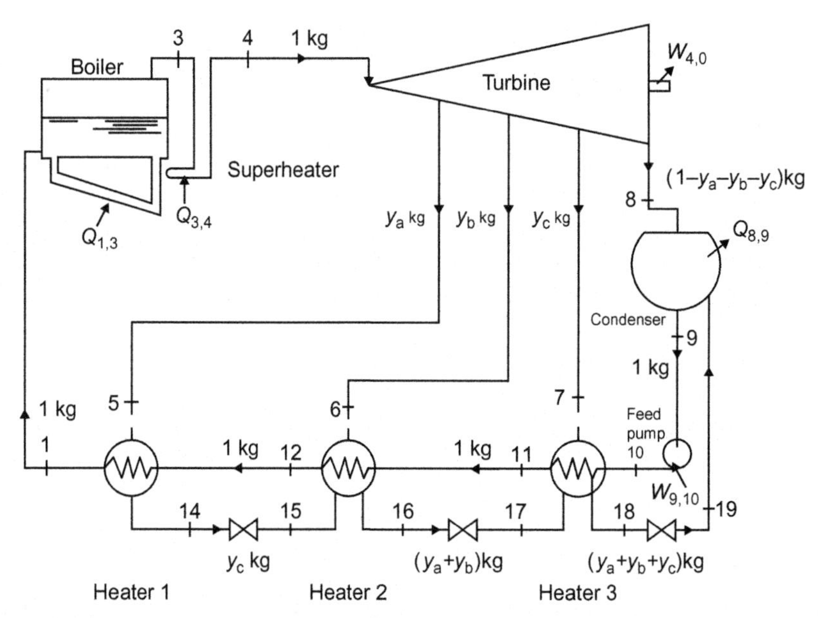

Figure 1.14 Triple extraction regenerative cycle.

feedheating, i.e., the heat addition to the cycle is reduced by the area 4′-4-4″-3-3′-4′, keeping the output unchanged, thereby reducing the cost of power generation.

Figure 1.14 and Figure 1.15 show a typical triple-extraction three-stage feedheating regenerative cycle [2].

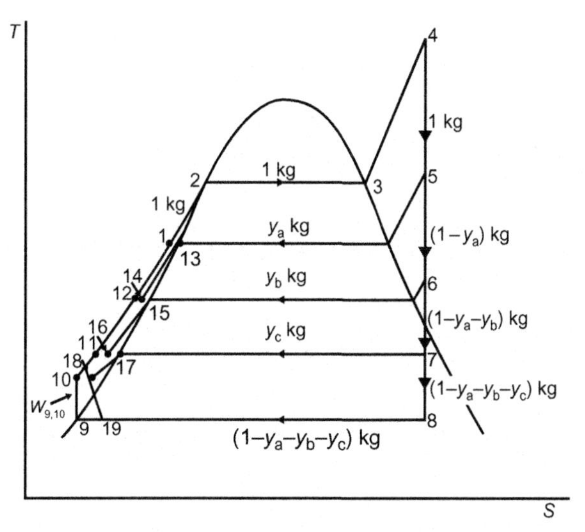

Figure 1.15 Triple extraction regenerative cycle.

The ideal cycle efficiency of the above feedheating cycle (see Figure 1.15) is given by

$$\eta_{rgn} = \frac{[1 * (h_4 - h_5) + (1 - y_a) * (h_5 - h_6) + (1 - y_a - y_b) * (h_6 - h_7) + (1 - y_a - y_b - y_c) * (h_7 - h_8)}{(h_4 - h_1)} \tag{1.32}$$

Example 1.7

Calculate the ideal cycle efficiency and specific steam consumption (s.s.c.) of a regenerative cycle with triple extractions, as shown in Figure 1.14 and Figure 1.15, with boiler outlet steam pressure and superheat steam temperature are 8 MPa and 773 K, respectively, and condenser pressure 9.6 kPa. If the steam flowing through the turbine is 100 kg/s, also find the power output. Determine the heat rate of the turbine, which is defined as "the heat required at the turbine inlet per unit of power generated." (Assume the turbine extraction bleed steam pressures such that temperature difference between T_2 and T_9 (Figure 1.15) is approximately equally divided. Neglect work input to boiler feed pump. Also assume that on completion of regenerative feedheating, the enthalpies of both the compressed feedwater and condensed steam are equal to the enthalpies of the saturated liquid corresponding to the respective extraction bleed steam pressures.)

Solution: From the steam table it is noted that at 8 MPa, the temperature at T_2 is 568 K and corresponding to 9.6 kPa, temperature at T_9 is 318 K.

For approximately equal temperature difference ((568−318)/4 = 62.5 K) among (T_2−T_{13}), (T_{13}−T_{15}), (T_{15}−T_{17}) and (T_{17}−T_9), the turbine extraction bleed steam pressures are chosen as 2.8 MPa, 0.8 MPa, and 0.12 MPa.

In order to facilitate solving this problem the following table has been prepared based on the given data and assumptions:

Steam pressure	Temperature K		Inlet/ext. steam entropy kJ/kgK	Enthalpy kJ/kg			Dryness fraction %
	Inlet/ext. steam	Saturated fluid		Inlet/ext. steam	Sat. steam	Sat. water	
1	2	3	4	5	6	7	8
8.00 MPa	773.0	568.0	$s_4 = 6.7262$	$h_4 = 3398.80$	$h_3 = 2759.90$	$h_2 = 1317.1$	–
2.80 MPa	608.0	503.0	$s_5 = 6.7262$	$h_5 = h_{EX1} = 3086.20$	2802.00	$h_{13} = 990.48$	–
0.80 MPa	456.0	443.0	$s_6 = 6.7262$	$h_6 = h_{EX2} = 2797.40$	2767.50	$h_{15} = 720.94$	–
0.12 MPa	378.0	378.0	$s_7 = 6.7262$	$h_7 = h_{EX3} = 2467.2$	2683.40	$h_{17} = 439.36$	$x_7 = 90.363$
9.60 kPa	318.0	318.0	$s_8 = 6.7262$	$h_8 = 2125.4$	2583.40	$h_9 = 188.50$	$x_8 = 80.878$

This table is also reflects the following:

Column 2 shows temperature at different extraction bleed steam stages and at the exhaust of the turbine
Temperature of saturated liquid is shown in column 3
Isentropic expansion is given in column 4
Enthalpies of vapor at the inlet, extraction bleed steam stages and at the exhaust of the turbine are given in column 5
Enthalpies of saturated vapor and liquid at various pressures are presented in columns =6 and 7, respectively
Dryness fractions of wet vapor at the last extraction bleed steam stage and at the exhaust of the turbine are given in column 8

Let us denote the turbine extraction bleed steam flow quantity to feedheaters per kg of steam supplied to the turbine as y_a, y_b, and y_c and give the extraction steam from each feedwater heater as EX1, EX2, and EX3 and the states of condensed steam/drain from each feedwater heater as D1, D2, and D3, both in decreasing order of pressure.

As per the given conditions and assumptions, we have $h_1 = h_{13}$, $h_{D1} = h_{12} = h_{14} = h_{15}$, $h_{D2} = h_{11} = h_{16} = h_{17}$, and $h_{D3} = h_9 = h_{10} = h_{18}$.

Using the energy equation for each heater, we get the following relationships:

No. 1 heater

$$h_{13} - h_{15} = y_a \times (h_{\mathrm{EX1}} - h_{\mathrm{D1}}) = y_a \times (h_5 - h_{15})$$

$$y_a = (h_{13} - h_{15})/(h_5 - h_{15})$$

$$= (990.48 - 720.94)/(3086.20 - 720.94) = 0.114 \text{ kg}$$

No. 2 heater

$$h_{15} - (1 - y_a) \times h_{17} = y_a \times h_{15} + y_b \times (h_{EX2} - h_{D2})$$

$$h_{15} - (1 - y_a) \times h_{17} = y_a \times h_{15} + y_b \times (h_6 - h_{17})$$

$$y_b = \{(1 - y_a) \times (h_{15} - h_{17})\}/(h_6 - h_{17})$$

$$= \{(1 - 0.114) \times (720.94 - 439.36)\}/(2797.40 - 439.36)$$

$$= 0.106 \text{ kg}$$

No. 3 heater

$$(1 - y_a) \times h_{17} - (1 - y_a - y_b) \times h_9 = y_b \times h_{D2} + y_c \times (h_7 - h_{D3})$$

$$(1 - y_a) \times h_{17} - (1 - y_a - y_b) \times h_9 = y_b \times h_{17} + y_c \times (h_7 - h_9)$$

$$y_c = \{(1 - y_a - y_b) \times (h_{17} - h_9)\}/(h_7 - h_9)$$

$$y_c = \{(1 - 0.114 - 0.106) \times (439.36 - 188.50)\}/(2467.2 - 188.50) = 0.086 \text{ kg}$$

Having found the turbine extraction bleed steam flows, it is possible to calculate the heat and work transfers for the cycle.

Heat added to the boiler is

$$Q_{4-13} = h_4 - h_{13} = 3398.80 - 990.48 = 2408.32 \text{ kJ/kg}$$

Heat rejected in the condenser is

$$Q_{8-9} = (1 - y_a - y_b - y_c) * (h_8 - h_9)$$
$$= (1 - 0.114 - 0.106 - 0.0860) * (2125.40 - 188.50)$$
$$= 1344.21 \text{ kJ/kg}$$

Net work output is

$$W = Q_{4-13} - Q_{8-9} = 1064.11 \text{ kJ/kg}$$

Therefore, the cycle efficiency is

$$\eta_r = 1064.11/2408.32 * 100 = 44.18\%$$

The specific steam consumption is

$$\text{s.s.c.} = 3600/1064.11 = 3.38 \text{ kg/kWh}$$

The power output is

$$P = 100 \times W = 100 * 1064.11 = 106411 \text{ kJ/s}$$
$$= 106411 \text{ kW}$$

Neglecting pressure drop in steam pipeline, it may be assumed that "heat required at the turbine inlet" is same as "heat added to the boiler."

The heat rate of the turbine is

$$\text{HR} = 100 * 2408.32 * 3600/106411 = 8147.61 \text{ kJ/kWh}$$

(Notes:

1. There are two types of feedwater heaters: open or direct contact feedwater heater and closed (contact less) feedwater heater. Two terms that are always used for heat transfer in closed feedwater heater are *terminal temperature difference* (TTD) and *drain cooler approach* (DCA) as defined below:

   TTD = Saturation temperature of extraction bleed steam to the heater − Feedwater temperature at heater exit.

   DCA = Heater extraction bleed steam condensate drain temperature − Feedwater temperature at heater inlet.

2. The results of Examples 1.4, 1.5, 1.6, and 1.7, as presented in the following table, compare the performance of the Ideal Saturated Rankine, Real Saturated Rankine, Ideal Superheat Reheat Rankine, and Ideal Superheat Regenerative Rankine cycles, with steam operating between 8 MPa and 9.6 kPa pressures.

Parameters	Ideal saturated rankine cycle	Real saturated rankine cycle	Ideal superheat reheat rankine cycle**	Ideal superheat regenerative rankine cycle**
Heat Added, kJ/kg	2561.58	2560.16	3931.18	2408.32
Cycle Efficiency, %	36.61	29.19	40.86	44.18
Specific Steam Consumption, kg/kWh	3.84	4.82	2.24	3.38

** In these cycles "work input to boiler feed pump" as well as "isentropic efficiencies of expansion and compression processes" have been neglected.)

Example 1.8

For the same cycle of Example 1.7 consider the following and calculate the cycle efficiency, specific steam consumption, power output, and heat rate of the turbine.

Isentropic efficiencies of expansion and compression processes as 0.9 and 0.8, respectively
Work input to boiler feed pump to be taken into consideration
Assume TTD for the lowest pressure heater as 3 K and for the remaining heaters as −2 K. DCA for all the heaters as 5 K
Neglect pressure drop in feedwater pipe, steam pipe, and extraction pipelines

Solution: For 0.9 isentropic efficiency for the expansion process and for 0.8 isentropic efficiency for the compression process all states on the vertical line of Figure 1.15 show that states 5, 6, 7, 8 and 10 would move toward the right. Hence, various thermodynamic parameters of these states will have new values as presented in the following tables:

States	Steam press.	Enthalpy, kJ/kg				Inlet/ext. steam entropy kJ/kgK	Specific volume m^3/kg	Dryness fraction x %
		Isentropic inlet/ext. steam	Actual inlet/ext. steam	Sat. steam	Sat. water			
4	$p_2 = 8.00$ MPa	$h_4 = 3398.80$	$h_4 = 3398.80$	$h_3 = 2759.90$	$h_2 = 1317.1$	$s_4 = 6.7262$	–	–
5	2.80 MPa	3086.20	$h_5 = h_{EX1} = 3117.46$	2802.00	$h_{13} = 990.48$	$s_5 = 6.7771$	–	–
6	0.80 MPa	2797.40	$h_6 = h_{EX2} = 2857.54$	2767.50	$h_{15} = 720.94$	$s_6 = 6.8546$	–	–
7	0.12 MPa	2467.2	$h_7 = h_{EX3} = 2560.36$	2683.40	$h_{17} = 439.36$	$s_7 = 6.9729$	–	$x_7 = 94.515$
8	$p_9 = 9.60$ kPa	2125.4	$h_8 = 2252.74$	2583.40	$h_9 = 188.50$	$s_8 = 7.1263$ $s_9 = s_{10} = 0.6388$	$v_{f9} = 0.00101$	$x_8 = 86.195$

States	Steam press.	Temperature K				Enthalpy, kJ/kg	
		Inlet/Ext. steam	Sat. fluid	Applying given TTD	Applying given DCA	Feedwater at 8.00 MPa	Heater drain
4	$P_2 = 8.00$ MPa	$T_4 = 773.0$	$T_2 = 568.0$	–	–	$h_2 = 1317.1$ $h_{10} = 196.56$	–
5	2.80 MPa	$T_5 = 621.0$	$T_{13} = 503.0$	$T_1 = 501.0$	$T_{14} = T_{12} + 5 = 446.0$	$h_1 = 982.09$	$h_{14} = 733.33$
6	0.80 MPa	$T_6 = 481$	$T_{15} = 443.0$	$T_{12} = 441.0$	$T_{16} = T_{11} + 5 = 386.0$	$h_{12} = 714.43$	$h_{16} = 474.48$
7	0.12 MPa	$T_7 = 378$	$T_7 = T_{17} = 378.0$	$T_{11} = 381.0$	$T_{18} = T_{10} + 5 = 323.0$	$h_{11} = 458.62$	$h_{18} = 209.35$
8	$p_9 = 9.60$ kPa	–	$T_8 = T_9 = 318$	–	–	$h_9 = 188.50$	–

Using the energy equation for each heater as in Example 1.6, we get the following results:

No.1 heater

$$y_a = (h_1 - h_{12})/(h_5 - h_{14}) = (982.09 - 714.43)/(3117.46 - 733.33)$$
$$= 0.112 \text{ kg}$$

No.2 heater

$$y_b = (1 - y_a) * (h_{12} - h_{11})/(h_6 - h_{16})$$
$$= (1 - 0.112) * (714.43 - 458.62)/(2857.54 - 474.48)$$
$$= 0.095 \text{ kg}$$

No.3 heater

$$y_c = (1 - y_a - y_b) * (h_{11} - h_{10})/(h_7 - h_{18})$$
$$= (1 - 0.112 - 0.095) * (458.62 - 196.56)/(2560.36 - 209.35)$$
$$= 0.088 \text{ kg}$$

Having found the turbine extraction bleed steam flows, it is possible to calculate the heat and work transfers for the cycle.

Heat added to the boiler is

$$Q_{1-4} = h_4 - h_1 = 3398.80 - 982.09 = 2416.71 \text{ kJ/kg}$$

Heat rejected in the condenser is

$$Q_{8-9} = (1 - y_a - y_b - y_c) * (h_8 - h_9) = (1 - 0.112 - 0.095 - 0.088) * (2252.74 - 188.50)$$
$$= 1455.29 \text{ kJ/kg}$$

Work output of turbine

$$W_T = 1 * (h_4 - h_5) + (1 - y_a) * (h_5 - h_6) + (1 - y_a - y_b) * (h_6 - h_7)$$
$$+ (1 - y_a - y_b - y_c) * (h_7 - h_8)$$
$$= (3398.80 - 3117.46) + (1 - 0.112) * (3117.46 - 2857.54)$$
$$+ (1 - 0.112 - 0.095) * (2857.54 - 2560.36)$$
$$+ (1 - 0.112 - 0.095 - 0.088) * (2560.36 - 2252.74)$$
$$= 281.34 + 230.81 + 235.66 + 216.87 = 964.68 \text{ kJ/kg}$$

Work input to feed pump is

$$W_P = v_{f9} * (p_B - p_{A'})$$
$$= 0.00101 * (81.600 - 0.098) \times 98/0.8$$
$$= 10.08 \text{ kJ/kg}$$

(since 1 kPa = 0.0102 kg/cm^2, 1 MPa = 10.2 kg/cm^2, 1 m^2 = 10000 cm^2 and 1000 mkgf = 9.8 kJ).
Net work output is

$$W = W_T - W_P = 954.60 \text{ kJ/kg}$$

Therefore, the cycle efficiency is

$$\eta_r = (954.60/2416.71) * 100 = 39.50\%$$

The specific steam consumption is

$$\text{s.s.c.} = 3600/954.60 = 3.77 \text{ kg/kWh}$$

The power output is

$$P = 100 * W = 100 * 954.60 = 95460 \text{ kJ/s} = 95460 \text{ kW}$$

The heat rate of the turbine is

$$\text{HR} = 100 * 2416.71 * 3600/95460 = 9113.93 \text{ kJ/kWh}$$

(Note: The following table gives a comparison of the results of Examples 1.7 and 1.8:

Example no.	F_{EX1} kg	F_{EX2} kg	F_{EX3} kg	η_r %	s.s.c. kg/kWh	P kW	HR kJ/kWh
1.7	0.114	0.106	0.086	44.18	3.38	106411	8147.61
1.8	0.112	0.095	0.088	39.50	3.77	95460	9113.93

The above table shows that the performance of the real cycle, i.e., Example 1.8 is influenced by the factors "work input to boiler feed pump" and "isentropic efficiencies of expansion and compression processes," resulting in reduction in cycle efficiency and power output and increase in specific steam consumption and turbine heat rate. Nevertheless, performance of this cycle does indicate substantial improvement in performance compared with the performance of the ideal saturated Rankine cycle.)

Figure 1.16 shows a typical steam-water flow diagram of a superheat-reheat-regenerative cycle.

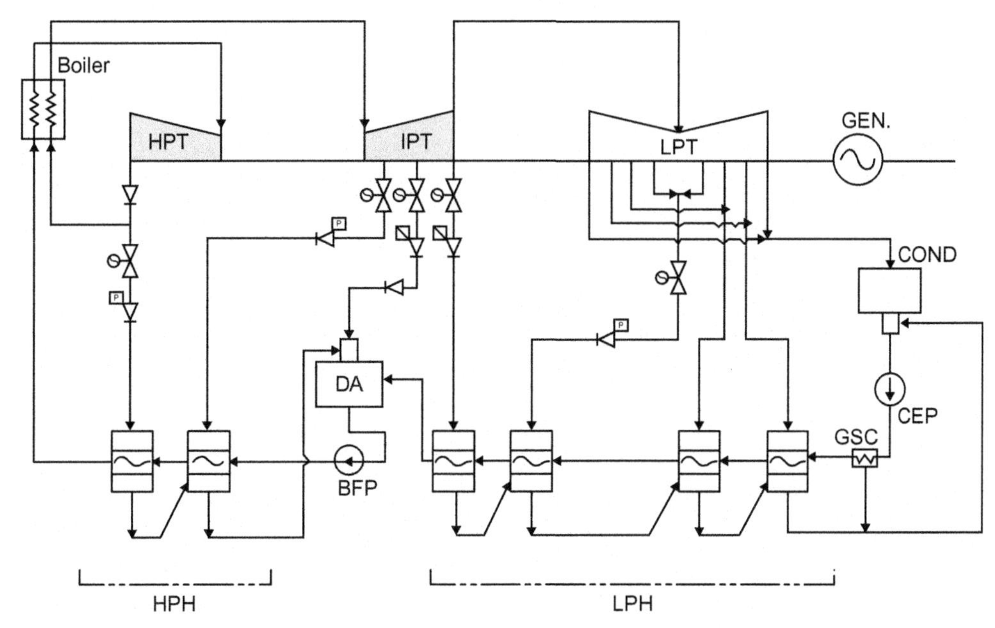

Figure 1.16 Typical steam water flow diagram.

1.7 KALINA CYCLE

The Kalina cycle, named after the Russian engineer *Alexander Kalina*, is a thermodynamic cycle for converting thermal energy to mechanical power. The working fluid in this cycle is a mixture of at least two different fluids (typically water and ammonia). Ammonia has a lower boiling point compared with water. Hence, when the temperature of the mixture increases, the ammonia will boil first. In contrast to this when the mixture is cooled, water will condense first (Figure 1.17) [13,14]. When the mixture begins to boil at 550 kPa (state 3) the concentration of ammonia is 70% and water is 30%. As the boiling of mixture continues, the temperature increases and at state 4 the concentration of the remaining fluid and vapor formed are given by states 6 and 5, respectively. On further boiling of the mixture, when it reaches state 7, the mixture is saturated vapor. The concentration of vapor at this point is the same as the concentration of the liquid at the beginning of the evaporation process.

Unlike the Rankine cycle, where considerable heat energy is lost in the isothermal vaporization of water to steam, the binary mixture in the Kalina cycle vaporizes non-isothermally, resulting in better performance. Variable temperature boiling permits the working fluid to maintain a temperature closer to that of the combustion gases in the boiler. The use of mixture as the working fluid allows manipulating pressure in the system by changing composition of the mixture. At given cooling conditions the

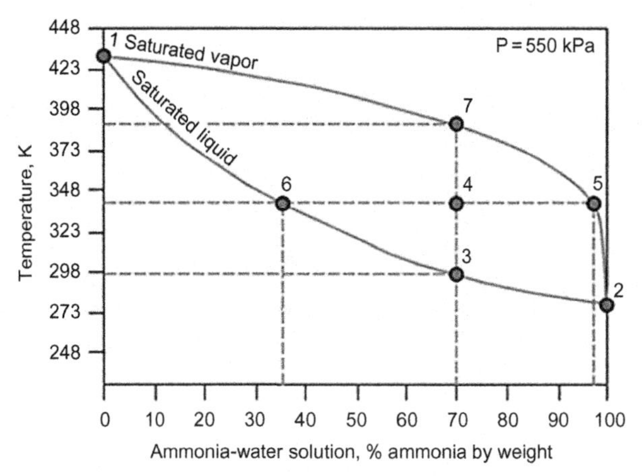

Figure 1.17 Kalina cycle.

pressure in the condenser can be reduced to slightly above atmospheric pressure by decreasing the ammonia concentration of the condensing fluid. Increasing the ammonia concentration may raise the evaporation pressure. A high ammonia concentration causes a steep increase in pressure. The demands of high ammonia concentration in the working fluid during evaporation and expansion and low ammonia concentration during condensation are met by the distillation and mixing processes.

High temperature exhaust from any process/equipment, e.g., a gas turbine, is the heat source of the ammonia-water mixture for recovering heat [15]. By absorbing heat the mixture vaporizes and then drives the turbine. On partial completion of expansion in the turbine, the mixture gets condensed. The condensed mixture is further diluted to lower its pressure and extract more work from the turbine. Finally, this diluted working fluid is re-concentrated and sent back to the heat source. However, this cycle is not commercially viable, since it is difficult to handle large quantities of ammonia safely and reliably.

Properties of Ammonia:	
Boiling Point at 101.3 kPa	239.5 K
Melting Point	195 K
Critical Point	11.28 MPa and 405.4 K
Latent Heat of Vaporization at 101.3 kPa	1371.2 kJ/kg
Vapor Pressure at 294 K	888 kPa
Liquid Density at Boiling Point	682 kg/m^3
Gas Density at Boiling Point	0.86 kg/m^3
Gas Density at 101.3 kPa and 288 K	0.73 kg/m^3
Liquid/Gas Equivalent at 101.3 kPa and 288 K	947 vol/vol
Gas Specific Gravity at 101.3 kPa and 294 K	0.597
Gas Specific Volume at 101.3 kPa and 294 K	1.411 m^3/kg
Liquid Specific Heat Capacity (constant pressure) at 300 K	4.75 kJ/kg.K
Gas Specific Heat Capacity (constant pressure) at 101.3 kPa and 288 K	0.037 kg/mol.K

Gas Specific Heat Capacity (constant volume) at 101.3 kPa and 288 K	0.028 kg/mol.K
Liquid Thermal Conductivity at 300 K	477×10^6 kW/m.K
Gas Thermal Conductivity at 101.3 kPa and 273 K	22.19 mW/m.K
Gas Solubility in Water at 101.3 kPa and 273 K	862 vol/vol
Gas Auto-ignition Temperature	903 K

1.8 BINARY VAPOR CYCLE

It is known that the higher the temperature of the heat source, the greater the cycle efficiency. Thus, a considerable improvement in cycle efficiency could be achieved provided evaporation takes place at a temperature corresponding to the metallurgical limit of the turbine. There is, however, a constraint. Although steam has a relatively low critical temperature (647.14 K), its critical pressure is very high (22.12 MPa). As a result, the design becomes critical. The process could be simplified if there is a working fluid whose critical temperature is higher than the metallurgical limit of about 873 K, but the vapor pressure of the fluid is moderate [2]. One such working fluid is mercury, with a vapor temperature of 873 K at a saturation pressure of only 2.3 MPa. It is evident that at a comparable temperature the saturation pressure of mercury is far below that of water.

Figure 1.18 and Figure 1.19 [2] show a cycle where mercury in conjunction with steam is used in a binary cycle to harness the benefit of the whole temperature range

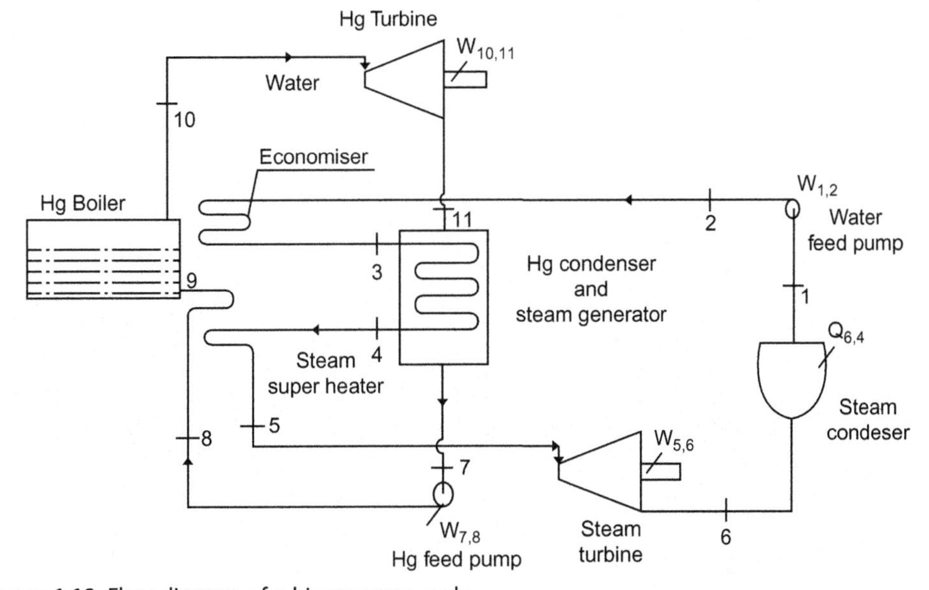

Figure 1.18 Flow diagram of a binary vapor cycle.

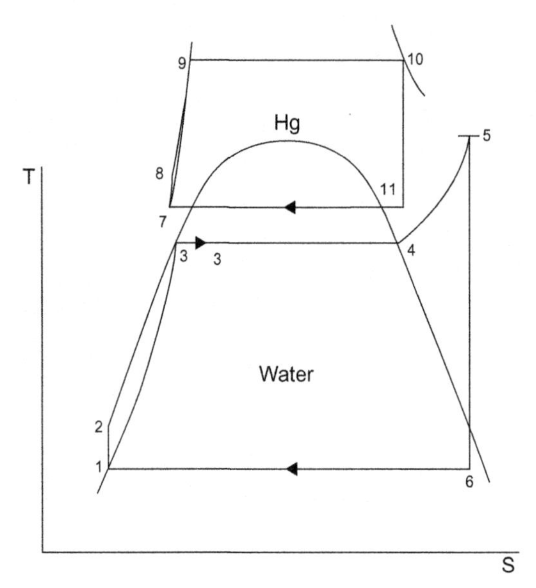

Figure 1.19 Binary vapor cycle.

beginning with the metallurgical limit down to the atmosphere. In a mercury boiler mercury is evaporated and then it expands in the mercury turbine to a pressure of about 10 kPa. The mercury condenser then serves the dual purpose of condensing mercury and acting as a steam generator. Feedwater heating and superheating of steam after evaporation are performed by the heat of condensation of mercury vapor.

From Figure 1.19 it is observed that dry saturated vapor of mercury leaves the mercury boiler at state 10 and expands isentropically in the mercury turbine to state 11. A two-phase mixture of mercury vapor-liquid then enters the "mercury condenser-water boiler" heat exchanger to exchange its heat of condensation to water steam and reaches state 7. Then the mercury pump raises its pressure to state 8 for further evaporation in the mercury boiler.

Water becomes dry saturated vapor in a "mercury condenser-water boiler" at state 4 and then passes through the superheater in the mercury boiler furnace to state 5. Superheated steam then enters the steam turbine, expands isentropically to state 6, and condenses to state 1. Saturated water thereafter is pushed to the boiler to repeat the cycle.

Although quite attractive, this cycle is not commercially viable since it suffers from the following disadvantages:

i. Mercury vapor is toxic and extremely hazardous.
ii. At the normal heat sink temperature of about 300 K, the vapor pressure of mercury is extremely low and its specific volume is very large. Thus, the vacuum to

be maintained in the mercury condenser is exorbitantly high, causing the size of the condenser to be impracticably large.

iii. Mercury has low latent heat (about one-eighth of the latent heat of steam), therefore, specific mercury consumption is high.

iv. Capital expenditure of this plant is exorbitant.

v. Its maintenance cost is also high.

Properties of Mercury:	
Boiling Point at 101.3 kPa	630 K
Freezing Point	234 K
Latent Heat of Vaporization at 101.3 kPa	272.1 kJ/kg
Latent Heat of Fusion at 101.3 kPa	11.3 kJ/kg
Vapor Pressure at 373 K	0.0364 kPa
Density at 373 K	13352.2 kg/m^3
Specific Heat Capacity (constant pressure) at 273 K	0.139 kJ/kg.K
Thermal Conductivity at 273 K	7.82 W/m.K

1.9 PROBLEMS

1.1 Steam expands in a Carnot cycle from 6 MPa pressure to 10.13 kPa pressure in the condenser. Calculate the cycle efficiency and specific steam consumption of the process.

(**Ans.:** $\eta_c = 41.83$ %; s.s.c. = 5.48 kg/kWh.)

1.2 Assume isentropic efficiencies of expansion and compression as 0.9 and 0.8, respectively, and calculate the cycle efficiency and specific steam consumption of the process of problem 1.1.

(**Ans.:** $\eta_c = 31.81$ %; s.s.c. = 7.20 kg/kWh.}

1.3 Calculate the heat and work transfers of a Carnot cycle, using steam between pressures 3 MPa and 8.67 kPa.

(**Ans.:**
Heat supplied to the cycle, $Q_{2\text{-}3} = 1793.90$ kJ/kg
Heat rejected from the cycle, $Q_{4\text{-}1} = 1118.22$ kJ/kg.s
Net work output of the system, $W_o = 675.68$ kJ/kg)

1.4 Calculate the heat and work transfers of an ideal saturated Rankine cycle, using steam between pressures 3 MPa and 8.67 kPa.

(**Ans.:**
Heat added in the steam generator, $Q_1 = 2619.00$ kJ/kg
Heat rejected to the condenser, $Q_2 = 1761.32$ kJ/kg
Work output of turbine, $W_T = 860.70$ kJ/kg

Work input to feed pump, $W_P = 3.02$ kJ/kg
Net work output of the system, $W_o = 857.68$ kJ/kg
Efficiency of the cycle, $\eta_r = 32.75$ %)

1.5 Steam at a rate of 80 kg/s expands through an ideal saturated Rankine cycle turbine at temperature 623 K and condenses to a temperature of 313 K. Calculate the efficiency of the cycle.
(**Ans.:** $\eta_r = 38.94$ %)

1.6 Assume isentropic efficiencies of expansion and compression as 0.85 and 0.75, respectively, and calculate the cycle efficiency and specific steam consumption of the process of problem 1.5.
(**Ans.:** $\eta_r = 32.84$%; s.s.c. = 4.61 kg/kWh)

1.7 Steam is generated at a pressure of 2 MPa and is then superheated to a temperature of 623 K. If this superheated steam is used in an engine working on ideal Rankine cycle, and expanded to a pressure of 10.33 kPa, calculate the thermal efficiency of the cycle and the dryness fraction of the steam after expansion.
(**Ans.:** $\boldsymbol{x_2} = 0.842$**;** $\eta_r = 31.55$ %)

1.8 Calculate the power output and the heat rate of a real superheat-reheat Rankine cycle turbine, through which 200 kg/s of steam at pressure and temperature of 13 MPa and 723 K, respectively, expands to a pressure of 3.5 MPa with an isentropic efficiency of 0.90, gets reheated to temperature of 753 K, and thereafter expands further with an isentropic efficiency of 0.85 to condenser pressure of 11.87 kPa. The saturated water is then compressed to pressure 13 MPa with an isentropic compression efficiency of 0.80.
(**Ans.:** $P = 248144$ kW; $HR = 10188.70$ kJ/kWh)

1.9 In an ideal superheat steam Rankine cycle turbine operating with four stages of feedheating, steam enters at pressure 10 MPa and temperature 673 K and condenses at pressure 8.66 kPa. Feedwater heating extraction bleed steam pressures are such that the difference between the saturated temperature at 10 MPa and the saturated temperature at 8.66 kPa is approximately equally divided. Assume isentropic efficiency of expansion as 0.8. Assume any other conditions that may be required to find the cycle efficiency, power output, and heat rate of the turbine. Neglect the work input to the boiler feed pump.
(**Ans.:** Assumptions:

i. Steam flow through turbine 100 kg/s
ii. Isentropic efficiency of compression as 1.0
iii. On completion of regenerative feedheating, the enthalpies of both compressed feedwater and condensed steam are equal to the enthalpies of saturated liquid corresponding to respective extraction bleed steam pressures.

$\eta_r = 36.22$%; $P = 71614$ kW; $HR = 9940.29$ kJ/kWh)

1.10 A steam cycle operates between pressures 3.0 MPa and 4.0 kPa, and uses a superheat temperature of 723 K. The mercury cycle operates between pressures 1.4 MPa and 10 kPa, and the mercury is dry saturated prior to entering the turbine. Determine the cycle efficiency of this binary vapor cycle.

(Assume quantity of steam in circulation is 1 kg and quantity of mercury in circulation is y kg. Neglect the work supplied to the feed pumps.

Given: Referring to Figure 1.19

$h_{10} = 362.55$ kJ/kg, $h_8 = h_7 = 34.33$ kJ/kg, $x_{11} = 0.7286$, $h_{11} = 248.93$ kJ/kg)

(**CLUE:** Sequence of calculations is to be as below:

i. First, calculate the heat and work transfers in the steam-water cycle.

ii. Perform the same calculations for the mercury cycle.

iii. Use the energy equation of the mercury condenser-steam generator as $y^*(h_7 - h_{11}) = 1^*(h_4 - h_3)$.)

(**Ans.:** $\eta_B = 51.79$ %)

BIBLIOGRAPHY

[1] Francis Weston Sears. Thermodynamics. Addison-Wesley Publishing Company, Inc.; 1966.

[2] Rogers, G.F.C., Mayhew, Y.R., Engineering Thermodynamics Work and Heat Transfer. The English Language Book Society; 1967.

[3] El-Wakil, M.M. Power Plant Technology. McGraw-Hill Book Company; 1984.

[4] Theodore Baumeister, Marks L.S., (Ed.), Mechanical Engineers' Handbook, sixth ed. McGraw-Hill Book Company, Inc.; 1958.

[5] Nag, P.K., Power Plant Engineering. Tata McGraw-Hill Publishing Company Limited; 2008.

[6] Goodall, P.M., (General ed.), The Efficient Use of Steam. Westbury House; 1980.

[7] Lyle O., The Efficient Use of Steam. Her Majesty's Stationery Office; 1968.

[8] Newton, R.G., A History of Physics. Universities Press (India) Private Limited; 2008.

[9] Wrangham D.A., Theory and Practice of HEAT ENGINE. The English Language Book Society; 1962.

[10] Black, Veatch., Power Plant Engineering. CBS Publishers and Distributors; 2001.

[11] Gaffert, G.A. Steam Power Stations. McGraw-Hill Book Company, Inc.; 1952.

[12] Skrotzki, B.G.A., Vopat, W.A., Power Station Engineering and Technology. McGraw-Hill Book Company, Inc.; 1988.

[13] Carlos Eymel Campos Rodriguez, et al., Exergetic and Economic Analysis of Kalina Cycle for Low Temperature Geothermal Sources in Brazil. Proceedings of ECOS 2012 – The 25th International Conference, Perugia, Italy; 2012.

[14] Kohler, S., Saadat, A., 2003. Thermodynamic Modelling of Binary Cycles Looking for Best Scenarios. International Geothermal Conference, Reykjavik; 2003.

[15] Wall, G., Chuang, C.-C., Ishida, M., 1989. Exergy Study of the Kalina Cycle. Presented at ASME Winter Annual Meeting, San Francisco, California; 1989.

CHAPTER 2

Steam Generators

2.1 INTRODUCTION

Man has used the technique of converting water (liquid) into steam (vapor) and using the expansive force of this vapor as far back as 150 B.C. Yet it took human beings more than 1800 years to harness the practical qualities of steam to do useful work. It was around the year 1711 that the first commercial piston-operated mine pump was invented [1], but it wasn't until the latter half of the nineteenth century that the first steam-operated central power station was introduced. It was in the year 1882 that Holborn Viaduct power station in London and Pearl Street Station in New York had been put into service using reciprocating steam engines.

The main objective of a steam power station is to generate electrical power. In a steam power station, the electrical energy is produced according to the principle of "external combustion," where the "heat of combustion" of the fuel is transferred to a prime mover by a "working medium." In the steam generator, low-temperature water is the working medium that receives the heat of combustion of fuel and becomes high-energy steam. The heat of steam is converted to mechanical energy in the steam turbine and then to electrical energy in the generator. The sequence of these activities is shown in Figure 2.1.

The chemical energy available in fossil fuel (i.e., coal, fuel oil, gas) is converted to heat energy by combustion in a steam generator. The heat thus liberated is absorbed by continuously feeding water in a combination of heat-transfer surfaces, resulting in a continuous generation of steam. The water fed into a steam generator is called feedwater. Steam and feedwater together is called working fluid. The name "steam generator" is also still called "boiler," but modern steam generators in the supercritical class do not involve the "boiling" phenomenon.

The fuel-firing equipment of a steam generator should completely burn the fuel used in the furnace to release as much energy as possible. Air is fed into the furnace for combustion of fuel-forming products of combustion or flue gas. The heat released by burning fuel is absorbed in different heat-transfer surfaces to the maximum level possible practically and economically to keep the loss of heat to a minimum. In the heating surfaces, the flue gas transfers its heat to the working fluid. Thus, the feed-water is pre-heated to the saturation temperature and vaporized. The saturated steam thus formed is further superheated. After passing over the heating surfaces at various zones the flue gas is cooled and discharged to the atmosphere through a stack.

Thermal Power Plant
DOI: http://dx.doi.org/10.1016/B978-0-12-801575-9.00002-0

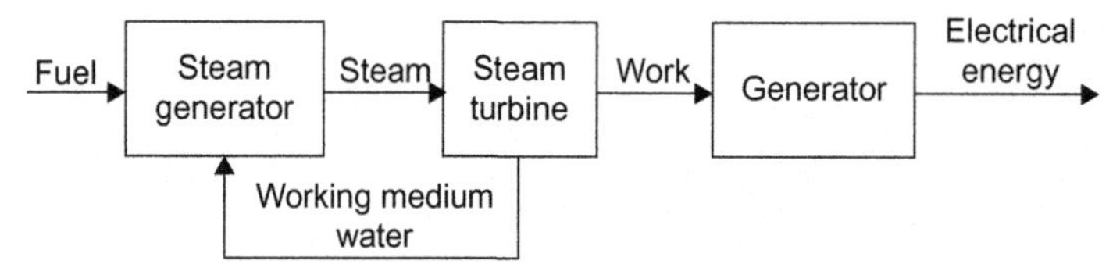

Figure 2.1 Schematic arrangement of a steam power station.

The primary function of a steam generator is to generate steam under pressure, but modern steam generators are also required to do the following:

i. Ensure generation of exceptionally high-purity steam by eliminating all impurities from saturated steam.
ii. Raise the degree of superheat of supplied steam as specified and maintain the same temperature over a defined range of load.
iii. In large power stations after partial expansion in the turbine steam is returned to the steam generator for further superheating and then transmitted to the turbine for complete expansion. This new degree of superheat is called "reheat" and should also be maintained constant over a defined range of load.
iv. While executing the above duties, a steam generator must utilize the heat of combustion of fuel as efficiently as possible.

2.2 BOILING AND CIRCULATION

Boiling of a liquid refers to a condition where vapor bubbles form on the heating surface of the liquid. Formation of bubbles depends on the fluid properties and the operating and surface conditions. When heat is added to a liquid, its temperature does not increase beyond its saturation temperature corresponding to its pressure, instead, the energy used results in a change of phase from a liquid to a gaseous state, e.g., from water to steam. This process, which takes place at constant pressure and constant temperature, is known as boiling.

The boiling point of a liquid is defined as the temperature at which its vapor pressure is equal to the pressure of the gas above it. This temperature is also called saturation temperature. In an open vessel the boiling point of a liquid refers to a temperature at which its vapor pressure is equal to the external pressure at one atmosphere. In the event that the external pressure becomes less than one atmosphere, the boiling point of the liquid gets reduced. As the external pressure rises above one atmosphere, the boiling point of the liquid also starts rising. Figure 2.2 shows the boiling point of water as a function of the external pressure. Note the boiling temperature of 373 K corresponds to a pressure of 101.3 kPa and temperature of 473 K to a pressure of 1554.0 kPa.

Boiling occurs when heat is added to the liquid at such a rate that its temperature is at least equal to its saturation temperature corresponding to the total pressure over the free surface of the liquid. If the vessel is open to atmosphere, the vapor displaces the air from its

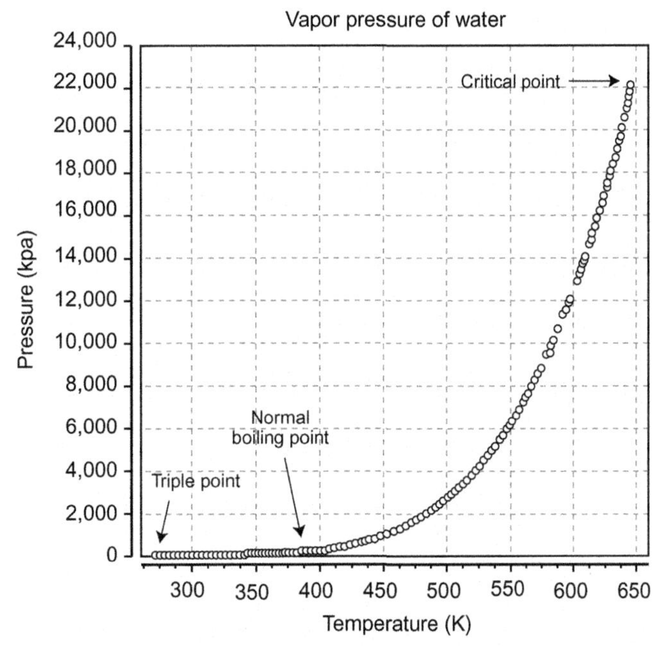

Figure 2.2 Saturated water/steam pressure versus temperature. *Source: http://upload.wikimedia.org/wikipedia/commons/thumb/e/e1/Vapor_Pressure_of_Water.png/800px-Vapor_Pressure_of_Water.png.*

surface of the liquid entirely, whereas with evaporation the vapor is removed by diffusion. The heat, which changes water to steam, is called the *heat of evaporation* or *heat of vaporization*.

The heat transfer that takes place from the wall to the liquid during boiling is given by the convective heat transfer equation:

$$Q = hA\Delta T \tag{2.1}$$

where

Q: Rate of heat flow, W
h: Heat transfer coefficient, W/m^2K
A: Area across which heat flow takes place, m^2
ΔT: Difference in temperature between the heating surface and the temperature of the fluid, K
(Log Mean Temperature Difference (lmtd))

The "heat flux" at the heating surface is found from Eq. 2.1 as:

$$q = \frac{Q}{A} = h\Delta T \tag{2.2}$$

2.2.1 Pool boiling

When a large volume of liquid is heated by a submerged heating surface and the motion is caused by free convection currents stimulated by agitation of the rising vapor bubbles,

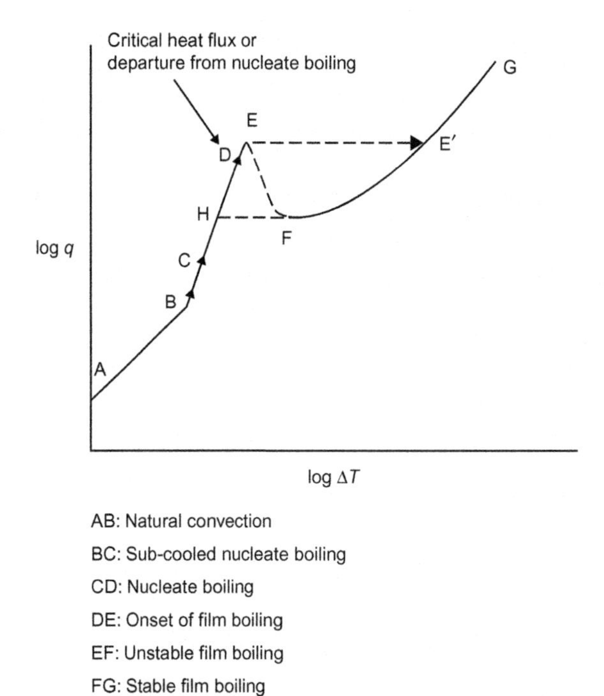

Figure 2.3 Heat flux versus temperature difference. *Source: P5-2. CH5. Boiling Heat Transfer, Two-Phase Flow and Circulation. STEAM ITS GENERATION AND USE. 41st Edition. The Babcock And Wilcox Company.*

it is called *pool boiling* [2]. During the course of heating it is noted that a general correlation lies between the gradually increasing heat flux, q, from the immersion heater and the corresponding temperature difference between the surface of the heater and the bulk of liquid, ΔT. The result of such correlation is presented in Figure 2.3 [1].

At point A since the heat flux is very low, no boiling occurs and no bubbles are formed, hence movement of liquid is by natural convection. In the region A-B, phase change takes place only as evaporation at the free surface of liquid that is at the liquid-vapor interface, and the corresponding heat flux is proportional to $\Delta T^{4/3}$ [2].

A further increase in heat flux will result in vapor bubble formation at the hot surface, yet the bulk of the liquid may still be below saturation temperature. The bubbles, which form in the hot boundary layer, condense as they rise into the colder liquid, giving up their latent heat to raise the temperature of the water. This phase change still occurs because of evaporation at the free surface. Bubble agitation in this region, B-C, is insignificant, since there is only a minor change in the slope of the curve. The formation and release of the steam bubbles at the metal water interface, with water still wetting the surface, cause ΔT to remain low. This region, wherein heat transfer takes place by natural convection, is called *subcooled nucleate boiling* [3].

Further rising of heat flux results in rapid formation of a large number of bubbles on the hot surface. As a result, strong local velocity is encountered within the liquid, causing a further increase in heat transfer. This is the region of *nucleate boiling*, where the bulk of the fluid is at saturation temperature and bubbles travel through bulk of the liquid to emerge at the free surface. Beyond point C bubbles do not collapse. This region, C-D, is one of stable boiling, and has a very steep slope. The heat flux is proportional to ΔT^{m}, where "m" lies between 2 and 4 [3]. In this region, as the surface temperature is increased, the heat flux also increases up to a maximum value at point D.

Nucleate boiling is incipient boiling, which is characterized by the growth of bubbles on the heated surface that rise from discrete points on a surface, whose temperature is only slightly above the liquid's saturation temperature. In general, the number of nucleation sites is increased by an increase in surface temperature. An irregular surface of the boiling vessel (i.e., increased surface roughness) can create additional nucleation sites, while an exceptionally smooth surface, such as glass, lends itself to superheating. Under these conditions, a heated liquid may experience boiling delay and the temperature may go somewhat above the boiling point and fail to boil.

Beyond the nucleate boiling region, the bubbles of the steam forming on the hot surface begin to interfere with the flow of water to the surface and eventually coalesce to form a film of superheated steam over part or all of the heating surface. This thin layer or film of vapor has low thermal conductivity and thus insulates the surface and reduces heat transfer and the heat flux, which in turn results in higher metal temperature. When vapor film insulates the surface from the liquid it is called *film boiling*. Region D-E shows the "onset of film boiling." At point E bubbles become so large and numerous that liquid has difficulty approaching the surface as the bubbles rise and thus starve the surface of liquid. Consequently there is a sudden jump in heating surface temperature, as shown at point E′. With water at atmospheric pressure, ΔT might be 25K at E and more than 1000K at E′ [2]. At this temperature there is a real danger of heating surface burning out or melting. Hence, point E is known as the *burn out point*. It is therefore imperative to operate as close as possible to point E that is at point D, without the risk of burn-out. If at point E a situation arises when heat flux would be dependent on the surface temperature, as when a fluid is condensing at variable pressure, it will be possible to reach point F from point E.

Region E-F in Figure 2.3 shows unstable film boiling where the vapor blanket alternately breaks down and recovers. As a result there is a decline in heat flux with increase in surface temperature.

Beyond point F, in the region F-G, the vapor film becomes thicker and film boiling becomes stable, where heat flux again increases with an increase in ΔT. Heat is transferred by conduction, convection and radiation across the layer of vapor that blankets the heating surface. At this region only evaporation takes place without formation of any bubble. This region thus is identified as *stable film boiling* one. If now the heat flux is reduced below the value at point G, the conditions do not jump back to E, but follow a path to point F. At this point the vapor film suddenly collapses and conditions revert to those at point H on the nucleate boiling curve.

Point D, which is very close to point E, is known as the point of *departure from nucleate boiling* (DNB) or the "critical heat flux." Heat flux at D is extremely high, as for example with water at atmospheric pressure heat flux is about 1500 kW/m^2 [2]. The temperature difference, ΔT, between the boiling liquid and the heating surface at which the critical heat flux occurs, is known as the *critical temperature difference*.

During operation of the boiler it is difficult to theoretically predict the heat transfer coefficients of nucleate boiling that would ensure ΔT remains below the critical temperature difference. Nevertheless, the ability to predict the value of the maximum heat flux, $q_{\max}$, i.e., point E in Figure 2.3, is also useful as this represents an upper limit in the nucleate boiling heat transfer. The following formula, from Roshenow and Griffith, gives the value of $q_{\max}$ in SI units, i.e., W/m^2 [3].

$$q_{\max}/(\rho_V * \lambda) = 0.0121 * \{(\rho_L - \rho_V)/\rho_V\}^{0.6} \tag{2.3}$$

where

ρ_L: Density of liquid, kg/m^3
ρ_V: Density of vapor, kg/m^3
λ: Latent heat of the fluid, kJ/kg

Once the critical heat flux data is determined it is compared with the required heat flux data. For the design to be acceptable, the critical heat flux, which is dependent on the following parameters, must always be greater than the heat flux generated in the course of boiling.

- Operating pressure
- Steam quality, i.e., surface tension, subcooling, etc.
- Type of tube: rifled or plain bore
- Tube diameter
- Flux profile around the tube
- Steam/water flow in the tube
- Angle of inclination of tube

2.2.2 Forced-convection boiling

Besides pool boiling, vapor is also generated by passing liquid through a tube heated either by firing fuel, as in a once-through boiler, or by condensing steam, as is commonly used in process industries. This method of generating vapor is known as forced-convection boiling.

The nature of boiling in forced convection is similar to pool boiling in many ways like the regimes of nucleate or film boiling. The most important difference is in the cause of "burn-out," since the vapor and liquid must travel simultaneously through the tube. As a result both heat transfer and pressure drop are affected by the behavior of the two-phase flow, which keeps on changing along the tube due to gradual evaporation of liquid. In a once-through tube, liquid below or at the saturation point

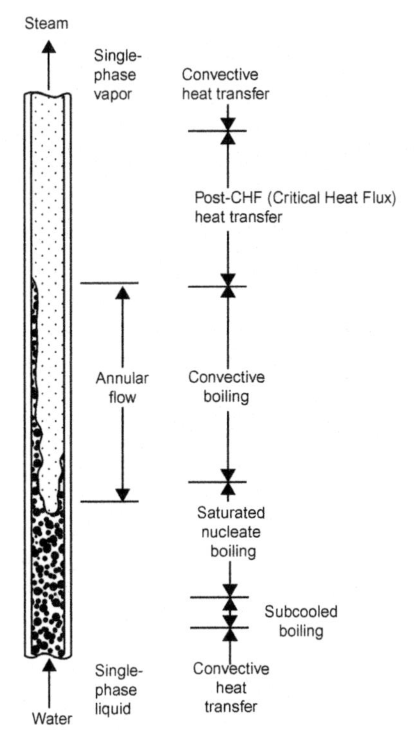

Figure 2.4 Evaporation of liquid in once-through tube. *Source: P5-3. CH5. Boiling Heat Transfer, Two-Phase Flow and Circulation. STEAM ITS GENERATION AND USE. 41st Edition. The Babcock And Wilcox Company.*

enters at the bottom and gradually evaporates until a dry or superheated vapor leaves at the top. In the process, some or all of the following regions are encountered by the flow system (Figure 2.4):

1. A liquid heating region or a subcooled boiling region: in this region bubbles are formed at the tube surface but subsequently collapse within the liquid. The heat transfer to the liquid will be by natural convection.
2. A saturated nucleate boiling region: where forced convection dominates. This is the region in which the local tube surface temperature is sufficiently superheated corresponding to the local pressure, causing bubbles to form at the surface.
3. An annular region: where the vapor travels along the tube as a continuous stream carrying with it some droplets of liquid, while the remaining liquid flows as a thin film. Heat transfer in this region takes place by conduction and forced convection. Following the annular region tube surface no longer remains wetted resulting in sudden jump in wall temperature; eventually 'dry out' condition or 'critical heat flux' is arrived at.
4. A region of pure vapor: in this region heat transfer takes place by natural convection.

The first and the fourth regions are extremes consisting of liquid only and vapor only regions, respectively. In these regions in turbulent flow the phenomena of heat transfer to water and to superheated steam in pipes is described by the Dittus-Boelter equation as follows [3].

For turbulent flow through tubes the phenomenon of heat transfer to water is described by

$$Nu = 0.023 * (Re)^{0.8}(Pr)^{0.4} \tag{2.4}$$

For superheated steam through pipes, the equation is

$$Nu = 0.0133 * (Re)^{0.84}(Pr)^{0.33} \tag{2.5}$$

where

Nu: Nusselt Number = hd/k
Re: Reynolds Number = ρVd/μ
Pr: Prandtl Number = c_pμ/k
h: Convective heat transfer coefficient, W/m^{2}K
ρ: Density of fluid, kg/m^3
V: Velocity of fluid, m/s
d: Internal diameter of pipe, m
μ: Dynamic viscosity of fluid, kg/m.s
c_p: Specific heat of fluid, kJ/kg.K
k: Thermal conductivity, W/m.K

From Eq. 2.4, the heat transfer coefficient of water is found as

$$h_{water} = 1063 * (1 + 2.93 * 10^{-3} * T_f) * (V_f^{0.8}/d^{0.2}) \tag{2.6}$$

From Eq. 2.5, the heat transfer coefficient of superheated steam is

$$h_{steam} = 4.07 * (1 + 7.69 * 10^{-4} * T_g) * \{V_g^{0.79}/(d^{0.16} * l^{0.5})\} \tag{2.7}$$

where

T_f: Absolute temperature of bulk water, K
T_g: Absolute temperature of superheated steam, K
V_f: Velocity of water through tube, m/s
V_g: Velocity of superheated steam through tube, m/s
d: Internal diameter of pipe, m
l: Length of pipe, m

Example 2.1

In a once-through boiler, feedwater enters through the bottom of tubes, receives heat from combustion of coal, and escapes from the top of the tubes as superheated steam. Calculate the heat transfer coefficients of water and superheated steam using the following data:

Pressure of feedwater at tube inlet = 27.0 MPa
Absolute temperature of feedwater, $T_f = 563$ K
Pressure of superheated steam at tube outlet = 24.5 MPa
Absolute Temperature of superheated steam, $T_g = 838$ K
Velocity of Feedwater through each tube, $V_f = 6$ m/s
Internal diameter of each tube = 63.5 mm
Length of pipe = 30 m

Solution: The density of feedwater at 27.0 MPa and 563 K is

$$\rho_f = 763.942 \text{ kg/m}^3$$

The density of superheated steam at 24.5 MPa and 838 K is

$$\rho_g = 74.532 \text{ kg/m}^3$$

For constant mass flow through each tube,

$$\rho_f * V_f * A = \rho_g * V_g * A$$

Hence, the velocity of superheated steam through each tube is

$$V_g = \rho_f * V_f / \rho_g = 763.942 * 6/74.532 = 61.5 \text{ m/s}$$

Applying Eq. 2.6, the heat transfer coefficient of water is

$$h_f = 1063 * (1 + 2.93 * 10^{-3} * 563) * (6^{0.8}/0.0635^{0.2}) = 20496.43 \text{ W/m}^2 \text{ K}$$

Applying Eq. 2.7, the heat transfer coefficient of superheated steam is

$$h_g = 4.07 * (1 + 7.69 * 10^{-4} * 838) * \{61.5^{0.79}/(0.0635^{0.16} * 30^{0.5})\} = 49.18 \text{ W/m}^2 \text{ K}$$

2.2.3 DNB (departure from nucleate boiling)

At the onset of boiling, metal temperature remains marginally above the saturation temperature of fluid. This is accomplished by maintaining adequate supply of liquid to avoid the occurrence of DNB. As heating continues, the liquid phase changes to vapor and the DNB point (point D in Figure 2.3) is reached, which is also the end point of the nucleate boiling. The metal temperature starts increasing at this point until point E is reached. In the region E-F, the metal temperature decreases, but increases again in the superheat region beyond point F. In the event the applied heat flux exceeds the *critical heat flux*, which is a function of pressure, the DNB would occur at low steam quality causing the metal temperature to be high enough to melt the tube.

In an optimistic design, in most boilers, ΔT is of the order of 3–6 K, which is well below the critical temperature difference of water at point E, i.e., about 25 K at

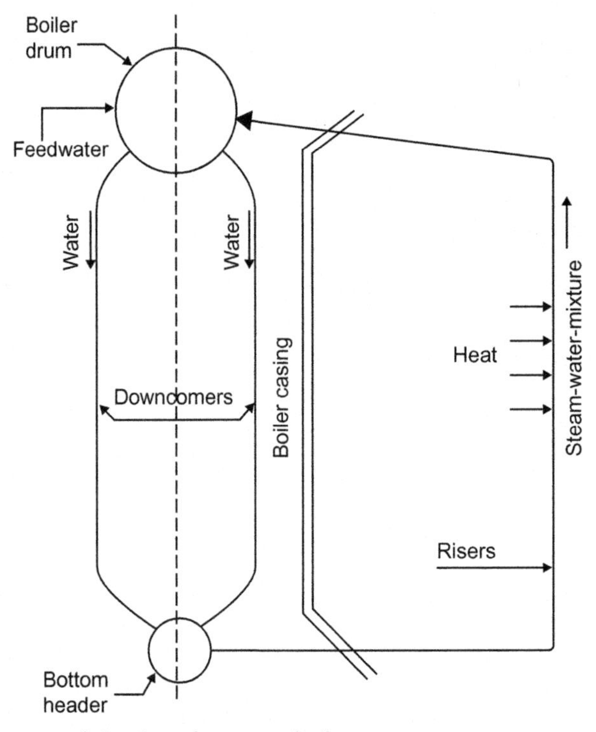

Figure 2.5 Downcomer and riser in a drum type boiler.

atmospheric pressure. Thus ensuring the design value of heat flux is much lower than the maximum value or the critical heat flux.

2.2.4 Circulation

In drum-type subcritical boilers, water flows from the drum through *downcomers* to the bottom of the furnace, then moves up through *risers* or evaporator tubes and returns to the drum. From Figure 2.5 it can be seen that the water in the downcomers does not receive any heat while risers, which form furnace walls, absorb the heat of combustion of the fuel. As a result, the water inside the riser tubes is heated and gradually steam bubbles form in the water. This mixture of water and steam in the riser tubes has a much lower density than the density of water alone in the downcomers. A density difference thus generated results in a static head difference causing thermo-siphon effect, which provides the driving force for a downward flow in downcomers and an upward flow in risers. Since the flow of fluid is generated by density difference alone, this is known as *natural circulation* in boilers [1,4,5].

For maintaining continuous circulation of fluid, the fluid flow must overcome friction losses in downcomer and riser tubes, headers, bends, etc. One way of overcoming resistance due to friction is by increasing heat input to the furnace. As the heat absorption

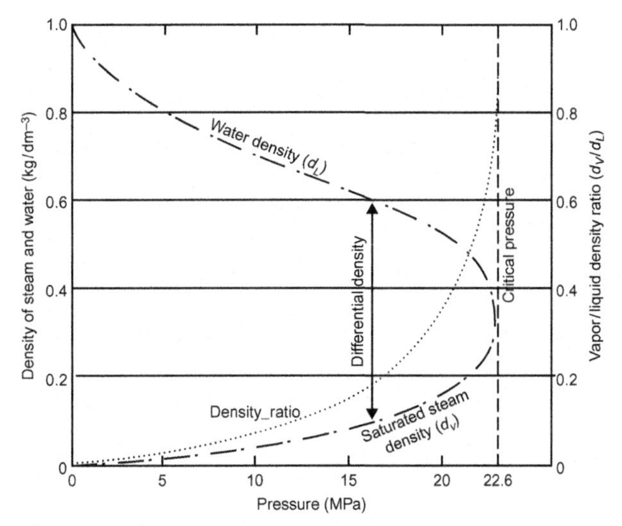

Figure 2.6 Density of steam and water versus pressure.

increases the density difference between the fluid in the downcomers and the fluid in the risers also increases, resulting in an increase in fluid flow. However, an increase in fluid flow in turn also causes an increase in friction loss. As a result, a point is reached when the rate of increase in friction loss surpasses the benefit of the enhanced density difference, where the rate of fluid flow starts to "drop." Hence, all natural circulation boilers are designed to operate below the point of "drop." The phenomenon of rising flow-rising heat input tends to make a natural circulation boiler self-compensating in the case of sudden overloads, changes in heat-absorbing surface condition, non-uniform burner disposition, etc.

As the operating pressure of fluid is increased, the density difference between the saturated liquid and the saturated vapor starts diminishing and reduces to zero at critical pressure, as is evident from Figure 2.6. Thus it becomes imperative to raise the height of the boiler higher to maintain adequate circulation with higher-pressure boilers.

In practice, at 18 MPa or higher fluid pressure the density difference is so small that it is difficult to maintain natural circulation, even after substantially increasing the height of the boiler. In such cases the fluid flow in the boiler tubes is ensured with the help of *assisted or forced circulation.*

In assisted or forced circulation, pumps are incorporated in the downcomer circuit to help natural circulation overcome friction losses. Downcomer tubes from the boiler drum are connected to a common header that serves as the suction header of the pumps. Discharge piping of these pumps is then connected to another header from which riser tubes are issued. Since forced circulation can overcome higher-resistance circuits smaller diameter riser tubes, than the tube diameter of natural circulation boilers, are used to economize the cost of tube materials. To reap the economic benefits of assisted circulation, this type of circulation is also adopted in some boilers operating at a pressure below 18 MPa.

The process of circulation, be it natural or forced, is dominated by what is called the *circulation ratio*, which is defined as the ratio of the "quantity of water in circulation" to the "quantity of steam produced." For a natural circulation boiler (up to 18 MPa) the circulation ratio may range from 4 to 30, while for forced circulation boiler the circulation ratio varies from 3 to 10 [6].

2.3 DESIGN

As discussed earlier, the chemical energy in fuel is converted to heat energy by combustion, and low-temperature water in the steam generator receives this heat of combustion and becomes high-energy steam for use in the steam turbine. The salient features of this heat-transfer process encompasses the following circuits, systems, and/or areas:

a. Water and Steam Circuit
b. Furnace
c. Fuel-burning System
d. Draft System
e. Heat Recovery System

2.3.1 Water and steam circuit

At operating pressure, temperature of the feedwater, prior to entering the steam generator, is less than that at the saturated condition, i.e., the entering feedwater is under the subcooled condition. Within the steam generator, the feedwater temperature is elevated to the near-saturation temperature in the economizer. Thereafter, the sensible heat for bringing the feedwater to saturation and the latent heat for evaporation of saturated feedwater is added in the *water-wall* for generating steam. This steam is separated from the steam-water mixture coming out from the water-wall and purified in the boiler drum to ensure the supply of saturated steam. The saturated steam is further superheated to attain the desired energy level.

The heat-transfer surfaces discussed so far, e.g., economizer, water-wall, superheaters/reheaters, etc., are located in the steam generator in the flow path of products of combustion so that heat is absorbed efficiently in proper proportions in various heat transfer zones.

The evaporation zones are comprised of the boiler drum, downcomers, and riser tubes or simply risers. While the boiler drum and downcomers are located outside the heat transfer zone, the riser tubes are exposed to heat transfer. While flowing through the riser tubes the feedwater receives the heat of combustion for evaporation. The riser tubes in modern steam generators are arranged such that they form the enclosure of the combustion chamber or furnace and receive the heat. Hence, riser tubes are usually called water-walls.

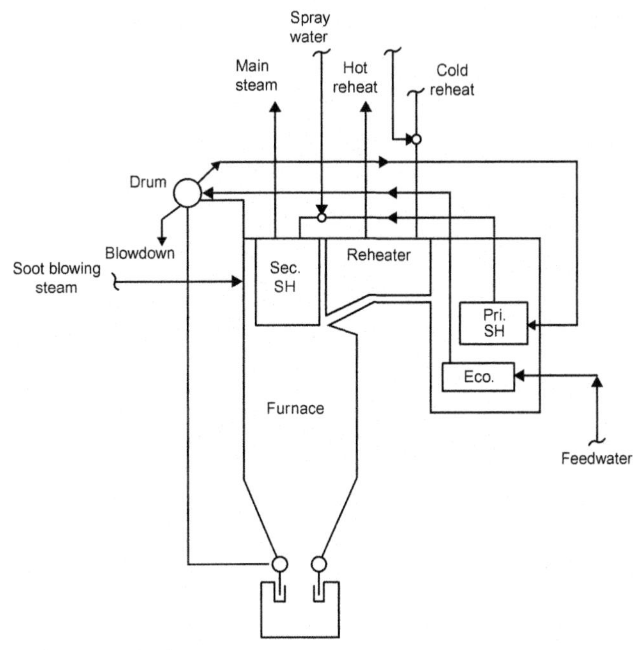

Figure 2.7 Water and steam circuit.

The feedwater from the economizer enters the boiler drum, flows down through the downcomers, passes through the pipes to the water-wall bottom header and rises through the riser tubes. In the riser tubes, the feedwater absorbs the heat, part of which is converted to steam and re-enters the boiler drum as a steam-water mixture. From this mixture, steam is separated in the boiler drum and purified. Dry saturated steam from the boiler drum is lead out through the saturated steam pipes to the superheaters. The complete flow path of working fluid is shown in Figure 2.7.

The process of separation and purification of steam in the boiler drum is accomplished by drum internals, e.g., cyclones, baffles, etc., chemical and feedwater admission piping, blow-down lines, etc. (Figure 2.8). The process includes three steps: separation, steam washing, and scrubbing.

Separation is the process of removing the bulk mass of water from steam and is accomplished by any one of following means:

- Gravity
- Abrupt change in flow direction
- Centrifugal action
- Impact against a plate
- Use of baffles

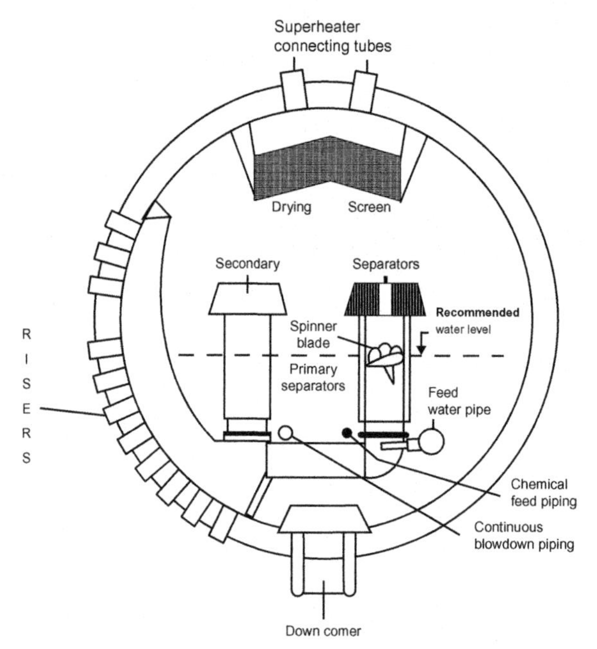

Figure 2.8 Drum internals.

Steam washing, scrubbing, or steam drying is the process of passing steam between closely fitted corrugated plates or screens on which mists are deposited in a way similar to that occurs during filtration. To avoid re-entrainment of water, velocity is kept low. The collected water drips to the boiler water below by gravity.

The arrangement of drum internals must be such as to ensure the following [1,3]:

i. Provide high-purity steam to prevent salt deposition in superheaters
ii. Provide steam-free water to the downcomers for maximum circulation
iii. Assure minimum pressure drop through the separators
iv. Provide simple design and easy accessibility for drum inspection

The function of the superheaters and reheaters is to raise the total heat content of the steam. There are two fundamental advantages of providing superheaters and reheaters: first, gain in efficiency of the thermodynamic cycle and second, enhancement of life in the last stage of the turbine by reduction of moisture content of the flowing steam.

In low-temperature boilers only convection superheaters are used, but large, high-temperature steam generators use both radiant and convection superheaters along with the reheaters to achieve the required superheat and reheat final steam temperatures. Today, modern steam power plants run at lower load at rated conditions during off-peak hours. Hence, the superheater/reheater elements are designed to achieve a normal operating temperature at about 60% of the boiler maximum continuous rating

(BMCR) load for better steam turbine performance. The location of superheaters and reheaters in the flue-gas path are normally split into primary and secondary sections to facilitate reasonably unchanged steam temperature control.

2.3.2 Furnace

In the furnace of a steam generator combustion of fuel takes place with atmospheric air to release heat. Hence, it is essential to ensure complete combustion of fuel under all operating conditions to harness the maximum heat potential of the fuel under combustion. (Detailed treatment of this process is included in Chapter 3, Fuels and Combustion.)

2.3.3 Fuel-burning system

The primary function of a fuel-burning system is to provide controlled and efficient conversion of chemical energy of the fossil fuel into heat energy. The heat energy thus generated is transferred to the heat-adsorbing surfaces of the steam generator. The main equipment for firing fossil fuel in a furnace is the burner.

An ideal fuel-burning system in a furnace is characterized by the following:

a. No excess oxygen or unburned fuel in the product of combustion
b. Low rate of auxiliary ignition energy input to initiate and sustain continuity of combustion reaction
c. Wide and stable firing range
d. Fast response to demand for a change in firing rate
e. Uniform distribution of flue-gas weight and temperature in relation to parallel circuits of heat-absorbing surface
f. High degree of equipment availability with low maintenance requirement
g. Safety must be the main requirement under all operating conditions

2.3.4 Draft system

For sustaining a healthy combustion process in a furnace it needs to receive a steady flow of air and at the same time remove products of combustion from the furnace without any interruption. When only a chimney is used for the release of products of combustion, the system is called a *natural draft* system. When the chimney is augmented with forced draft (FD) fans and/or induced draft (ID) fans the system is called a *mechanical draft* system. Small steam generators use natural draft, while large steam generators need mechanical draft to move large volumes of air and gas against flow resistance. A furnace utilizing an FD fan only is subjected to "positive draft" (pressurized), while one using an ID fan only is subjected to "negative draft" (subatmospheric pressure). Normally large coal-fired steam generators utilize *balanced draft*, employing both FD fans and ID fans. Figure 2.9 presents the flow path of a typical balanced draft furnace.

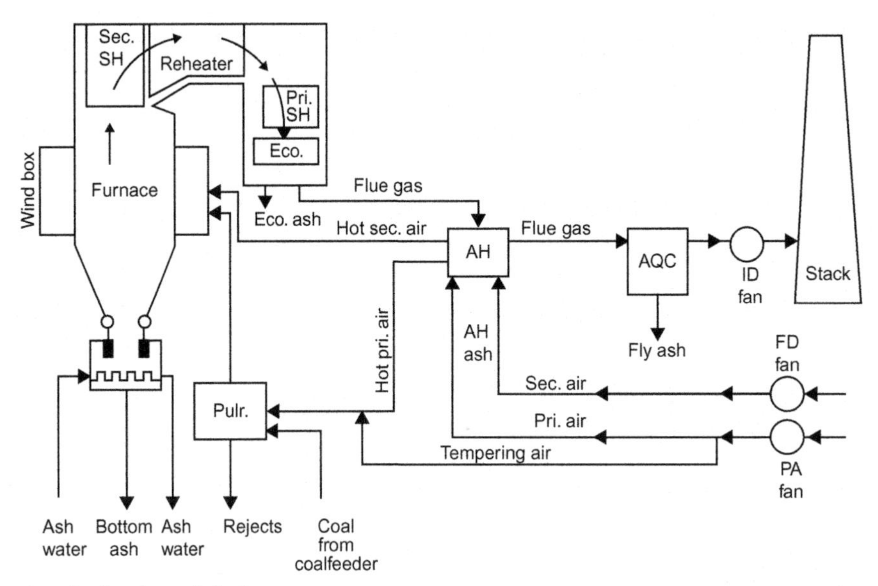

Figure 2.9 Fuel, ash, and draft system.

2.3.5 Heat recovery system

The flue gas or products of combustion leaving the convection heat-adsorbing surfaces in large steam generators still carry considerable heat energy at a temperature much above the steam saturation temperature. This heat, if exhausted to atmosphere without further heat transfer, will result in considerable losses. As a natural consequence of economic operation of a steam generator, a major part of this heat is recovered by placing the economizer thereafter the air heater in the exit flue-gas path.

Theoretically it is possible to reduce the flue-gas temperature to that of the incoming air to the air heater, but for technical and economic considerations, as described below, such reduction in exit flue-gas temperature is prohibited.

i. All fossil fuels are sulfur-bearing fuels. As such, if the exit flue-gas temperature from the air heater falls below the acid dew point of the as-fired fuel, sulfurous or sulfuric acid will form, causing back-end corrosion in the air heater. As a consequence, both maintenance cost and plant outages will increase.

ii. Capital costs incurred for installing additional heat recovery surfaces may offset the savings obtained by reducing flue-gas temperature.

In an *economizer* the incoming feedwater absorbs heat from the exiting flue gases, raising the temperature of the feedwater as close as possible but lower than its saturation temperature prior to entering into the evaporating circuit.

The remaining heat in flue gas downstream of the economizer is recovered in *air heater* to raise the temperature of combustion air before it enters the furnace. The energy thus recovered returns to the furnace and reduces the amount of heat that must be released by the fuel to maintain the desired furnace temperature, thereby ensuring reduction in the amount of fuel to be burned. The hot air is also used to dry coal in a pulverized coal-fired boiler.

2.4 CLASSIFICATION AND UTILIZATION

The function of a steam generator is to provide controlled release of heat in the fuel and to transfer this heat safely, reliably, and efficiently to the feedwater and steam. In a boiler no work is done on or by the fluid as it passes through the system. Hence, based on its use, location of fire and water spaces, type of circulation, arrangement of steam and water spaces, etc., steam generators are classified as follows [7]:

a. As per location of fire and water spaces –
- Fire Tube Boiler
- Water Tube Boiler

b. Based on circulation –
- Natural Circulation Boiler
- Forced/Assisted Circulation Boiler

c. Based on pressure requirement –
- Subcritical Boiler
- Supercritical Boiler

d. As per arrangement of steam and water spaces –
- Drum Type Boiler
- Once-through Boiler

e. Based on type of firing/heat transfer –
- Stoker-fired Boiler
- Fossil Fuel (Gas/Oil/Coal)-fired Boiler
- Fluidized-bed Boiler
- Waste Heat Recovery Boiler

Although boilers are classified in various categories, in practice all boilers fall into two or all of the above categories. Hence, selection of the type and size of boiler depends on the following factors:

- Required quantity of steam to be generated
- Boiler outlet steam pressure and temperature to be maintained
- Type of fuel to be used
- Availability of water
- Desired load factor
- Economy of steam/power generation
- Geographical location of the plant
- Ash disposal arrangement

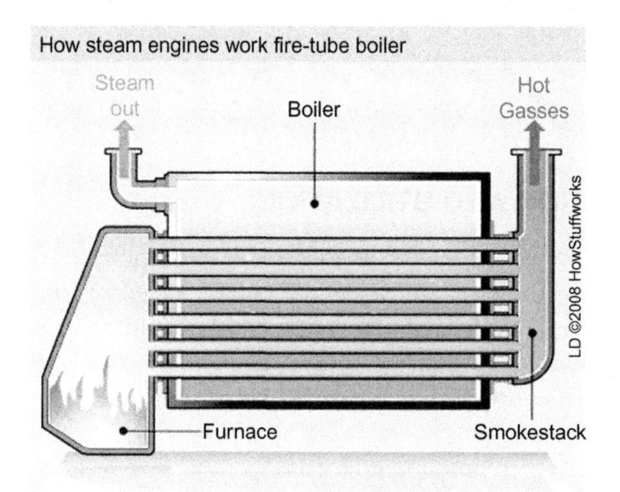

Figure 2.10 Fire-Tube Boiler. *Source: http://science.howstuffworks.com/transport/engines-equipment/steam2.htm.*

2.4.1 Fire-tube boiler

In this type of boiler hot gases pass through the tubes and water is contained in the shell. As the name suggests, its general construction is as a tank of water perforated by tubes that carry hot flue gases from the fire. The tank is usually cylindrical in shape to realize the maximum strength from simple structural geometry. The tank may be installed either horizontally or vertically.

Typically the furnace and the grate, on which fuel is burnt, are located underneath the front end of the shell (Figure 2.10). The gases pass horizontally to the rear, then either release through the stack at the rear or reverse directions and pass through the horizontal tubes to stack at the front. The fire tubes can be placed horizontally or vertically or at an inclined plane in a furnace.

In this type of boiler boiling takes place in the same compartment where water is stored. As a result, only saturated steam used to be produced in older designs of this type of boiler, but today, a fire-tube boiler may generate superheated steam as well.

A fire-tube boiler is sometimes called a "smoke-tube boiler" or "shell boiler." This type of boiler was used on virtually all steam locomotives in horizontal "locomotive" form. It has a cylindrical barrel containing the fire tubes, but also has an extension at one end to house the "firebox." This firebox has an open base to provide a large grate area and often extends beyond the cylindrical barrel to form a rectangular or tapered enclosure.

Fire-tube boilers are also typical of early marine applications and small vessels. Today, they are used extensively in the stationary engineering field, typically for low-pressure steam use such as for heating a building. However, the steam-generating capacity and outlet-steam pressure of these boilers are limited, therefore, they are unable to meet the needs of larger units. Another disadvantage of these boilers is that they are susceptible to explosions.

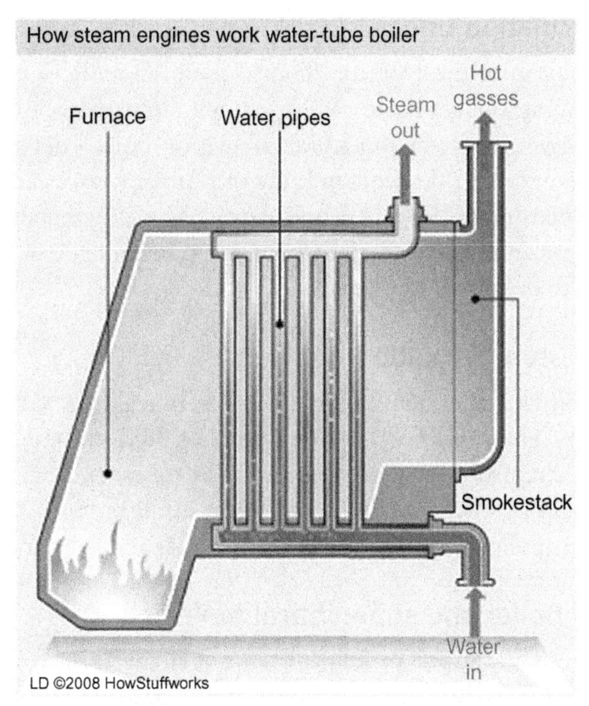

Figure 2.11 Water-tube boiler. *Source: http://science.howstuffworks.com/transport/engines-equipment/steam2.htm.*

2.4.2 Water-Tube Boiler (Figure 2.11)

To generate high evaporation rate accompanied by high steam pressure, the fire-tube boiler becomes exorbitantly heavy; therefore, the size and weight become extremely difficult to manage. In regards to the size of the shell, these shortcomings are circumvented by passing flue gases outside the tubes, instead of inside, and water is circulated through the tubes for evaporation. Baffles are installed across the tubes to allow cross flow of flue gases to ensure maximum exposure of the tubes. On the basis of configuration of tubes inside the furnace, this boiler is further classified as "straight-tube boiler" and "bent-tube boiler."

In a water-tube boiler water circulates in tubes heated externally by the hot flue gas. Fuel is burned inside the furnace, creating hot gas that heats up the water in the steam-generating tubes. Cool water at the bottom of the steam drum returns to the feedwater drum of small boilers via large-bore "downcomer tubes," where it helps pre-heat the feedwater supply. In large utility boilers, feedwater is supplied to the steam drum and the downcomers supply water to the bottom of the water-walls. The heated water then rises into the steam drum. Here, saturated steam is drawn off the top of the drum. In large utility boilers water-filled tubes form the walls of the furnace to generate steam and saturated steam coming out of the boiler drum re-enter the furnace through a superheater to become superheated. The superheated steam is used in driving turbines.

2.4.3 Natural circulation boiler

Boilers in which motion of the working fluid in the evaporator is caused by thermo-siphon effect on heating the tubes are called "natural circulation boilers." As discussed in section 2.2 circulation of water in natural circulation boilers depends on the difference between the density of a descending body of relatively cool and steam-free water and an ascending mixture of hot water and steam. The difference in density occurs because the water expands as it is heated, and thus, becomes less dense. All natural circulation boilers are drum-type boilers.

2.4.4 Forced/assisted circulation boiler

As noted in section 2.2 the density difference between the saturated liquid and saturated vapor starts diminishing at 18 MPa or higher fluid pressure, thus it is difficult to maintain natural circulation of fluid flow in boiler tubes. In such cases fluid flow is ensured with the help of forced/assisted circulation using pumps. The forced/assisted circulation principle applies equally in both supercritical and subcritical ranges.

2.4.5 Subcritical boiler and supercritical boiler

The critical pressure is the vapor pressure of a fluid at the critical temperature above which distinct liquid and gas phases do not exist. As the critical temperature is approached, the properties of the gas and liquid phases become the same, resulting in only one phase. The point at which the critical temperature and critical pressure is met is called the *critical point*. The critical pressure and critical temperature of water and steam are 22.12 MPa and 647.14 K, respectively. Any boiler that operates below the critical point is called a subcritical boiler, and one that operates above the critical point is known as a supercritical boiler. (Supercritical boilers are covered in section 2.10.)

2.4.6 Drum-type boiler

In a drum-type boiler the drum acts as a reservoir for the working fluid. These boilers have one or more water drums, depending on the size and steam-generating capacity. The drum is connected to cold downcomers and hot riser tubes through which circulation of working media takes place. The lower portion of the drum with feed-water is called the water space and the upper portion of the drum occupied by steam is called the steam space. A drum-type boiler can be either the natural circulation type or forced/assisted circulation type. Drum-type boilers are essentially subcritical boilers; they operate below the critical pressure of the working fluid. The economic design pressure limit of fluid in a drum-type boiler is around 18 MPa.

2.4.7 Once-through boiler

This type of boiler does not have a drum. Simply put, a once-through boiler is merely a length of tube through which water is pumped, heat is applied, and the water is converted into steam. In actual practice, the single tube is replaced by numerous small tubes arranged to provide effective heat transfer, similar to the arrangement in a drum-type boiler. The fundamental difference lies in the heat-absorbing circuit. Feedwater in this type of boiler enters the bottom of each tube and discharges as steam from the top of the tube. The working fluid passes through each tube only once and water is continuously converted to steam. As a result there is no distinct boundary between the economizing, evaporating, and superheating zones. The circulation ratio of this type boiler is *unity* [6]. These boilers can be operated either at subcritical or at supercritical pressures.

2.4.8 Stoker-fired boiler

Prior to the commercial use of "fluidized-bed boilers" (Chapter 5), stoker-fired boilers provided the most economical method for burning coal in almost all industrial boilers rated less than 28 kg/s (100 tph) of steam. This type of boiler was capable of burning a wide range of coals, from bituminous to lignite, as well as byproducts of waste fuels. However, over the years this type of boiler has become less popular due to more efficient technological advancement.

In stoker-fired boilers coal is pushed, dropped, or thrown on to a grate to form a fuel-bed. Stokers are divided into two general classes: *overfeed*, in which fuel is fed from above, and *underfeed*, wherein fuel is fed from below. Under the active fuel-bed there is a layer of fuel ash, which along with air flow through the grate keeps metal parts at allowable operating temperatures. Figure 2.12 and Figure 2.13 show a chain-grate overfeed stoker and a traveling grate overfeed stoker, respectively.

Figure 2.12 Chain-grate overfeed stoker. *Source: P 9-18, SECTION 9, POWER GENERATION Editors, Theodore Baumeister and Lionel S. Marks. Mechanical Engineers' Handbook (6th Edition), 1958. McGraw-Hill Book Company, Inc.*

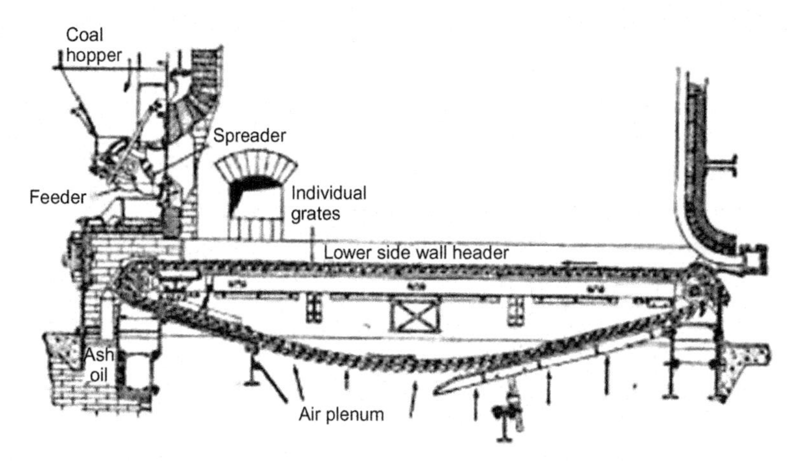

Figure 2.13 Traveling-grate overfeed stoker. *Source: P 9-18, SECTION 9, POWER GENERATION Editors, Theodore Baumeister and Lionel S. Marks. Mechanical Engineers' Handbook (6th Edition). McGraw-Hill Book Company, Inc.*

Table 2.1 Comparison of 'traveling grate stoker' and 'spreader stoker'

Traveling grate stoker	Spreader stoker
Less fly ash generation	Responds more quickly to load changes
Smokeless at low load	
Should not be used with caking coals	Burns a greater variety of coal
10:1 turn-down ratio without smoking	3:1 or 4:1 turn-down ratio without smoking
Maximum continuous burning rate:	Maximum continuous burning rate:
$4.80*10^6$ kJ/m^2 of grate surface,	$7.95*10^6$ kJ/m^2 of grate surface
$24.90*10^6$ kJ/front m of stoker width	$34.60*10^6$ kJ/front m of stoker width

A comparison of the traveling grate stoker and the spreader stoker is given in Table 2.1.

2.4.9 Fossil fuel-fired boiler

Coal, fuel oil, and natural gas are the main types of fossil fuel [8]. These fuels can generate a substantial quantity of heat by reacting with oxygen. These fuels consist of a large number of complex compounds comprised of five principal elements: carbon (C), hydrogen (H), oxygen (O), sulfur (S), and nitrogen (N). Any of these three fuels, i.e., coal, fuel oil, and natural gas, can be used in steam power stations, but coal plays a large role in power generation because of the enormous reserves of coal in many countries around the world, e.g., the United States, China, Mongolia, India, Indonesia, Russia, Poland, South Africa, Australia, etc. Most of the large steam-generating stations in these countries are coal based, and coal is expected to remain a dominant fuel for this purpose for many years. Coal-fired power plants at present account for about 41% of power produced globally [9].

Coal-based steam-generating stations also dominate in those countries where natural gas and fuel oil are abundantly available, since the cost of natural gas and fuel oil is exorbitantly high compared to the cost of coal. Additionally, natural gas and fuel oil are more economical in other industries. In modern, large-capacity, coal-fired steam generators coal is burnt in suspension. (Pulverized coal-fired boilers are covered in more depth in Chapter 4.)

2.4.10 Fluidized-bed boiler

Fluidized-bed combustion ensures burning of solid fuel in suspension, in a hot inert solid-bed material of sand, limestone, refractory, or ash, with high heat transfer to the furnace and low combustion temperatures (1075–1225 K). The combustor-bed material consists of only 3–5% coal. Fluidized-bed combustion is comprised of a mixture of particles suspended in an upwardly flowing gas stream, the combination of which exhibits fluid-like properties.

Fluidized-bed combustors are capable of firing a wide range of solid fuels with varying heating value, ash content, and moisture content. In this type of boiler, pollutants in products of combustion are reduced concurrently with combustion – much of the ash and hence the particulate matter as well as sulfur is removed during the combustion process. Further, lower temperature combustion in the fluidized bed results in lower production of NO_X and obviates any slagging problem. (Fluidized-bed boilers are covered in Chapter 5.)

2.4.11 Waste-heat recovery boiler

A *waste-heat recovery boiler* (WHRB) or a *heat-recovery steam generator* (HRSG) is a heat exchanger that recovers heat from a gas stream and in turn produces steam that can be used in a process or to drive steam turbines. A common application for HRSGs is in a combined cycle power plant (CCPP), where hot exhaust gas from a gas turbine is fed to the HRSG to generate steam. (These types of boilers are discussed in Chapter 7.)

2.5 BOILER MOUNTING AND ACCESSORIES

Equipment that are directly attached to, or within, the boiler are generally called boiler mountings. They are essential for safety, economics, and convenience. These mountings include water-level gauges, safety or relief valves, drain and blow-down valves, vent valves, water and steam sample connections, stop-check valves, soot blowers, etc. Accessories for measuring boiler-operating conditions include pressure gauges, water-level gauges, thermometers, thermocouples, water and steam flow meters, alarms, etc. There are also combustion control equipment and measuring devices. Interlock and protection devices protect the boiler from abnormal operating conditions such as low drum water level, high steam pressure, high steam temperature, and other off-normal conditions [10].

In many countries of the world it is mandatory that each and every boiler be provided with at least the following:

i. Two safety valves on the boiler drum.
ii. In case of boilers fitted with integral superheaters, an additional safety valve at the end of superheater outlet header.
iii. Two means of indicating water level in boiler drum. These indicators should be fitted with shut-off valves and drain cocks.
iv. A steam-pressure gauge.
v. A steam-stop valve connecting the boiler or superheater to the steam delivery pipe. This valve should be located as near the outlet from the final superheater as is convenient and practical. In the case of boilers without superheaters the valve location should be as close as practical to the drum.
vi. A non-return type of feed-check valve.
vii. Two independent feed apparatus; each such apparatus should have a capacity of no less than the maximum continuous rating of the boiler. In case of battery of boilers two independent feed apparatus should be considered, provided a total supply of feedwater, no less than the combined maximum continuous rating of all active boilers, can be maintained even if any one of the two sources of power supply should fail.
viii. A blow-down cock or valve placed at or as near as practical to the lowest point of the drum. The cock or valve should be fitted with a device that indicates its open and closed positions. (Blow-down is required to reduce the concentration of soluble salts in the boiler water, which results from the hardness of feedwater.)
ix. Low-water alarms, equipped with automatic tripping device, to disconnect fuel supply and start the feed pump simultaneously in the event of low water in the boilers.
x. Every boiler should be fitted with a valve or cock with a receiving screw for attachment of the Boiler Inspector's pressure gauge.
xi. A manhole for inspection and effective cleaning of drum internals.

Every boiler that is provided with automatic water-level and/or firing control should also comply with the following additional requirements:

1. In the event of automatic-control failure, the boiler should be capable of being brought under immediate manual control.
2. Means of manual resetting should be provided to turn off the fuel and air supplies to the boilers should there be a failure of electricity supply or any fault in electrical circuits to water-level and/or firing control equipment.

3. Where the evaporating capacity of the boiler is greater than 1.03 kg/s (3.7 tph), it should have automatic water-level alarms and firing controls and can be tested regularly without altering the level of water in the boiler.

The control equipment for automatic water-level controls should be such as to regulate the feedwater supply to the boiler to effectively maintain the level of water in the boiler drum according to the pre-determined limits.

Automatic firing controls should effectively control the supply of fuel and air to the combustion equipment at all times. It is necessary to maintain complete control over the combustion process based on the quantity of CO, CO_2, or O_2 in the flue gases. The combustion equipment should shut off the fuel supply to burners or the fuel and air supply to the furnace under the following conditions [11]:

a. Flame failure or pilot flame failure in cases of gas, oil, or pulverized fuel-fired boilers
b. Failure to ignite fuel at the burner within a pre-determined time
c. Fall in drum water level below a pre-determined safe level
d. Failure of forced draft or induced draft fans or any automatic flue damper
e. Increase or decrease in furnace pressure to a pre-determined value

Each boiler is also provided with steam/water/air soot blowers to dislodge soot and fly ash from the fireside passages of the boiler.

2.6 SUPERHEATERS AND REHEATERS

The superheater is one of the most critical elements of a boiler plant. Since it is located in the high-temperature zone it acquires close to the highest value of steam temperature as allowed by the metallurgical limit. The average temperature at which heat is supplied is increased by the superheating and hence the ideal cycle efficiency is increased and accordingly specific steam consumption (kg/kW.s) decreases.

The efficiency of an un-superheated cycle does not increase continuously with boiler pressure up to the critical pressure, but reaches the maximum when the pressure is about 16 MPa; thereafter, it drops with an increase in pressure. This is due to the fact that the latent heat and the quantity of heat transferred at higher temperature becomes smaller as the pressure increases. Eventually, the effective average temperature at which heat is added starts to decrease and the efficiency falls. In constrast to the efficiency of an un-superheated cycle, the efficiency of the superheated cycle increases continuously with pressure.

Furthermore, the dryness fraction at the steam turbine exhaust of an un-superheated cycle reduces as the boiler pressure is increased, resulting in the formation of droplets in the steam that erode the turbine blades and also reduce the turbine isentropic efficiency. To avoid these problems, the dryness fraction of steam is not allowed to fall below about 88% when passing the steam through superheater [12].

Even though superheating increases the dryness fraction, with the current operating steam temperature of about 810 K or above it is sometimes difficult to maintain the dryness fraction at or above 88%, particularly in large units. This problem is circumvented by adding another heat transfer surface called a "reheater" in the flue-gas path. With a combined superheat-reheat arrangement, the steam expands in the turbine at two different pressure stages. The steam coming from the superheater expands in the high-pressure turbine to some intermediate pressure. The low-temperature exhaust steam from the high-pressure turbine is then passed through the reheater in the steam generator to raise the steam temperature at a constant pressure to a level equal or higher than the superheat steam temperature. The reheated steam is then fed to the intermediate pressure turbine.

The superheater and reheater heat transfer surfaces absorb heat from the products of combustion or flue gases either through radiation, i.e., they receive heat through radiation from the furnace or through convection, which means these surfaces receive heat from the hot flue gas from the furnace by convection heat transfer [13]. Accordingly, these heat transfer surfaces are called "radiant type" or "convection type."

2.6.1 Convection superheater/reheater

Convection superheater/reheater heating surfaces are placed above or behind the banks of water tubes at the rear of the radiant heating surfaces to protect them from combustion flames and high temperatures. The main mode of heat transfer between the flue gases and the convection superheater/reheater tubes is convection. With an increase in steam demand, both fuel and air flow increase, resulting in an increase in flue-gas flow. As a result, convective heat transfer coefficients increase both inside and outside the tubes. The overall heat transfer coefficient between gas and steam hence increases faster than the increase in mass flow rate of steam. Thus, steam receives greater heat transfer per unit mass flow rate and the steam temperature increases with the load as shown in Figure 2.14.

2.6.2 Radiant superheater/reheater

Some sections of superheaters and reheaters are placed closer to the higher-temperature zone of the combustion flames for greater heat absorption. With this placement of the superheater/reheater sections, the heat transfer between the hot flue gases and the flame and the outer surface of the tubes takes place by radiation. Hence, these heating surfaces are called "radiant superheater/reheaters" [14].

From fundamental heat transfer theory it is known that radiation heat transfer is proportional to $(T_{\text{flame}}^4 - T_{\text{tube}}^4)$, where T_{flame} is the flame absolute temperature and T_{tube} is the tube surface absolute temperature. However, T_{flame} is much greater than T_{tube} and is also not dependent on load. As a result, the heat transfer is distinctly proportional to

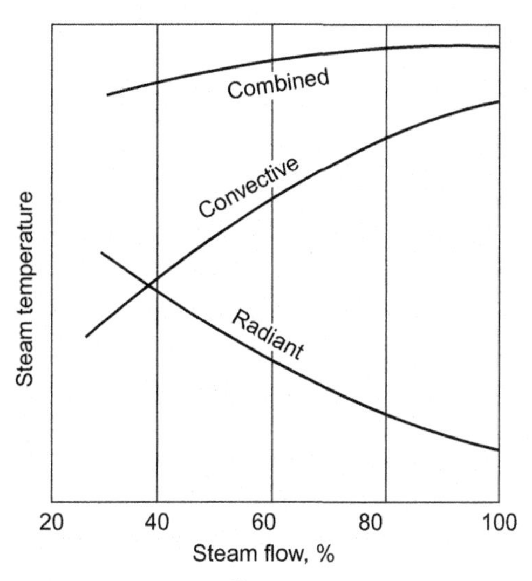

Figure 2.14 Steam temperature versus steam flow.

T_{flame}^4 and the resulting heat transfer per unit mass of flow decreases as the steam flow increases. Hence, an increase in steam flow due to enhanced load demand would cause the final steam temperature to fall, which is opposite of that experienced by the convection superheater/reheater (Figure 2.14). To allow radiation through radiant heating surfaces they are arranged in flat panels or platen sections with wide spacing.

2.7 ECONOMIZERS

In a subcritical boiler, the temperature of flue gases that leave steam-generating surfaces is determined by the saturation temperature of boiling water. The higher the steam pressure the higher the saturation temperature. Therefore, to permit the exchange of heat from flue gases to steam-generating surfaces, the temperature of flue gases must be above the saturation temperature of the water and is generally maintained at around 648–818 K. At this high temperature if the flue gases are exhausted to atmosphere, the loss in boiler efficiency will be exorbitantly high and unacceptable. Hence, there is an opportunity to absorb the residual heat of flue gases leaving the superheater and reheater by the incoming subcooled feedwater in a separate heat transfer surface known as the "economizer." With this type of additional heat absorption the working fluid efficiency of the boiler also is enhanced substantially. For each 5.5 K increase in the feedwater temperature, the boiler efficiency rises by about 1%. For each 22K drop in flue gas temperature, there is an increase in boiler efficiency by 1% in a conventional boiler [1].

When flue gases travel across water-walls and superheaters only the latent heat and superheat of steam are supplied by the flue gases. The primary state of this latent heat is accomplished in the economizer, which is placed in series with the feedwater heaters of the regenerative system. The function of the economizer is, therefore, to raise the temperature of feedwater downstream of the highest pressure feedwater heater to the saturation temperature corresponding to the boiler pressure.

The economizer, comprised of a bank of tubes over which flue gases pass, is a forced flow convection heat exchanger through which feedwater is supplied at a pressure higher than the pressure at the evaporating section. The size of the economizer is dictated by a comparative analysis of its costs and savings of fuel due to improvement in thermal performance. Economizer size is also governed by the temperature of the feedwater at the economizer inlet and the temperature of the exit flue gas from the economizer. With very low feedwater temperature, there is a possibility of external corrosion of economizer tubes due to the condensation of flue-gas constituents passing over them if the surface temperature of the tubes falls below the acid or water dew point of the flue-gas stream.

In the event that the feedwater contains a significant amount of dissolved oxygen (>7 ppb (parts per billion) in high-pressure boilers) pitting and corrosion may take place inside the tubes. This problem is overcome by deaerating the feedwater and maintaining a pH of around 9.

An economizer may be designed as follows [5]:

a. According to geometrical arrangement –
- Horizontal Tubes (Figure 2.15)
- Vertical Tubes (Figure 2.16)

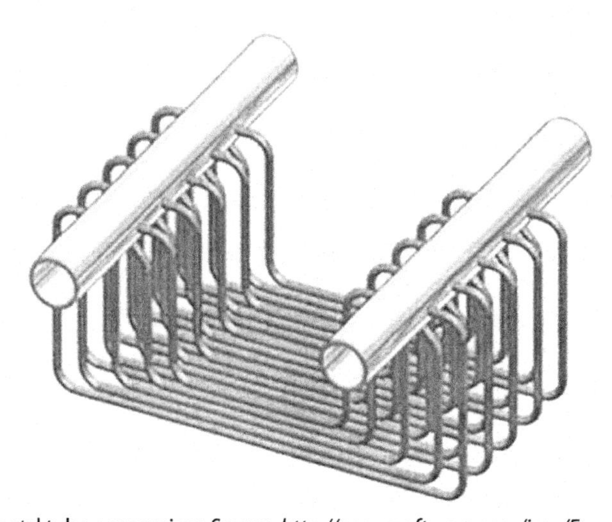

Figure 2.15 Horizontal tube economizer. *Source: http://genussoftware.com/img/Economizer5x5web.png.*

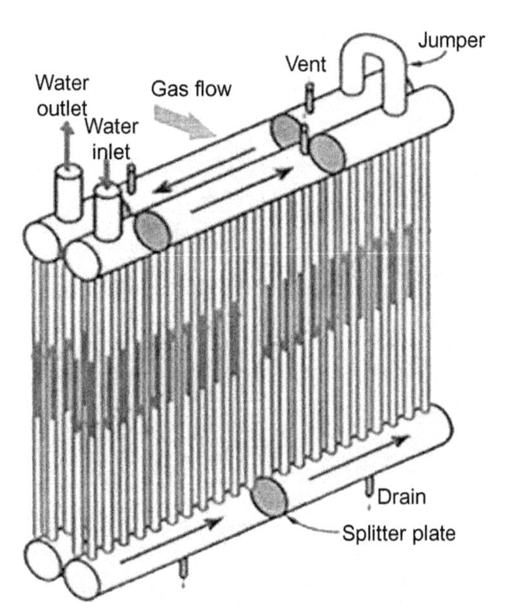

Figure 2.16 Vertical tube economizer. *Source: http://www.ccj-online.com/wp-content/uploads/2012/08/4-51-1024x813.jpg.*

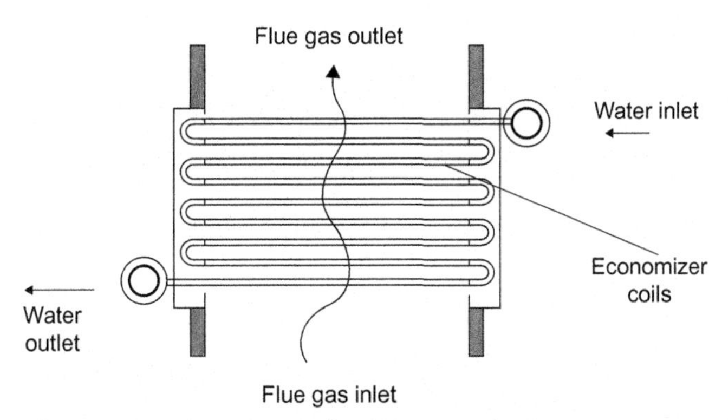

Figure 2.17 Crossflow economizer. *Source: http://www.mechnol.com/wp-content/uploads/2014/12/Economizer.jpg.*

b. Based on the direction of flue gases with respect to tubes in the bank –
- Longitudinal Flow
- Cross Flow (Figure 2.17)

c. According to the relative direction of flue-gases flow and feedwater flow –
- Parallel Flow
- Counter Flow (Figure 2.18)

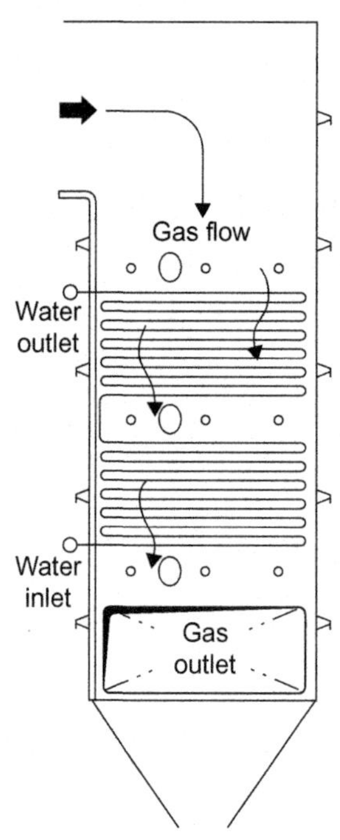

Figure 2.18 Counter flow economizer.

Figure 2.19 Finned-tube economizer. *Source: http://www.bossgoo.com/product-detail/carbon-steel-titanium-spiral-finned-tube-1788088.html.*

d. With regard to thermal performance –
- Steaming
- Non-steaming

e. Based on the details of design and form of heating surface –
- Plain Tube
- Finned Tube (Figure 2.19)

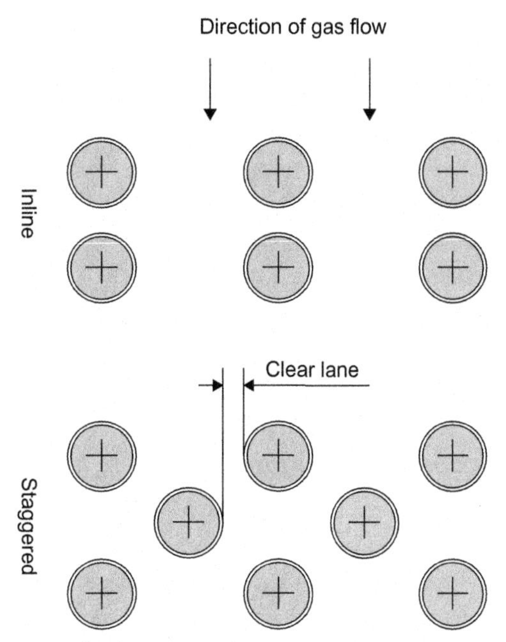

Figure 2.20 Inline and staggered tube economizer.

- Return-bend Tube
- Continuous Tube

f. According to the spacing of tubes and pattern of tubes (Figure 2.20) –
- Staggered Tube
- In-line Arrangement

2.7.1 Counter-flow economizer (Figure 2.18)

It is always best for flue gases from the boiler to flow down across the economizer tubes and for water to enter at the bottom and flow up through the tubes, thus reducing both surface and draft losses. Up-flow of water eliminates unstable water flow, gives the most uniform gas distribution, and makes the proper use of a steaming economizer possible.

2.7.2 Steaming economizer

A steaming economizer may be defined as an economizer in which the temperature rise in the economizer is equal to or more than two-thirds of the difference between the economizer inlet feedwater temperature and the saturation temperature of the fluid at that pressure. This may be expressed mathematically as $(T_O - T_I) = (2/3)^* (T_S - T_I)$, where T_O is the economizer outlet feedwater temperature, T_I the economizer inlet feedwater temperature, and T_S is the saturation temperature of feedwater at its prevailing pressure. The steam content of water at the economizer outlet should not exceed 20% of the feedwater flow at full load, and should be lower at part loads.

When feedwater make-up is low, it is advantageous to produce part of the steam in an economizer rather than in the evaporating section. From commercial point of view steaming is not allowed in the economizer, since there is the possibility of tube failure due to two phase flows through the tubes.

2.7.3 Plain-tube economizer

In some designs of boilers a continuous loop-type economizer is used. Tubes of this type of economizer have relatively small diameter to escalate the magnitude of heat transfer and to minimize fouling.

2.7.4 Finned-tube Economizer (Figure 2.19)

Heat absorption in the economizer tubes may also be intensified by welding fins to the plain tubes on the flue-gas side. This type of arrangement helps make the economizer compact, and it also increases the use of metal per unit heating surface area, providing much more gain in heat absorption and requiring smaller space. These types of economizers are more suited to gaseous and liquid fuel-fired boilers and in combined cycle plants. In coal-fired boilers, finned tubes often get plugged with ash and are usually not used.

2.7.5 Return-bend tube economizer

When it is essential to clean the inside of the tubes mechanically this design is best. This type of economizer is basically a horizontal steel-tube economizer, one end of which is bent 180° and the other end is equipped with a special flange bolted with a return-bend fitting.

2.7.6 Staggered-tube economizer (Figure 2.20)

A staggered arrangement of economizer tubes is preferable for gas and oil-fired boilers. This arrangement provides ascending motion of the feedwater in the economizer, resulting in free release of flue gases entering from the top of the economizer tubes and flowing down across the staggered tubes leaving at the bottom. This economizer is not suitable for coal-fired boilers. In a coal-fired boiler tubes must be in line to allow free flow of ash chunks through the bank of tubes.

2.8 AIR HEATERS

The temperature of flue gases downstream of the economizer is still quite high (588–698 K). The remaining heat in these flue gases is trapped in another heat-absorbing surface called the air heater. The air heater uses a colder heat-receiving fluid (ambient air) and thus the outlet flue-gas temperature from the air heater is reduced to about 393–433 K, while the temperature of ambient air is raised from atmospheric to about 523–613 K prior to using this hot air in combustion.

The heat recovered is recycled to the furnace by the combustion air and is added to the energy released from the fuel. This higher heat energy is further absorbed in the evaporating and/or the superheating surfaces, resulting in gain in boiler efficiency and less fuel consumption for the same output. The fuel savings are almost directly proportional to the rise in air-heater outlet air temperature. For a 110 K air temperature rise, typical fuel savings are about 4% and for a 280 K air temperature rise, fuel savings are about 11% [14]. Beyond fuel savings, air pre-heating is also required for drying coal in the pulverizer. The typical air temperature for drying coal ranges from 423–588 K.

When considering the principle of operation there are two types of air heaters: recuperative or regenerative. Both recuperative and regenerative air heaters may be arranged in vertical or horizontal configuration. The heat transfer efficiency of regenerative air heaters is high compared to the efficiency of recuperative air heaters. However, the disadvantages of regenerative air heaters are leakage of air into the gas space and transport of fly ash into the combustion air system.

2.8.1 Recuperative air heater (Figure 2.21)

In a recuperative air heater, heat from the flue gases is transferred continuously to air through a heating surface. In this type of air heater the metal parts are stationary and heat is transferred by conduction through the metal wall. A commonly used recuperative air heater is the tubular type, but for lower air and flue-gas pressures the air heater can also be the plate type.

Tubular air heaters are normally the counter-flow shell-and-tube type where hot flue gases pass through the inside of the tubes and air flows outside. Baffles are provided to ensure maximum air contact with hot tubes. These heaters, however, consume much metal and occupy a large space.

2.8.2 Regenerative air heater

In a regenerative air heater, the heating surface is swept alternately by the flue gas and the air undergoes alternate heating and cooling cycles and transferring of heat by thermal storage capacity of the heat transfer surfaces. The gas and air counter-current flows move through the rotor separately and continuously. Thus, when half of the heating elements are exposed to flue gases at any instant, the other half is exposed to air. This air heater may include either rotating heat transfer surfaces with stationary air/gas distribution hoods (Ljungstrom design: Figure 2.22) or rotating air/gas distribution hoods, where the heat transfer surfaces are stationary (Rothemuhle design: Figure 2.23).

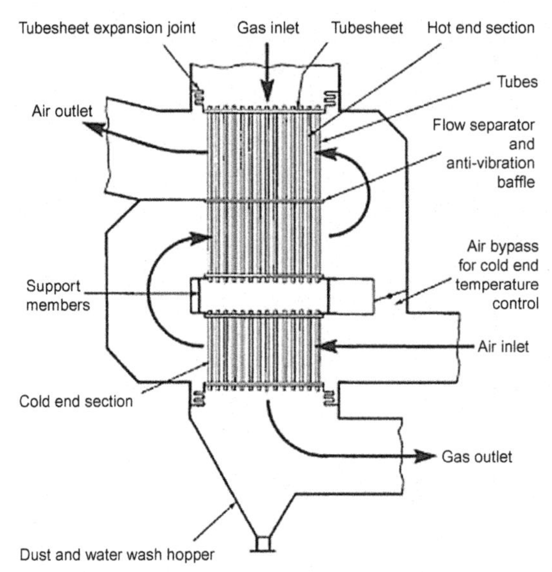

Figure 2.21 Recuperative air heater. *Source: Fig. 13, Page 20-7, STEAM Its Generation and Use (41st Edition). The Babcock and Wilcox Company.*

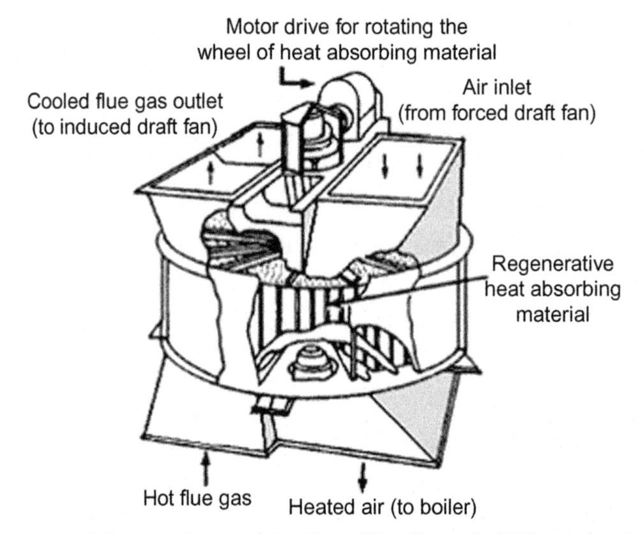

Figure 2.22 Ljungstrom air heater. *Source: http://en.wikipedia.org/wiki/Air_preheater.*

The rotor of a Ljungstrom air heater is driven by a motor and has several radial members that form sectors. Two opposite sectors are provided with stationary seal covers. The heat transfer surfaces are comprised of corrugated, notched, or undulated ribbing steel sheets that form baskets. They constitute the heat-storage medium of the air heater. As the

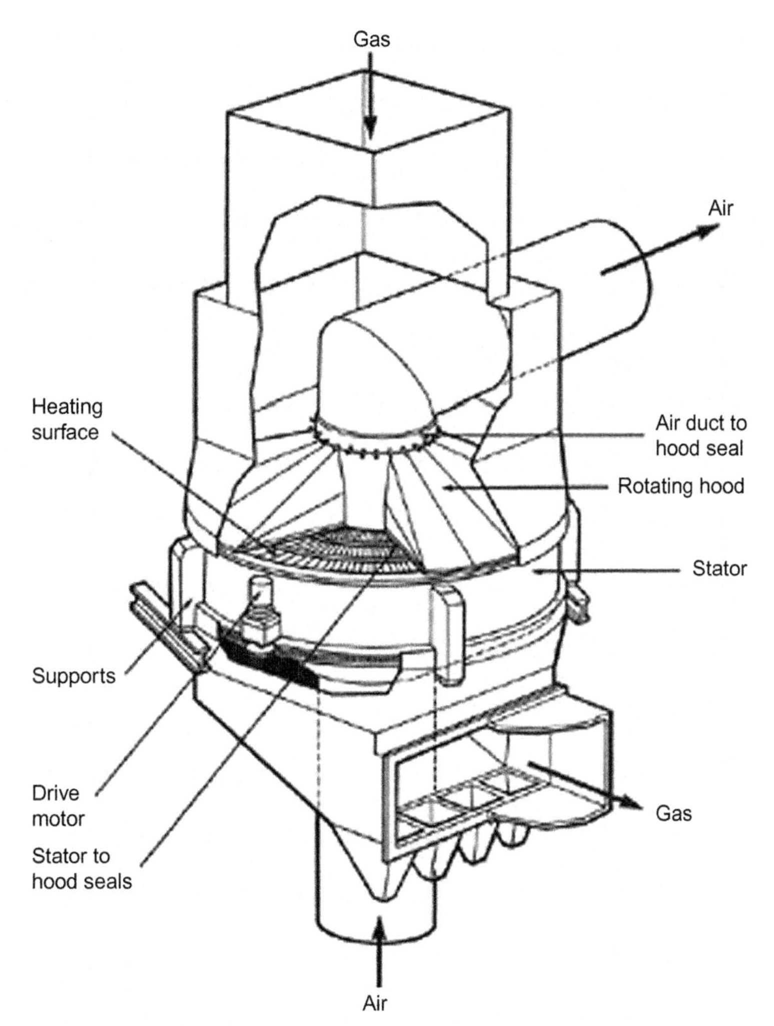

Figure 2.23 Rothemuhle air heater. *Source: Fig. 20, Page 20-11, STEAM Its Generation and Use (41st Edition). The Babcock and Wilcox Company.*

rotating sectors enter the hot gas zone they are heated by the gas, and they store the heat as sensible heat. When they enter the air zone, they progressively release this heat to the air.

In a Rothemuhle air heater the storage elements are stationary and the gas and air ducts are connected to two rotating segments each [1]. The principle of heat transfer from hot gas to cold air is identical to that of the Ljungstrom air heater. Ljungstrom air heaters, however, are better for commercial use due to better heat transfer and lower leakage, sometimes less than 5%.

While they provide substantial fuel savings, air heaters are prone to common problems as discussed in the following.

2.8.2.1 Leakage

Leakage is a normal phenomenon in all air heaters. While leakage in a new recuperative air heater is essentially zero, leakage occurs with the passage of time and due to accumulation of thermal cycles. Nevertheless, leakage in a recuperative air heater can be restricted to 3% with regular maintenance.

With a rotary regenerative air heater air leakage is inherent and its design value ranges from 5–10%, which increases further over time as seals wear. When the higher-pressure air side passes to the lower-pressure gas side leakage occurs through the gaps between rotating and stationary parts.

The rate of such leakage is given by the following expression [1]:

$$W_L = KA(2g\Delta P\rho)^{0.5} \tag{2.8}$$

where

W_L = Leakage flow rate, kg/s
K = Discharge coefficient (generally 0.4–1.0)
A = Flow area, m^2
g = Acceleration due to gravity (1 kgm/Ns^2)
ΔP = Pressure differential across gap, kg/m^2
ρ = Density of leaking air, kg/m^3

Ambient air infiltration through holes, doors, gaskets, expansion joints, etc., into lower-pressure gas streams is another form of air-heater leakage. The air-heater leakage can be determined by analyzing the $\%O_2$ content of flue gases entering and leaving the air heater (dry basis) and is given by

$$\%Leakage = \frac{\%O_2 Leaving - \%O_2 Entering}{21 - \%O_2 Leaving} * 0.9 * 100\ [1] \tag{2.9}$$

Example 2.2

While evaluating the performance of a boiler it is observed from flue-gas analysis that $\%O_2$ content of flue gases entering and leaving a rotary air heater are 5.0% and 7.5%, respectively. Determine the percentage infiltration of ambient air.

Solution: Applying Eq. 2.9, we get

$$\%Leakage = (7.5 - 5.0)/(21.0 - 7.5) * 90 = 16.67\%$$

Hence, the percentage of infiltration of ambient air in the boiler is 16.67%.

(NOTE: In a rotary air heater, whose seals are in good condition, leakage of air should not exceed 10%. The above result reveals that the seals of the air heater are in bad shape and replacement and proper adjustment of seals should be done.)

2.8.2.2 Corrosion

The air heater operates in the lowest temperature zone of the flue-gas path; as a result, the cold end of the air heater may be subjected to a temperature equal to or less than the dew point of the flue gases, resulting in a moisture film at the cold end. The formation of moisture in the cold end may cause corrosion and fouling. These adverse effects are exacerbated in units firing sulfur-bearing fuels. Upon combustion of the fuel, its sulfur content is oxidized to form SO_2, a portion of which is converted to SO_3 that combines with the moisture to form sulfuric acid vapor. This vapor condenses on surfaces at temperatures below its dew point of 393−423 K, resulting in potential corrosion in the cold end of the air heater.

The obvious solution would be to operate at cold-end metal temperatures above the acid dew point, but this results in unacceptable boiler heat losses. In practice, air heaters are designed to operate the same at minimum metal temperature (MMT), therefore, the resulting increase in boiler efficiency offsets the additional maintenance cost needed to prevent low-temperature gas corrosion. Some typical recommendations of leading boiler manufacturers regarding limiting MMTs while burning sulfur-bearing fuels are given in Figure 2.24 and Figure 2.25.

A generally accepted method of preventing gas corrosion is to raise the inlet temperature of the air in the steam coil air heater. Another method of overcoming corrosion is to apply corrosion-resistant coatings to low-temperature elements of an

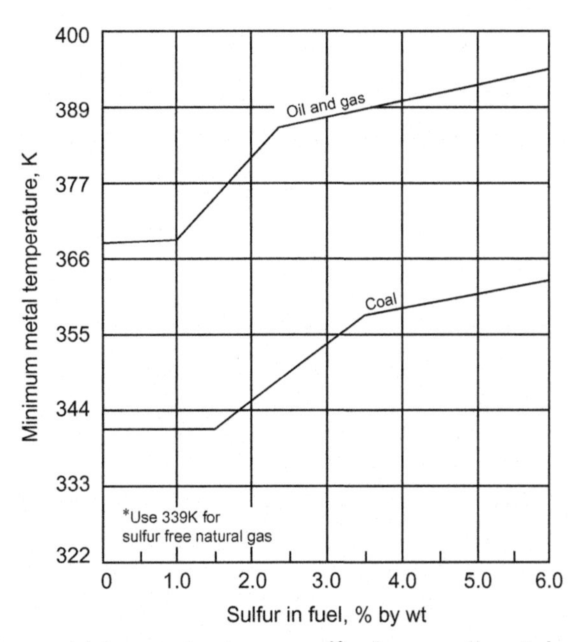

Figure 2.24 Minimum metal temperature versus sulfur (regenerative air heater). *Source: Fig. 23, Page 20-14, STEAM Its Generation and Use (41st Edition). The Babcock and Wilcox Company.*

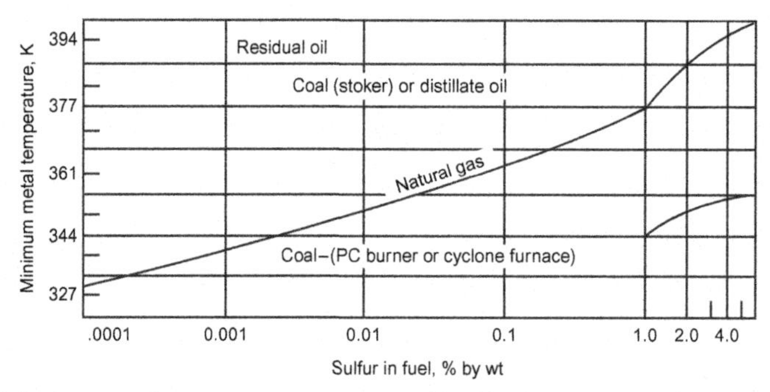

Figure 2.25 Minimum metal temperature versus sulfur (recuperative air heater). *Source: Fig. 22, Page 20-13, STEAM Its Generation and Use (41st Edition). The Babcock and Wilcox Company.*

air heater. Cold-air bypass of air heaters or hot-air recirculation to combustion-air inlet to air heaters are other methods used to prevent gas corrosion.

2.8.2.3 Fouling

Fouling in an air heater is done by the flow of gas-entrained ash and corrosion products, which is controlled and removed by soot blowing, cold-end temperature control, and off-line cleaning.

2.8.2.4 Fire

Air heater fires, although not common, sometimes occur, especially in regenerative air heaters. Most air-heater fires occur during start-up when unburned fuel oil, deposited on ash-fouled heating surfaces, is ignited. Fire can be prevented by maintaining a clean air heater surface by resorting to soot blowing during start-up as well as prior to shut-down of the boiler. Today air heaters are also provided with automatic fire-extinguishing systems.

2.9 INSULATION

All hot surfaces at temperatures above 333 K must be insulated for personal protection. The thickness of insulation is chosen to ensure a maximum external surface temperature of 328 K for metal surfaces and 333 K for non-metallic surfaces. In addition to personal protection, insulation is also applied to flue-gas paths and flue-gas ducts to ensure the flue-gas temperature entering the chimney remains above its acid dew point to prevent low-temperature gas corrosion. Additionally, thermal insulation is applied to a steam generator to ensure the steam is supplied to its point of application at the correct temperature without any drop. Various heat-traced piping and vessels are also applied with insulation to ensure maximum transfer of heat and minimum heat loss from surfaces.

2.9.1 Thermal insulation materials

There are four types of thermal insulation materials: granular, fibrous, cellular, and reflective. Granular materials, such as calcium silicate, contain air entrained in the matrix. Fibrous materials, such as mineral wool, contain air between fibers. Cellular materials, e.g., cellular glass and foamed plastics, contain small air or gas cells sealed or partly sealed from each other. Reflective insulation materials consist of numerous layers of spaced thin-sheet material of low emissivity, such as aluminum foil, stainless-steel foil, etc. In practice, a combination of two or more of the above four types are used.

2.9.1.1 Properties of thermal insulation materials

Heat from a hot surface is transferred or lost through an insulation material to the surrounding area by radiation and convection. Thermal insulation should limit the heat loss to a minimum. The main requirement of thermal insulation materials is "low thermal conductivity" since heat is transferred through insulation material by conduction (at 373 K, thermal conductivity of mineral wool is 0.047 W/m.K, that of glass fiber is 0.038 W/m.K, of calcium silicate is 0.057 W/m.K and of magnesia is 0.062 W/m.K).

Thermal insulation material should have the following properties to ensure satisfactory performance of the material for the life of a plant:

- Material must be suitable for continuous use at maximum operating temperatures without degradation of its physical properties
- It must be non-corrosive to plant and pipe work if wet by rain or leakage of water/steam
- It should not be permanently damaged if contaminated with water
- It should have adequate compressive strength to resist local loads, such as foot traffic, ladders, etc.
- It must have adequate flexural strength and impact resistance to permit transportation and application without breakages
- Material must be non-combustible
- It should not cause any discomfort or health hazards to operating personnel
- It should be tight, free of voids, and well anchored
- Removal and replacement should be required only during maintenance or modification of the plant

2.9.1.2 Design of thermal insulation

From basic heat transfer theory [13] it is known that under steady state conditions the heat loss from a hot flat surface through thermal insulation to the ambient air is expressed as

$$Q = (T_h - T_c)/(L/k) \tag{2.10}$$

or

$$Q = (T_c - T_a) * f \tag{2.11}$$

Combining Eq. 2.10 and Eq. 2.11 we get

$$Q = (T_h - T_a)/(L/k + 1/f) \tag{2.12}$$

where

Q = Heat loss, W/m^2
T_h = Hot surface temperature, K
T_c = Cold surface temperature, K
T_a = Ambient air temperature, K
L = Thickness of insulation, m
k = Thermal conductivity of insulation material, W/m.K
f = Surface coefficient of insulation material, W/m^2.K
(combined effect of radiation and convection heat transfer)

The temperature of a cold surface or the surface of an insulation material can then be calculated from Eq. 2.11 as

$$T_c = (Q/f) + T_a \tag{2.13}$$

For steady flow of heat through composite walls, e.g., wall of a vessel lined with thermal insulation, Eq. 2.12 takes the shape of

$$Q = (T_h - T_a)/(L1/k1 + L2/k2 + 1/f) \tag{2.14}$$

where

Q = Heat loss, W/m^2
T_h = Hot surface temperature, K
T_a = Ambient air temperature, K
L_1 = Thickness of wall, m
k_1 = Thermal conductivity of material of wall, W/m.K
L_2 = Thickness of insulation, m
k_2 = Thermal conductivity of insulation material, W/m.K
f = Surface coefficient of insulation material, W/m^2.K

The expression of cold-surface temperature, T_c, remains the same as Eq. 2.13.

The heat loss from insulated cylindrical surfaces (e.g., pipes, tubes, small diameter vessels) can be written as

$$Q = (T_h - T_a)/[\{r_i * \ln(r_o/r_i) * (1/k)\}(r_i/r_o) * 1/f] \quad (2.15)$$

where

Q = Heat loss, W/m^2
T_h = Hot surface temperature, K
T_a = Ambient air temperature, K
r_i = Inner radius of insulation, m
r_o = Outer radius of insulation, m
k = Thermal conductivity of insulation material, W/m.K
f = Surface coefficient of insulation material, $W/m^2.K$

The temperature of cold surface or the surface of insulation material can then be calculated from

$$T_c = (Q/f) * (r_i/r_o) + T_a \quad (2.16)$$

For steady flow of heat through composite cylindrical surfaces, Eq. 2.15 changes to

$$Q = (T_h - T_a)/[\{(r_i/k_1) * \ln(r_c/r_i)\} + \{(r_c/k_2) * \ln(r_o/r_c)\} + (r_i/r_o) * 1/f] \quad (2.17)$$

where

Q = Heat loss, W/m^2
T_h = Hot surface temperature, K
T_a = Ambient air temperature, K
r_i = Inner radius of wall, m
r_c = Outer radius of wall/Inner radius of insulation, m
r_o = Outer radius of insulation, m
k_1 = Thermal conductivity of material of wall, W/m.K
k_2 = Thermal conductivity of insulation material, W/m.K
f = Surface coefficient, $W/m^2.K$

The expression of the cold-surface temperature remains the same as Eq. 2.16.

Example 2.3

Hot-surface temperature of a 150-mm thick fire brick wall is maintained at 1088 K. If the average thermal conductivity of the wall material is 0.196 W/m.K determine the maximum heat loss through the wall to ensure its cold-surface temperature does not exceed 333 K. For an ambient air temperature of 294 K, what is the surface coefficient of the wall material?

Solution: Applying Eq. 2.10, the maximum heat loss through the brick wall may be determined as

$$Q = (1088 - 333)/(0.15/0.196) = 986.53 \ W/m^2$$

The surface coefficient of the wall material can be found from Eq. 2.11 as

$$f = 986.53/(333 - 294) = 25.296\ W/m^2.K$$

Example 2.4

Steam at a temperature of 623 K is flowing through a pipe with an outer diameter of 100 mm. The pipe is insulated with 20-mm thick mineral wool, the average thermal conductivity of which is 0.055 W/m.K. Determine the outside surface temperature of mineral wool if the ambient air temperature is 300 K. Assume the hot-surface temperature of the mineral wool is the same as that of the flowing steam and the value of the surface coefficient of the insulating material is 30 $W/m^2.K$.

Solution: The inside diameter of the insulating material may be assumed to be the same as the outside diameter of the pipe. Hence, $r_i = 50$ mm and $r_o = 70$ mm. Applying Eq. 2.15 we get

$$\begin{aligned} Q &= (623 - 300)/[\{(0.05/0.055) * \ln(0.07/0.05)\} + (0.05/0.07) \times 1/30] \\ &= 323/(0.9091 \times 0.3365 + 0.0238) \\ &= 979.64\ W/m^2 \end{aligned}$$

The cold-surface temperature of insulating material is calculated from Eq. 2.16 as

$$\begin{aligned} T_c &= (979.64/30 \times 0.05/0.07) + 300 \\ &= 323.32\ K \end{aligned}$$

Example 2.5

One vessel having **a** carbon-steel wall of thickness 10 mm is carrying saturated steam and water at 441 K. The vessel is insulated with magnesia of thickness 50 mm. If the ambient air temperature is 303 K, determine the heat loss from the vessel.

Given:

- **i.** thermal conductivity of carbon steel is 52 W/m.K
- **ii.** thermal conductivity of magnesia is 0.05 W/m.K
- **iii.** surface coefficient of insulation surface is 10 $W/m^2.K$

Solution: Assuming no temperature drop across the metal and applying Eq. 2.14 we get

$$\begin{aligned} Q &= (441 - 303)/(0.01/52 + 0.05/0.05 + 1/10) \\ &= 151.83\ W/m^2 \end{aligned}$$

2.10 SUPERCRITICAL BOILERS

2.10.1 Introduction

From Carnot cycle we know that the efficiency of a thermodynamic cycle depends on the temperature of the heat source and the temperature of the heat sink and is independent of the type of working fluid. The higher the temperature of the heat source the lower the temperature of the heat sink and the more efficient the cycle. Since a temperature below 288 K for the heat sink and cooling water is rarely available, the efficiency can be increased by raising the temperature of the heat source. This gain in efficiency is substantially raised by using fossil fuel-fired plants in the thermodynamic supercritical zone.

The supercritical condition of a steam-water cycle is a state at which its temperature and pressure are above its thermodynamic critical point, where the pressure of the steam water is 22.12 MPa, the temperature is 647.14 K, and the density is 324 kg/m^3. At the critical point, liquid-vapor phase shows the following special phenomenon:

i. Latent heat of vaporization is zero, or there is no boiling
ii. Density difference between liquid and vapor is zero
iii. Specific enthalpy difference between liquid and vapor is zero

As heat is applied to water above the critical pressure, the temperature rises, water molecules are gradually agitated, inter-molecular space increases uniformly, and fluid becomes less dense. Thus, at the critical point the liquid and gas phases of any fluid become a single supercritical phase. The transition from the dense-phase water with a compact molecular arrangement to a wide-spaced random arrangement of vapor is uniform. No internal bubbles are formed. Specific enthalpy changes uniformly and all other physical properties change continuously from liquid to vapor state with a gradual rise in temperature. Near to the critical point, small changes in pressure or temperature result in large changes in density. Thus, the density difference between the liquid and vapor phases reduces sharply.

The benefit of adopting supercritical technology in fossil fuel-fired power plants is not restricted to improvement in generating efficiency alone. Efficient operation of supercritical power plants leads to burning less fuel and consequently less SO_X, NO_X, and CO_2 emissions, which would in turn reduce global warming as well. The relative improvement in efficiency that can be achieved with a supercritical plant compared to a subcritical plant is presented in Table 2.2.

2.10.2 Steam-water circuit

Supercritical boilers do not have boiler drums. This type of boiler consists of a number of parallel circuits connected by inlet and outlet headers. Pressurized water enters the circuit at one end and leaves as superheated steam at the other end (Figure 2.26).

Table 2.2 Relative efficiency improvement of supercritical plant over a subcritical plant [15]

Parameter	Increase in superheat steam pressureby 9.0 MPa	Increase in superheat steam temperature by 28 K	Increase in reheat steam temperature by 28 K	Increase in feedwater temperature by 28 K
GAIN IN EFFICIENCY, %	2.2	1.0	0.7	0.8

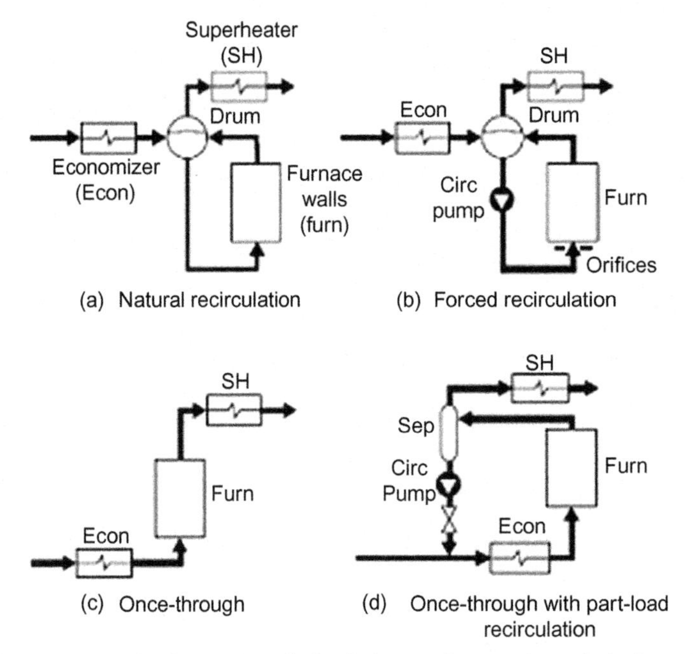

Figure 2.26 Steam water circuit – natural circulation and once-through boilers. *Source: Fig. 26, Page 5-18, STEAM Its Generation and Use (41st Edition). The Babcock and Wilcox Company.*

As water absorbs energy it gradually expands in volume without internal bubbling and acts as extremely dense gas. Feedwater enters the bottom header of the economizer and passes upward through the economizer elements and then enters the evaporator, located downstream of the economizer.

In a supercritical boiler the evaporator circuit is the once-through type, but this type of boiler starts operating in the once-through mode beyond a particular minimum load of about 30–40%. Therefore, regardless of load it is necessary to ensure adequate flow of fluid within individual tubes so that the wall temperature does not exceed the allowable metal temperature limit. To meet this requirement below the minimum load, the supercritical boiler operates in circulation mode and needs a separator and circulation system for water-steam separation. Fluid coming from the evaporator enters the separator vessel, and the separated water is circulated back to the boiler.

There are two types of circulation systems in use today. In the first type, the separated water from the separator is led to the deaerator and is circulated to the feedwater system through the boiler feed pump. This system is simple and relatively inexpensive but involves loss of heat. In the other type of system a circulation pump is provided to circulate the water from the separator directly to the economizer (Figure 2.27). This prevents heat loss but increases cost.

From the separator vessel all fluid is routed through the pipes to the convection enclosure, similar to a subcritical design. There may be two to three stages of superheaters with interstage desuperheater for temperature control.

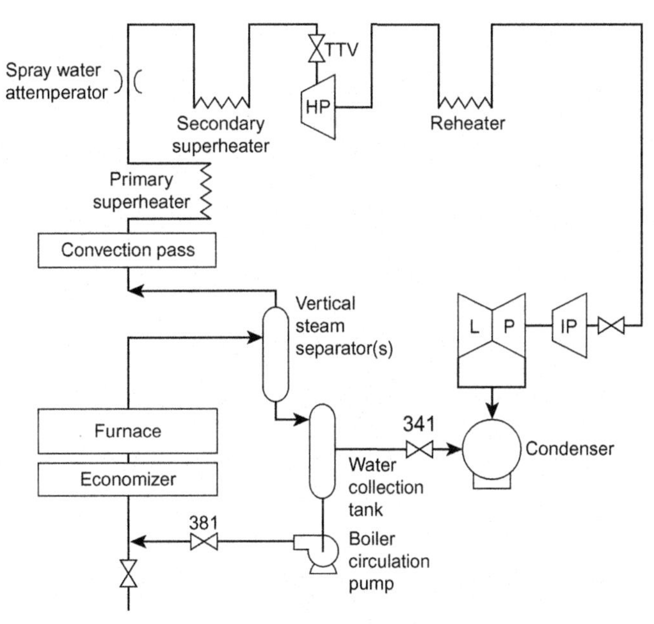

Figure 2.27 Separator drain circulation system of a supercritical boiler. *Source: P17. Paper on STEAM GENERATORS USC AND A-USC STEAM CONDITIONS by Paul Weitzel, Niranjan Pawgi & Aaron Virgo at Partnership to Advance Clean Energy – Deployment. Babcock & Wilcox Company 2013. November 21–23, 2013. New Delhi. (US AID & Ministry of Power, Govt. of India).*

Another way of ensuring each tube of evaporator receives uniform total heat input above the minimum load is to deploy either inclined smooth tubes spiraled around the furnace enclosure (Figure 2.28) or a vertical wall furnace with rifled tubes (Figure 2.29).

Spiral construction of tubes provides the following benefits:

1. For the same cross-sectional heat receiving area the number of inclined tubes is less than the number of vertical tubes to envelope the furnace without increasing the spacing between the tubes.
2. Uniform heat absorption in each tube renders the spiral wall system less sensitive to changes in the heat absorption profile in the furnace.
3. Increased mass flow rate of water in the tubes to a value that will prevent the DNB at partial loads, so there is no risk of tube starvation, and overheating is limited due to the critical heat flux/dry-out. It also equalizes the heat absorbed by water-wall tubes and thus limits the tube-to-tube temperature unbalance at the outlet of the furnace water-wall circuit (Figure 2.30).

However, spiral-tube construction involves a complex support structure, which raises fabrication costs, and it is also relatively difficult to erect and repair spiral tubes.

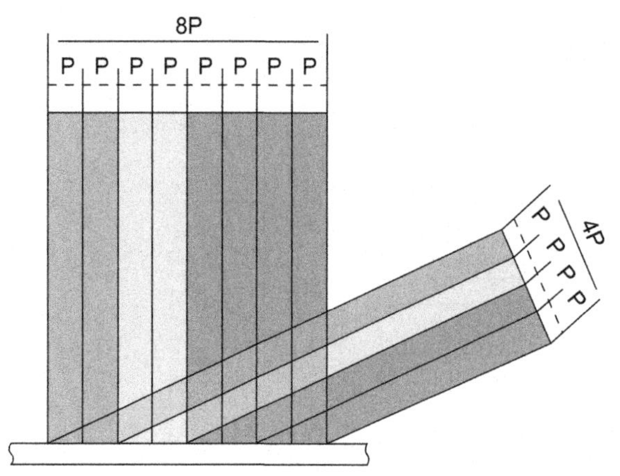

Figure 2.28 Spiral water tubes. *Source: ALSTOM Presentation to The Kuljian Corporation, Philadelphia. August 25, 2010.*

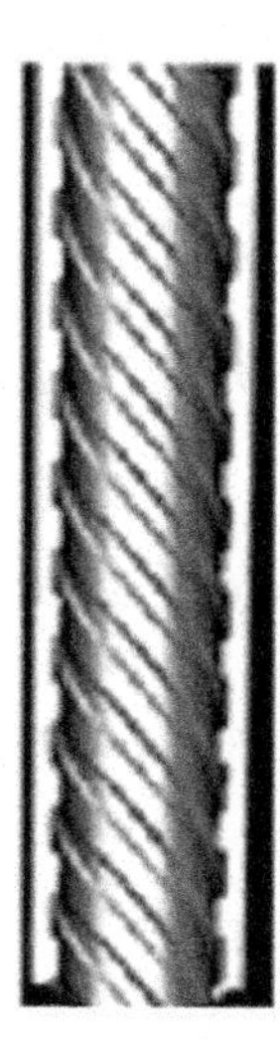

Figure 2.29 Rifled water tubes.

The vertical water-wall design uses rifled tubing for improved mass flow without DNB or dry-out at part loads. Its advantages lie in ease of maintenance, along with the potential to lower costs, and it is much easier to erect, support, and repair. The main disadvantage of the rifled tube water-wall design is the overall height of the boiler is excessive.

2.10.3 Steam-water cycle

A typical steam-water cycle of a supercritical plant is shown in Figure 2.31, which shows that the expansion of steam from state 1 to the condenser pressure would cause the exhaust vapor to become very wet and may cause severe damage to the

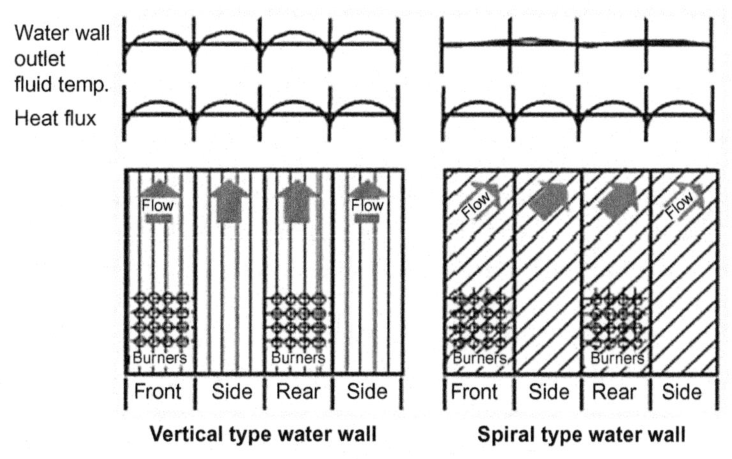

Figure 2.30 Furnace water wall circuit.

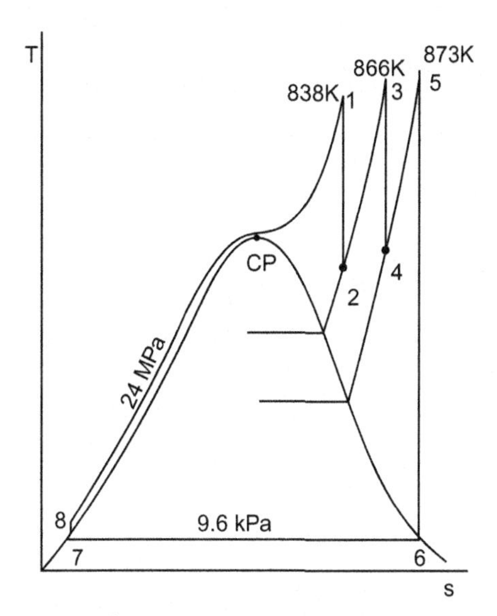

Figure 2.31 Typical steam-water cycle.

low-pressure turbine stages. Hence, this cycle has to essentially use single reheat, and if the high-pressure turbine inlet temperature is low (813 K or less) and pressure is high (above 26 MPa) double reheat may also be required.

2.10.4 Special features

1. Supercritical boilers are essentially once-through boilers because of a gradual change in density from the liquid phase to the vapor phase. Hence, storage vessel of subcritical boiler, i.e., boiler drum, is no longer valid for the supercritical one.

2. The volume of the evaporator system of a once-through boiler is much smaller compared with a natural circulation boiler. Due to smaller inventory of stored water and steam, the theoretical rate of response on the load changes of a supercritical unit is much faster than for the drum unit.
3. These boilers are better suited for frequent load variations than drum-type boilers, since the drum is a component with a high wall thickness, requiring controlled heating. Therefore, pressure changes in a supercritical boiler can be achieved more easily, and these boilers require 15–20% less cold start-up time compared with that needed for subcritical boilers of the same size.
4. A once-through boiler does not have any boiler blow-down, therefore, less waste water is disposed of and eventually condensate make-up is usually significantly less.
5. In a supercritical unit the main steam temperature is controlled by the water-fuel ratio control with back-up spray attemperation. As a result, the rated steam outlet temperature can be achieved at all loads with all types of fuel.
6. The wall-thickness of HP turbine section of a supercritical plant should be high enough to withstand elevated pressure and temperature, yet in order to increase the thermal flexibility and fast load changes the thickness has to be as low as possible to avoid large amount of material that would require controlled heating.

2.10.5 Advantages

1. Reduced fuel costs due to higher plant efficiency
2. Significant reduction in CO_2 emissions
3. Reduced NO_x, SO_x, and particulate emissions
4. Excellent availability, equal to or higher than that of a subcritical plant
5. Plant costs comparable with subcritical technology and less than other clean coal technologies
6. Maintenance costs of supercritical units are comparable to subcritical plants, if not better
7. Compatible with biomass co-firing
8. Can be fully integrated with appropriate CO_2 capture technology
9. More amenable to sliding-pressure operation
10. Adaptable to load swing and operational flexibility

2.10.6 Disadvantages

1. Boiler pressure drop is higher than that with a subcritical unit. As a result, boiler feed pump power consumption is higher than that of equivalent size drum type boiler.

2. Cost of high-temperature and high-pressure superheater and reheater alloy steel is exorbitant.
3. Superior water quality essentially requires a 100% condensate polishing unit, increasing overall costs.
4. Proven performance with high-ash Indian coal has yet to be established.
5. Not suitable for sustained constant pressure operation at part load.
6. Design of furnace sizing is critical. Improper sizing may cause slagging problems.
7. Inappropriate flow through evaporator tubes results in reduced mass flux that may lead to tube failure.

Nonetheless, power plants operating with supercritical steam conditions show lower emissions of CO_2, SO_X and NO_X, higher efficiency, flexibility and operating reliability, reduced fuel costs, and investment costs compared to subcritical plants.

2.10.7 History

The first application of supercritical technology in a power plant took place in 1957 in the United States. It was the Philo Unit 6 of the American Electric Power with a capacity of 125 MW designed with the superheater outlet steam conditions of 31 MPa and 883 K along with a double-reheat steam temperature of 839/811 K. Within two years, the 325 MW Eddystone Unit 1 of the Philadelphia Electric Company started operating with superheater outlet steam conditions of 34 MPa and 922 K along with a double-reheat steam temperature of 839/839 K.

The performance of these plants, however, was unreliable, expensive, and they had poor operational flexibility, mostly related to thermal effects. As a result, most countries stayed with subcritical technology. However, development continued in the 1970s and 1980s in inland Europe and Japan to explore the benefits of supercritical technology.

It should also be noted that the former USSR was not far behind the United States in the use of supercritical technology. The first supercritical power plant with a capacity of 300 MW was commissioned in the USSR in 1963, and the first 800 MW unit was put into operation in 1968. Unlike the United States, the USSR approach toward adopting supercritical steam conditions was quite modest, while steam pressure was raised to 25.5 MPa, superheat steam temperature was kept the same as that of the subcritical plants, i.e., 818 K, and the reheat steam temperature was either the same as the superheat temperature or slightly higher, at 818–838 K. The world's largest supercritical unit with a capacity of 1200 MW has been operating in the USSR since 1980 [16].

From the 1980s on, the technology got so matured that currently there are more than 600 supercritical power plants operating in the United States, Russia, Denmark, Germany, Poland, Japan, China, Korea, South Africa, India, Australia, and Taiwan. Power producers in other countries are also embarking on supercritical technology.

When considering supercritical steam conditions research shows that the industry is predominantly adopting a superheat steam pressure in the range of 24–31 MPa, with superheat/reheat steam temperatures in the range of 838–883 K [17]. In the near future the focus will likely be on adopting steam conditions of 31–32 MPa and 893 K/893 K with the ultimate goal of reaching ultra-supercritical steam conditions of 34–36 MPa pressure and 973/993 K temperature [17].

2.11 PROBLEMS

2.1 Steam is generated at a pressure of 10 MPa with 150 K superheat. If the rate of steam generation is 100 kg/s and the feedwater temperature at the economizer inlet is 463 K, determine the rate at which the heat must be supplied. (Assume enthalpy of compressed liquid is the same as the enthalpy of saturated liquid.)

(Ans.: 246595 kJ/s**)**

2.2 In a pressurized vessel water at 8 MPa pressure is heated by an immersion-heating coil. Find the maximum heat flux at the limit of nucleate boiling. Also determine the difference in temperature between the heating surface and the temperature of the fluid, if the heat transfer coefficient of the fluid is 160 W/m^2 K.

(Ans.: qmax = 1088 W/m^2; $\Delta T = 6.8$ K**)**

2.3 A 15-m long carbon-steel tube of 25 mm diameter is heated externally to 423 K. Water enters this tube at 2 MPa pressure with an average velocity of 3.06 m/s. While passing through the tube water is heated from 313 K to 393 K.

Find the "mean film temperature, Tm." Calculate the convective heat transfer coefficient and the heat transfer to the water by forced convection. (Neglect the thickness of the tube and friction drop inside the tube.)

(Ans.: $T_m = 388$ K; h = 17.73 kW/m^2 K; **Q** = 1285.95 kW**)**

2.4 A furnace wall is comprised of a 200-mm thick fire clay wall with an average thermal conductivity of 0.26 W/m.K. Determine the heat loss through the wall when the inside temperature of the furnace is 1253 K and temperature outside the wall is 328 K. If the ambient air temperature is 300 K, what is the surface coefficient of the wall material?

(Ans.: Q = 1202.5 W/m^2; f = 42.95 W/m^2.K**)**

2.5 An 80 mm outer diameter pipe carrying steam at 508 K is lagged with glass fiber of 25-mm thickness. Find the heat loss to ambient air if its average temperature is 395 K. The thermal conductivity of glass fiber may be taken as 0.045 W/m.K and the average surface coefficient as 40 W/m^2.K.

(Ans.: Q = 252.77 W/m^2**)**

2.6 A carbon-steel pipe has an internal diameter of 250 mm and an outside diameter of 275 mm. The outside of the pipe is covered with a layer of mineral wool that is 25 mm thick and has a thermal conductivity of 0.037 W/m.K. The inside temperature of the pipe is 623 K and the ambient air temperature is 298 K. Calculate the heat loss through the pipe and the surface coefficient of the insulating surface if the cold surface temperature is maintained at 333 K. Given: the thermal conductivity of carbon steel is 48 W/m.K.

(Ans.: $Q = 466.98\ W/m^2$; $f = 10.2633\ W/m^2.K$**)**

BIBLIOGRAPHY

[1] Stultz SC, Kitto JB (Ed.). Steam its Generation and Use, 41st ed. The Babcock And Wilcox Company; 2005.
[2] Rogers GFC, Mayhew YR. Engineering Thermodynamics Work and Heat Transfer, The English Language Book Society; 1967.
[3] Goodall PM (General Ed.). Efficient Use of Steam, Wetsbury House; 1980.
[4] Bozzuto C (Ed.). Clean Combustion Technologies, fifth ed. Alstom; 2009.
[5] Baumeister T, Marks LS (Ed.). Mechanical Engineers Handbook, sixth ed. Mcgraw-Hill Book Company, Inc.; 1958.
[6] Reznikov MI, Lipov YuM. Steam Boilers of Thermal Power Stations, Mir Publishers, Moscow; 1985.
[7] Shields CD. Boilers – Types, Characteristics and Functions, F. W. Dodge Corporation, New York; 1961.
[8] Sarkar S. Fuels and Combustion, Orient Longman; 2009.
[9] World Coal Association. Coal Facts; 2014.
[10] Central Boilers Board. Indian Boiler Regulations (Ibr), New Delhi.
[11] National Fire Protection Association, USA Nfpa 85 – 2004 Boiler and Combustion Systems Hazards Code; 2004.
[12] Gaffert GA. Steam power stations. McGraw-Hill Book Company, Inc.; 1952.
[13] Kern DQ. Process heat transfer. McGraw-Hill Book Company, Inc.; 1997.
[14] El-Wakil MM. Power plant technology. McGraw-Hill Book Company; 1984.
[15] A Clean and Efficient Supercritical Circulating Fluidised Bed Power Plant by Patrick Laffont, Jacques Barthelemy, Brendon Scarlin & Christian Karvenec.
[16] Advanced Russian Technologies For Raising The Operating Performances of Large Power Steam Turbines by Dr. Alexander S. Leyzerovich Actinium Corp., Mountain View, CA. In: Proceedings of 2000 International Joint Power Generation Conference Miami Beach, Florida. July 23–26, 2000.
[17] International Conference on Advanced Technologies and Best Practices for Supercritical Thermal Plants, New Delhi, India. November 21–23; 2013. (US AID & Ministry of Power, Govt. of India).

CHAPTER 3

Fuels and Combustion

3.1 INTRODUCTION

Any matter that is a source of heat is called fuel [1]. Fuel releases its energy either through a chemical reaction, such as during combustion, or via nuclear fission or fusion. An important property of a useful fuel is that its energy can be stored and released only when needed, and that the release is controlled in such a way that the energy can be harnessed to produce the desired work. Fuels are broadly classified as fossil or organic or chemical fuel, nuclear fuel, and rocket fuel.

Fossil fuels are hydrocarbons, primarily coal and petroleum (liquid petroleum or natural gas), formed from the fossilized remains of dead plants and animals by exposure to heat and pressure in the earth's crust over hundreds of millions of years. Fossil fuels generate substantial quantities of heat per unit of mass or volume by reacting with an oxidant in a combustion process. In most practical applications including combustion in a steam generator, air is used as the oxidant, although in certain processes oxygen, oxygen-enriched air and other chemicals are used as the oxidant.

Energy generation by a *nuclear fuel* takes place either by the process of nuclear fission of heavy fissile elements in a nuclear reactor resulting in chain reactions or by the process of nuclear fusion, in which simple atomic nuclei are fused together to form complex nuclei, as in case of fusion of hydrogen isotopes to form helium. The process of nuclear fusion is also known as a thermonuclear reaction, which is difficult to control even today. As a result, the main source of nuclear energy presently available is mainly from nuclear fission. The most common fissile radioactive heavy metals are naturally occurring isotope of uranium, U^{235}, artificial isotope of uranium, U^{233}, and artificial element plutonium, P^{239}. In a nuclear reactor plutonium, P^{239}, is produced from naturally occurring isotope of uranium, U^{238}, and U^{233} is produced from the naturally occurring element thorium, Th^{232}. Nuclear fission of 1 kg of U^{235} generates about $85*10^6$ MJ of heat, which is equivalent to the heat generated by combustion of about $5*10^6$ kg of coal with a high heating value (HHV) of 17 MJ/kg. In nuclear reactions, the product is either isotopes of the reactants or other nuclei.

Unlike fossil fuel, *rocket fuel* does not depend on its surroundings for the oxidant. The oxidant is carried by the rocket itself. The propellant is the chemical mixture burned to produce thrust in rockets and consists of a fuel and an oxidizer. Propellants are classified according to their state – liquid, solid, or hybrid. In a liquid-propellant

Thermal Power Plant
DOI: http://dx.doi.org/10.1016/B978-0-12-801575-9.00003-2

rocket, the fuel and oxidizer are stored in separate tanks, and are fed to a combustion chamber where they are combined and burned to produce thrust. Liquid propellants used in rocketry can be highly refined kerosene, liquid hydrogen (LH_2), liquid methane, etc. Solid propellant motors are the simplest of all rocket designs. Unlike liquid propellant engines, solid propellant motors cannot be shut down. Once ignited, they will burn until all the propellant is exhausted. The fuel is generally aluminum and the propellant is ammonium perchlorate. In hybrid propellant engines fuel is generally solid and the oxidizer is liquid.

Besides availability of energy from various form of fuels, energy is also available from sun, wind, tides, geothermal, biomass, etc., which are broadly called renewable sources of energy. Commercial use of such energy at present, however, is not large scale in many countries.

3.2 SOURCES OF CHEMICAL ENERGY

In steam power plants, fossil fuels that produce large quantities of heat economically have been used for combustion. These fuels are abundantly available in nature and their extraction process is reasonably simple. Most widely used fossil fuels for the production of steam in relation to power are available in all the states of matter, e.g., gaseous—natural gas, blast furnace gas, coke-oven gas, liquefied petroleum gas or (LPG), etc., liquid – high-speed diesel (HSD), light diesel oil (LDO), heavy fuel oil (HFO), furnace oil (FO), low sulfur heavy stock (LSHS), naphtha, liquefied natural gas (LNG), etc., and solid—bituminous coal, anthracite, lignite, peat, oil shale, biomass, etc.

Fossil fuels consist of a large number of complex compounds of five elements: carbon (C), hydrogen (H), oxygen (O), sulfur (S), and nitrogen (N). Besides these elements, all fuels contain mineral matter (A) and moisture (M) to some extent. However, there are just three combustible elements of significance in a fuel, e.g., carbon, hydrogen, and sulfur, of which carbon is the principal combustible element with a HHV of 32.780 MJ/kg. Hydrogen has a very high HHV of 141.953 MJ/kg, but its content in solid fuel is quite low, about 2–4%. The HHV of sulfur is only 9.257 MJ/kg, hence as a source of heat its presence is insignificant, although it is more so since its presence in coal is small in quantity. Major concern regarding sulfur is that it promotes corrosion and creates atmospheric pollution problems.

Contrary to solid and liquid fuels, gaseous fuels are mixtures of combustible and non-combustible gases. NFPA 85 defines natural gas as "A gaseous fuel occurring in nature and consisting mostly of organic compounds, normally methane (CH_4), ethane (C_2H_6), propane (C_3H_8), and butane (C_4H_{10}) [2]. The calorific value of natural gas varies between 26.1 and 55.9 MJ/m^3, the majority averaging 37.3 MJ/m^3." Natural gas on average contains 80–90% methane, 6–9% ethane, and 2–5% propane. Noncombustible gases present in minor quantities in natural gases are nitrogen (0.5–2.0%) and carbon dioxide (0.1–1.0%).

LPG, as per NFPA 85, is "A material composed predominantly of the following hydrocarbons or mixtures of them: propane, propylene, n-butane, isobutene, and butylenes" [2]. The mixture is liquefied at room temperature at very high pressure. An average mixture of LPG is comprised of about 80% butane and about 20% propane.

In addition to these fuels, other fuels have also been used as sources of chemical energy for combustion. Some of these fuels are coke, peat, bagasse, tars and char, wood, rice husk, producer gas (product received by burning coal or coke in air deficient environment in presence of controlled amount of moisture), water gas (produced by passing steam through a bed of hot coke; the product contains mainly hydrogen and carbon monoxide), etc.

3.2.1 Heating value

The *heating value* of a fuel is the amount of heat recovered when the products of complete combustion of a unit quantity of fuel are cooled to the initial temperature (298 K) of the air and fuel. As the heating value of fuel increases, the heat content delivered to the burners increases. The heat of combustion of a fuel is also called its potential heat.

When a fuel is burned in oxygen saturated with water vapor, the quantity of heat released is known as the high heating value (HHV), or gross calorific value (GCV), of fuel. When the latent heat of water vapor contained in the combustion products is subtracted from the HHV we get the low heating value (LHV) or net calorific value (NCV) of fuel. In a laboratory, the HHVs of solid and liquid fuels are determined at constant volume and those of gaseous fuels are determined at constant pressure. Combustion in a furnace, however, takes place at constant pressure.

For bituminous coal and anthracite, the HHV can be calculated approximately by Dulong's formula, as follows, which will be within 2–3% of the value determined by the calorimeter.

$$HHV_{COAL} = 33.823C + 144.251(H_2 - O_2/7.937) + 9.419S, \mathrm{MJ/kg} \tag{3.1}$$

Once the value of HHV_{COAL} is known, either from laboratory determination or from Dulong's formula, the LHV_{COAL} is then calculated as follows:

$$LHV_{COAL} = HHV_{COAL} - 2.44(H * 8.937 + M), \mathrm{MJ/kg} \tag{3.2}$$

where C, H, O, S, and M correspond to carbon, hydrogen, oxygen, sulfur, and moisture content of coal, respectively, expressed in parts by weight of each constituent, and heat of condensation of water vapor at 298 K is 2.44 MJ/kg.

3.3 AVAILABILITY OF FUELS

Fossil solid fuels are available in almost all countries but are distributed unevenly. The majority of countries is not rich in coal production. There are around 70 countries that contain recoverable coal reserves. The largest producer of hard coal is the People's Republic of China (about $3561*10^9$ kg was produced in the year 2013) [3], followed by the United States, India, Indonesia, Australia, Russia, and South Africa. Germany, Poland, and Kazakhstan also contribute to global coal production to some extent.

The largest producer of lignite (brown coal) is Germany (about $183*10^9$ kg produced in the year 2013). Russia, the United States, Poland, Turkey, Australia, Greece, and India also produce lignite. Coal is the major fuel used to generate over 40% of the world's electricity demand and provides 30% of global primary energy needs. Coal is also used in the production of 70% of the world's steel [3]. Almost all (93% or more) electricity-generating stations in Mongolia and South Africa are coal-fired [3]. Other countries that depend heavily on coal for electricity generation are Poland, the People's Republic of China, India, Australia, Israel, Indonesia, Germany, the United States, the United Kingdom, and Japan. Globally about $7823*10^9$ kg of hard coal and brown coal were produced in the year 2013. At the current production level, proven coal reserves are estimated to last 134.5 years as estimated by the German Federal Institute for Geosciences and Natural Resources (BGR), and the World Energy Council (WEC) reports that the coal reserves left are equivalent to 113 years of coal output [3].

Petroleum is a term that includes a wide variety of liquid hydrocarbons. The most familiar types of petroleum are tar, oil, and natural gas. Petroleum is found in porous rock formations in the upper strata of some areas of the earth's crust. There is also petroleum in oil sands (tar sands). Petroleum deposits can be found in almost all parts of the world, but commercial exploration stretches from Indonesia throughout Mayanmar, India, the Middle East, Central Europe, Africa, to North and South America. Over 50% of the world's oil reserves are in the Middle East. As of 2012, the known reserves of petroleum are typically estimated to be around $210.5*10^9$ m^3.

Oil accounts for a large percentage of the world's energy consumption, ranging from a low of 32% for Europe and Asia, up to a high of 53% for the Middle East. Other geographic region consumption patterns are as follows: South and Central America (44%), Africa (41%), and North America (40%).

More than 110 countries produce petroleum globally. According to *Businessinsider*, the largest supplier of petroleum oil in the year 2012 was Saudi Arabia, followed by the United States, Russia, Iran, People's Republic of China, Canada, United Arab Emirates, Brazil, Kuwait, Iraq, Venezuela, Nigeria, Qatar, Kazakhstan, and Libya [4]. Other countries that produce a substantial amount of petroleum are Angola and Algeria.

Crude oil is also found in semi-solid form mixed with sand as in the Athabasca oil sands in Canada, where it is usually referred to as crude bitumen. Venezuela also has large amounts of oil in the Orinoco oil sands. Canada and Venezuela contain an estimated $570*10^9$ m^3 of bitumen and extra-heavy oil.

Oil shales are found in many countries, but the United States has the world's largest deposits. Oil shales can be converted into crude oil using heat and pressure in a process called destructive distillation.

The world at large consumes $4.8*10^9$ m^3 of oil per year, and the top oil consumers largely consist of developed nations. In fact, 21% of global petroleum consumption is from the United States. Other countries that are large consumers of petroleum are the European Union, People's Republic of China, Japan, Russia, Germany, and India. (**Note**: Algeria, Angola, Ecuador, Iran, Iraq, Kuwait, Libya, Nigeria, Qatar, Saudi Arabia, United Arab Emirates, and Venezuela are jointly known as OPEC: Organization of Petroleum Exporting Countries.)

Natural gas is a major source of electricity generation through the use of gas turbines and steam turbines in combined cycle modes. Natural gas burns cleaner than other fossil fuels, such as oil and coal, and produces less carbon dioxide per unit energy released. For an equivalent amount of heat, burning natural gas produces about 30% less carbon dioxide than burning petroleum and about 45% less than burning coal.

Natural gas is commercially produced in oil and natural gas fields. Gas produced from oil wells is called casinghead gas or associated gas. Over 66% of gas reserves are concentrated in the Middle East and Russia. As of 2014, the total proven reserves of natural gas in the world as reported by the *International Energy Statistics* are more than $197*10^{12}$ m^3 [5]. The world's largest gas field by far is the offshore South Pars/North Dome gas-condensate field, shared by Iran and Qatar, estimated to have 58×10^{12} m^3 of natural gas in place. The second largest natural gas field is in Russia.

According to the *International Energy Agency* global production of natural gas in the year 2013 was about $3.48*10^{12}$ m^3. More than 90 countries produce natural gas. The United States and Russia together is the largest producer, with around $1.35*10^{12}$ m^3 per annum. Other main contributors are European Union, Iran, Canada, Qatar, Norway, People's Republic of China, Saudi Arabia, Algeria, The Netherlands, Indonesia, Malaysia, Uzbekistan, Egypt, Turkmenistan, Mexico, United Arab Emirates, Bolivia, Australia, UK, Trinidad and Tobago, and India.

3.4 CHARACTERISTICS OF COAL

Coal is not another form of carbon. It is a fossil fuel or an organic sedimentary rock, formed by the result of temperature and pressure on plant debris; it is far more plentiful than oil or gas. Coal is comprised of a complex mixture of organic chemical substances containing carbon, hydrogen, oxygen, together with small amounts of nitrogen, sulfur, and trace elements such as mercury, selenium, arsenic, fluorine, etc. It also contains various amounts of moisture and minerals.

The different grades of coal include peat, lignite, sub-bituminous, bituminous, semi-anthracite, and anthracite. From a geological perspective anthracite is the oldest form of coal and lignite is the youngest. In the lower ranked group (peat and lignite) coal is low in carbon and high in oxygen. At the upper end, the reverse is true. The amount of hydrogen remains reasonably constant. All ranks of coal oxidize in a normal storage environment.

Peat is the earliest formation of coal from decomposition of organic matter. With increasing depth and increasing temperature this peat changes into lignite. These low rank coals gradually become mature with further increase in temperature and more time.

Mature forms of coal – such as "bituminous" (soft) and "anthracite" (hard) – contain more carbon and less moisture, while, the most immature forms of coal –"peat" and "lignite"– contain the high moisture and lowest carbon.

Anthracite is a hard coal comprised of mainly carbon with little volatile matter. Its moisture content is negligibly small. On the other side of the scale is lignite, which is a very soft coal whose constituents are mainly moisture and volatile matter with very little presence of fixed carbon.

Because of the wide variation in constituents of different forms of coal it is important to know the various characteristics of coal, as discussed in the following, since they affect the burning of coal in a power plant [1]. It is also important to remember that burning an unsuitable coal can reduce the efficiency of a power plant with consequent increase in pollutant emissions.

Caking, coking An essential property of coal that is important for the production of coke is *caking*. When caking coals are heated they soften, swell, become plastic, then form lightweight, porous coke particles. These coals are also called agglomerating coals (free swelling index of 1 or greater). They can easily float out of the furnace before they are burned, unless pulverization is sufficiently fine. [NOTE: The 'free swelling index' is determined by heating 1 gm of coal sample for a specified time and temperature. The shape of the button formed by the sample coal is then compared to a set of standard buttons. Larger formed button indicates higher swelling index. Oxidized coal tends to have lower swelling index. This index can be used to show caking properties of coal.] [6].

Coal that shows little or no fusing action is called *free-burning coal*. Free-burning coals do not have the same swelling characteristic (free swelling index of 0.5 or less) and, hence, do not require the same degree of pulverization. Sub-bituminous coals and lignite are classified as free-burning coals. Both caking and free-burning coals can be burned without difficulty in pulverized coal-fired boilers.

When heated at uniformly increasing temperature in an air-less furnace or oven or in an atmosphere very deficient in oxygen, the volatile matter of coal is driven off, leaving behind a residue of fixed carbon and ash. During heating in the range of 620–670 K coal softens and the residue fuses into a solid mass. This is called *coke*, which is suitable for metallurgical processes.

Caking and coking coals generally have lower oxygen content than free-burning and non-coking coals. All coking coals are caking, but not all caking coals are suitable as coke.

Reactivity of coal The reactivity or rate of combustion of coal is defined as the rate at which coal combines with oxygen at temperatures above the ignition point. During the progress of heating, a portion of the inherent oxygen of coal also becomes available to the oxidation process. As the oxygen level increases, the fuel reactivity increases. Reactivity is influenced by the particle size as well as by the surface area to mass ratio of the particle.

Ash fusibility or ash-fusion temperature Ash fusibility is the means of measuring the performance of coals related to slagging and deposit build-up. In general, high fusion temperatures result in low slagging potential in dry bottom furnaces, while low fusion temperatures are considered mandatory for wet-bottom (slag tap) furnaces.

When coal ash is heated, it becomes soft and sticky, as the temperature continues to raise it becomes fluid. Ash-fusibility temperature is measured by heating cones of ash in a furnace arranged to produce either oxidizing or reducing atmosphere. The temperature at which the tip of the cone starts to deform is known as the *initial-deformation temperature* (IDT). The temperature at which the cone fuses down into a round lump, in which the height is equal to the width at the base, is called the *softening temperature* (ST). The temperature at which the cone has fused down to a hemispherical lump, where the height of the cone equals one half the width of the base, is known as the *hemispherical temperature.* When the melted cone spreads out in a flat layer, the temperature is called the *fluid temperature.*

The softening temperature serves as the single best indicator of clinkering and slagging tendencies under given fuel-bed and furnace conditions. If ash arrives at a heat-absorbing surface at its softening temperature, the resulting deposit is likely to be porous in structure, and may either fall off metal surfaces by its own weight or be removed by soot blowing. If such a deposit builds up in a high-temperature zone, its surface can reach its melting point and then run down the wall surfaces.

The temperature differential between initial deformation and fluid temperatures, if small, indicates that the wall slag will be thin, runny, adhesive, and sticky. This type of slag is extremely difficult to control by soot blowing. As the differential temperature increases, the resulting slag deposit will build up to thicker proportions, which is less adhesive and therefore responds better to removal by soot blowing.

Sulfur Sulfur in coal exists in three forms: organic, pyritic, and sulfate. Sulfur that is an inherent constituent of coal is organic sulfur, which is considered as a non-removable impurity. Pyritic sulfur occurs primarily as pyrite or mercasite. Sulfate sulfur usually exists as calcium sulfate or iron sulfate.

When coal is burned, sulfur oxides form and emit through the stack. However, the quantity of sulfur oxides emission to atmosphere should comply with local air pollution control regulations.

In addition to its air-polluting properties, sulfur also plays a role in promoting corrosion of air heaters, economizers, and stacks. Sulfur oxides, when combined with moisture in combustion air form acids that may be deposited when the combustion gas is cooled below its dew-point temperature. Sulfur also contributes to clinkering and slagging, and to spontaneous combustion of stored coal.

3.4.1 Analysis of coal

The two different methods of analyzing coal are known as *ultimate* and *proximate.* Ultimate analysis determines the proportion of the main chemical elements contained in the coal, i.e., carbon, hydrogen, nitrogen, sulfur, and carbonate contents. The oxygen may be determined chemically but usually is estimated by difference. Thus, the results of ultimate analysis are important since they facilitate combustion calculations. This analysis provides information regarding the primary combustible and non-combustible components that are used to compute combustion air requirements, flue-gas volumes, and the losses associated with the combustion of hydrogen whose latent heat of evaporation is lost with the gas. The results of ultimate analysis are given in various categories, e.g., on a "as received," "air dried," "dry," "dry ash-free," or "dry mineral matter free" basis. Conversion from one analysis report to another is presented in Table 3.1.

The results of proximate analysis provide information regarding moisture, ash, volatile matter, and by difference, the fixed carbon content of coal. The higher the value of volatile matter the easier the ignition of fuel. On the other hand, the fixed carbon is the main heat generator during the burning of coal. The major contributor to the heating value of coal is fixed carbon together with the amount of volatile matter content of coal.

Proximate analysis of coal in combination with its calorific value facilitates the design of the furnace including burner locations along with determining the quantum of coal required for generating a quantity of steam.

Moisture All coal contains some natural moisture, since all coals are mined wet. This natural moisture is called *inherent moisture* and it lies in the pores of the coal and remains within the coal after it is air-dried. *Surface moisture*, on the other hand, depends on conditions in the mine and the weather during transit. The surface moisture may be removed from coal by heating it to 373–378 K.

Moisture is generally determined quantitatively in two steps: by air-drying and oven-drying. The air-dried component of total moisture is required in the design and selection of coal-handling and coal-preparation equipment. High moisture content may cause serious difficulties in the fuel combustion process. It reduces the heating value of fuel, increases fuel consumption, and the volume of products of combustion. It also results in higher heat losses with flue gases and raises power consumption of induced draft fans. Surface moisture adversely affects pulverizer performance. In the event coal dust is over-dried, it is liable to self-ignite in places where it is stored or accumulates and may become explosive when mixed with air.

Table 3.1 Conversion of ultimate analysis report of coal from one category to another category

Given	Required				
	As received (ar)	**Air dried (ad)**	**Dry (db)**	**Dry ash-free (daf)**	**Dry mineral matter free (dmmf)**
As Received (ar)	–	(100 – Mad)/(100 – Mar)	100/(100 – Mar)	100/(100 – (Mar + Aar))	100/(100 – (Mar + MMar))
Air Dried (ad)	(100 – Mar)/(100 – Mad)	–	100/(100 – Mad)	100/(100 – (Mad + Aad))	100/(100 – (Mad + MMad))
Dry (db)	(100 – Mar)/100	(100 – Mad)/100	–	100/(100 – Aad)	100/(100 – MMdb)
Dry Ash-Free (daf)	(100 – (Mar + Aar))/100	(100 – (Mad + Aad))/100	(100 – Adb)/100	–	(100 – Adb)/(100 – MMdb)
Dry Mineral Matter Free (dmmf)	(100 – (Mar + MMar))/100	(100 – (Mad + MMad))/100	(100 – MMdb)/100	(100 – MMdb)/(100 – Adb)	–

M = Moisture Content, MM = Mineral Matter Content, A = Ash Content.

Ash The ash content of coal is the incombustible mineral matter residue that is left behind after coal burns completely. Ash is chiefly comprised of SiO_2, Al_2O_3, Fe_2O_3, CaO, with smaller amounts of TiO_2, MgO, K_2O, Na_2O, SO_3, P_2O_5, etc. While ferric oxide, lime, magnesia, potassium oxide and sodium oxide are basic components of ash, silica, alumina and titania are acidic components.

The ash-fusion temperature increases as the "percentage of acidic components" increases or the "percentage of basic components" becomes very high or low. As discussed earlier high fusion temperatures result in low slagging potential, while high slagging potential is the consequence of low fusion temperatures. A low ash fusion temperature is not suitable for pulverized coal firing as the "furnace exit gas temperature" (FEGT) will have to be too low, resulting in a slagging problem in the furnace. The mineral matter in the ash plays a significant role in the slagging, fouling, erosion, and corrosion of components exposed to the combustion gases. The slagging index of ash can be determined by using the following formula:

$$Slagging\ Index = \frac{(Sum\ of\ Basic\ Components), \%}{(Sum\ of\ Acidic\ Components), \%} \tag{3.3}$$

Ash affects fuel selection and the design and sizing of the furnace and other coal quality-dependent components. The properties of ash also play an important part in the operation of a steam generator. Gas velocity in convective heat transfer areas needs to be reduced to prevent erosion and wear of steam and water tubes as well as flue-gas ducts. High quartz content in ash can exacerbate the wearing of grinding parts of pulverizers.

Ash must be removed from the furnace and plant using special equipment to prevent air pollution, hence increasing costs. Ash also increases shipping and handling costs.

Volatile matter When dry coal is heated gradually in an inert medium in the absence of air, the total loss of weight is called the volatile matter. During the process of combustion it is driven off in gaseous form from coal. The yield of volatile matter from coal takes place at a temperature range of 383–1373 K, and the highest yield takes place at around 1073 K. This is the principal indicator of the reactivity of a coal, and for predicting ignitability and flame stability.

Volatile matter is comprised of combustible gases, such as methane and other hydrocarbons, hydrogen, oxygen and carbon monoxide, and non-combustible gases. It affects the firing mechanism, furnace volume, and arrangement of heating surfaces. High-volatile coals ignite more readily than low-volatile coals. High-volatile coals need less-fine pulverization than do low-volatile coals. Low-volatile coals, except anthracite, have higher grindabilities, because they are softer.

Fixed carbon The fixed carbon is that portion of coal that remains as residue after volatile matter distills off, after the sum of moisture and ash content in the coal is subtracted. It is essentially carbon, but contains minor quantity of hydrogen, oxygen, nitrogen, and sulfur not driven off with the gases. In a combustion process this is the combustible residue left after the volatile matter distills off. In general, the fixed carbon represents the portion of the coal that must be burned in a solid state. Knowledge of fixed carbon helps in the selection of combustion equipment, since its form and hardness are an indication of the caking properties of a fuel.

3.5 OTHER SOLID FUELS

Some of the solid fuels that find wide industrial applications are described in the following. None of these fuels, however, can be used as a source of energy in large utility power plants.

3.5.1 Biomass

Biomass is an organic matter that has been in use since human beings started burning wood to make fires. It is a source of renewable energy derived from plant material, urban garbage, and animal waste. It can regrow over a relatively short period of time. Biomass can be used directly through combustion or indirectly by converting it to biofuel.

Biomass is produced from solar energy by photosynthesis. Through the process of photosynthesis plants convert CO_2 from the air and water from the ground into carbohydrates, which on combustion gets back to CO_2 and water along with the release of solar energy absorbed during photosynthesis.

In many developing countries biomass is the only source of domestic fuel. Wood is the largest source of biomass energy. Rotting garbage and agricultural and animal waste all release methane gas. The potential of algae and water hyacinth as a source of biomass is under research. It is believed that the use of biomass as an alternative source of energy will reduce greenhouse gas emissions because the carbon dioxide released into atmosphere by using biomass is recovered again by the growth of new biomass. In contrast, the use of fossil fuels increases greenhouse gas emission without any recovery.

Some of the biomass used to generate bio-power includes the following:

- **i.** Agricultural wastes, e.g., wheat straw, rice husk, jute stick, etc.
- **ii.** Energy crops, e.g., bagasse, bamboo, special type of grass, e.g., switchgrass, etc.
- **iii.** Wood and forest residues, e.g., dry leaves, twigs, etc.
- **iv.** Wood wastes, e.g., sawdust, wood shavings
- **v.** Clean industrial and municipal wastes
- **vi.** Cattle dung, poultry litter, etc.

3.5.2 Peat

Peat is the earliest formation of coal from decomposition and disintegration of wetland vegetation, e.g., bogs, mosses, sedges, shrubs, and other plants, yet it is not considered a coal or a fossil fuel. Peat forms when plant material, in wetlands, where flooding obstructs the flow of oxygen from the atmosphere, is prevented from decaying fully. It is partially carbonized vegetable matter saturated with water. The moisture content of peat is more than 75%. Peatland features can include ponds, ridges, and raised bogs. Large deposits of peat are found in Canada, People's Republic of China, Russia, Finland, Ireland, Sweden, and the United States.

Peat is friable and its quality is variable. It can be ignited easily, it burns freely releasing intense heat, and its color ranges from yellowish to dark-brown to black. Peat is not considered a renewable source of energy. With increasing depth and increasing temperature peat changes into lignite.

Bog is wet spongy ground of decomposing vegetation, which can be cut and dried and used as fuel. Sedge is grass like plant having solid stems and narrow leaves that grows in wetlands.

3.5.3 Charcoal

Charcoal is the light-black residue left on heating wood or other organic substances such as saw dust, coconut shell, bark, bamboo, etc., in the absence of air. Charcoal is comprised of mainly pure carbon with varying amounts of hydrogen, oxygen, and ash. It can be ignited easily. The heating value of charcoal is higher than that of wood. It is also lighter than wood, thus transport costs are lower. It is used as a cooking fuel, in smelting metal ores, in explosives, and as an absorbent of gases and liquids from solutions. Wood charcoal is also used as a component of gun powder.

3.6 PETROLEUM AND NATURAL GAS

In its natural state *petroleum* is generally a brownish-green to black liquid with a density of 0.80–0.95 at 288 K. Although petroleum is liquid, from its formation it may also include natural gas. Petroleum is comprised of a mixture of innumerable hydrocarbons of differing molecular weight and structure that may be classified into three groups as follows:

- GROUP A: *Paraffins* (C_nH_{2n+2}, where $n = 1$ to 35)
  When $n = 1$ to 4, constituents are gases
  (e.g., Methane: CH_4, Ethane: C_2H_6, Propane: C_3H_8 and Butane: C_4H_{10})
  When $n = 5$ to 15, constituents are liquids
  (e.g., Pentane: C_5H_{12}, Hexane: C_6H_{14}, Heptane: C_7H_{14}, ...)
  When $n = 16$ and above, constituents are solids/semisolids
  (e.g., Wax: $C_{17}H_{36}$, ...)

- GROUP B: *Naphthenes* (C_nH_{2n}, where n ≥ 5)
  (e.g., Cyclopentane: C_5H_{10}, Cyclohexane: C_6H_{12}, ...)
- GROUP C: *Aromatics* (C_nH_{2n-6}, where n ≥ 6)
  (e.g., Benzene: C_6H_6, Toluene: C_7H_8, Xylene: C_8H_{10}, ...)

Olefins (C_nH_{2n}, e.g., Ethylene: C_2H_4, Propylene: C_3H_6, Butylene: C_4H_8) and *Acetylenes* (C_nH_{2n-2}, e.g., Acetylene: C_2H_2) were not included in the above list. Olefins are formed in the processing of petroleum.

From the ultimate analysis of crude petroleum it is observed that its composition varies within a narrow band and is composed of mainly carbon (83–87% by weight) and hydrogen (11–14% by weight) along with elements like oxygen (2–3% by weight), nitrogen (0.1–1.0% by weight), and sulfur (0.5–3.0% by weight) in combination. Ash content of crude is about 0.1% or less that contains metals like vanadium, nickel, silicon, aluminium, iron, calcium, magnesium, and sodium to some extent.

Various constituents of crude oil are separated by fractional distillation. In the initial stage of distillation methane, ethane, propane, butane, and light gasoline are separated by flash evaporation. This is the primary or pressure stage. Crude from this stage distills further in secondary or atmospheric stage to separate heavy gasoline, naphtha, kerosene, light gas oil, and heavy gas oil. Residue from the atmospheric stage is put in the third or vacuum stage, thereby separating fuel oil blends, distillates for lubricating oil, bitumen, etc. Distillation of coal tar that is rich in aromatic hydrocarbons yields benzene, toluene, xylene, etc. Coal tar is thick, dark, and oily liquid and is obtained by coal carbonization. Modern civilization depends heavily on petroleum for the production of chemicals, fertilizers, dyestuffs, detergents, fibres, plastics, etc. Almost all internal combustion engines run on petroleum yield fuels. Petroleum products are also used in ovens and furnaces. Replacement of solid fuels by liquid and gaseous fuels improves the operating conditions of power stations and reduces the cost of the equipment substantially and increases efficiency of the stations. However, these benefits are achieved at the expense of exorbitantly high running cost.

3.6.1 Liquid fuels

Different types of liquid fuels obtained as products from petroleum refineries include the following.

Gasoline Liquid fuel that is used in a reciprocating spark-ignition internal combustion engine is called gasoline. Gasoline is broadly classified as motor gasoline and aviation gasoline. Motor gasoline is a complex mixture of low boiling hydrocarbons. Low boiling hydrocarbons usually cause vapor lock in engines and hence are not suitable for use in aircraft engines. Motor gasoline has a lower octane rating than aviation gasoline. Because of its use in aircraft engines, volatility of aviation gasoline is higher than that of motor gasoline.

Jet fuel Because of its low freezing point and wider boiling point, commercial jet airlines run on this fuel. The vapor pressure of jet fuel is quite low (14–20 kPa at 311 K). Antioxidants, metal deactivators, and corrosion inhibitors are used as additives in jet fuel. This fuel has some bearing with kerosene.

Kerosene The volatility of kerosene is lower than that of gasoline. Kerosene is basically paraffin used for burning in oil lamps and stoves. It is obtained as a straight-run distillate from crude petroleum after gasoline is recovered.

Diesel fuel The characteristics of diesel fuel vary from heavy kerosene to residual fuels. Diesel fuel is rich in aromatics and iso-paraffins. Ignition quality, cleanliness, and viscosity are the important properties of diesel fuel that determine its use in high-speed engines. Use of diesel fuel in diesel engines is attractive because of its low cost, even though the maintenance cost is high.

Fuel oil There are five commercial grades of fuel oil, also known as bunker fuel oil, or furnace oil, viz. high speed diesel (HSD), light diesel oil (LDO), heavy fuel oil (HFO), furnace oil (FO), low sulfar heavy stock (LSHS). Any oil used for generation of power or heat is identified as fuel oil. It ranges from light diesel oil (LDO) to heavy fuel oil (HFO). LDO is used in burners without preheating, while elaborate heating arrangement is provided to HFO for its use in burners. Equipment used for burning HFO is designed to inhibit formation of carbon deposits. Burners are also provided with an easy removal facility to quickly clean carbon deposits from burner nozzles.

3.6.1.1 Properties

As discussed above liquid fuels cover a wide range of products. These products have different uses based on these properties:

Specific gravity The specific gravity of a fuel determines the carbon-hydrogen contents of that grade of fuel. For example, a higher specific gravity denotes a higher concentration of carbon than that of hydrogen in that fuel. Eventually the gross calorific value per unit weight of a heavier fuel will be lower than that of a lighter fuel. Knowledge of specific gravity also helps to convert the volume of a liquid to its weight.

The specific gravity of petroleum liquid is specified as *Degree API* by the American Petroleum Institute (API) to denote the relative density of petroleum liquid to the density of water at the same temperature. If the API gravity is greater than 10, the fluid is lighter and floats on water; if it is less than 10, it is heavier and sinks. On a hydrometer API gravity is graduated in degrees such that most values will fall between 10 and 70 API gravity degrees, defined as follows:

$$^{o}\mathrm{API} = (141.5/G) - 131.5 \tag{3.4}$$

where G stands for specific gravity of a liquid at 288.5 K in relation to water at 288.5 K.

Viscosity The viscosity of a fluid is a measure of its resistance to flow. It is the most important property in furnace-oil specification. It influences the degree of pre-heating required for handling, storage, and satisfactory atomization. If the oil is too viscous it may become difficult to pump, the burner may be hard to light, and operation may be erratic. Poor atomization may result in carbon deposits on the burner tips or on the walls. The upper viscosity limit for furnace oil is such that it can be handled without heating in the storage tank except under severe cold conditions.

Knowledge of viscosity of a fuel oil also facilitates the design of fuel burners to ensure optimum performance of burners for achieving the highest combustion efficiency. Pre-heating may be necessary for proper atomization. For ease of atomization in conventional fuel oil burners the desired maximum viscosity is 25 centistokes. The maximum viscosity for easy pumping of oil in pipes is 1200 centistokes, for which the temperature of the fuel oil should be maintained at around 333−343 K.

The viscosity of fuel oils falls as the temperature rises but becomes nearly constant above about 393 K. The density of fuel oils also decreases as the temperature rises, resulting in less weight of oil delivered to the burner. Hence, beyond 393 K additional preheating actually lowers the burner capacity. Viscosity of oils is measured conventionally as *kinematic viscosity* and is expressed as stokes, the unit being m^2/s, or centistokes. Other standards to denote viscosity include Redwood No. 1, Redwood No. 2, Saybolt Universal, and Engler Degrees.

Flash point and fire point The flash point of a flammable liquid is the lowest temperature to which it must be heated to give off sufficient fuel vapor to form an ignitable mixture with air. When a certain quantity of oil is heated slowly and a small flame is passed over the oil surface the temperature at which the first flash occurs is the *flash point* of that oil sample. At this temperature the vapor may cease to burn when the source of ignition is removed. At a slightly higher temperature, called the *fire point*, the vapor continues to burn after being ignited. This is the lowest temperature at which the liquid gives off vapor fast enough to support continuous burning even after the ignition flame is removed. Neither of these parameters is related to temperatures of the ignition source, which are much higher. For the same liquid the flash point is lower than the fire point.

A flammable liquid with high volatility will have a low flash point, and one having low volatility will have a high flash point. Every flammable liquid has a vapor pressure; one with high vapor pressure will have a high flash point, and low vapor pressure denotes a low flash point.

Flash points indicate comparatively the degree of safety in storage, transportation, and use of liquid petroleum products, either in closed or open containers. They are not directly related to the fire hazards involved. An oil having flash point below 294 K (e.g., gasoline) is called a Class "I" petroleum. Flammable liquids with flash points between 294 K and 328 K (e.g., kerosene) are considered as Class "II" petroleum.

A Class "III" petroleum liquid has a flash point above 328 K (e.g., gas oils and fuel oils). The vapor space of a tank or vessel, which contains any of the above three classes of petroleum liquids, is classified as a Zone 0 hazardous area (discussed in the following).

There are two ways to measure flash points: open-cup testers and closed-cup testers. In the open-cup type (e.g., Cleveland Open Cup (COC) and Pensky-Martens Open Cup), the sample is contained in an open cup, which is heated, and at intervals a flame is brought over the surface. The main difference between the two types is that the Cleveland Open Cup is heated from below, while the Pensky-Martens Open Cup is heated both from the sides as well as below. Closed-cup testers are sealed with a lid through which the ignition source can be introduced periodically. The oil is heated in the closed cup under specified conditions and a pilot flame is introduced in the vapor space by opening a shutter in the lid. The vapor above the liquid is assumed to be in reasonable equilibrium with the liquid. For oils with flash points above 323 K, e.g., fuel oils, gas oils, diesel oils, etc., a Pensky-Martens Closed Cup tester is used, while a Abel Closed Cup tester is used for oils below this temperature, e.g., kerosene. For the same sample, closed-cup testers give lower values for the flash point (typically 5–10 K lower) than the open-cup tester value.

Pour point and cloud point The pour point of oil is the lowest temperature at which it will remain still fluid or can be poured under prescribed conditions. This temperature is 2.8 K higher than the temperature at which oil ceases to flow when cooled. Petroleum products having temperatures 8–11 K above their pour points are readily pumpable. Heavy fuel oils having significantly high pour points require additional heating to be pumpable. The cloud point of a fluid is the temperature at which dissolved solids are no longer completely soluble, precipitating as a second phase, giving the fluid a cloudy appearance. In the petroleum industry, cloud point refers to the temperature below which wax in liquid form has a cloudy appearance. The presence of solidified waxes thickens the oil and clogs fuel filters and injectors in engines. The wax also accumulates on cold surfaces (e.g., pipelines or heat exchangers) and forms an emulsion with water. Therefore, the cloud point indicates the tendency of oil to plug filters or small orifices at cold operating temperatures. The cloud-point temperature is 5–6 K higher than the pour-point temperature.

Carbon residue Fuel oils tend to form carbonaceous deposits when they are burned. The carbon residue value of a fuel gives an approximate indication of the combustibility and deposit-forming tendencies of the fuel. There are two methods for the determination of carbon residue: Conradson's test and Ramsbottom's test. Atomizing burners are practically insensitive to the carbon residue of a fuel.

Sulfur Crude oils in general contain sulfur in some form or another, which increases with an increase in the boiling range of the oil. Sulfur cannot economically be completely removed, nor is it required to remove it completely. The sulfur content in petroleum products is classified either as corrosive sulfur or total sulfur.

Heavy fuel oil may contain 2–4% sulfur. On combustion, sulfur releases foul gases, and when it comes across moisture it promotes corrosion.

Moisture Moisture in oil is not desirable since it interferes with combustion. Moisture may be present in free or emulsified form and can on combustion cause damage to the inside of furnace surfaces, especially if it contains dissolved salts. It can also cause sputtering of the flame at the burner tip. The water content of furnace oil when supplied is normally very low.

Ash Ash content represents the incombustible component remaining after a sample of the furnace oil is completely burned. The ash content of petroleum products is generally low. Typically, the ash value is in the range of 0.03–0.07% by weight, although in certain oils higher ash content may be found. Ash consists of extraneous solids, residues of organometallic compounds in solution, and salts dissolved in water present in the fuel. These salts may be compounds of sodium, vanadium, aluminium, nickel, calcium, magnesium, silicon, iron, etc. Sodium and vanadium content varies widely in fuel oils depending on the crude oil source or crude oil mixes and ranges up to 200 ppm (parts per million) and 600 ppm, respectively. The ratio of sodium (Na) to vanadium (V) in fuel oil greatly influences the melting point and thereby the corrosive and slagging effect. The critical range of Na/V ratios is 0.08–0.45 of which 0.15–0.30 are particularly destructive. The sodium content of fuel oil causes bonding of ash constituents on boiler superheater surfaces.

Vanadium in fuel oil combines with oxygen in the combustion process and creates pentavanadate (V_2O_5) droplets that melt at 573–773 K. Over time, these pentavanadate droplets accumulate and form a hard crust in the interior of the equipment. The thick shield prevents heat transfer that in turn reduces the efficiency of boiler. To mitigate these problems magnesium oxide is introduced, creating a new molecule that will not stick to the interior walls and pipes of boilers and furnaces, allowing better heat transfer thereby increasing combustion efficiency.

In internal combustion (IC) engines vanadium is responsible for forming slag on exhaust valves and seats on 4-cycle engines, and piston crowns on both 2- and 4-cycle engines, causing localized hot spots, which eventually lead to burning away of exhaust valves, seats, and piston crowns. Vanadium can be neutralized during combustion in IC engines by the use of chemical inhibitors (such as magnesium or silicon).

Sediment As a blend of residues furnace oil contains some quantity of sediments. These have adverse effect on the burners and cause blockage of filters, etc. However, typical values of sediments are normally much lower than the stipulated value of maximum 0.25 percent, by mass.

Calorific value Calorific value or heat of combustion of a fuel is the amount of heat developed by a fuel when completely burned. Gross calorific value of fuel oils varies within a comparatively narrow range, highest value of 48 MJ/kg is for light distillate to the lowest value of 42 MJ/kg for heavy fuel oil. Fuel oils are high in hydrogen content, 11.8−14.5%; thus, their net calorific values are less than the respective gross calorific values by as much as 2.60−3.18 MJ/kg.

Empirically, the gross calorific value of fuel oils can be determined as follows:

$$HHV_{OIL} = 51.91 - 8.79G^2, \mathrm{MJ/kg} \quad (3.5)$$

where G stands for specific gravity of liquid at 288.5 K in relation to water at 288.5 K.

Octane number The octane number of a gasoline is a measure of its anti-knock value or its ability to resist knock during combustion in an engine. In spark-ignition internal combustion engines combustion of air/fuel mixture in the cylinder may begin smoothly in response to ignition by the spark plug and then one or more pockets of unburned fuel may explode outside the envelope of the normal combustion front. This results in *knocking* of the engine.

The octane number of a fuel is defined as the percentage of iso-octane in a mixture of n-heptane and iso-octane. The higher the octane number of gasoline the better its anti-knock capability. Iso-octane has excellent anti-knock quality and has been assigned as 100 octane number, while n-heptane has a poor anti-knock quality with an octane number of zero.

Higher-compression-ratio engines need higher octane number gasoline for smooth operation.

Cetane number Cetane number is a measure of the ignition quality of diesel engine fuels as it indicates the comparative ease with which a diesel fuel will ignite in a diesel engine cylinder. The time interval between the start of injection and the start of combustion (ignition) of a fuel in a diesel engine, i.e., the fuel's ignition delay, is measured by the cetane number. This number is a measurement of the combustion quality of diesel fuel during compression ignition. In a particular diesel engine, higher cetane fuels will have shorter ignition delay periods than lower cetane fuels. Hydrocarbons in decreasing order of ignition quality, i.e., increasing ignition delay, are: n-paraffins, olefins, naphthenes, iso-paraffins and aromatics. Cetane number of 100 is assigned to n-paraffin ($C_{16}H_{34}$) while an aromatic, e.g., α-methyl naphthalene, is assigned with zero cetane number.

Hazardous areas Areas wherein petroleum fluid is handled or stored may be enveloped with a flammable atmosphere. To ensure safety in the application of electrical circuit design, instrumentation, and all aspects of power engineering, these areas need to be classified as dangerous or safe. To assess the extent of areas that are hazardous, the Institute of Petroleum Electrical Code (IPEC) has defined the following:

i. Non-hazardous area: An area in which explosive gas-air mixtures (flammable) are not expected to be present in quantities such as to require special precautions for the construction and use of electrical apparatus.

ii. Hazardous area: An area in which explosive gas-air mixtures (flammable) are or may be expected to be present in quantities such as to require special precautions for the construction and use of electrical apparatus.

In hazardous areas three types of zones are recognized in order of decreasing probability of explosive gas-air mixtures (flammable) being present:

- Zone 0: A zone in which a flammable atmosphere is present continuously, or present for long periods.
- Zone 1: A zone in which a flammable atmosphere is likely to occur under normal operating conditions.
- Zone 2: A zone in which a flammable atmosphere is likely to occur under abnormal operating conditions, and, if it does, will only exist for a short time.

3.6.2 Gaseous fuels

Gaseous fuels may be classified as follows [2]:

- Fuels naturally found in nature, e.g., natural gas, methane from coal mines, etc.
- Fuel gases made from solid fuels
  - Gases derived from coal (coal gas)
  - Gases derived from waste and biomass
  - From other industrial processes (blast furnace gas)
  - Gases produced by blowing air and sometimes steam through an incandescent fuel bed (Producer gas)
  - Gases produced in a similar manner to above but allows the production of a higher calorific value fuel by intermittently blasting the incandescent bed with air and steam such that the overall heat balance is maintained (water gas)
  - Gases from other gasification processes, including substitute natural gas (SNG).
- Gases made from petroleum, e.g., liquefied petroleum gas (LPG), refinery gases, gases from oil gasification
- Gases from some fermentation process

3.6.2.1 Properties

Knowledge of the following three factors helps determine which of the gases can be used in an appliance:

a. Is the heat release roughly the same as for the same pressure drop?
b. Is the flame shape the same as for the same air and fuel flows?
c. Are pollutants within a specified tolerance for the same heat-release conditions?

Wobbe index or wobbe number This index or number gives an indication of the interchangability of the gases. The Wobbe Index (Wo) is found by dividing the calorific value (CV) of the gas by the square root of its specific gravity and is denoted by

$$Wo = CV/\sqrt{G} \tag{3.6}$$

where G stands for specific gravity of gas in relation to air.

Table 3.2 Gross calorific value of gaseous fuels

Type	GCV, MJ/ Nm^3
Coal gas coke oven (debenzolized)	20
Coal gas continuous vertical retort (steaming)	18
Coal gas low temperature	34
Commercial butane	118
Commercial propane	94
North Sea gas natural	39
Producer gas coal	6
Producer gas coke	5
Water gas carbureted	19
Water gas blue	11

The Wobbe Index is used to compare the combustion energy output of different composition fuel gases in an appliance. If two fuels have identical Wobbe indices then for a given condition the energy output will also be identical.

The Wobbe Index of natural gas ranges from 35.8–71.5 MJ/Nm^3 and of liquefied petroleum gas (LPG) from 71.5–87.2 MJ/Nm^3. These two gases are mostly used for steam or power production.

Weaver flame speed factor This factor is used to define the probability of the gas to react. It is defined as the ratio between the laminar flame speed of a particular gas in relation to the laminar flame speed of hydrogen, which is arbitrarily given a value of 100. The lower the value of this factor the lower the flame speed. The Weaver flame speed factor is greatly influenced by the amount of hydrogen in the mixture.

The Weaver factor (We) for gases may be classified into three groups:

i. High flame speed gases, We = 32–45, having calorific value between 17–21 MJ/Nm^3

ii. Intermediate flame speed gases, We = 25–32, having calorific value between 21–31 MJ/Nm^3

iii. Low flame speed gases, We = 13–25, having calorific value between 31–42 MJ/Nm^3

If both the Wobbe Index and Weaver factor are identical for two gases they are completely interchangeable.

Calorific value Gross calorific values of some gaseous fuels at 288 K temperature and 101.33 kPa pressure dry are given in Table 3.2.

3.6.3 Natural gas

Natural gas in pure form is obtained from gas fields and is also extracted in association with crude petroleum from oil fields. Its principal heat-producing constituents are methane (CH_4) and hydrogen. Additionally, natural gas also contains ethane, propane, butane, and pentane in varying proportions along with the presence of iso-paraffins and naphthenes in small quantities. Some gases also contain hydrogen sulphide.

It is the cheapest and most efficient of all fuels. However, to ensure complete combustion natural gas requires a large amount of air and special burners. Its calorific value is high, specific gravity is moderate, and flame speed is low. Terms used to describe natural gases are:

i. Dry or lean natural gas—this gas contains high methane and $<15\ g/m^3$ recoverable condensate
ii. Wet natural gas—it is comprised of high concentration of higher hydrocarbons (C5—C10) and recoverable condensate $>50\ g/m^3$
iii. Sour or foul—gas containing H_2S
iv. Sweet—gases free from H_2S

Natural gases can be liquefied for distribution by tankers. Liquefied natural gas (LNG) contains mostly methane, and LPG (liquefied petroleum gas) mostly butane and propane.

3.7 PRINCIPLES OF COMBUSTION

Combustion is a form of oxidation. When oxygen is combined rapidly with fuels a substantial amount of heat is released. Combustion usually takes place when heat is applied to a fuel from an external source. The potential energy stored in fuels is released by combustion and made available in the form of heat and power. However, nothing burns properly until it becomes a gas. Hence, the degree of flammability of any matter is characterized by how quickly it turns into a gas. In burning this gas it is necessary to have a gas-air mixture that will ignite, after which the temperature of this mixture is raised to its ignition temperature and kept at this condition. The initial stage of fuel burning takes place at a high concentration of the combustible substance and oxidant and an elevated turbulence of the flow, which is formed by the burner.

The combustible elements in solid and liquid fuels are carbon, hydrogen, and sulfur, while combustible components of gaseous fuels are hydrogen, carbon monoxide, methane, and other unsaturated hydrocarbons as discussed earlier. Chemical reactions that ensure complete combustion of these combustible elements are as follows:

Solid and liquid fuels:

$$C + O_2 = CO_2, \quad H_2 + \frac{1}{2}O_2 = H_2O, \quad S + O_2 = SO_2$$

Gaseous fuels:

$$H_2 + \frac{1}{2}O_2 = H_2O, \quad CO + \frac{1}{2}O_2 = CO_2, \quad CH_4 + 2O_2 = CO_2 + 2H_2O,$$

$$C_nH_{2n+2} + \frac{3n+1}{2}O_2 = nCO_2 + (n+1)H_2O$$

These reactions are exothermic, i.e., heat gets released when these chemical reactions happen. From this it may be concluded that combustion is a chemical reaction in which the chemical energy in the combustibles is converted to heat energy.

In boiler furnaces, oxygen needed for combustion is supplied by air that contains 23% by weight (21% by volume) of oxygen. The air supplied also contains a large amount of nitrogen, about 77% by weight (about 79% by volume) of nitrogen, which performs no useful action. On the other hand, during the process of combustion this nitrogen combines with oxygen to form NO_X (NO and NO_2) and creates an air-pollution problem. In addition to oxygen and nitrogen, air also contains a certain amount of humidity that also does not perform any useful function.

When just the right amount of air is supplied to completely burn a fuel of given composition (other than hydrogen), the flue-gas analysis will reveal only CO_2 and N_2. If excess air is supplied, O_2 will also be found in the flue gas, while an air-deficient combustion or incomplete combustion will show CO_2, CH_4, CO, etc., in the flue-gas analysis. The amount of oxygen in flue gases indicates a meaningful completeness of combustion process. Little presence of oxygen in flue gases reveals reasonably correct supply of excess air and low dry flue-gas heat loss, while a higher value of oxygen will mean higher flue-gas heat loss. The quantum of excess air supplied for combustion can be easily found from the following equation provided analysis of flue-gas constituent gases are known:

$$\%\text{ExcessAir} = 100 * \left(O_2 - \frac{CO}{2}\right) / \left\{0.264N_2 - \left(O_2 - \frac{CO}{2}\right)\right\} \tag{3.7}$$

Once analysis of flue-gas constituent gases is done, it is also possible to find out from the following equation, the maximum CO_2 content of flue gas for complete combustion of fuel:

$$\%CO_2(\max) = \frac{100 * CO_2}{100 - 4.76 * CO_2} \tag{3.8}$$

The objective of reasonably perfect combustion is to release all of the heat released during the combustion reaction, keeping losses from imperfect combustion and supply of excessive air to a minimum. Once a proper fuel-air ratio is obtained complete combustion of the fuel is assured by three factors, e.g., temperature, turbulence, and time, usually referred to as the 3 Ts [6].

Temperature To initiate combustion, first the combustible elements need to be raised to a specific temperature, known as the ignition temperature of the concerned combustible/s. At this temperature the rate of chemical reaction is sufficiently accelerated to produce more heat than is lost to the surroundings so that the combustion process becomes self-sustaining.

Turbulence After reaching the ignition temperature, combustibles and air must be brought into close contact for the reaction to occur. In the first step fuel should be so spread out to expose as much surface as practicable to the air stream. This creates enough turbulence to bring air into close contact with the combustible elements/components of the fuel to initiate rubbing action and expose new surface to oxygen for continued combustion.

Time To ensure complete combustion over and above the temperature and turbulence, time is another criterion. The mixture of combustible elements/components and air must remain in contact sufficiently long to effect complete combustion before their temperature is quenched by coming into contact with the heat-absorbing surfaces of the boiler.

3.8 COMBUSTION CALCULATIONS

The combustion calculations facilitate design and performance determination of boilers and associated components. To achieve complete combustion, air and fuel must combine in exact proportions commensurate with combustion stoichiometry. The following equations delineate the chemical reaction of some common combustible elements, i.e. carbon (C), hydrogen (H) and sulfur (S):

$$C + O_2 = CO_2 + 32.780\ \text{MJ/kg of C} \tag{3.9}$$

$$2H_2 + O_2 = 2H_2O + 141.953\ \text{MJ/kg of } H_2 \tag{3.10}$$

$$S + O_2 = SO_2 + 9.257\ \text{MJ/kg of S}, \tag{3.11}$$

From Eq. 3.9 it may appear that combustion of carbon in the furnace takes place in one step to form CO_2, but this is not true all the time. In certain cases, combustion of carbon may take place in two steps: first to form CO then further combustion of CO will yield CO_2 [6]. Note that the heat of combustion released in the first step is much smaller than the heat of combustion released in the latter step as is evident from the following reactions:

$$2C + O_2 \approx 2CO + 9.227\ \text{MJ/kg of C}, \tag{3.12}$$

$$2CO + O_2 \approx 2CO_2 + 23.553\ \text{MJ/kg of C}, \tag{3.13}$$

Before undertaking the above combustion calculations it is important to understand the following points:

i. For solid and liquid fuels combustibles are expressed as a percentage of elements by weight.

ii. For gaseous fuels combustibles are expressed as a percentage of components by volume or mole (mass of a substance in *gm* equal to its molecular weight is known as *gm-mole* or simply *mole*).

iii. In combustion calculations involving gaseous mixtures, it follows from Avogadro's law that the weights of equal volumes of gases are proportional to their molecular weights, i.e.,

1 mole of $O_2 = 32$ gm of O_2

1mole of $H_2 = 2$ gm of H_2

1 mole of $CH_4 = 16$ gm of CH_4

1 mole of $CO_2 = 44$ gm of CO_2

1 mole of $SO_2 = 64$ gm of SO_2

iv. After a combustion calculation the composition of products of combustion (flue gas) is expressed on a dry basis.

v. During the process of combustion, substances combine on a molar basis, but they are usually measured on a mass basis [1].

vi. The amount of air required for stoichiometrically ensuring complete combustion of combustibles is called *theoretical air.* The resulting products of combustion or flue gas are called *theoretical products of combustion.* Per ASME PTC 4–2008, the value of calculated theoretical air for typical fossil fuels should fall within the ranges of theoretical air as follows:

Coal (VMdaf >30%) ~0.316–0.333 kg/MJ (735–775 lbm/MBtu)

Oil ~ 0.316–0.324 kg/MJ (735–755 lbm/MBtu)

Natural Gas ~0.307–0.316 kg/MJ (715–735 lbm/MBtu)

(Note: VMdaf–Volatile matter dry-ash-free basis.)

vii. The practical maximum and minimum values of theoretical air for hydrocarbon fuels are denoted as follows:

For combustion of carbon theoretical air required: 0.351 kg/MJ

For combustion of hydrogen theoretical air required: 0.222 kg/MJ

viii. During the process of actual combustion it is observed that some additional amount of air beyond theoretical air is required to ensure complete combustion of any fuel. This additional air is called "excess air," [6] or EA, which is expressed as

$$EA, \% = 100 * \frac{ActualAirSupplied - TheoreticalAir}{TheoreticalAir} \quad (3.14)$$

In a pulverized coal-fired boiler the amount of excess air is usually considered as 20% of theoretical air. While in oil or gas fired boiler excess air is less, e.g., about 15–20% for oil and 10–15% for gas.

ix. Eq. 3.9 reveals that when pure carbon is burned, since CO_2 replaces O_2 in air, the percentage of CO_2 in flue gas can never exceed 21% by volume. However, when excess air is supplied this percentage gets reduced, thus 16% by volume of CO_2 in flue gas indicates an efficient combustion system.

x. For convenience of calculation, amount of solid and liquid fuels is considered to be 100 kg and amount of gaseous fuels and flue gas is 100 kmole [1].

xi. In combustion calculations air is considered to be dry and carbon dioxide free. The average molecular weight of such air is 29.0.

xii. The volume or mole ratios of components in dry air are as follows [1,7]:

Oxygen/Nitrogen = 0.27	Nitrogen/Oxygen = 3.76
Oxygen/Air = 0.21	Air/Oxygen = 4.76
Nitrogen/Air = 0.79	Air/Nitrogen = 1.27

xiii. The weight ratios of components in dry air are as follows:

Oxygen/Nitrogen = 0.30	Nitrogen/Oxygen = 3.32
Oxygen/Air = 0.23	Air/Oxygen = 4.32
Nitrogen/Air = 0.77	Air/Nitrogen = 1.30

xiv. The weight and volume relationships of common fuel and oxygen in combustion calculations can be determined with the help of Table 3.3 (Chapter 10, Steam, 41st Edition) [6]:

Table 3.3 may also be used in combustion reactions between combustible elements/components while using air in lieu of oxygen by adding 3.76 moles of nitrogen per mole of oxygen, e.g.,

For burning carbon in air

$$C + O_2 + 3.76N_2 = CO_2 + 3.76N_2 \tag{3.15}$$

For burning methane in air

$$CH_4 + 2O_2 + 2 * 3.76N_2 = CO_2 + 2H_2O + 7.52N_2 \tag{3.16}$$

xv. Before carrying out combustion calculations of a coal-fired boiler, it is essential to know the percentage contents of various constituents of the coal, e.g., carbon (C), hydrogen (H), oxygen (O), sulfur (S), moisture (M), and ash (A). Then using the following frequently used common formulae (Source: ASME PTC 4–2008) combustion calculations are done:

Theoretical Air (kg/kg of fuel)

$$TA = \left\{11.51C + 34.30 * \left(H_2 - \frac{O_2}{7.94}\right) + 4.31S\right\}/100 \tag{3.17}$$

Total Dry Air (kg/kg of fuel)

$$DA = TA(1 + EA/100) \tag{3.18}$$

Table 3.3 Essential data for combustion calculation

Combustible	Reaction	Moles	Weight kg	High heat of combustion MJ/kg of Fuel	Air required for combustion kg/kg of combustibles	Products of combustion (flue gas) kg/kg of combustibles
Carbon (to CO)	$2C + O_2 = 2CO$	2 + 1 = 2	24 + 32 = 56	9.227 C		
Carbon (to CO_2)	$C + O_2 = CO_2$	1 + 1 = 2	12 + 32 = 44	32.780 C	11.510	3.664 (CO_2), 8.846 (N_2)
CO to CO_2	$2CO + O_2 = 2CO_2$	2 + 1 = 2	56 + 32 = 88	23.553C/10.099 CO	2.468	1.571 (CO_2), 1.900 (N_2)
Hydrogen	$2H_2 + O_2 = 2H_2O$	2 + 1 = 2	4 + 32 = 36	141.953 H_2	34.290	8.937 (H_2O), 26.353 (N_2)
Sulfur (to SO_2)	$S + O_2 = SO_2$	1 + 1 = 2	32 + 32 = 64	9.257 S	4.310	1.998 (SO_2), 3.320 (N_2)
Methane	$CH_4 + 2O_2 = CO_2 + 2H_2O$	1 + 2 = 1 + 2	16 + 64 = 80	55.570 CH_4	17.235	2.743 (CO_2), 2.246 (H_2O), 13.246 (N_2)
Ethane	$2C_2H_6 + 7O_2 = 4CO_2 + 6H_2O$	2 + 7 = 4 + 6	60 + 224 = 284	51.949 C_2H_6	16.092	2.927 (CO_2), 1.797 (H_2O), 12.367 (N_2)
Ethylene	$C_2H_4 + 3O_2 = 2CO_2 + 2H_2O$	1 + 3 = 2 + 2	28 + 96 = 124	50.341 C_2H_4	14.784	3.138 (CO_2), 1.284 (H_2O), 11.362 (N_2)
Hydrogen Sulphide	$2H_2S + 3O_2 = 2SO_2 + 2H_2O$	2 + 3 = 2 + 2	68 + 96 = 164	16.500 H_2S	6.093	1.880 (CO_2), 0.529 (H_2O), 4.682 (N_2)

Total Wet Air (kg/kg of fuel)

$$WA = DA * (1 + 0.013) \tag{3.19}$$

Weight of Wet Gas from Fuel (kg/kg of fuel)

$$WF = 1 - \frac{A}{100} \tag{3.20}$$

Weight of Wet Flue Gas (kg/kg of fuel)

$$W_{FGW} = WA + WF \tag{3.21}$$

Weight of Dry Flue Gas (kg/kg of fuel)

$$W_{FGD} = W_{FGW} - \left\{ \left(8.94H_2 + \frac{M}{100} + 0.013DA \right) \right\} \tag{3.22}$$

Moles of Dry Air (moles/kg of fuel)

$$M_{DA} = DA/29 \tag{3.23}$$

Moles of Dry Flue Gas (moles/kg of fuel)

$$M_{FGD} = \left(\frac{C}{12} + \frac{S}{32} + \frac{N_2}{28} \right)/100 + 0.79 * M_{DA} + 0.23 \left(\frac{(DA - TA)}{32} \right) \tag{3.24}$$

Moles of Wet Flue Gas (moles/kg of fuel)

$$M_{FGW} = M_{FGD} + \left(\frac{H_2}{2} + \frac{M}{18} \right)/100 + (M_{DA} * 29 * 0.013)/18 \tag{3.25}$$

Molecular Weight of Dry Flue Gas

$$MW_D = W_{FGD}/M_{FGD} \tag{3.26}$$

Molecular Weight of Wet Flue Gas

$$MW_W = W_{FGW}/M_{FGW} \tag{3.27}$$

Specific Volume of Dry Flue Gas (Nm^3/kg)

$$SV_D = 22.4/MW_D \tag{3.28}$$

Specific Volume of Wet Flue Gas (Nm^3/kg)

$$SV_W = 22.4/MW_W \tag{3.29}$$

Volume of Dry Flue Gas (Nm^3/kg of fuel)

$$V_{FGD} = SV_D * W_{FGD} \tag{3.30}$$

Volume of Wet Flue Gas (Nm^3/kg of fuel)

$$V_{FGW} = SV_W * W_{FGW} \tag{3.31}$$

xvi. Heat release rate in the furnace firing coal may be calculated using the following equation [8]:

$$QRR = \{HHV + 0.25 * 10^{-3} * WA(T_F - T_A) - 1.1 * (8.937 * H) - 15.4 * UC\} * WF \tag{3.32}$$

where

QRR = Furnace Heat Release Rate, MJ/h
HHV = Higher Heating Value of Fuel, MJ/kg
WA = Total Air, kg/kg of fuel
T_F = Furnace Inlet Air Temperature, K
T_A = Airheater Inlet Air Temperature, K
H = Hydrogen Content in Fuel, kg/kg of fuel
UC = Unburned Carbon in Refuse, kg/kg of fuel
WF = Fuel Consumption, kg/h

(Notes: In Eq. 3.17, the factor $O_2/7.94$ is a correction for the hydrogen already combined with oxygen in the fuel to form water vapor.
The term EA, in Eq. 3.18, stands for "excess air" (kg/kg of fuel).
In Eq. 3.19, the factor 0.013 (kg//kg dry air) is the moisture content of (standard) air at 60% RH and 300 K temperature.)

Example 3.1

The dry-ash-free ultimate analysis of a coal in percentages is as follows:

C = 77.5
H = 6.0
O = 13.2
N = 2.6
S = 0.7

Air-dried proximate analysis of the same coal in percentages is as follows:

FC = 35.0
VM = 22.0
M = 8.0
A = 35.0

Determine the air-dried ultimate analysis and theoretical oxygen requirement for complete combustion of this coal. (The HHV of the coal is given as 18.63 MJ/kg.)

Solution:

Alternate – 1: Ultimate analysis of coal is known

Per the conversion factors given in Table 3.1, air-dried analysis may be obtained by multiplying dry-ash-free ultimate analysis by the factor {100–(Mad + Aad)}/100.

For the given coal, above multiplying factor is {100−(8.0 + 35.0)}/100 = 0.57.
Hence, the air-dried ultimate analysis of this coal is

C = 77.5 * 0.57 = 44.2
H = 6.0 * 0.57 = 3.4
O = 13.2 * 0.57 = 7.5
N = 2.6 * 0.57 = 1.5
S = 0.7 * 0.57 = 0.4
M = 8.0
A = 35.0

Following Eq. 3.17 the theoretical air required for complete combustion of the given coal is

$$TA = \{11.51 * 44.2 + 34.30(3.4 - 7.5/7.94) + 4.31 * 0.4\}/100$$
$$= 5.95 \text{ kg/kg of fuel}$$

Therefore, the theoretical oxygen requirement for complete combustion of coal is calculated as

$$TO = 0.23 * 5.95 = 1.37 \text{ kg/kg of fuel}$$

Alternate − 2: Proximate analysis of coal is known.
Volatile matter on a dry-ash-free basis, VM = 22/(100−8 − 35) = 38.6%.
For VM >30% the theoretical air required is as follows (Table 8, page 10-9, Steam 41st Edition) [6].

$$TA = 0.325 \text{ kg/MJ} = 0.325 * 18.63 = 6.06 \text{ kg/kg of fuel}$$

(Notes:

i. The above value of TA is within 2% of the actual TA.
ii. 1 kg/MJ = 23.26 lbm/10^4 Btu.)

Example 3.2

Ultimate analysis of a sample of anthracite was found to contain percentage by weight of the following elements:

C = 90.0
H = 3.3
O = 3.0
N = 0.8
S = 0.9
A = 2.0

Calculate the weight of the theoretical air for the complete combustion of 1 kg of fuel. If the excess air supplied is 20%, find the percentage composition of dry flue gas by volume.

Solution: Following Eq. 3.17 the theoretical air required for complete combustion of anthracite may be calculated as:

$$TA = \{11.51 * 90.0 + 34.30(3.3 - 3.0/7.94) + 4.31 * 0.9\}/100 = 11.40 \text{ kg/kg of fuel}$$

For calculating the percentage composition of various constituents of flue gas, take the following steps:

i. Total Dry Air (Eq. 3.18)

$$DA = TA(1 + EA/100) = 11.40\ (1 + 20/100)$$
$$= 13.68 \text{ kg/kg of fuel,}$$

ii. Moles of Dry Air (Eq. 3.23)

$$MDA = DA/29 = 0.472 \text{ moles/kg of fuel,}$$

iii. Moles of various constituents of dry flue gas (Eq. 3.24)

$$CO_2 = C/1200 = 0.075 \text{ moles/kg of fuel}$$
$$SO_2 = S/3200 = 2.813 * 10^{-4} \text{ moles/kg of fuel}$$
$$N_2 = N/2800 = 2.857 * 10^{-4} \text{ moles/kg of fuel}$$

Nitrogen in total air

$$N_2 = 0.79\, M_{DA} = 0.373 \text{ moles/kg of fuel}$$

Excess oxygen in total air

$$EO = 0.23(DA - TA)/(32 \times 100)$$
$$= 0.23(13.68 - 11.40)/32 = 0.0164 \text{ moles/kg of fuel}$$
$$M_{FGD} = 0.075 + 2.813 * 10^{-4} + 2.857 * 10^{-4} + 0.373 + 0.0164$$
$$= 0.465 \text{ moles/kg of fuel}$$

iv. Hence, the percentage composition of the various constituents of flue gas are

$$CO_2 = 100 * 0.075/0.465 = 16.13\%,$$
$$SO_2 = 100 * 2.813 * 10^{-4}/0.465 = 0.06\%,$$
$$N_2 = 100 * (0.373 + 2.857 * 10^{-4})/0.465 = 80.28\%,$$
$$O_2 = 100 * 0.0164/0.465 = 3.53\%$$

Example 3.3

For the following air-dried ultimate analysis of coal of Example 3.1 find the flue-gas analysis and the air-fuel ratio by weight. (Assume 50% excess air.)

$C = 58.320$
$H = 3.312$
$O = 8.640$
$N = 1.296$
$S = 0.432$
$M = 8.000$
$A = 20.000$

Solution: The following approach to calculation is different from the one followed in Example 3.2.

i. Moles of theoretical oxygen required for complete combustion of given fuel:

$$M_O = (C/12 + H/4 + S/32 - O/32)/100$$
$$= (58.320/12 + 3.312/4 + 0.432/32 - 8.640/32)/100$$
$$= (4.8600 + 0.8280 + 0.0135 - 0.2700)/100$$
$$= 0.054115 \text{ moles/kg of fuel}$$

ii. Moles of dry air required

$$M_{DA} = 4.76 * M_O * 1.5 = 0.3864 \text{ moles/kg of fuel}$$

iii. Analysis of Flue Gas

Products of combustion	Formula	Moles per kg of fuel	Volume, %
CO_2	C/1200	0.048600	11.836
H_2O	$(H_2/2 + M/18)/100 + (MDA*29*0.013)/18$	0.029097	7.086
SO_2	S/3200	0.000135	0.033
N_2	$0.79*MDA + N_2/2800$	0.305719	74.455
Excess O_2	0.5*MO	0.027058	6.590
TOTAL		0.410609	100.00

iv. Air-Fuel Ratio, AFR = MDA*29 = 11.21 kg/kg of fuel

Example 3.4

Assuming 32% excess air and 94 kg/s of coal burned, calculate the following.

Quantity of theoretical air in kg/MJ
Quantity of Excess oxygen
Weight of wet and dry flue gas
Moles of dry air
Moles of dry and wet flue gas
Molecular weight of dry and wet flue gas
Specific volume of dry and wet flue gas at NTP
Volume of dry and wet flue gas at 300 K
Volume of dry and wet flue gas at 273 K.

Given ultimate analysis of coal as

C = 44.1%
H = 3.0%
O = 8.1%
N = 0.9%
S = 0.4%
M = 8.5%
A = 35.0%
GCV = 17.80 MJ/kg

Solution: Using the various equations presented earlier the desired values may be calculated as follows:

a. Quantity of theoretical air (Eq. 3.17)

$$\begin{aligned} TA &= \{11.51 * 44.1 + 34.30(3.0 - 8.1/7.94) + 4.31 * 0.4\}/100 \\ &= (507.591 + 67.909 + 1.724)/100 = 5.772 \text{ kg/kg of fuel} \\ &= 5.772/17.80 = 0.324 \text{ kg/MJ} \end{aligned}$$

b. Quantity of total dry air (Eq. 3.18)

$$DA = 5.772(1 + 32/100) = 7.619 \text{kg/kg of fuel}$$

Quantity of excess oxygen

$$EO = 0.23(DA - TA) = 0.425 \text{ kg/kg of fuel}$$

c. Total Wet Air (Eq. 3.19)

$$WA = DA + 0.013DA = 7.718 \text{ kg/kg of fuel}$$

Weight of Wet Gas from Fuel (Eq. 3.20)

$$WF = 1 - A/100 = 0.65 \text{ kg/kg of fuel}$$

Weight of Wet Flue Gas (Eq. 3.21)

$$W_{FGW} = WA + WF = 8.368 \text{ kg/kg of fuel}$$

Weight of Dry Flue Gas (Eq. 3.22)

$$W_{FGD} = 8.368 - \{(8.94 * 3.0 + 8.5)/100 + 0.013 * 7.619\}$$
$$= 8.368 - 0.452 = 7.916 \text{ kg/kg of fuel}$$

An alternate method of calculating Weight of Dry Flue Gas (Table 3.3) is

$$W_{FGD} = (3.664 * C + 1.998 * S + N)/100 + DA - 0.23 * TA$$
$$W_{FGD} = (3.664 * 44.1 + 1.998 * 0.4 + 0.9)/100 + 7.619 - 0.23 * 5.772$$
$$= 7.924 \text{ kg/kg of fuel}$$

The difference between the two calculating values of W_{FGD} cropped up due to the approximation in calculation as well as in selecting coefficients used for different elements.

d. Moles of Dry Air (Eq. 3.23)

$$M_{DA} = DA/29 = 0.263 \text{ moles/kg of fuel}$$

e. Moles of Dry Flue Gas (Eq. 3.24)

$$M_{FGD} = (44.1/12 + 0.4/32 + 0.9/28)/100 + 0.79 * 0.263 + 0.425/32$$
$$= 0.258 \text{ moles/kg of fuel}$$

Moles of Wet Flue Gas (Eq. 3.25)

$$M_{FGW} = 0.258 + (3.0/2 + 8.5/18)/100 + (0.263 * 29 * 0.013)\}/18$$
$$= 0.283 \text{ moles/kg of fuel}$$

f. Molecular Weight of Dry Flue Gas (Eq. 3.26)

$$MW_D = 7.916/0.258 = 30.68$$

Molecular Weight of Wet Flue Gas (Eq. 3.27)

$$W_W = 8.368/0.283 = 29.57$$

g. Specific Volume of Dry Flue Gas (Eq. 3.28)

$$SV_D = 22.4/30.68 = 0.730 \text{ Nm}^3/\text{kg}$$

Specific Volume of Wet Flue Gas (Eq. 3.29)

$$SV_W = 22.4/29.57 = 0.758 \text{ Nm}^3/\text{kg}$$

h. Volume of Dry Flue Gas at 300 K (Eq. 3.30)

$$V_{FGD} = (0.730 * 300/273) * 7.916 * 94 = 597 \text{ m}^3/\text{s}$$

Volume of Wet Flue Gas at 300 K (Eq. 3.31)

$$V_{FGW} = (0.758 * 300/273) * 8.368 * 94 = 655\ m^3/s$$

i. Volume of Dry Flue Gas at 423 K

$$V_{FGD} = 597 * 423/300 = 842\ m^3/s$$

Volume of Wet Flue Gas at 423 K

$$V_{FGW} = 655 * 423/300 = 924\ m^3/s$$

Example 3.5

For the following Gandhar natural gas composition by volume, assuming 15% excess air, determine:

Flue-gas analysis
Moles of dry and wet flue gas
Weight of wet and dry flue gas in kg/MJ
Specific volume of wet flue gas at NTP
Volume of wet flue gas at 423 K

Methane CH_4 = 80.9%
Ethane C_2H_6 = 8.8%
Propane C_3H_8 = 5.2%
i-Butane C_4H_{10} = 0.6%
n-Butane C_4H_{10} = 1.2%
i-Pentane C_5H_{12} = 0.2%
n-Pentane C_5H_{12} = 0.3%
CO_2 0.9%
N_2 1.9%
Average LCV = 40.80 MJ/Nm^3

Solution: First find out the "molecular weight" of the given gas, then calculate the "specific volume" of this fuel, followed by elemental breakdown of each constituent of the fuel and finally analysis of flue gas.

a. Molecular weight i.e., weight of one mole of the fuel.

Constituent	Moler weight of each constituent
Methane CH_4	0.809*16 = 12.944
Ethane C_2H_6	0.088*30 = 2.640
Propane C_3H_8	0.052*44 = 2.288
Butane C_4H_{10}	0.018*58 = 1.044
Pentane C_5H_{12}	0.005*72 = 0.360
CO_2	0.009*44 = 0.396
N_2	0.019*28 = 0.532
	= 20.204

Hence, specific volume of the fuel, SV_{NG} = 22.4/20.204 = 1.109 Nm^3/kg
and calorific value of fuel per kg, LCV_{NG} = 40.80*1.109 = 45.25 MJ/kg

b. The elemental breakdown of each constituent may be expressed mole per 100 moles of the fuel as follows:

Constituent	C	H_2
CH_4 =	80.9*1 = 80.9	80.9*2 = 161.8
C_2H_6 =	8.8*2 = 17.6	8.8*3 = 26.4
C_3H_8 =	5.2*3 = 15.6	5.2*4 = 20.8
C_4H_{10} =	1.8*4 = 7.2	1.8*5 = 9.0
C_5H_{12} =	0.5*5 = 2.5	0.5*6 = 3.0
TOTAL	123.8	221.0

c. Analysis of Flue Gas

Fuel constituents	Moles of fuel constituents	Mol. WT. of fuel constituents	Weight of constituent, kg/kg of fuel	O_2 multiplier	Required theoretical O_2 moles	Flue-gas composition, moles/kg of fuel			
						CO_2	O_2	N_2	H_2O
C to CO_2	123.8	12	1485.6	1	123.8	123.80			
H_2	221.0	2	442.0	0.5	110.5				221.00
CO_2	0.9	44	39.6	–	–	0.90			
N_2	1.9	28	53.2	–	–			1.90	
SUM	347.6	–	2020.4	–	234.3				
Excess O_2 supplied, 234.3*0.15 = 35.15							35.15		
Total O_2 supplied 234.3 + 35.145) = 269.45									
Total N_2 supplied, 269.445*79/21 = 1013.63								1013.63	
Dry Air supplied, DA = N_2 + O_2 = 1283.08									
Moisture in Air, H_2O = 0.0212*DA = 27.20									27.20
Wet Air supplied, WA = 1283.08 + 27.20 = 1310.28									
Flue-gas Constituents						124.70	35.15	1015.53	248.20

Moles of Wet Flue Gas, MFGW = 1423.58 moles/kg of fuel,
Moles of Dry Flue Gas, MFGD = 1175.38 moles/kg of fuel.
Above moles of flue gas results from burning100 moles of fuel.
Hence,
Weight of wet flue gas, W_{FGW} = 1423.58/100 = 14.24 kg/kg of fuel.
Weight of wet flue gas per unit of heat input, W_{FWH} = 14.24/45.25 = 0.315 kg/MJ
Weight of dry flue gas, W_{FGD} = 1175.38/100 = 11.75 kg/kg of fuel.
Weight of dry flue gas per unit of heat input, W_{FDH} = 11.75/45.25 = 0.260 kg/MJ

d. Volume of Flue Gas

Flue-gas constituents	Moles/kg of fuel	Mol. Wt. of flue-gas constituents	kg/kg of fuel
CO_2	124.70	44	124.70*44/1423.58 = 3.85
O_2	35.15	32	35.15*32/1423.58 = 0.79
N_2	1015.53	28	1015.53*28/1423.58 = 19.97
H_2O	248.20	18	248.20*18/1423.58 = 3.14
SUM	1423.58	–	27.75

Therefore, Specific Volume of Wet Flue Gas, $SV_W = 22.40/27.75 = 0.807\ Nm^3/kg$.
Volume of Wet Flue Gas at 423 K, $V_{FGW} = SV_W * W_{FGH} * 423/273 = 0.394\ m^3/MJ$.

Example 3.6

For the following percentage analysis by volume of dry flue gas, calculate the air drawn through the furnace per kg of coal-fired, given the carbon content of coal is 58.3% by weight.

Products of combustion	Volume, %
CO_2	12.74
SO_2	0.04
N_2	80.13
O_2	7.09

Solution: Relative composition by weight of the flue gas is

$$12.74 * 44 : 0.04 * 64 : 80.13 * 28 : 7.09 * 32$$

12.74 * 44 kg CO_2 requires 12.74*44*12/44 kg of carbon for formation.
Hence, the total amount of carbon in flue gas, C = 12.74*12 = 152.88 kg.
Relative weight of nitrogen to carbon in flue gas, N_2/C is (80.13*28)/152.88 = 14.68.
Now each kg of coal contains 0.583 kg of carbon.
Therefore, for combustion of each kg of coal amount of N_2 in flue gas is N_2 = 14.68*0.583 kg, and the amount of air required per kg of coal is DA = 14.68*0.583*100/77 = 11.11 kg.

3.9 DESIGN ASPECT OF BURNER

The purpose of a fuel burner is to mix the fuel and air so that rapid ignition and complete combustion result. Burners do not ignite the fuel [9–12]. Their function is to provide thorough and complete mixing of fuel and air, to permit stable ignition and to control flame shape and travel effectively. The primary stage of fuel burning takes place at a high concentration of the combustible substance and oxidant at elevated turbulence of the flow that is formed by the burner. The fuel-air mixture from a burner enters the furnace in the form of straight or swirled jets, whose progression in the furnace space determines the conditions of ignition and combustion intensity.

In a pulverized coal burner, only a part of the air, called the "primary air," is initially mixed with fuel to obtain rapid ignition. The remaining air, called "secondary air," is introduced outside the primary-air ports. Primary air is usually 20–30% of the total air.

(Pulverized coal burners are discussed in Chapter 4, Pulverized Coal-Fired Boiler.)

In the fuel oil burner, oil is atomized either utilizing fuel pressure or with the help of compressed air or steam. The former type is called the *mechanical* type or pressure type atomized burners. They use 1.5–2.0 MPa oil pressure at maximum flow. Here a high velocity swirl is imparted to oil, which is released through the orifice in the form of a conical mist. Regulation of burner output is done by varying the supply oil pressure, but since the pressure determines efficiency of atomization, there is not much of range for regulating the burner output. Therefore, the turn-down ratio (operating range) of these burners is generally low, e.g., 2:1 to 4:1. Redwood No.1 viscosity of oil at pressure atomized burner is in the range of 80–120 sec.

The limitation of the turn-down ratio of the pressure-atomized burner is largely overcome by air- or steam-atomized burner. Burner output in *air or steam atomized* burners may be varied by varying the oil pressure and air/steam pressure correspondingly. Air/steam is fed into a central tube (Figure 3.1) at a pressure of 0.5–1.2 MPa to

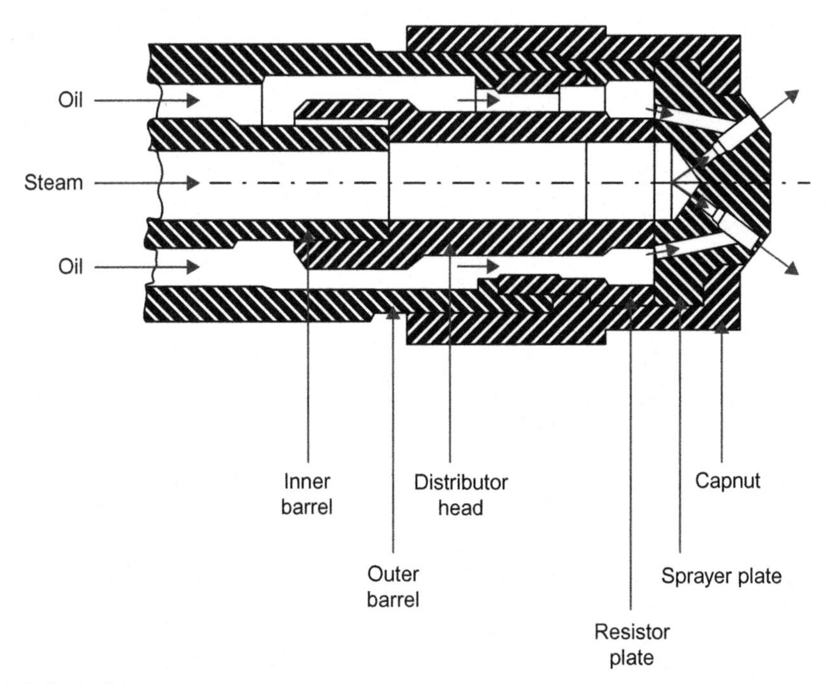

Figure 3.1 Fuel oil burner.

a perforated plate where it meets the oil, which has passed along an annular space between the central air/steam tube and the concentric outer tube.

This burner produces a fine spray and for this reason it may be used to light a burner in a cold furnace. The turn-down ratio of an atomized oil burner is as high as 10:1 and requires a maximum oil pressure of about 0.7 MPa. The details of a typical oil-burner connection are shown in Figure 3.2.

Gas burners use either a perforated ring or a gun to admit fuel to the burner. Cleaning these burners is generally simple.

Depending on the type of fuel, amount of volatile matter, etc., burners may be the turbulent or vortex type and straight-flow type [12]. In a turbulent burner, the fuel-air mixture and secondary air are fed as swirled jets. They are widely used for low-volatile fuels. In a straight-flow burner, the fuel-air mixture and secondary air are supplied as parallel jets. These burners are employed with high-volatile coals, brown coals, peat, oil-shale, etc.

Arrangement of Burners: Burners may be arranged in one or more tiers based on the heat requirement for the desired steam generating capacity. A typical arrangement of turbulent pulverized-coal burners are the front wall, opposed (front and rear) wall, and opposed side wall. A corner or tangential arrangement is used for straight-flow burners. In this arrangement flame jets are directed tangentially to an imaginary circle in the furnace center that ensures complete combustion.

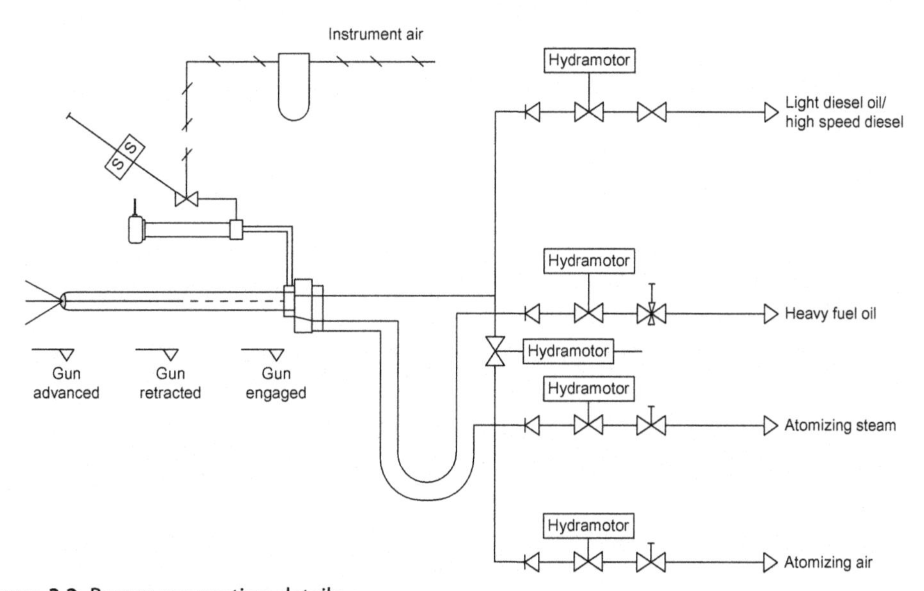

Figure 3.2 Burner connection details.

In a pulverized coal-fired boiler, the location of the fuel oil burner is governed by the type and arrangement of the main coal burner. For horizontal turbulent burners, the oil burner is concentric to the coal burner. In a horizontal lignite-fired boiler the oil burner is located in the side walls. While in a tangential corner-fired boiler the oil burner is located between two tiers/elevations of the coal burner.

In addition to these burners, other burners may be placed in the furnace in the following configurations:

Horizontal Burner: This burner has a central coal nozzle with internal ribs in the form of rifling. A central pipe allows insertion of ignition torch. The nozzle of this burner is surrounded with a housing that is provided with adjustable vanes for controlling air turbulence as well as flame shape.

Circular Register Burner: Gas, oil, or pulverized coal or any combination of these fuels can be burned in this burner. However, combination of oil and pulverized coal is seldom recommended, since coal may form coke, thus reducing burner performance.

Intervane Burner: Vanes contribute a swirling action to the coal-air mixture in a central nozzle. Thus, considerable turbulence results, which ensures efficient combustion.

Directional Burner: In this burner air and fuel are introduced through vertical openings between tubes of furnace walls. Each opening has directional vanes that can be adjusted to obtain optimum placement of fuel for efficient combustion.

Fuel is generally fired in burners in any one of the following basic configurations:

a. Horizontal Firing—through fixed burners from one wall or opposed wall
b. Vertical or Down-shot Firing—with single-U or double-U flame shape
c. Tangential or Corner Firing—usually with tilting tip burner assembly for steam temperature control

3.9.1 Horizontal firing (Figure 3.3 and Figure 3.4)

For medium and high volatile coal horizontal firing using short flame turbulent type burners is suitable. Burners may be arranged in the front wall, front and rear wall, or on both sidewalls based on convenience of design and layout. This type of burner can accommodate any type of fuel, e.g., gas, oil, and pulverized fuel.

By virtue of their construction a composite burner burning any combination of the above fuels is also easily accomplished. Each one of this type of burner produces a short, turbulent, stable flame around itself, and the combustion is more or less complete within its burning zone. This feature makes each burner self-sufficient, and adjacent burners have only a slight effect on its performance. A certain amount of regulation permits flame length adjustment by secondary air, impeller position, etc., but care is needed to ensure that the point of ignition is not shifted too far from the burner throat.

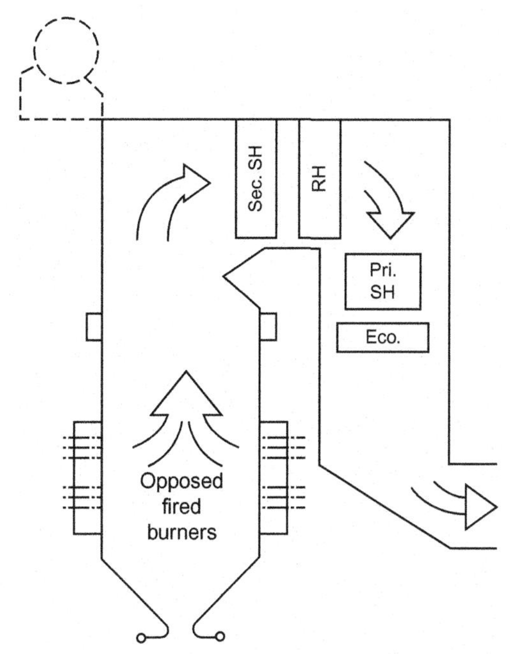

Figure 3.3 Horizontal firing (front and rear).

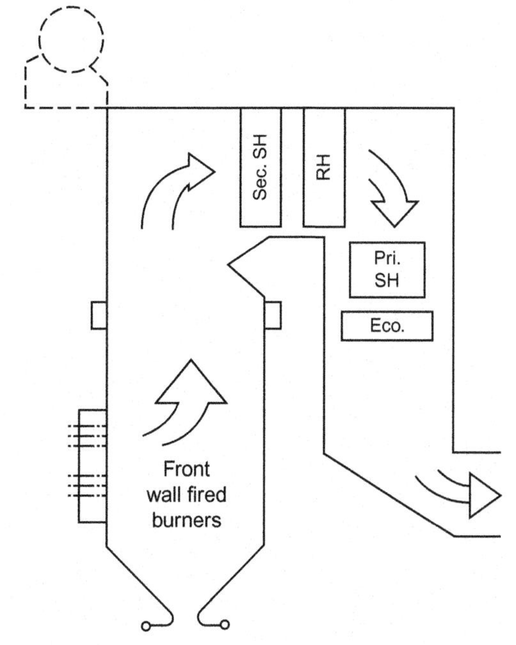

Figure 3.4 Horizontal firing (front).

3.9.2 Vertical firing (Figure 3.5)

Vertical down-shot firing is used where volatile matter of coal is low, usually 18% or below. Low volatile coal takes longer to complete combustion and does not require high turbulence as in horizontal firing. The coal and primary air is fed vertically downward from the burners near one side of the furnace roof. The gas outlet is through another side of the roof and it follows a U-shaped path. Secondary air may be introduced either through rows of ports on the sidewall adjacent to the burners or concentrically with primary air jets. A down-shot furnace may also be fired from opposite ends of the furnace roof, the gas outlet being at the center that results in W-shaped flame.

Coal and secondary air are injected in narrow parallel streams and get mixed thoroughly during their downward travel. About 40% of total combustion air, i.e., primary and secondary, is provided at the burners, and remaining 60% are provided progressively at right angles to the flame path, through the wall.

By progressive addition of combustion air a controlled amount of turbulence is created to ensure complete combustion and products of combustion finally pass upwards through the furnace.

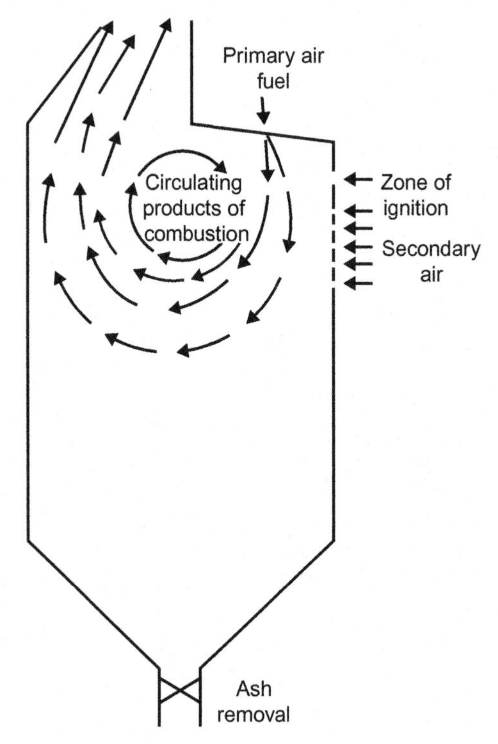

Figure 3.5 Vertical firing.

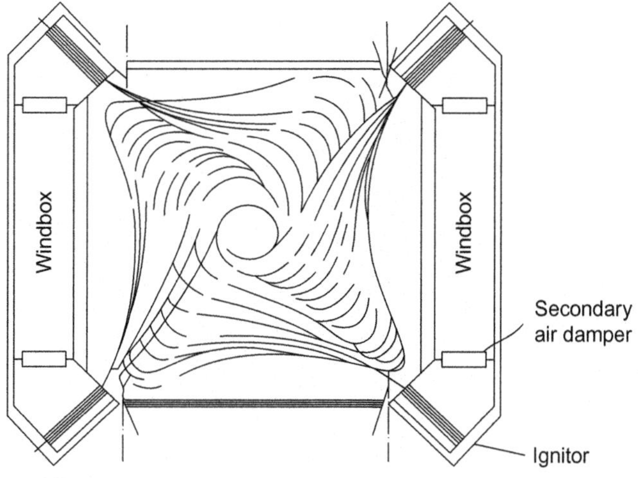

Figure 3.6 Tangential firing.

3.9.3 Tangential firing (Figure 3.6)

In this type of firing, the furnace itself acts as the burner. Admission of fuel and air into the furnace is through four windbox (airbox) assemblies located in four corners of the furnace. The fuel and air streams from the windbox nozzles are directed to an imaginary firing circle in the furnace center. The swirling action of this type of firing is most effective in turbulent mixing of the burning fuel in a constantly changing air and gas atmosphere. This burner system has the advantage of a single-flame envelope for burning wide range of fuels.

Fuel streams are injected into the furnace between air streams. Burners are arranged vertically and the fuel is normally fired on a level basis (Figure 3.7) so that proper interaction of the separate streams is obtained. Even if one corner is blanked off entirely, combustion can be stable and efficient.

3.10 FLAME STABILITY

A burner sustains a stable flame when the flame front close to the burner experiences no backlash or blow-off of flame under various operating conditions. The rate of flame propagation or flame velocity is of paramount importance in determining flame stability. When the flame velocity balances the velocity of the fuel-air mixture at the flame front the flame is said to be stable [12]. Changes in fuel composition and the ratio of fuel-air mixture along with its temperature have considerable effect on flame stability. The inside diameter of the burner also greatly affects the flame stability. Too high a diameter will cause burner backlash while too low a diameter will result in burner blow-off.

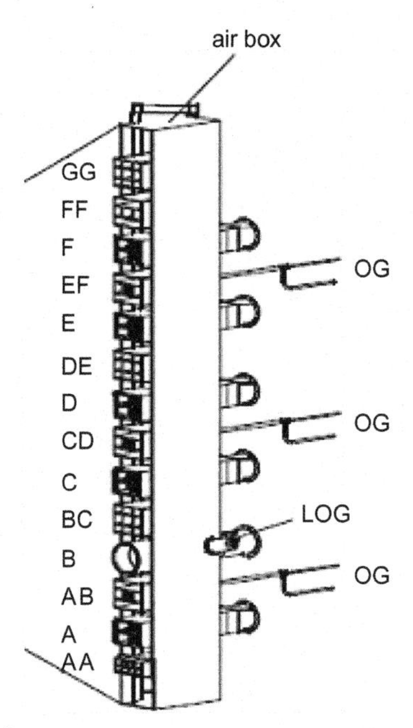

Figure 3.7 Typical corner fired burners.
Coal burners are located at tiers "A, B, C, D, E, and F." Oil burners are located between coal burners at tiers "AB, CD, and EF." Air is supplied at tiers "AA, AB, BC, CD, DE, EF, and FF." Over-fire air is supplied at tier "GG."

3.11 DESIGN ASPECTS OF FURNACE

The furnace of a steam generator is where fuel is ignited and burnt completely to release heat. Modern, large-capacity steam-generating units employ membrane/finned tube waterwalls for furnace enclosure with an external layer of insulation. These watercooled furnaces are designed to comply with the following requirements:

1. Sufficient height in a drum type steam generator to ensure adequate fluid circulation in the furnace tubes
2. Sufficient surface area such that the furnace gas temperature can be reduced to an acceptable level of superheating requirement; the furnace exit gas temperature should be slightly lower than the ash fusion temperature
3. Sufficient furnace depth to obviate flame impingement on opposite wall
4. Sufficient furnace width to accommodate all burners on acceptable pitching to avoid interference of flames and impingement on side-walls

5. Sufficient overall dimension and shape to ensure a gas path that will "fill" the furnace and provide optimum heat absorption to all parts
6. A fuel ash residence time sufficient enough to burn-out and at the same time to restrict furnace gas temperature to remain below the ash melting state
7. Dry bottom that employs water impounded bottom ash hopper

In a furnace the oxygen needed for combustion of combustible elements in a fuel is supplied by atmospheric air, where chemical energy in the combustible elements is converted to heat energy. Hence, to obviate loss of high-energy heat released during combustion it is essential to ensure complete combustion of fuel under all operating conditions [6].

The heat liberation rate and heat release rate in a furnace depends on the type of firing in the furnace, e.g., tangential firing, front-wall firing, opposed-wall firing, down-shot firing, etc., analysis of coal, and the type of steam generator (subcritical/supercritical). Some typical values of heat liberation rate and heat release rate are given in Table 3.4. Typical heat absorption profile along risers is shown in Figure 3.8.

The furnace enclosure should be designed to be capable of withstanding a transient pressure excursion without permanent deformation due to yield or buckling of any support member. The structural stability of the furnace enclosure along with the air

Table 3.4 Typical values of heat liberation rate and heat release rate

Heat Liberation Rate (Heat Input Divided by Furnace Volume)	120 (dry bottom furnace)–210 (slag bottom furnace)	kW/m^3
EPRS (Effective Projected Radiant Surface) Heat Release Rate (Heat Available by Furnace Area)	150–235	kW/m^2
Heat Input Divided by Furnace Horizontal Plan Area	3500 (clinkering coal) – 6400 (brown coal)	kW/m^2
Burner Zone Heat Release Rate	900–2300	kW/m^2
Typical values for once-through steam generator are:		
Heat Input Divided by Furnace Horizontal Plan Area	5700	kW/m^2
Burner Zone Heat Release Rate	1600	kW/m^2

(NOTES:
Heat Input: The total heat in the fuel actually burned.
Heat Available: GCV of fuel actually burned plus the sensible heat in combustion air minus unburned combustible heat minus latent heat of moisture in fuel minus latent heat of water formed by combustion of hydrogen in fuel minus one half of the radiation and unaccounted for losses above 300 K.
EPRS/Furnace Area: Flat projected area of all water-cooled surface including walls, roofs, floor plus radiant superheater surface, and the plane of the furnace exit measured at the point of flue-gas entrance to convection superheater surface with centerline to centerline spacing less than 375 mm.
Furnace Horizontal Plan Area: Width*Depth.
Burner Zone Area: 2*(Width + Depth)*(Bottom to Top Burner Height + 3 m)
Furnace Volume: The space enclosed by the boundaries described in EPRS.)

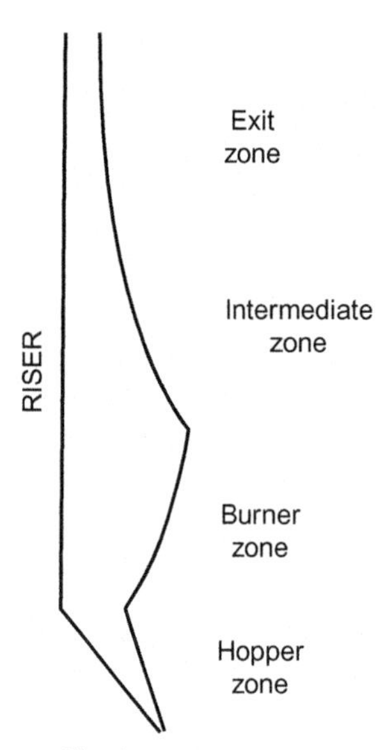

Figure 3.8 Typical heat absorption profile along risers.

supply and flue-gas removal system should be designed to comply with the following recommendations of NFPA 85 to minimize the risk of furnace pressure excursions [2]:

1. Applicable to all boilers (except fluidized-bed boilers):
   - **i.** The positive transient design pressure should be at least, but should not be required to exceed, +8.7 kPa (+890 mmwc).
   - **ii.** If the test block capability of the forced draft fan at ambient temperature is less than +8.7 kPa, the positive transient design pressure should be at least, but should not be required to exceed, the test block capability of the forced draft fan.
   - **iii.** The negative transient design pressure should be at least as negative as, but should not be required to be more negative than, −8.7 kPa (−890 mmwc).
   - **iv.** If the test block capability of the induced draft fan at ambient temperature is less negative than −8.7 kPa, the negative transient design pressure should be at least as negative as, but should not be required to be more negative than, the test block capability of the induced draft fan.
2. Applicable to fluidized-bed boiler:
   - **i.** The positive transient design pressure should be 1.67 times the predicted operating pressure of the component or +8.7 kPa (+890 mmwc), whichever is greater, but should not be in excess of the maximum head capability of the air supply fan at ambient temperature.
   - **ii.** The negative transient design pressure should be the maximum head capability of the induced draft fan at ambient temperature, but not more than −8.7 kPa (−890 mmwc).

3.12 PROBLEMS

3.1 Ultimate analysis of a sample of coal was found to contain percentage by weight of following elements:

C = 86.80
H = 4.25
O = 3.06
N = 0.80
S = 0.83
A = 4.26

Calculate the weight of theoretical air for complete combustion of 1 kg of fuel and the weight of products of combustion per kg of fuel burnt.

(**Ans.**: 11.35 kg of air, 12.31 kg of products of combustion)

3.2 A boiler is fired with a coal having following analysis by percentage weight. The coal is burned using 40% excess air and produces ash containing 25% unburned carbon. Determine

- Volume of air required to burn 1 kg of coal
- Actual volume of air
- Percentage composition of dry products of combustion

C = 54
H = 4
O = 12
N = 1
S = 4
A = 22
M = 3

(**Ans.**: 5.60 Nm^3/kg, 7.84 Nm^3/kg, CO_2: 15.02%, SO_2: 0.48%, O_2: 8.11% and N_2: 76.39%)

3.3 In a boiler, air-dried coal with the following percentage ultimate analysis by weight has been used with 50% excess air:

C = 58.4
H = 3.3
O = 8.6
N = 1.3
S = 0.4
A = 20.0
M = 8.0

Determine

Moles of oxygen per 100 kg of coal
Moles of dry air per kg of coal
Percentage analysis of products of combustion by volume

(**Ans.**: 5.44 moles of O_2 per 100 kg of coal, 0.39 mole of dry air per kg of coal, CO_2: 11.76%, H_2O: 7.04%, SO_2: 0.03%, O_2: 6.58% and N_2: 74.59%)

3.4 The following is a percentage analysis by volume of dry flue gas. Calculate the air drawn through the furnace per kg of coal fired, given the carbon content of coal is 86.8% by weight.

Products of combustion	Volume, %
CO_2	11.9
CO	0.1
N_2	81.8
O_2	6.2

(**Ans.**: 17.94 kg of air)

3.5 Convert the following percentage analysis by volume of dry flue gas to gravimetric analysis.

Products of combustion	Volume, %
CO_2	10.9
CO	1.0
N_2	81.0
O_2	7.1

(**Ans.**: CO_2: 15.97%, CO: 0.93%, N_2: 75.53%, and O_2: 7.57%)

3.6 For the following gas composition by volume, find out the volume of air required for the combustion of 100 m^3 of gas. Determine also the volumetric analysis of the products of combustion.

Gas composition	Volume, %
CH_4	39.5%
H_2	46.0%
CO	7.5%
CO_2	4.5%
H_2O	2.0%
N_2	0.5%

(**Ans.**: 504 m^3 of air, H_2O: 22.63%, CO_2: 8.86% and N_2: 68.51%)

BIBLIOGRAPHY

[1] Sarkar, S., Fuels and Combustion. Orient Longman; 2009.
[2] National Fire Protection Association, U.S.A., NFPA 85–2004 Boiler and Combustion Systems Hazards Code; 2004.
[3] World Coal Association. Coal Facts–2014 Edition.

[4] 15 Countries Sitting On Gigantic Oceans Of Oil. www.businessinsider.com.15-countries-with-the-biggest-oil-reserves-2012.
[5] International Energy Agency. The World Factbook. Natural gas - production—2014.
[6] Stultz, S.C., Kitto, J.B. (Eds.), Steam Its Generation and Use (41st Edition). The Babcock and Wilcox Company; 2005.
[7] Basu Mallik, A.R., Solution of Problems on Thermodynamics, Steam and Other Heat Engines. Basu Mallik and Co.; 1950.
[8] Li, K.W., Priddy, A.P., Power Plant System Design. John Wiley and Sons; 1985.
[9] The Editors of the Power Magazine. The Power Handbook 2nd Edition. Platts, A division of the McGraw-Hill Companies, Inc.; 2010.
[10] Bozzuto, C., (Ed.), Clean Combustion Technologies, fifth ed. Alstom; 2009.
[11] Baumeister, T., Marks, L.S. (Eds.), Mechanical Engineers Handbook, sixth ed. McGraw-Hill Book Company, Inc.; 1958.
[12] Reznikov, M.I., Lipov, YM., Steam Boilers of Thermal Power Stations. Mir Publishers, Moscow; 1985.

CHAPTER 4

Pulverized Coal-Fired Boilers

4.1 BRIEF HISTORY

Steam-operated central power stations started operating in the late nineteenth century (1881–1900). Within a couple of decades suspension firing of pulverized coal (finely ground coal particles) was in use at the Oneida Street Power Station, owned by Milwaukee Electric Railway and Light Company, for the first time in a utility [1]. The concept of suspension firing evolved from the belief that if coal were ground to the fineness of flour, it would flow through coal pipes like oil [2] and would burn in the furnace space as easily and efficiently as gas.

By the 1920s, pulverized coal firing became so developed that it resulted in more complete coal combustion and higher system efficiencies. As a result, pulverized coal firing also became attractive to larger boilers. However, it took nearly 5 years for pulverized coal firing to become a dependable method of coal firing for commercial production and utilization of steam. By the 1950s, pulverized coal firing was the main method of coal firing, leading to the construction of large, efficient, and reliable steam generators and power plants.

In pulverized coal firing, fine particles of coal are easily moved by the flow of air and products of combustion through parts of the furnace. Combustion takes place in the furnace space within a very short time (1–2 s) of the presence of particles in the furnace [3]. The process of pulverization includes two stages. In the first stage, raw coal is crushed to a size of not more than 15–25 mm. The crushed coal is then delivered into raw-coal bunkers and transferred to pulverizers, where it is ground to a fine particle size.

4.2 COMBUSTION OF PULVERIZED COAL

Burning a coarse ground coal may reduce the efficiency of the boiler, increase emissions of pollutants, and in certain cases may damage the boiler or its auxiliaries. Efficient combustion of pulverized coal depends on following characteristics of coal [4]:

- **i.** Grinding and abrasion properties
- **ii.** Volatile matter yield
- **iii.** Ignitability and flame stability
- **iv.** Reactivity and burnout

Thermal Power Plant
DOI: http://dx.doi.org/10.1016/B978-0-12-801575-9.00004-4

v. Fouling and slagging due to the presence of ash
vi. SO_X and NO_X emission

As mentioned, when coal is ground to the fineness of flour, the ground coal will flow in pipes like oil and burn in furnace like gas. To arrive at such fineness, it is essential to dry and pulverize the coal using special equipment. Thereafter, other equipment is provided for transporting this dry and pulverized coal to the furnace in an air stream. In addition, another set of equipment is required for injecting this coal along with the air needed for combustion to the furnace.

When combined with air in a furnace, pulverized coal first passes through the stage of thermal preparation, which consists of the evaporation of residual moisture and separation of volatiles. Once these minute particles enter the furnace at a temperature of 1200–1500 K (depending on the type of fuel) and are exposed to heat, their temperature rises and the volatile matter distills off within a fraction of a second. To facilitate transport of pulverized coal, adequate quantity of primary air (20–30% of total combustion air) is introduced at the burner to dry and carry the pulverized coal from pulverizers to the burner or the bin. This air intimately mixes with the stream of coal particles to burn gas resulting from distillation of each particle in the stream [2].

Fuel particles are heated to a temperature 673–873 K at which volatiles are evolved intensively in a few tenths of a second. Volatile matter, mostly hydrocarbons, ignites more easily than the carbon components of the coal, and heating the latter produces coke. The intensive burning of the volatiles takes 0.2–0.5 s [3]. A high yield of volatiles produces enough heat through combustion to ignite coke particles. When the yield of volatiles is low, the coke particles must be heated additionally from an external source.

The final stage is the combustion of coke particles at a temperature above 1073–1273 K. This is a heterogeneous process whose rate is determined by oxygen supply to the reacting surface. Secondary air (70–80% of total combustion air), introduced around the burner, sweeps past and scrubs the hot carbon particles and gradually burns them. The burning of a coke particle takes up the greater portion of the total time of combustion, which may be of the order of 1.0–2.0 s, depending on the kind of fuel and the initial size of particles.

For efficient combustion of pulverized coal two basic factors need to be considered:

a. Pulverized coal must be fed without segregation, and
b. the mixture of pulverized coal and air fed to the burners should permit stable ignition.

For better flame stability the required coal-primary air ratio generally increases with decreasing grindability and decreases with decreasing volatile matter and falling pulverizer loading. Complete burnout of pulverized coal particles depends on physical and chemical properties of the resultant char, individual boiler design, and boiler operating conditions. NO_X production from a pulverized coal-fired boiler is dependent on coal characteristics, boiler and burner design, and operating conditions. During combustion of coal particles the reacting mechanism between carbon and oxygen takes place in two stages. Oxygen is adsorbed on the surface of particles and reacts chemically with carbon to form CO. Carbon monoxide then reacts with oxygen within the boundary gas film to get oxidized to CO_2 [1].

(The principle of combustion is discussed in Chapter 3, Fuels and Combustion.)

4.3 PULVERIZER PERFORMANCE

The factors that affect pulverizer-grinding performance along with its processing capacity are discussed as follows.

Raw-Coal Size: Normally plants receive run-of-mine coal, although in certain plants washed coal is also supplied. While using run-of-mine coal, it is important to make feed size as uniform as possible, since pulverizer performance is strongly influenced by raw coal size. The smaller the size the better the performance. For small mills top size usually falls between 15 and 25 mm. For medium and large mills, the desired top size of coal may be up to 30 mm and 55 mm, respectively. However, for a particular type of mill, the size of the coal to be used as mill input should conform to the coal size recommended by the manufacturer.

Fineness: Fineness of pulverized coal is extremely important in system design. The higher the fineness of pulverized coal, the lower the pulverizer processing capacity (Figure 4.1).

The degree of fineness of pulverized coal depends to a large extent on coal characteristics. The desired fineness is also determined by the way it affects the coal combustion in the furnace. For pulverized coal testing, the percentage of pulverized coal passing

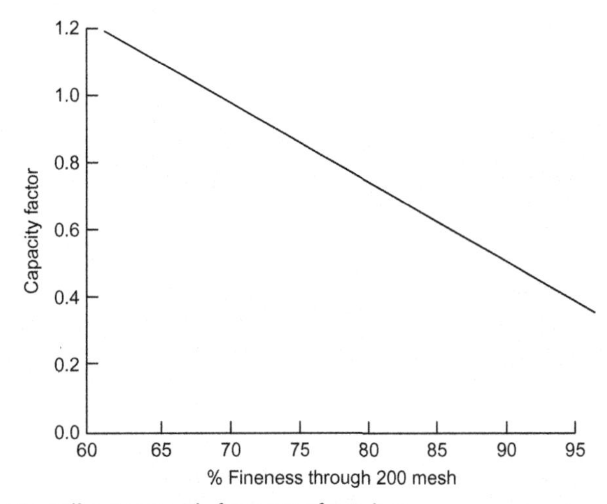

Figure 4.1 Variation in mill output with fineness of product.

through a sieve of size 50 mesh (which corresponds to a 297 μ opening) is classified as oversize and through a sieve of size 200 mesh (which corresponds to a 74 μ opening) is considered as coal dust [2, 5]. For caking coals, pulverization should produce sufficiently fine coal particles to prevent lightweight porous coke particles from floating out of the furnace before completing effective combustion. Free-burning coal, however, does not require the same degree of fineness [2].

To burn pulverized coal successfully in a furnace, the following two requirements must be met:

a. Large quantities of very fine particles of coal, typically 70% of coal that will pass through a 200-mesh screen to ensure ignition because of their large surface-to volume ratios.
b. Minimum quantity of coarser particles, at least 98% of coal that will pass through a 50-mesh screen, to ensure high combustion efficiency.

Note that the percentage of larger coarse particles, of size larger than 50 mesh (297 μ), should not be more than 2% [6], since they cause slagging and loss of combustion.

Volatile Matter: It is known that high-volatile coal ignite more readily than low-volatile coal. Hence, high-volatile coal requires less pulverization than low-volatile coal. Contrary to this, low-volatile coal, except for anthracite, has higher grindability, because it is softer [2]. As a result, the balance between these two characteristics determines the pulverizer output. In practice, with volatile content above 22%, the fineness can be kept between 65 and 70% passing through 200 mesh to ensure ignition stability; while with less than 18% volatile matter content the fineness has to be increased to 80% or above through 200 mesh.

Grindability: This term is used to measure the ease of pulverizing a coal. Unlike moisture, ash, or heating value, this index is not an inherent property of coal [7]. Grindability should not be considered as synonymous with the hardness of coal [7]. Anthracite is a very hard coal, whereas lignite is soft; however, both are difficult to grind.

The hardness of coal is generally in the range 10–70 kg/mm^2, depending on rank, with a minimum near 20% volatile matter. Its index is generally given as the *Hardgrove Grindability Index* (HGI). The higher the index of coal the softer it is to pulverize. It is generally considered that for a desired coal fineness and selected feed size, the mill capacity has a direct relationship with grindability; however, this capacity is significantly influenced by the moisture content of the coal as is evident from Figure 4.2 [8]. If the coal-drying operation is inadequate, the mill capacity is dictated by moisture content and not by grinding. Thus, more output may be available from a mill with a dry coal of lower grindability than with a wet coal of higher grindability.

From the above discussion it is easy to argue that to maintain the same steam-generating capacity, if the moisture content of coal, coal fineness, and feed size remain unchanged, the higher the HGI value, the lower the required pulverizer output. Thus, a higher correction factor should be applied to arrive at correct pulverizer

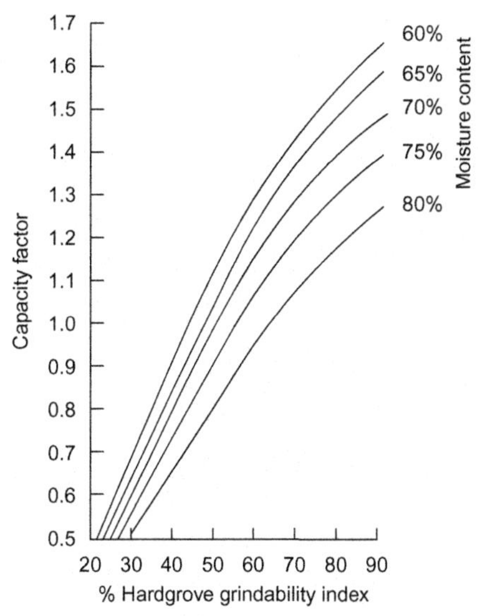

Figure 4.2 Pulverizer capacity versus grindability [ref.: ASME PTC PM-1993]. *Source: ASME PTC PM-1993. Performance monitoring Guidelines for Steam Power Plants.*

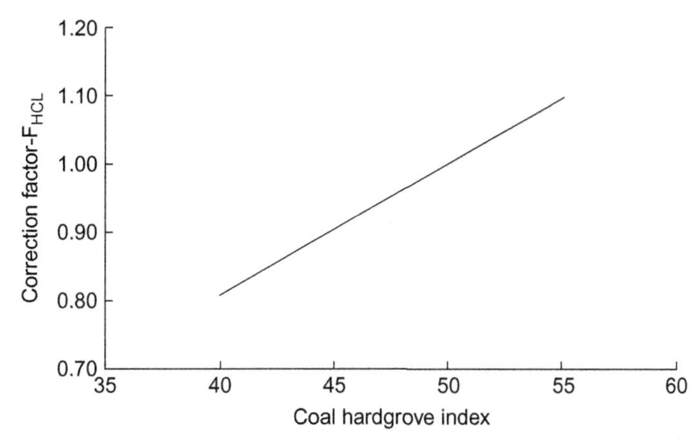

Figure 4.3 Pulverizer capacity versus coal hardgrove index.

output (Figure 4.3). Reason being pulverizer design capacity is guaranteed under specified HGI value. Deviation from this value will cause a change in design capacity. Hence for deviations from specified HGI value, it is necessary to correct the measured capacity using the correction factor to get the design capacity of the pulverizer.

Moisture: As discussed above the moisture content of coal significantly influences pulverizer performance. It is the *surface moisture* (SM) that affects pulverization the most; *inherent moisture* (IM) has little or no effect. If the temperature of the surface moisture in the coal fed to the pulverizer is below its dew point, agglomeration of fines in the pulverizing zone will take place that adversely affect pulverizer capacity.

Agglomeration of fines has the same effect as coarse coal during the combustion process. Hence, during the process of pulverization coal must be dried enough to make it free of surface moisture [1].

Hot air is normally supplied for drying of coal inside the pulverizer. If the surface moisture is excessive, the pulverizer must be capable of handling large volumes of high-temperature air. In the event that there is deficiency of hot air or if the pulverizer is unable to handle additional hot air, transport of the pulverized coal to the burners will be reduced. Figure 4.4 shows the relationship between the weight of the air per unit of fuel and the temperature of the coal-air mixture leaving the pulverizer with varying surface moisture for a particular coal.

For high-volatile coal to prevent pre-ignition of dry coal inside the pulverizer, the coal-air mixture temperature at the pulverizer outlet has to be maintained at around 338–348 K, while for low-volatile coal this temperature may be safely raised to 355–373 K. Thus, the supplied air may be pre-heated to a temperature of 600 K or above, as recommended by the manufacturer.

GCV (HHV) of Coal: The heating value of coal also influences the pulverizer output. Pulverizer design output is guaranteed under certain specified GCV. Any deviation in GCV will cause a change in design output. Hence for deviations in GCV, it is imperative to correct the measured output using the correction factor to get the design output. For maintaining the same steam generating capacity, the higher the heating value, the lower the required pulverizer output. Hence, a higher correction factor should be applied to arrive at the correct pulverizer output as is evident from Figure 4.5.

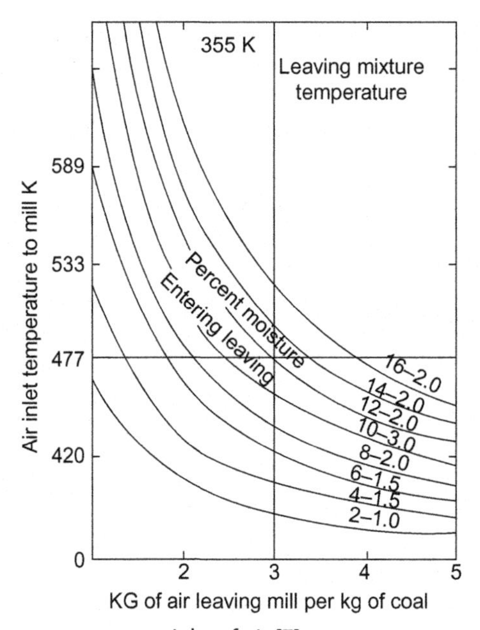

Figure 4.4 Air inlet temperature versus weight of air [7].

Abrasive Index (YGP – Yancy, Geer, and Price – Index): Pulverizing results in an eventual loss of grinding-element material. Balls, rolls, rings, races, and liners gradually erode and wear out as a result of abrasion and metal displacement in the grinding process. The wear rate is linearly related to the mineral content of coal. Two basic mechanisms of mechanically induced wear are present in the grinding zone. The first mechanism is abrasion and the second basic mechanism is impaction erosion wear, the effect of which is considered to be negligible. The *YGP Index* typically indicates the abrasive characteristic of coal. Figure 4.6 presents a typical "wear life curve" of a pulverizer.

Extraneous Materials: The wear life of pulverizer-grinding rings, balls, etc., is influenced by coal-feed size, HGI, and the presence of extraneous materials such as rock, slate, sand, stone, pyrite, quartz, etc., which are quite abrasive. The effect of coal-feed size (larger than 25 mm or so) on wear rate is small. Wear rate is greatly influenced

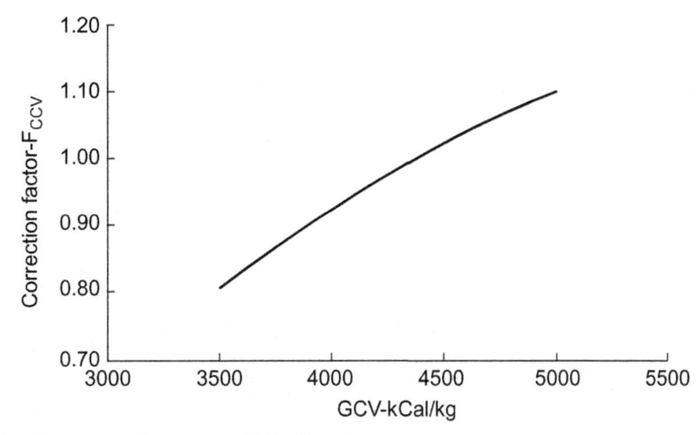

Figure 4.5 Pulverizer capacity versus GCV of coal.

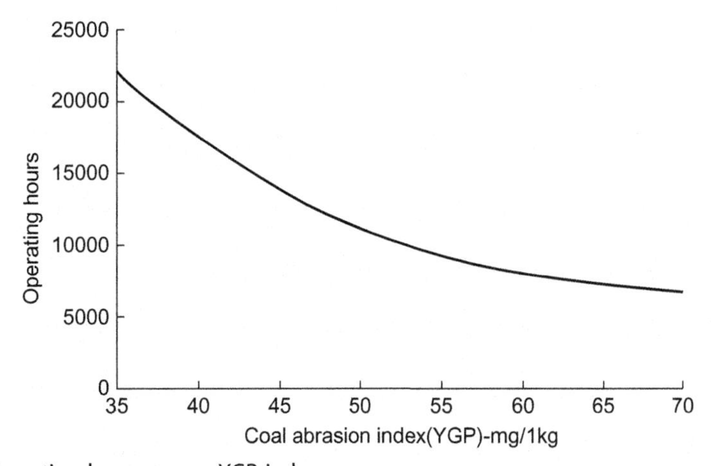

Figure 4.6 Operating hours versus YGP index.

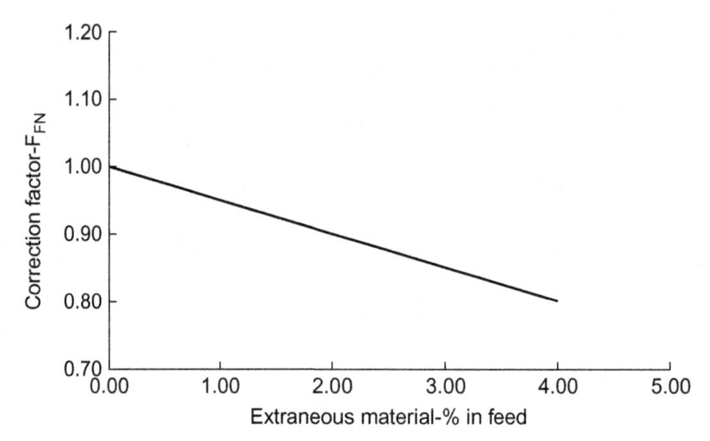

Figure 4.7 Pulverizer capacity versus extraneous material.

by *quartz* and *pyrite particles*. High quartz content in ash would exacerbate the wearing characteristics of grinding parts but not mill capacity. The larger the size of these particles, the longer they remain in circulation in the mill, hence wear rates are increased. Quartz, with a density of approximately twice that of coal, has to be ground to a finer size when compared to the coal particles before it is released out of the pulverizer by the primary air. Rounded quartz particles cause higher wear than that resulting from angular particles. Quartz causes a three- to four-fold increase in abrasion over *siderite* (a mineral composed of $FeCO_3$), which in turn is about three times more abrasive than *calcite* (a carbonate mineral and the most stable polymorph of $CaCO_3$).

The wear life of the pulverizer-grinding components is adversely affected by the presence of rock and stones (typical Moh's scale of hardness 6), which cause five times the wear caused by coal (typical Moh's scale of hardness 3). Hence, a lower correction factor should be applied to arrive at correct pulverizer output for a coal containing a higher quantity of extraneous materials as presented in Figure 4.7.

Pulverizers also suffer from chemical wear, i.e., corrosion, in a moist atmosphere. Sulphur (oxidized pyrite) and chlorine (chloride and sodium ion) compounds and organic acids are comprised of an acidic and corrosive medium in moist coal. Furthermore, moisture in the absence of acids can enhance abrasion wear.

4.4 COAL BURNERS (FIGURE 4.8)

After pulverization, coal is transported pneumatically to the burners through pipes. Air and fuel are supplied to the furnace in a manner that permits:

i. Stable ignition
ii. Effective control of flame shape and travel
iii. Thorough and complete mixing of fuel and air

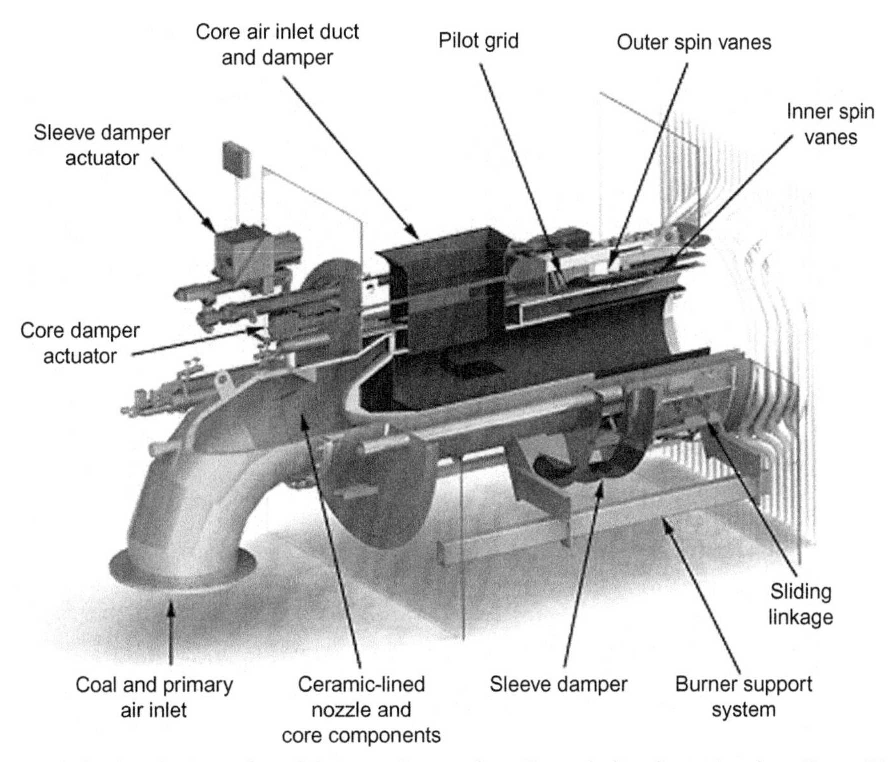

Figure 4.8 Sectional view of coal burner. *Source: http://www.babcock.com/products/Pages/AireJet-Low-NOx-Coal-Burner.aspx.*

The air that transports coal to the burner is called the primary air, while secondary air is introduced around or near the burner. The burners impart a rotary motion to the coal-air mixture in a central zone, and the secondary air around the nozzle – all within the burner. The rotary motion provides pre-mixing for the coal and air along with some turbulence.

In steam generators burners may be installed on a front wall or on front and rear (opposed) walls or on furnace corners. Front-wall burners are provided with either internal ribs or circular registers that impart a rotary motion to coal-air mixture that produces turbulence. Opposed-wall burners throw their flames against each other to increase turbulence. Tangential or corner-fired burners produce inherently turbulent flame.

The total air-fuel ratio in coal burners is much greater than the stoichiometric requirement but just enough to ensure complete combustion without wasting energy by adding too much sensible heat to the air. The initial ignition of coal burners is accomplished with the help of a sustained flame of light fuel-oil burner. The control equipment of coal burners ranges from manual to a remotely operated programmed sequence. (Detailed treatment on the design aspects of burners is given in Chapter 3, Fuels and Combustion.)

4.5 PULVERIZER SYSTEM DESIGN

NFPA 85 recommends the pulverizer system arrangement should be such as to provide only one direction of flow, i.e., from the points of entrance of fuel and air to the points of discharge. The system should be designed to resist the passage of air and gas from the pulverizer through the coal feeder into the coal bunker. To withstand pulverizer-operating pressures and to resist percolation of hot air/gas, a vertical or cylindrical column of fuel at least the size of three coal-pipe diameters should be provided between the coal-bunker outlet and the coal-feeder inlet as well as between coal-feeder outlet and the pulverizer inlet. Within these cylindrical columns there will be accumulation of coal that will resist percolation of hot air/gas from the pulverizer to the coal bunker. All components of the pulverized coal system should be designed to withstand an internal explosion gauge pressure of 344 kPa [9].

Besides the above, design specifications of a pulverizer system must also consider the following:

Number of Spare Pulverizers: To overcome forced outage and consequent availability of a number of operating pulverizers it is generally considered that while firing the worst coal one spare pulverizer should be provided under the TMCR (Turbine Maximum Continuous Rating) operating condition. In certain utilities one spare pulverizer is also provided even while firing design coal, but under the BMCR (Boiler Maximum Continuous Rating) operating condition. Practice followed in the United States generally is to provide one spare pulverizer for firing design coal, in larger units two spare pulverizers are provided. However, provision of any spare pulverizer is not considered in current European design [5].

Pulverizer Design Coal: The pulverizer system should be designed to accommodate the fuel with the worst combination of properties that will still allow the steam generator to achieve the design steam flow. Three fuel properties that affect pulverizer-processing capacity are moisture, heating value, and HGI, as discussed earlier.

Pulverizer Product Fineness: Fineness of pulverized fuel is one of the pulverizer sizing criteria, as discussed earlier.

Unit Turndown: The design of a pulverizer system determines the turndown capability of the steam generator. The minimum stable load for an individual pulverizer firing coal is 50% of the rated pulverizer capacity. Normally in utility boilers, the operating procedure is to operate at least two pulverizers to sustain a self-supported minimum boiler load. Thus, the minimum steam generator load when firing coal without supporting fuel is equal to the full capacity of one pulverizer. Therefore, a loss of one of the two running pulverizers will not trip the steam generator because of "loss of fuel" and/or "loss of flame."

Pulverizer Wear Allowance: A final factor affecting pulverizer system design is a "capacity margin" that would compensate for loss of grinding capacity as a result of wear between overhauls of the pulverizer (Figure 4.6). A typical pulverizer-sizing criterion is 10% capacity loss due to wear.

4.6 CLASSIFICATION OF PULVERIZERS

A pulverizer, also known as a grinding mill, is the main equipment associated with a pulverized coal-fired boiler. Grinding inside a pulverizer is realized by impact, attrition, crushing, or a combination of these. Based on their operating speed, pulverizers/mills are classified as "low," "medium," and "high" speed mills [10]. Depending on the pressure existing inside the grinding zone, pulverizers may be categorized as the suction type or the pressurized type. For both types, the pulverizing of coal is accomplished in the following two stages:

Feeding System: This system must automatically control fuel-feed rate according to the boiler-load demand and the air rates required for drying and transporting pulverized coal to the burner.

Drying: One important property of coal being prepared for pulverization is that it has to be dry and dusty.

4.6.1 Low-speed mill

Mills operating below 75 rpm are known as low-speed mills. Low-speed units include *ball or tube or drum mills*, which normally rotate at about 15–25 rpm. Other types of mills, e.g., ball-and-race and roll-and-race mills, that generally fall into the medium-speed category may also be included in this category provided their speed is less than 75 rpm.

Tube mills (Figure 4.9), also known as ball mills, are usually a drum-type construction or a hollow cylinder with conical ends and heavy-cast wear-resistant liners, less than half-filled with forged alloy-steel balls of mixed size. This is a very rugged piece

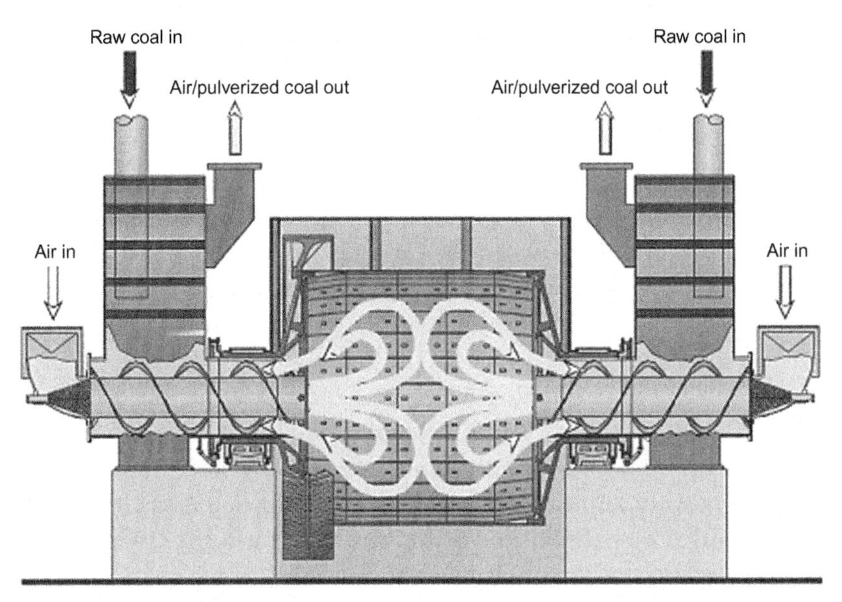

Figure 4.9 Sectional view of low speed mill. *Source: Foster Wheeler Corporation 'Hard-coal burn-up increase with adjustable classifier for Ball Mill', POWER-GEN International Conference, Orlando, Florida, USA, 2000.*

of equipment, where grinding is accomplished partly by impact, as the grinding balls and coal ascend and fall with cylinder rotation, and partly by attrition between coal lumps inside the drum.

Primary air is circulated over the charge to carry the pulverized coal to classifiers. In this type of mill pulverized coal exits from the same side of the mill that solid coal and air enter. In some designs entry of solid coal air and exit of pulverized coal are provided at each end of the mill. Both ends of the mill are symmetrical in nature. Consequently, each mill is served by two coal feeders.

Reliability of this type of mill is very high and it requires low maintenance. The disadvantages of this type of mill are high power consumption, larger and heavier construction, greater space requirement, etc. To pulverize coal of high rank and low grindability, ball/tube mills are preferred because they can achieve high fineness, required for proper burning, and maintain high availability.

4.6.2 Medium-speed mill

This type of pulverizer is usually one of two types: ball and race and roll and race. The speed of the grinding section of these mills is usually between 75 and 225 rpm. Medium-speed mills are smaller than low-speed units and are generally the vertical spindle type. They operate on the principles of crushing and attrition. Pulverization takes place between two surfaces, one rolling on top of the other. The rolling elements may be balls or rolls that roll between two races, in the manner of a ball bearing. Primary air causes coal feed to circulate between the grinding elements, and when it becomes fine enough, it becomes suspended in air and the finished product is conveyed to the burners or the classifier depending up on the applicable system. Configuration of vertical spindle mills may be either the pressurized type with a cold/hot primary air fan or the exhauster type. Medium-speed mills require medium to high maintenance, but their power consumption is low.

Ball and Race Mill (Figure 4.10): In the ball-and-race mill, balls are held between two races, much like a large ball bearing. The top race or grinding ring remains stationary while the bottom race rotates. As the coal is ground between large diameter balls and the ring, the balls are free to rotate on all axes and therefore remain spherical. Grinding pressure is adjusted by controlling the tension of springs.

Roll and Race Mill: The grinding elements of this type of mill consist of three equally spaced, spring-loaded heavy conical (Figure 4.11) or toroidal rolls (Figure 4.12), which, suitably suspended inside near the periphery, travel in a concave grinding ring or bowl (with heavy armoring). Force is applied to the rolls from above by a uniformly loaded thrust ring. The main drive shaft turns the table supporting the grinding ring, which in turn transmits the motion to the rollers.

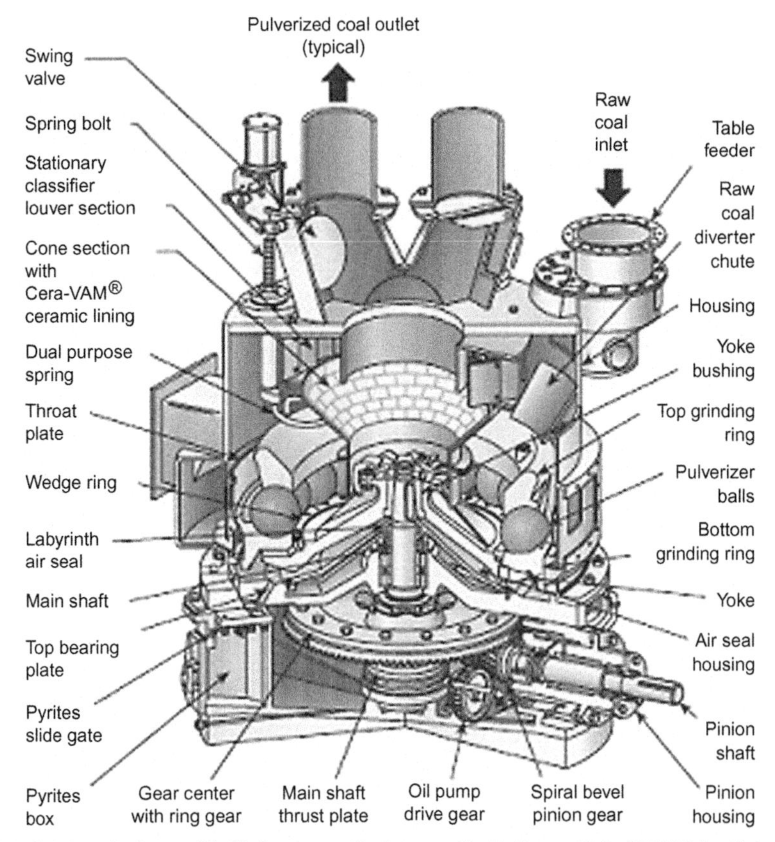

Figure 4.10 Sectional view of ball & race mill. *Source: Fig.5, Page 13-3, STEAM Its Generation and Use, 41st Edition. Babcock & Wilcox Company, 2005.*

Mills with conical rolls are also known as *bowl mills*. As the coal is ground between large diameter rolls and the bowl, rolls revolve about their own axes, and the grinding bowl revolves about the axis of the mill.

In the toroidal rolls type, grinding occurs only under the rollers in the replaceable grinding ring. There is no metal-to-metal contact between grinding elements, since each roller rests on thick layer of coal. Thus the maintenance is minimized.

4.6.3 High-speed Mill (Figure 4.13)

High-speed mills were developed during the early days of pulverized fuel firing. These mills are either *hammer beater* or *attrition mills*, which operate at speeds above 225 rpm. The beaters revolve in a chamber equipped with high-wear-resistant liners. Both impact and attrition is combined in this mill to pulverize coal, which is pulverized by the rubbing of coal on coal, by the impact of coal on impeller clips, and also by being rubbed between pegs. A classifier returns the coarse coal particles for further classification. Its

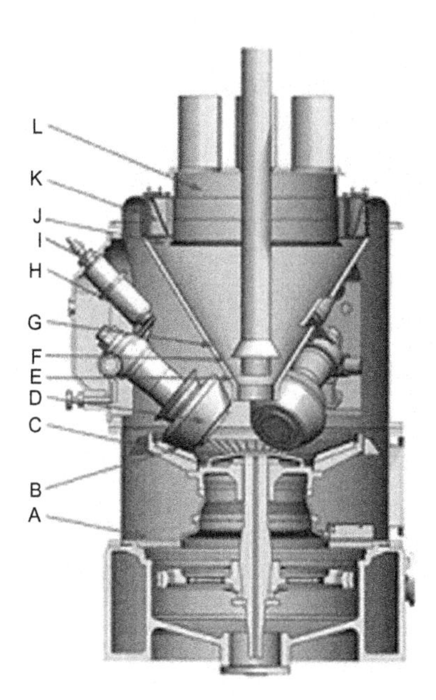

A. Pyrite sweep conditions/clearances
B. Grinding element condition/clearances
C. Throat dimensions/opening
D. Roll/journal condition
E. Feed pipe clearances
F. Inverted cone/conical baffle clearances
G. Classifier cone condition
H. Button clearance/spring height
I. Preload of spring canisters
J. Outlet cylinder height in relation to classifier blades
K. Classifier blade condition/ length/stroke synchronized angles
L. Outlet smooth, free of any obstructions or spin arresting protrusions into the spinning two-phase mixture of coal and air

Figure 4.11 Sectional view of conical roll and race mill. *Source: Page 4, Pulverizers 101: Part I by Dick Storm, PE. Storm Technologies Inc. POWER Magazine, 08/01/2011.*

capital cost per unit output is low, it requires minimum space, there is no speed reducer between the drive and the pulverizer, and its parts are lightweight to facilitate maintenance. Its maintenance cost, however, is very high, which is the reason this type of mill has long been discontinued from service. This type of mill is mostly used with low-rank coal with high moisture content, e.g., lignite, and uses flue gas for drying.

4.7 COAL PREPARATION SYSTEMS

There are two basic types of fuel preparation systems, the bin system and direct firing system, that have been used for the processing, distributing, and burning of pulverized coal.

4.7.1 Bin System (Figure 4.14)

In the bin system coal is processed at a location away from the furnace, and the coal is pulverized in pressurized or suction type mills. The drying and conveying of coal is done by hot air or hot flue gas extracted from the boiler. The resulting pulverized

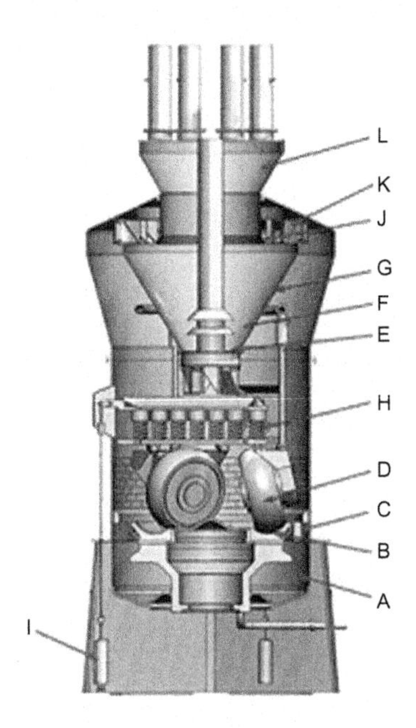

A. Pyrite sweep conditions/clearances
B. Grinding element condition/clearances
C. Throat dimensions/opening
D. Roll/journal condition
E. Feed pipe clearances
F. Inverted cone/conical baffle clearances
G. Classifier cone condition
H. Button clearance/spring height
I. Preload of spring canisters
J. Outlet cylinder height in relation to classifier blades
K. Classifier blade condition/ length/stroke synchronized angles
L. Outlet smooth, free of any obstructions or spin arresting protrusions into the spinning two-phase mixture of coal and air

Figure 4.12 Sectional view of toroidal roll and race mill. *Source: Page 4, Pulverizers 101: Part I by Dick Storm, PE. Storm Technologies Inc., POWER Magazine, 08/01/2011.*

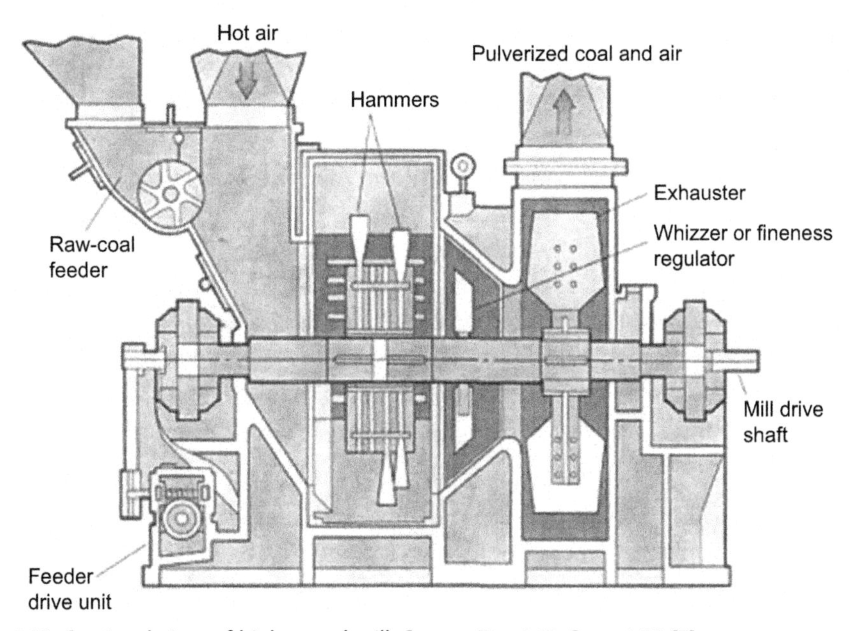

Figure 4.13 Sectional view of high speed mill. *Source: Fig. 6-18, Page 6-20 [7].*

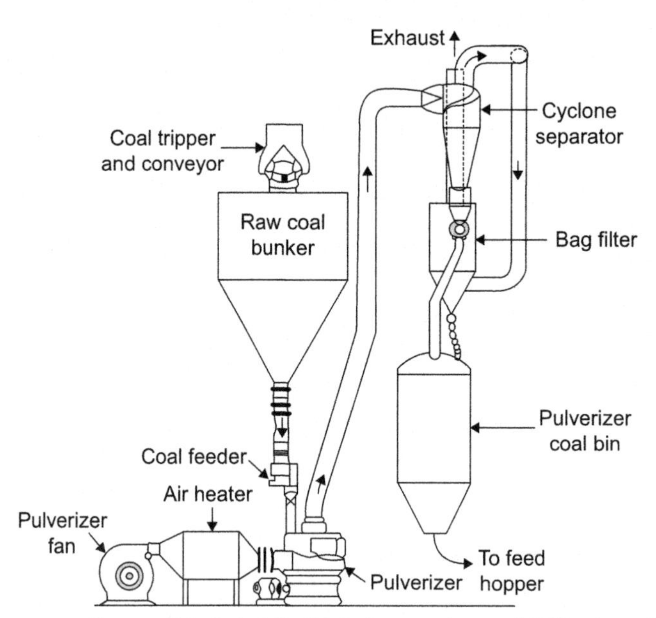

Figure 4.14 Bin system. *Source: FIG. 1. P 9-2. CH 9: Preparation and Utilization of Pulverized Coal. STEAM Its Generation and Use 39th Edition. The Babcock & Wilcox Company, 1978.*

coal-primary air mixture is pneumatically conveyed to a cyclone separator and/or fabric bag filter that separate and exhaust the moisture laden air to the atmosphere and discharge the pulverized coal to a pulverized coal bunker/bin near the furnace.

The stored coal from the pulverized coal bunker is fed to the burners through the respective pulverized coal feeders and conveyed by primary air. Additional vapor burners are provided in this system to burn the escaping coal dust (2–3%) after the cyclone separators. The basic advantage of this system is that the pulverization is not tied up with the minute-to-minute operations of the boiler and the system can withstand the outage of the pulverizer to a certain extent, depending upon the storage capacity available. The disadvantages are intrinsic dust nuisance of installation, fire risk of pulverized coal storage, higher cost of pulverization, etc. It is essential to incorporate an additional fire-prevention system and dust-coal disposal system to prevent unforeseen boiler outages.

The bin system was widely used before pulverizing equipment reached the stage where it was reliable enough for continuous steady operation.

For successful operation, the following guidelines need to be adhered to:

i. Surface moisture in the pulverized coal must not exceed 3%.
ii. Fineness should not be less than 90% through a 50-mesh sieve.

Because of the many stages of drying, storing, transporting, etc., the bin system is subject to fire hazards from spontaneous combustion. Nevertheless, it is still in use in many older plants.

4.7.2 Direct Firing System (Figure 4.15)

The bin system has been overshadowed by the direct-firing system because of greater simplicity, lower initial investment, lower operating costs, less space requirement, and greater plant cleanliness. The pulverizing equipment developed for direct-firing systems permits continuous utilization of raw coal directly from the bunkers through a feeder, pulverizer, and primary-air fan, to the furnace burners.

There are two direct-firing methods in use – the pressure type, where the primary-air fan is located at the pulverizer inlet, and the suction type, where the primary-air fan is located downstream the pulverizer. In either type, the coal is delivered to the burners with air as the transport medium.

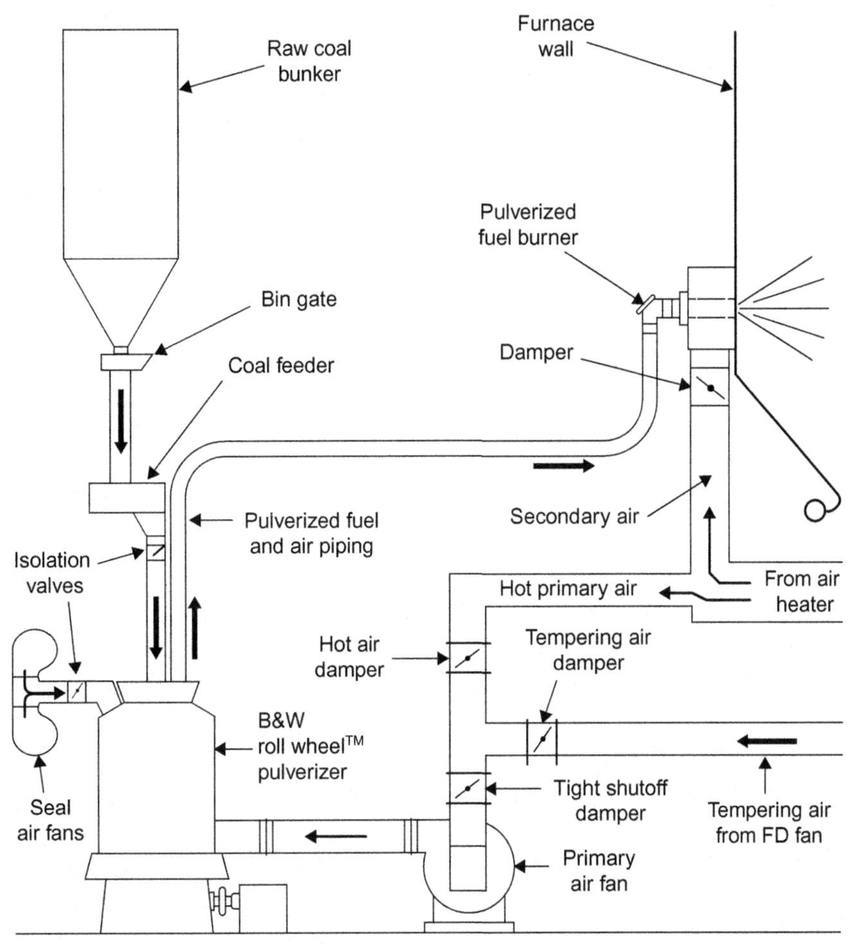

Figure 4.15 Direct firing system. *Source: Fig. 9, Page 13-6, STEAM Its Generation and Use, 41st Edition. Babcock & Wilcox Company, 2005.*

The advantages of the direct firing system are:

i. Low investment cost
ii. Low operating cost
iii. Convenience of installation
iv. Safety from fire hazard
v. Flexibility of operation
vi. Reliability of operation
vii. Less space requirement
viii. Greater plant cleanliness

The only disadvantage of this system is any outage in the pulverizer system immediately affects unit output, even though spare pulverizers are usually provided. Fuel feed is regulated to load demand by a combination of controls on the feeder and the primary-air fan to give fuel-air ratios in accordance with variations in steam generation. Large steam generators are provided with more than one pulverizer system, each feeding a number of burners, so that a wide control range is possible by varying the number of pulverizers and the load on each.

4.8 COAL FEEDERS

Coal from the raw-coal bunker is fed to the pulverizer through a raw-coal feeder or simply the coal feeder. The function of a coal feeder is to control the flow of coal to the pulverizer to comply with the demand of steam generation. In general, there are various kinds of raw-coal feeders in use, e.g., table type, overshot roll type, belt type, drag-link type, etc. Coal flows from raw coal bunker on to a moving belt through a vertical pipe, and from this belt coal flows to a pulverizer through another vertical pipe.

In most cases, coal feeders are driven by a constant speed electric motor through step-less variable speed drives to control the coal flow. The variable speed is generally achieved through positive infinitely variable (PIV) gears, variable hydraulic or electro-magnetic coupling, variable frequency drive, or some sort of variable stroke clutch mechanism.

Feeders may be the volumetric type, maintaining a constant speed vs. volume (for the same bed thickness) relationship or the gravimetric type with advanced electronic controls, capable of automatic control and recording of mass flow of coal feed.

The roll feeder and the belt feeder may be considered a highly efficient volumetric feeding device. In a volumetric system (Figure 4.16) the flow of coal is controlled by the boiler demand system with the help of a fixed position-leveling bar in combination with a variable speed belt, but the actual amount fed is estimated and the accuracy can vary greatly.

For efficient operation of the plant, however, it is essential to know the exact amount of fuel being fed. The gravimetric feeder (Figure 4.17) does this by determining the product of the fuel belt load multiplied by the speed to give the actual flow

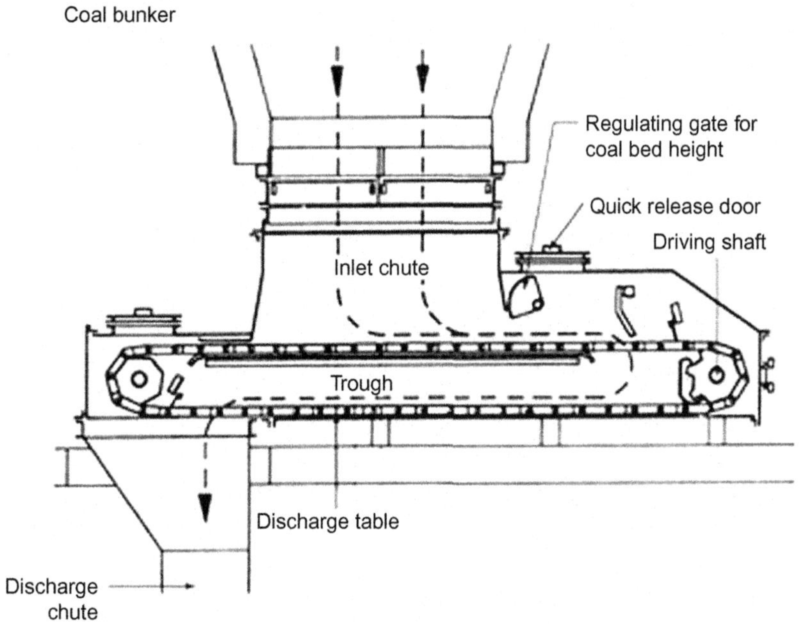

Coal bunker
Regulating gate for coal bed height
Quick release door
Driving shaft
Inlet chute
Trough
Discharge table
Discharge chute

Figure 4.16 Volumetric coal feeder. *Source: http://gazogenerator.com/boilers-for-power-and-process/milling-plant/.*

Cut-off gate
Crushed coal in
Fuel spout
Load cell
DC motor
Clean-out conveyor
Digital feed control
Speed signal
Transition chute
Crushed coal out

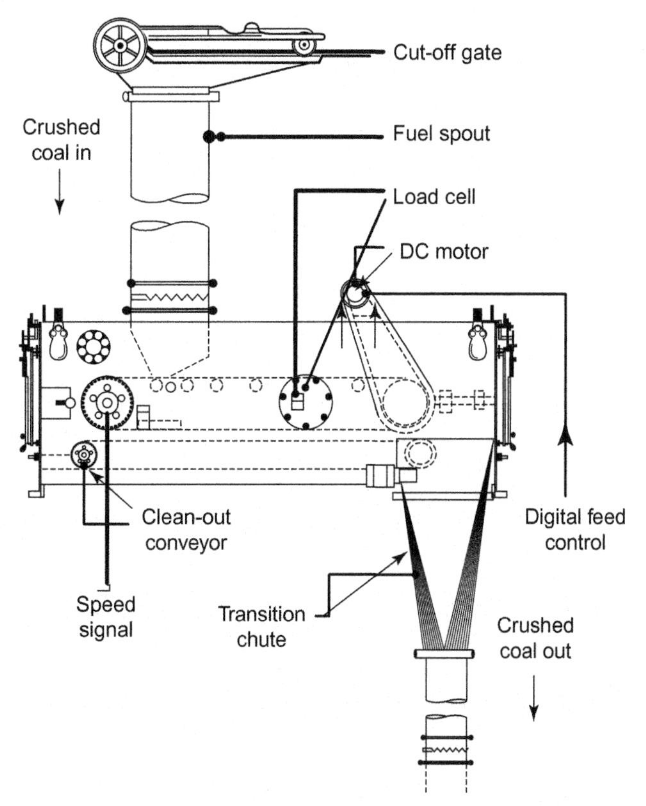

Figure 4.17 Gravimetric coal feeder. *Source: http://gazogenerator.com/boiler-for-power-and-process/milling-plant/.*

rate. This flow rate is compared to the boiler set point and the belt speed is varied accordingly. The first installation of gravimetric feeder was placed in continuous operation at Niagara Mohawk Power Corporation's Dunkirk Station in 1957 [11]. The benefits of gravimetric feeding are:

i. Exact and constant amount of coal feed to the mill
ii. Saving of coal due to exact control of excess air
iii. Fast and stable boiler load changes
iv. Accurate energy accounting

Two different types of gravimetric coal feeder generally used in the industry are the Stock® gravimetric feeder and the Pfister rotor weighfeeder. The Stock gravimetric coal feeder weighs material on a length of belt between two fixed rollers (span) precisely located in the feeder body. A third roller, located midway in the span and supported at each end by precision load cells, supports half the weight on the span. As material passes over the span, the load cell generates an electrical signal directly proportional to the weight supported by the center roller.

The Pfister rotor weighfeeder works like a horizontal star feeder. The gravimetric coal mass in the feeder is measured by a load cell. The speed of the feeder motor is controlled inverse to the coal content in the feeder. Less coal in the feeder results in higher speed, and more coal in the feeder results in lower speed. The outcome is a highly accurate feed rate at the outlet.

BIBLIOGRAPHY

[1] Stultz SC, Kitto JB, editors. STeam Its Generation and Use, 39th/41st ed. The Babcock And Wilcox Company; 1978/2005.
[2] Power (Special Issue). Controlling Combustion and Pollution in Industrial Plants, Mcgraw-Hill Book Company, Inc.; 1982.
[3] Reznikov MI, Lipov YuM. Steam Boilers of Thermal Power Stations, Mir Publishers, Moscow; 1985.
[4] British Electricity International. Modern Power Station Practice Volume "B" Boilers and Ancillary Plant, Pergamon Press, London; 1991.
[5] Black and Veatch. Power Plant Engineering, CBS Publishers and Distributors; 2001.
[6] Sarkar S. Fuels and Combustion, Orient Longman; 2009.
[7] Bozzuto C, editor. Clean Combustion Technologies, fifth ed. Alstom; 2009.
[8] The American Society of Mechanical Engineers. ASME PTC PM-1993 Performance Monitoring Guidelines for Power Plants.
[9] National Fire Protection Association, U.S.A. NFPA 85 – 2004 Boiler and Combustion Systems Hazards Code.
[10] Baumeister T, Marks LS, editors. Mechanical Engineers Handbook, sixth ed. Mcgraw-Hill Book Company, Inc.; 1958.
[11] The STOCK® Gravimetric Feeder. An International Historic Mechanical Engineering Landmark. Chagrin Falls, Ohio: Stock Equipment Company; May 4, 1995.

CHAPTER 5

Fluidized-Bed Combustion Boilers

5.1 BRIEF HISTORY

Fluidized-bed technology was initiated in the 1920s as a process for refining petroleum and for the production of chemical feedstock from coal. This technology has been applied for the following since the 1940s:

i. Calcination of Alumina
ii. Cracking of Hydrocarbon
iii. Roasting of Ore
iv. Incineration of Wastes

However, fluidized-bed technology as an alternative method for combustion of coal gained popularity in the 1950s when researchers started looking for ways to reduce atmospheric pollutants such as SO_2 and NO_X [1]. During the second half of the 1980s, fluidized-bed combustion (FBC) rapidly emerged as a viable option for stoker-fired and pulverized coal-fired units for the combustion of solid fuels due to following:

i. Rise in cost of conventional fuel and demand for multi-fuel firing
ii. Fuel flexibility
iii. Stringent SO_2 and NO_X emission regulation
iv. Availability of only low-grade coal for firing in boilers in some countries
v. High combustion efficiency
vi. Low combustion temperature
vii. High heat and mass transfer
viii. Low time of reaction
ix. Less capital investment compared to that required to install pulverized coal-fired boiler
x. Less operating costs leading to more profits
xi. Reduced auxiliary power consumption

The use of bubbling fluidized-bed (BFB) technology in the electric utility industry in the United States began in 1976 [2]. The first application of circulating fluidized-bed (CFB) technology in the United States occurred in 1981 and became operational by 1988. In Germany, although development of CFB combustors began in 1955, the CFB technology was first applied to the combustion of coal in 1982 when a co-generation

Thermal Power Plant
DOI: http://dx.doi.org/10.1016/B978-0-12-801575-9.00005-6

Table 5.1 List of some of the world's large CFBC boilers

Sl. No.	Unit size MW	Fuel	Location	Year of commissioning
1	4 × 125	Waste coal, Slurry	Emile Huchet, France	1990
2	2 × 125	Lignite	Akrimota, India	2003
3	2 × 220	Anthracite	Tonghae, Korea	1999
4	250	Sub-bituminous	Provence, France	1998
5	2 × 250	Brown coal	Red-Hill, USA	2001
6	2 × 250	Bituminous	Guyama, Puerto Rico	2003
7	300	Anthracite	Biama, China	2004
8	2 × 150	Coal and Bio	Tha Toom, Thailand	1999
9	202 (BFB)	Bio fuel	Power River, Canada	1999
10	220	Lignite	Chorzow, Poland	2003
11	6 × 235	Brown coal	Elektronia Turow, Poland	1998–2005
12	2 × 300	Bituminous	JEA Florida, USA	2002
13	2 × 330	Petcoke	Cleco, USA	2009

unit set up [2]. By that time FBC technology had reached commercial status and was well-established for generation of heat, power, as well as combined heat and power.

With the constant development and refinement of the technology the use of CFB technology has expanded from small industrial boilers of capacity equal to or less than 30 kg/s to large utility boilers. By the early 1990s, several electrical utility steam generators had been in operation globally in the range of 100–165 MW [2]. The world's largest sub-critical circulating fluidized bed combustion (CFBC) with a capacity of 250 MW was synchronized on October 29, 1995, and is in operation at Provence Power Station, France [3]. The largest supercritical CFBC plant is the 460 MW Lagisza plant in Poland. It started at the site in February 2006 and the project was commissioned in June 2009 [4,5]. Supercritical CFB boiler for which a detailed boiler design for up to 600 MWe CFB is now ready for first implementation with steam conditions 27 MPa/853 K/873 K and economizer inlet feedwater temperature 563 K [6]. Some of the world's large CFBC boilers successfully operating in various countries are listed in Table 5.1.

5.2 THE TECHNOLOGY

Fluidized-bed combustion is a process in which solid particles are made to exhibit fluid-like properties by suspending these particles in an upwardly flowing evenly distributed fluid (air or gas) stream [7]. Combustion takes place in the bed with high heat transfer to the furnace and low combustion temperatures.

The fluidized-bed process is shown schematically in Figures 5.1 through 5.10. When at rest the fluidized bed resembles a uniformly distributed bed of solid particles, e.g., sand stacked on a perforated plate of fine mesh at an intermediate position of an enclosed vessel (Figure 5.1). The perforated plate facilitates uniform distribution of fluid (air or gas) flow through the bed. At the bottom of the plate pressurized fluid is supplied.

When this pressurized fluid is allowed to pass upward through the bed of solid particles the bed tends to offer resistance to low fluid velocities such that the particles remain in contact and stay stagnant. This bed is called a *fixed bed* (Figure 5.2 and Figure 5.3).

With an increase in fluid velocity the particles offer less resistance and tend to expand. With further increase in fluid velocity a situation occurs in which particles are unable to remain in contact and start separating, resulting in bubble formation, vigorous turbulence, and rapid mixing. The motion of bubbles in this bed resembles the motion of bubbles in a liquid, and the bed of solid particles looks similar in

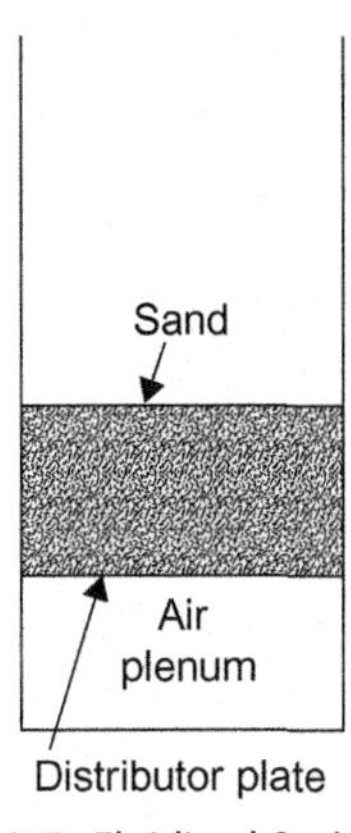

Figure 5.1 Bed at rest. *Source: P17-1. CH17. Fluidized-Bed Combustion STEAM ITS GENERATION AND USE. 41st Edition.*

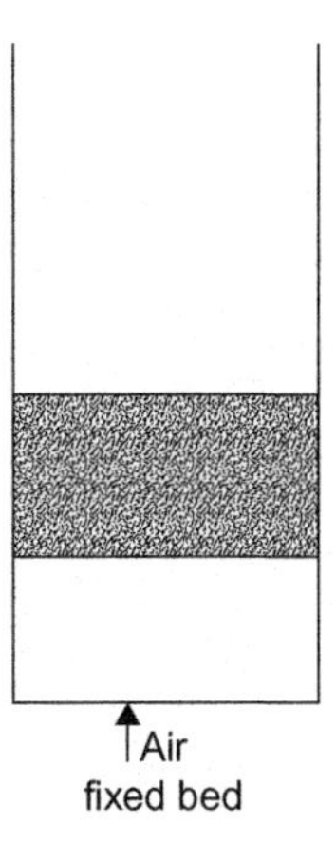

Figure 5.2 Fixed bed.

appearance to a boiling liquid. At this point the drag force exerted on particles counterbalances the gravity force of the particles, thereby causing the fluid-solid mixture to behave like a fluid [7]. This is called the *minimum fluidization condition* of the bed (Figure 5.4) and the bed is called a *fluidizing bed*. The increase in the fluidizing-bed volume is insignificant when compared with the fixed bed. The transition from fixed bed to fluidized bed caused by the changes in air/gas pressure with an increase in fluid velocity through the solid bed is shown in Figure 5.5.

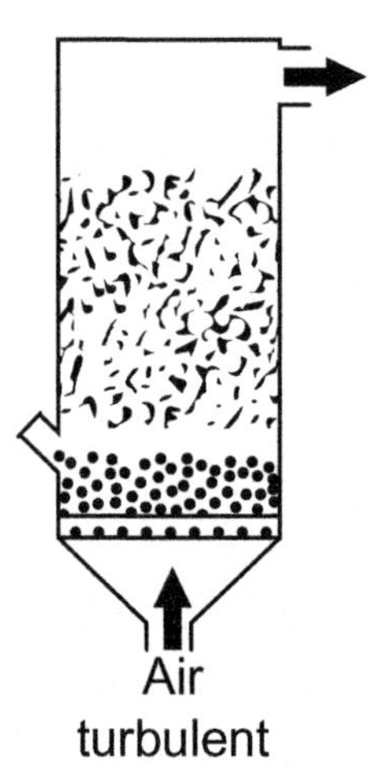

Figure 5.3 Fixed bed. *Source: P19. Article 2.1.2. Technical Study Report on Biomass Fired Fluidized Bed CombustionBoiler Technology for Cogeneration. UNEP-DTIE Energy Branch http://www.unep.org/climatechange/mitigation/Portals/93/documents/EnergyEfficiency/FBC_30_sep_2007.pdf.*

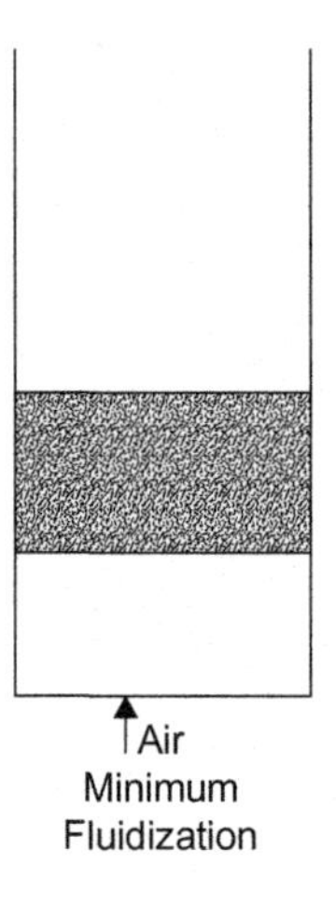

Figure 5.4 Minimum fluidization.

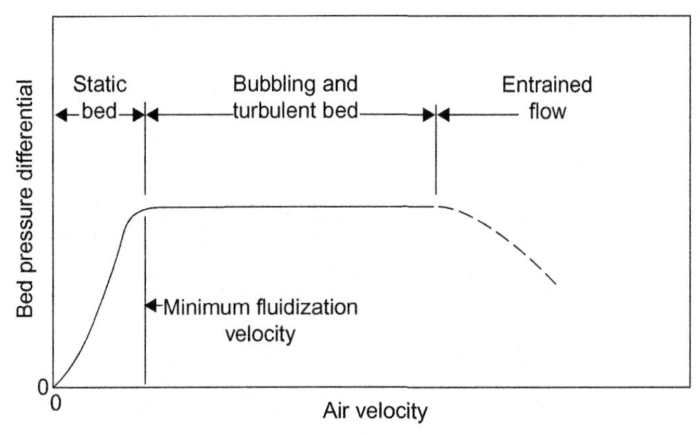

Figure 5.5 Gas pressure drop vs gas velocity. *Source: Fig. 2, Page 17-2, Fluidized-Bed Combustion. STEAM ITS GENERATION AND USE. 41st Edition. Babcock & Wilcox Company 2005.*

The velocity of the fluid at which the fixed bed transforms into the fluidizing bed is called the *minimum fluidization velocity*, denoted by v_m, and it depends on the following factors:

i. Particle diameter
ii. Particle density
iii. Particle shape
iv. Density of fluid
v. Viscosity of fluid
vi. Void fraction, etc.

The minimum fluidization velocity for spherical particles is given by [8]

$$v_m = \{\mu/(d_m * \rho_f)\}\{(C_1^2 + C_2 * Ar)^{0.5} - C_1\} \quad (5.1)$$

where

v_m: Minimum fluidization velocity, m/s
μ: Viscosity of the fluid, kg/m.s
d_m: Mean diameter of particles, m
ρ_f: Density of the fluid, kg/m^3
ρ_p: Particle density, kg/m^3
C_1: Empirical constant = 27.2
C_2: Empirical constant = 0.0408
Ar: Archimedes number

$$= \{\rho_f(\rho_p - \rho_f) * g * d_m^3\}/\mu^2 \quad (5.2)$$

If the viscosity term of the fluid is neglected, then equating the drag force with the gravity force the minimum fluidization velocity for spherical particles can be calculated from Eq. 5.3:

$$C_D \frac{\pi}{4} d_m^2 \rho_f \frac{v_m^2}{2} = \frac{\pi}{6} d_m^3 \rho_p g \tag{5.3}$$

where C_D is drag coefficient, a function of the particulate shape and the Reynolds number, which is dimensionless [1]. (Typical value = 0.6.)

Solving Eq. 5.3,

$$v_m = \sqrt{\left(\frac{4}{3} \frac{1}{C_D} \frac{\rho_p}{\rho_f} d_m g\right)} \tag{5.4}$$

As fluid flow increases further, particles accelerate. A physical condition is reached when the buoyant force and drag force of the particles balance the gravity force and relative velocity between the fluid and the particles reached a limiting value. This limiting relative velocity is called the *terminal velocity* (v_t), which is expressed mathematically for the range of Reynolds number 0.4–500 as [8]

$$v_t = \{\mu/(d_m * \rho_f)\}(Ar/7.5)^{0.666} \tag{5.5}$$

A further increase in the fluid flow causes the bed to become less uniform, bubbles of fluid to form, and the bed to become violent. This is called a *bubbling fluidized bed* (BFB), which is shown in Figures 5.6 and 5.7. The volume occupied by the fluid-solid mixture increases substantially. For this case, there is an easily seen bed level and a distinct transition between the bed and space above.

By increasing the fluid flow further, the bubbles become larger and begin to coalesce, forming large voids in the bed. The solids are present as interconnected groups

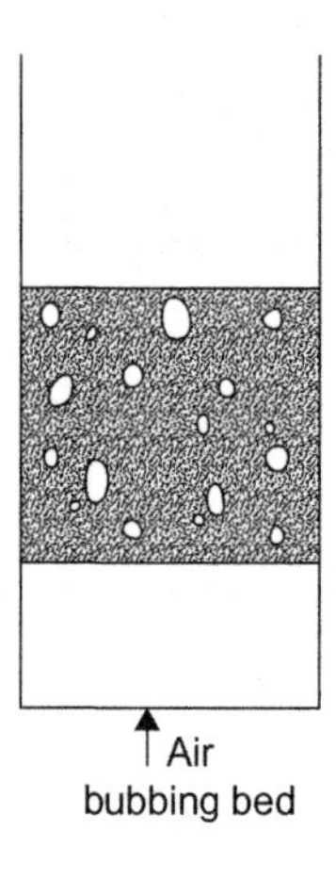

Figure 5.6 Bubbling fluidized bed.

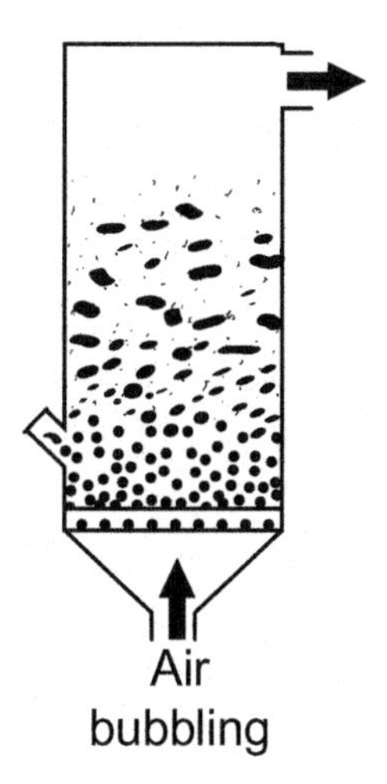

Figure 5.7 Bubbling fluidized bed.

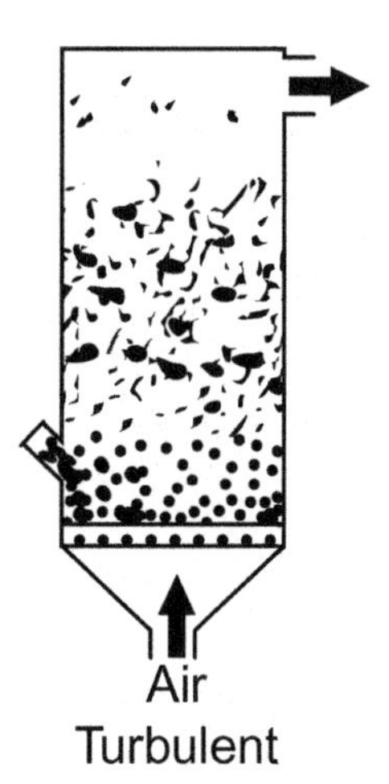

Figure 5.8 Turbulent fluidized bed.

of high solids concentration. This condition is called a turbulent fluidized bed (Figure 5.8).

If the solids are caught, separated from the fluid, and returned to the bed they will circulate around a loop [7]. This type of system is called a *circulating fluidized bed* (CFB) and is shown in Figures 5.9 and 5.10. Unlike the bubbling bed, there is no distinct transition between the dense bed in the bottom of the container and the dilute zone above. The solids concentration gradually decreases between these two zones.

A commercial fluidized bed is comprised of sand, limestone, or sorbent, refractory, or ash along with only 3–5% coal. Above the fluidized bed there is provision for

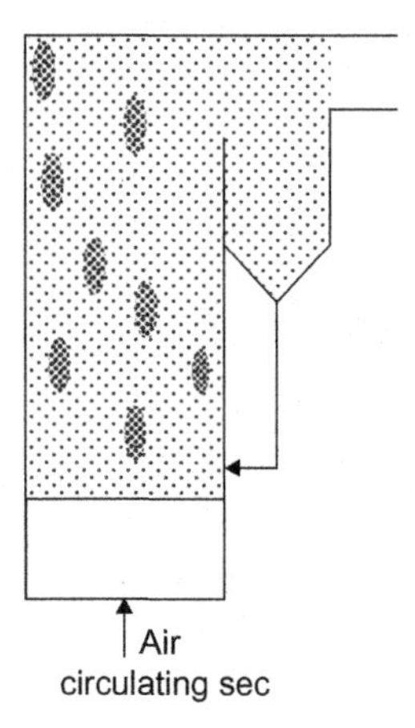

Figure 5.9 Circulating fluidized bed.

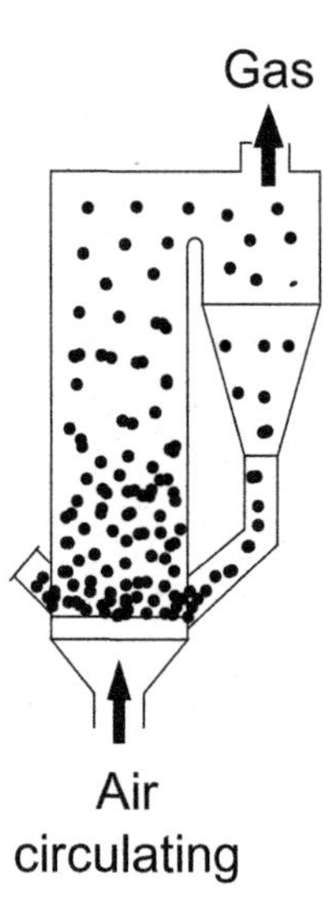

Figure 5.10 Circulating fluidized bed.

sufficient free-board height to prevent excessive elutriation of bed material or unburned carbon at the operating fluidizing velocity. The main feature of a fluidized bed is that throughout the combustor a homogeneous temperature is maintained at around 1073–1223 K. Thus, sintering of the bed is avoided, which takes place once the temperature exceeds 1223 K while burning coal. The above temperature range also suggests that combustion should take place at a minimum temperature of 1073 K below which combustion efficiency declines.

In most fluidized-bed boilers a low excess air level is achieved by providing an in-bed tube bundle designed to extract the heat released in excess of the heat removed by the gas. The majority of the heat in the gas is recovered in convective heat transfer surfaces. A fluidized bed of particles is capable of exchanging heat very effectively with fluidizing gas because of the very large surface area exposed by the particles. With its relatively low heat capacity the gas will rapidly approach the temperature of the solids it is fluidizing. It is the temperature of the solids that dominates the bed thermal behavior because of its high heat capacity, and it is the gas temperature that follows temperature of the particles.

Fuel flexibility and reduced emissions are the benefits utilities can gain by adopting FBC. FBC is appropriate for firing a variety of solid fuels, e.g., coal, petro-coke, biomass, wood, coal washery rejects, etc., with varying heating value, ash content, and moisture content in the same unit. The calorific value can be as low as 7.5 MJ/kg.

In FBC, pollutants in products of combustion are reduced concurrently with combustion – much of the ash and hence the particulate matter is removed during the combustion process, including sulfur. In addition, FBC occurs at much lower temperatures than the combustion temperatures of 1573–1773 K required in pulverized coal-fired boilers, thus resulting in lower production of NO_X. This reduced combustion temperature also helps avoid slagging problems with inferior grades of coal. The resulting ash of this boiler is less erosive in nature. Oil support requires a much lower load, about 25%, than that required in pulverized coal-fired boiler, which is about 40%.

The residues from FBC consist of mineral matter content of the as-fired fuel, as well as residual sorbent, i.e., $CaO/MgO/CaSO_4/CaCO_3$. High free lime may also be present in the residues. As a result strongly alkaline matter will leach from the bed.

5.3 CATEGORIES OF FBC

There are two basic categories of FBC: atmospheric fluidized-bed combustion (AFBC) and pressurized fluidized-bed combustion (PFBC). NFPA 85 defines AFBC as "a fuel-firing technique using fluidizing-bed operating at near-atmospheric pressure on the fire side" [9]. The AFBC option is further comprised of two types: bubbling fluidized-bed combustion (BFBC) and CFBC. In AFBC, steam is generated for power production using the conventional Rankine cycle.

A fluidizing bed in which the fluidizing velocity exceeds the terminal velocity of individual bed particles and in which part of the fluidizing gas passes through the bed as bubbles is called a *bubbling fluidized bed.* In a *circulating fluidized bed* the fluidizing velocities substantially exceed the terminal velocity of the individual bed particles where the bed and the free board above the bed cannot be distinguished and become a uniform fluidized bed.

PFBC units operate at 1.0–1.5 MPa pressure and offer the potential of smaller boilers in comparison with AFBC [10]. In a PFBC boiler, the combustion gas temperature at the exit of the boiler is in the range of 1088–1143 K, which is further used to drive a gas turbine. PFBC boilers are normally used in combined cycle units [11]. Combustion efficiency of PFBC is better than that of AFBC. As with AFBC, two configurations, i.e., bubbling bed and circulating bed, are also possible in PFBC.

Out of above four configurations of FBCs, atmospheric CFBC finds the widest application globally.

5.3.1 Advantages

The advantages of FBC boilers over conventional boilers are:

i. Combustion efficiency of FBC is comparable to that of conventional boilers because of effective contact among heated bed material, gas, and solid fuel even though the bed temperature is low.

ii. An environmentally attractive feature of FBC is that sulfur dioxide (SO_2) can be removed in the combustion process by adding sorbent (e.g., slaked lime (CaO) or limestone ($CaCO_3$) or dolomite ($CaCO_3 \cdot MgCO_3$)) to the fluidized bed, eliminating the need for an external scrubber or flue gas desulfurization (FGD) plant. The sorbent, in combination with oxygen from the supplied air, absorbs sulfur dioxide according to the following reactions:

$$CaO + SO_2 + \frac{1}{2}O_2 + 2H_2O = CaSO_4 \cdot 2H_2O \tag{5.6}$$

$$CaCO_3 + SO_2 + \frac{1}{2}O_2 + 2H_2O = CaSO_4 \cdot 2H_2O + CO_2 \tag{5.7}$$

$$CaCO_3 \cdot MgCO_3 + SO_2 + \frac{1}{2}O_2 = CaSO_4 \cdot MgO + 2CO_2 \tag{5.8}$$

The maximum rate of this reaction is achieved at bed temperatures between 1088 and 1143 K, even though a practical range of fluidized-bed operation of 1023–1223 K is common [1]. All lime and limestone units are designed for 90–95% SO_2 removal. Magnesium-enhanced lime systems are designed for 95–98% removal [12]. The compound $CaSO_4 \cdot 2H_2O$, produced in this process is gypsum, which is either regenerated or used in cement plants.

iii. Compared to the combustion temperature of conventional boilers, approximately 1573–1773 K, the combustion temperature of a FBC is quite low. Combustion at lower temperatures has several benefits:

1. Lower temperature minimizes sorbent requirements because the required Ca/S molar ratio for a given SO_2 removal efficiency is minimized in this temperature range (Figure 5.11) [13].

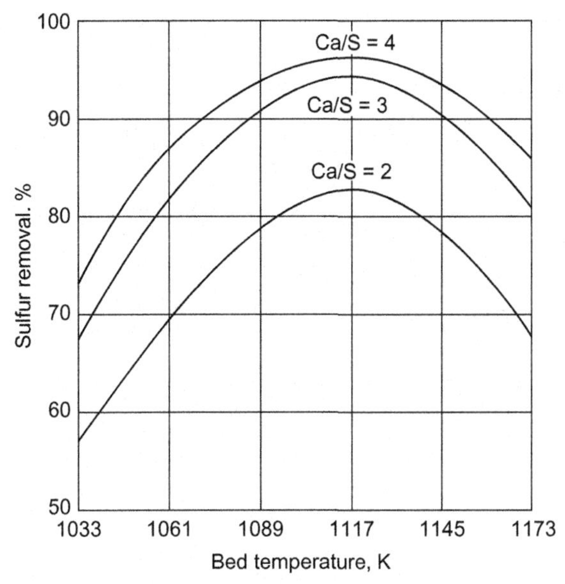

Figure 5.11 Sulfur removal vs bed temperature [13].

2. Bed temperature of 1023–1223 K is well below the ash-fusion temperature of most fuels so the fuel ash never reaches its melting point. The slagging and fouling characteristic of pulverized-coal units are significantly reduced, if not eliminated.
3. Lower temperature reduces NO_X emissions.

iv. FBC can be designed to incorporate the boiler within the bed, resulting in volumetric heat-transfer rates that are 10–15 times higher and surface heat-transfer rates that are 2–3 times higher than a conventional boiler. A fluidized-bed steam generator is therefore much more compact than a conventional one of the same capacity [1].

v. In addition, reduction in SO_2 (and SO_3) in the flue gas means that lower stack gas temperatures can be tolerated because less acid is formed as a result of the condensation of water vapor. Lower stack-gas temperatures result in an increase in overall plant efficiency.

vi. With a suitable ash cooler, bottom ash is available at a comparatively lower temperature (below 573 K), enhancing boiler efficiency.

vii. Variation in moisture content in fuel does not cause problems as in the case of conventional boilers. When introduced into the bed fuel gets immediately mixed with the bed material and the water of fuel gets vaporized and superheated on admission. As a result, the dried fuel particles reach the ignition temperature and burn in fluidizing air.

viii. Flexibility to use a wide range of fuels in the same boiler. Some of the fuels that can be successfully fired in FBC boilers are coal, peat, washery rejects, lignite, sludge, wood waste, bagasse, straw, husk, bark, paper waste, petroleum cake, biomass, etc.

ix. Even with low-grade fuel FBC boilers will generate rated output.

x. Fine coal of size below 6 mm [14], which are difficult to burn in conventional boilers, can be efficiently burnt in FBC boilers.

xi. Lower combustion temperature, softness of ash, and low particle velocity result in less corrosion and erosion effects.

xii. Due to high turbulence in the bed quick start-up and shut down are feasible.

xiii. Load following is more appropriate for automation because of inherent high thermal storage.

xiv. The boiler is more compact and thus less expensive.

xv. Maintenance cost is low since routine overhauls are infrequent.

5.3.2 Disadvantages

Some of the typical problems experienced during operation of FBC boilers include:

i. Size and shape of the particle plays a vital role. Over-size particles result in improper fluidization and impair the combustion process. Thus efficiency of boiler gets reduced. Too large a particle may lead to smoldering.

ii. In the event coal particles become under-sized, some of them may escape the free-board (firing) zone, resulting in burning of fine carbon particles around the cyclone, consequently there will be excursion of flue gas temperature beyond the furnace zone.

iii. Fluidizing nozzles sometimes get plugged, inhibiting proper fluidization of the fuel bed.

iv. It is very important to maintain optimum fluidizing velocity. In the event that velocity is inadequate incomplete combustion will result.

When the fluidizing velocity exceeds the limit, the heat transfer coefficient drops, causing the bed-wall heat transfer to fall. Erosion may occur in radiant superheater and convective heat-exchanging surfaces.

v. Gradual and undetected wear of boiler parts.

vi. At times, bed ash gets accumulated when removed manually, and bed height exceeds the permissible limit. A higher bed height may increase bed resistance, resulting in improper fluidization of the bed. As a result, the bed temperature may exceed the recommended upper limit, causing sintering of bed ash and associated problems. A higher bed height also enhances bed pressure.

vii. Bed temperature may exceed the recommended upper limit if the loop-seal air supply is inadequate and/or primary airflow is low. Low primary airflow may also cause the wind-box pressure to rise.

5.4 DESIGN VARIABLES

For the design of a fluidized-bed boiler the first step is to calculate the fuel-feed rate according to the required thermal output of the boiler with a targeted efficiency. Then the stoichiometric air supply requirement is calculated, which is further enhanced by the design excess air quantity. It is also essential to estimate the net heat release rate and its utilization to arrive at the temperature of the flue gas. The heat transfer from this hot gas to various heat-absorbing surfaces is then calculated. The next step is to find out the amount of sorbent requirement to ensure sulfur capture as per environment-protection standards.

The bed temperature is greatly influenced by fuel composition. For high-sulfur fuel, e.g., pet-coke, the bed temperature has to be maintained at around 1123 K for optimum sulfur capture. While low-sulfur low reactivity fuel, like anthracite culms, can be safely operated at a higher temperature to ensure high combustion efficiency. In the case of low-grade fuel less heat is absorbed in the fuel bed and bulk of the heat is absorbed in the convective surface.

Bed pressure drop resulting from static weight of solids in the bed is calculated as [1]

$$\Delta P = H(1-\alpha)\rho_p \tag{5.9}$$

where

H: Height of the fixed bed, m
α: Fraction not occupied by solids (voids) in fixed bed, dimensionless (the typical value is 0.40, for particles of equal diameter)
ρ_p: Particle density, kg/m^3

The air-distributor grate of CFBC is less critical than that of BFBC, since erosion of bed internals and heat exchanger tubes is a major concern for BFBC. In CFBC, air-distributor grates are narrow, deep, and operate at much higher velocity to distribute the fluidizing air uniformly over the cross-section of the bed.

Solid wastes produced from fluidized-bed boilers require special design attention, since the total quantity of solid waste is much higher than that produced from an equivalent-size pulverized coal-fired boiler.

The amount of sorbent, e.g., limestone, dolomite, etc., required depends on the sulfur content of the fuel, bed temperature, properties of ash content of fuel, emission restrictions, particle size distribution, reactivity of sorbent, etc. Figure 5.12 shows how the amount of sorbent is affected by the bed temperature and sulfur capture rate [15].

Considering all of the above aspects as well as ensuring optimum performance of the FBC boiler, the design variables, which primarily affect efficiency, sulfur capture, and operational flexibility, are as given in Table 5.2.

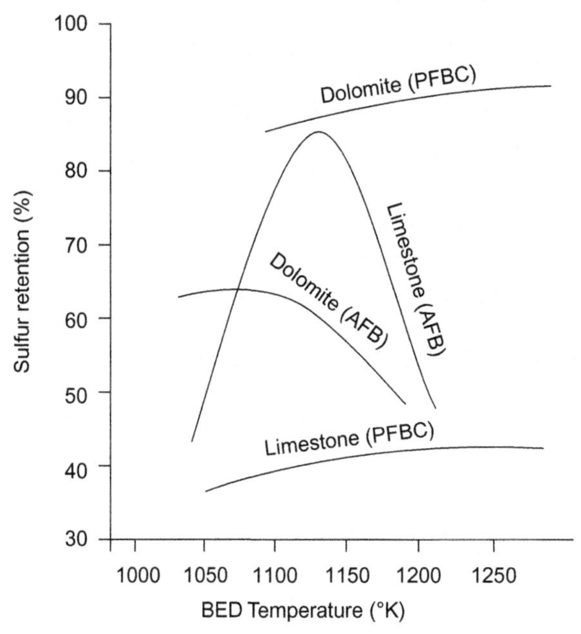

Figure 5.12 Sulfur retention vs bed temperature. *Source: Fig. 2, Page 45, P. Chattopadhyay. Emissions Control In CFB Boilers. Proceedings of the 4th National Workshop & Conference on CFB Technology and Revamping of Boilers in India. Mechanical Engineering Department – Bengal Engineering College (A Deemed University).*

Table 5.2 Design variables

Sl. No.	Parameters	Principal effect
1.	Fluidizing Air Velocity (Superficial Velocity)	Mixing of Bed Material, Elutriation of Fines, Erosion of Bed Tubes
2.	Bed Depth (BFBC)	Residence Time, Fan Power Consumption
3.	Free Board Height	Elutriated Fines, Staged Combustion
4.	Coal-feed Size	Combustor Design
5.	Excess Air	Combustion Efficiency, Heat Loss
6.	Calcium/Sulfur (Ca/S) Ratio	Desulfurization Efficiency, Combustion Efficiency
7.	Gas Recirculation	Turn–Down, Part Load Efficiency
8.	Grit/Ash Recirculation	Desulfurization Efficiency, Combustion Efficiency
9.	Bed Temperature	SO_2 and NO_X emission

5.4.1 Basic parameters

In addition to the design variables discussed in the previous section the basic parameters that need to be considered during the design stage of BFBC and CFBC boilers are given in Table 5.3.

Table 5.3 Basic parameters

Sl. No.	Description	BFBC boiler	CFBC boiler
1.	Fluidizing Velocity, m/s	1.2–3.7	3.7–9.0
2.	Coal-feed Particle Size (max), mm	Under Bed Feed: 6–10 Over Bed Feed: 20–32	6–10
3.	Mean Bed Particle Size, μ	500–1500	150–500
4.	Depth of Bed, m	0.9–1.5	No Distinguished Bed Level. Suspension Density Close to BFBC at Bottom and Gradually Thinning Toward Top
5.	Bed Pressure Drop, kPa	20–25	10–18
6.	Bed-Surface Heat Transfer Coefficient, W/m^2K	200–550	100–250
7.	Grate Heat Release Rate, $MWth/m^2$	0.5–1.5	3–5
8.	Entrainment Rate, kg/m^2s	0.1–1.0	10–40
9.	Combustion Efficiency, %	90–96%	More than 98%
10.	Boiler Efficiency (based on HHV), %	82–85%	86–87%
11.	NO_x Emission, ppm	300–400	100–200
12.	Calcium/Sulfur Ratio	3.0–4.5	1.5–2.0
13.	Excess Air, %	20–25	10–20
14.	Preferable Boiler Size, tph	Below 100	Above 50
15.	Turn – Down Ratio	With Slumping – 5:1 Without Slumping – 2:1	3:1–4:1
16.	Number of Coal-feed Points	Under Bed Feed – 1 Per m^2 of Bed Area Over Bed Feed – 1 Per Bed Compartment	1 Per 10–30 m^2 of Bed Area
17.	Moisture in Coal	Under Bed Feed – Coal Pipe Choking Over Bed Feed – No Problem	No Problem
18.	Excess Fines in Coal	Under Bed Feed – No Problem Over Bed Feed – Affects Efficiency	No Problem
19.	Agglomeration	Some	No problem
20.	Start-Up Time	High	Low
21.	Oil Support	Not Required	Required Below 25% Load

5.5 BUBBLING FLUIDIZED-BED COMBUSTION (BFBC) BOILERS

In a bubbling fluidized-bed unit the fluidizing velocity is low (1.2–3.7 m/s). As a result, the particles are held mainly in a bed that has a depth of 1 m or so and has a definable surface, as already discussed in section 5.2. The combustion of heat is absorbed from the gas by the heat transfer tubes immersed in the bed and by a conventional water-wall surface. The in-bed tubes can serve as either a steam-generation surface or a superheat surface and are used to control the bed temperature. Although an in-bed heat exchanger is a possible source of erosion problems, the close contact with the bed materials and excellent mixing provide high heat transfer. Heat transfer in the convection pass is similar to that for a conventional stoker-fired or pulverized coal-fired steam generator.

Figure 5.13 shows a bubbling-bed combustor, where the hot-bed material is in a state of suspension and behaves like a bubbling liquid by fluidized air. The bed temperature (1073–1173 K) is controlled by varying the amount of fuel and air within the bed.

As coal particle size decreases to about 250 micron, either as a result of combustion or attrition, the particle is elutriated from the bed and carried out of the combustor. These elutriated particles are then removed as fly ash. This loss of unburned carbon results in a combustion efficiency of 85–90%. By recycling grit and ash from the cyclone separator back into the bed, the combustion efficiency can be increased to 96%.

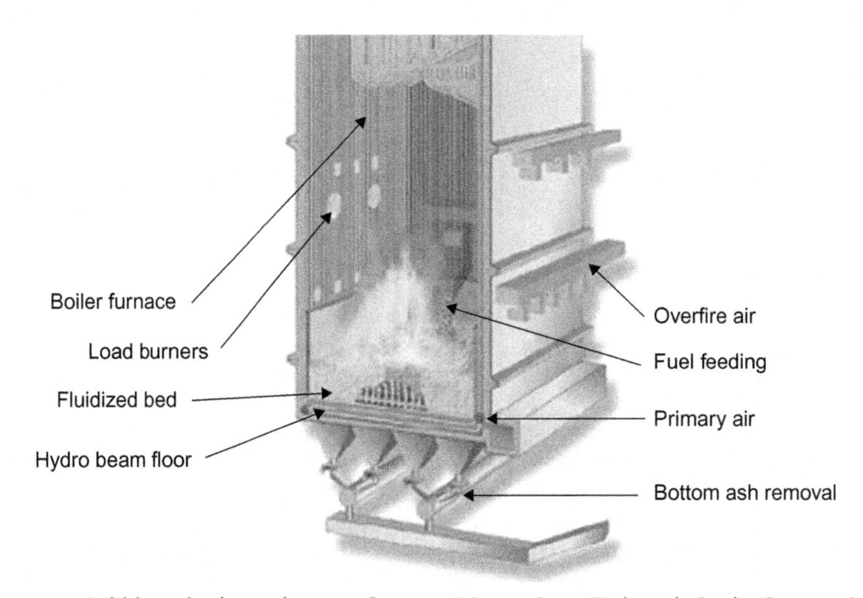

Figure 5.13 Bubbling bed combustor. *Source: FIG. 5, P20, Technical Study Report: Biomass Fired FBC Boiler for Cogeneration http://www.unep.org/climatechange/mitigation/Portals/93/documents/EnergyEfficiency/FBC_30_sep_2007.pdf.*

Sand is often used for bed stability, together with limestone or dolomite for SO_2 absorption. Since combustion takes place at 1073–1173 K, generation of NO_X is also low as compared to that of a pulverized coal-fired boiler. Flue gases are normally cleaned using cyclone, then gases pass through further heat exchangers. The fuel-feed size in this bed is restricted to 30 mm in top size, with 20% about 6 mm. The BFBC unit size is generally of the order of 25 MWe, although larger sizes are also of use for retrofitting existing units.

The unburned carbon level is higher than that of a pulverized coal-fired boiler. Another disadvantage of BFBC is that to remove SO_2, a much higher Ca/S ratio (resulting in higher limestone or dolomite consumption) is needed than for CFBC. This increases costs, in particular the cost of residue disposal. Other disadvantages are higher carbon loss, higher fan power consumption, erosion of bed tubes, etc. Chloride content in fuel accelerates erosion.

The bottom ash of this boiler comprises about 30–40% of the total ash, with the remaining being fly ash.

There are two types of fuel-feeding systems in a BFBC boiler, e.g., under-bed feed and over-bed feed (Figure 5.14) [12]. A comparison of these two types is given in Table 5.4.

BFBC boilers are mainly used in township heating systems or in industrial applications. Gas flow path of a typical BFBC boiler is given in Figure 5.15 [7].

Compared to a stoker-fired boiler a BFBC boiler has the following advantages:

- **i.** Reduction of furnace slagging
- **ii.** Reduction of convection pass fouling
- **iii.** Increased boiler availability
- **iv.** Increased boiler efficiency

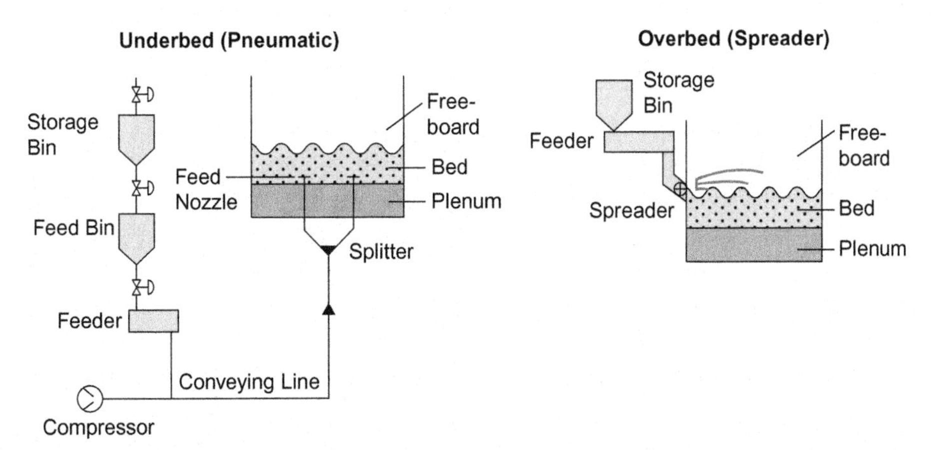

Figure 5.14 Fuel feeding in BFBC boiler. *Source: Fig. 3-49, Page 3-55, Boiler, Clean Combustion Technologies, 5th Edition ALSTOM, 2000.*

Table 5.4 Comparison of coal-feeding systems in BFBC boilers

Sl. No.	Description	Under-bed feed	Over-bed feed
1.	Coal-feed size (max), mm	6–10	20–32
2.	Number of coal-feed point	1 per m^2 of bed area	1 per compartment
3.	Coal pipe choke and erosion problem	Yes	No
4.	Sorbent consumption	Less	More
5.	Sensitivity on fines	Less	More
6.	Combustion efficiency without recycle	89%	86%
	Combustion efficiency with recycle	>95%	>95%
7.	Power consumption (crusher, fan)	More	Low
8.	Operation simplicity	Complex	Simple
9.	Maintenance on coal path	More	Less
10.	Plugging due to moisture in coal	More	Less
11.	Turn down	Low	High

5.6 CIRCULATING FLUIDIZED-BED COMBUSTION (CFBC) BOILERS

In a circulating fluidized-bed unit, the bed material is comprised of fuel, fuel ash, sorbent, and other inert bed materials. The bed is supported within the furnace by air flowing into the bed from the bottom of the furnace. The air flow supports the bed and ensures complete combustion by providing close mixing of fuel and air.

This type of boiler is capable of burning low volatile content, typically 8–9%, fuel, e.g., pet-coke, and fuels with low ash-melting temperature, e.g., wood, biomass. It can also burn fuels with ash content as high as 70%, e.g., coal washery rejects. Fuels with high moisture content, such as lignite, can also be burnt in this boiler.

In a circulating fluidized-bed unit, combustion of heat is absorbed from the gas by a conventional water-wall surface, by platens located in the upper region of the combustor, or by a heat transfer surface located in external heat exchangers.

The velocity of the gas in a CFBC boiler is relatively higher than that of the solids. The high slip velocity (the difference between the mean gas velocity and the mean solid velocity; see Figure 5.16) [2], in combination with long residence and contact times and intense mixing, results in higher heat and mass transfer rates and higher combustion efficiency in CFBC than are available with BFBC.

The circulating fluidized bed, as shown in Figure 5.17 [16], has been developed to improve combustion efficiency and reduce sorbent consumption for required level of desulfurization. In this boiler, higher fluidizing air velocity (3.7–9.0 m/s) results in the coal and inert bed material being entrained and carried through the combustor and over-board region into the hot cyclone. Thus, a distinct dense fluidized bed does not exist at

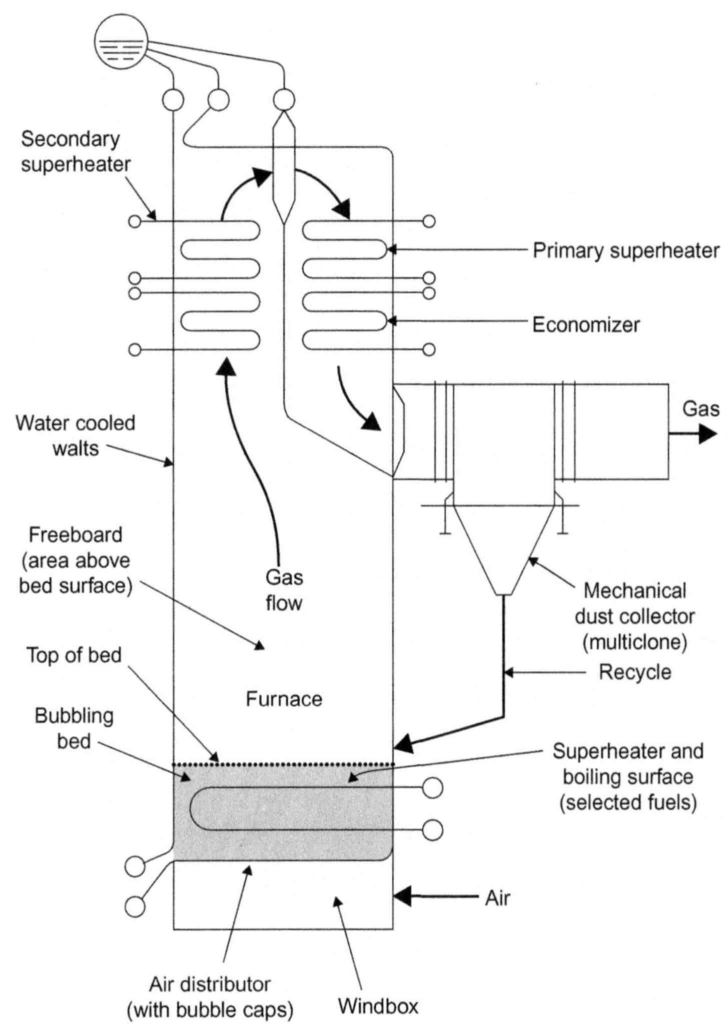

Figure 5.15 Gas flow path of a typical BFBC boiler. *Source: Fig. 3, Page 17-2, Fluidized-Bed Combustion. STEAM ITS GENERATION AND USE. 41st Edition. Babcock & Wilcox Company 2005.*

higher load in this case. The coarse particles of sorbent and unburned coal are recovered in the cyclone and are recycled to the combustor. Individual particles may recycle anything from 10–50 times, depending on their size, and how quickly the char burns away.

The primary air is injected below the air-distributor plate located at the furnace floor and the secondary air is injected at a certain height above the furnace grate to ensure complete combustion. Fuel particles are burnt in the furnace to release heat. Part of the heat released is absorbed in the in-bed water-steam surface and the bulk of the remaining heat is absorbed in the convective heat transfer area.

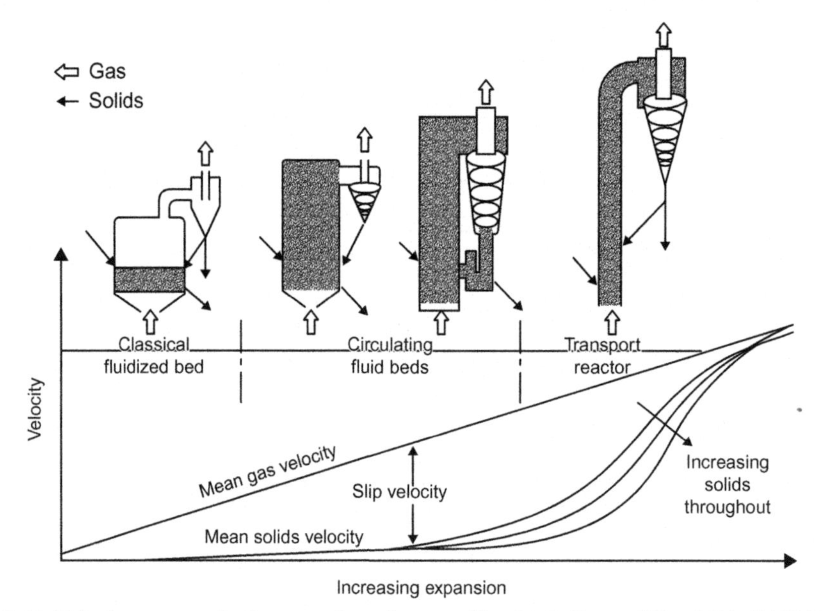

Figure 5.16 Velocity versus bed expansion. *Source: Fig. 21-1, Page 690, CH21 FLUIDIZED BED COMBUSTION. Black & Veatch. Power Plant Engineering. CBS Publishers & Distributors. Chapman & Hall, Inc. New York 1996.*

The combustion zone extends beyond the furnace into the hot cyclone. The furnace heat release rate in a CFBC boiler is of the order of 3.5–4.5 MW/m^2, which is compatible with that of a pulverized coal-fired boiler. For a bed burning bituminous coal, the carbon content of the bed is only 1%, while the rest of the bed is made up of ash, together with sand and/or lime and calcium sulphate. Because of the small particle size and recycling of solids back to the combustor, the CFBC boiler is able to achieve a combustion efficiency over 98% and a Ca/S ratio of 2.5 for 90% sulfur capture (Figure 5.11). The fuel-feed size in this boiler can be as low as 1.5 mm to 10 mm top size.

Contrary to a BFBC bed, a CFBC bed is not provided with steam-generating tubes. To capture and recycle large bed materials CFBC boilers are provided with big cyclone separators, requiring the boiler to be very tall (Figure 5.18.).

The CFBC boiler is used in number of units around 250–300 MW in size. In a CFBC unit, heat losses from the cyclone(s) are considerable. Thus, the thermal efficiency of a CFBC boiler is 3–4% lower than an equivalent-size pulverized coal-fired boiler. The operating performance of the CFBC boiler shows that whenever there is a need to increase load on the boiler it can be achieved by raising the bed temperature and keeping the bed material height unchanged (Figure 5.19) [15]. From the operating performance it is further noted that if the bed temperature is maintained the boiler load is linearly related with the bed material height (Figure 5.20) [15].

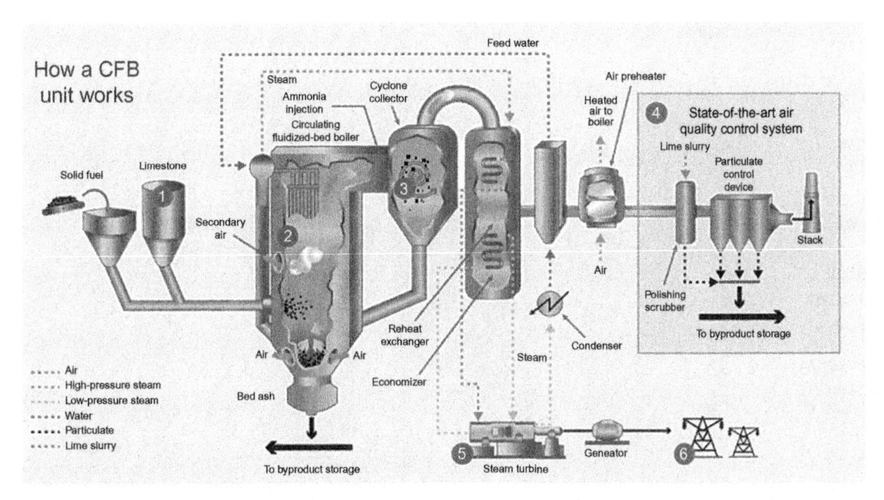

1. Fuel input

Fuel and limestone are fed into the combustion chamber of the boiler while air (primary and secondary) is blown in to "fluidize" the mixture. The fluidized mixture burns at a relatively low temperature and produces heat. The limestone absorbs sulfur dioxide (SO_2), and the low-burning temperature limits the formation of nitrogen oxide (NO_x)–two gases associated with the combustion of solide fuels.

2. CFB boiler

Heat from the combustion process boils the water in the water tubes turning it into high-energy steam. Ammonia is injected into the boiler outlet to further reduce NO_x emissions.

3. Cyclone collector

The cyclone is used to return ash and unburned fuel to the combustion chamber for re-burning, making the process more efficient.

4. State-of-the-art air quality control system

After combustion, lime is injected into the "polishing scrubber" to capture more of the SO_2. A "baghouse" (particulate control device) collects dust particles (particulate matter) that escape during the combustion process.

5. Steam turbine

The high-pressure steam spins the turbine connected to the generator, which converts mechanical energy into electricity.

6. Transmission lines

The electricity produced from the steam turbine/generator is routed through substations along transmission lines and delivered to distribution systems for customer use.

Figure 5.17 Circulating bed combustor. *Source: FIG. 7, P23, Technical Study Report: Biomass Fired FBC Boiler for Cogeneration http://www.unep.org/climatechange/mitigation/Portals/93/documents/EnergyEfficiency/FBC_30_sep_2007.pdf.*

The advantages of a CFBC boiler compared to a BFBC boiler are:

a. Higher combustion efficiency
b. Lower SO_2 and NO_X emission
c. Lower Ca/S ratio
d. Lower air requirement
e. No submerged bed tubes
f. No limitation on boiler size
g. Improved gas-solid contact
h. Reduced axial dispersion of gas
i. Reduced cross-sectional area due to higher fluidizing velocity
j. Potentially more control over suspension-to-wall heat transfer because of the ability to use the solids circulation flux as an additional variable

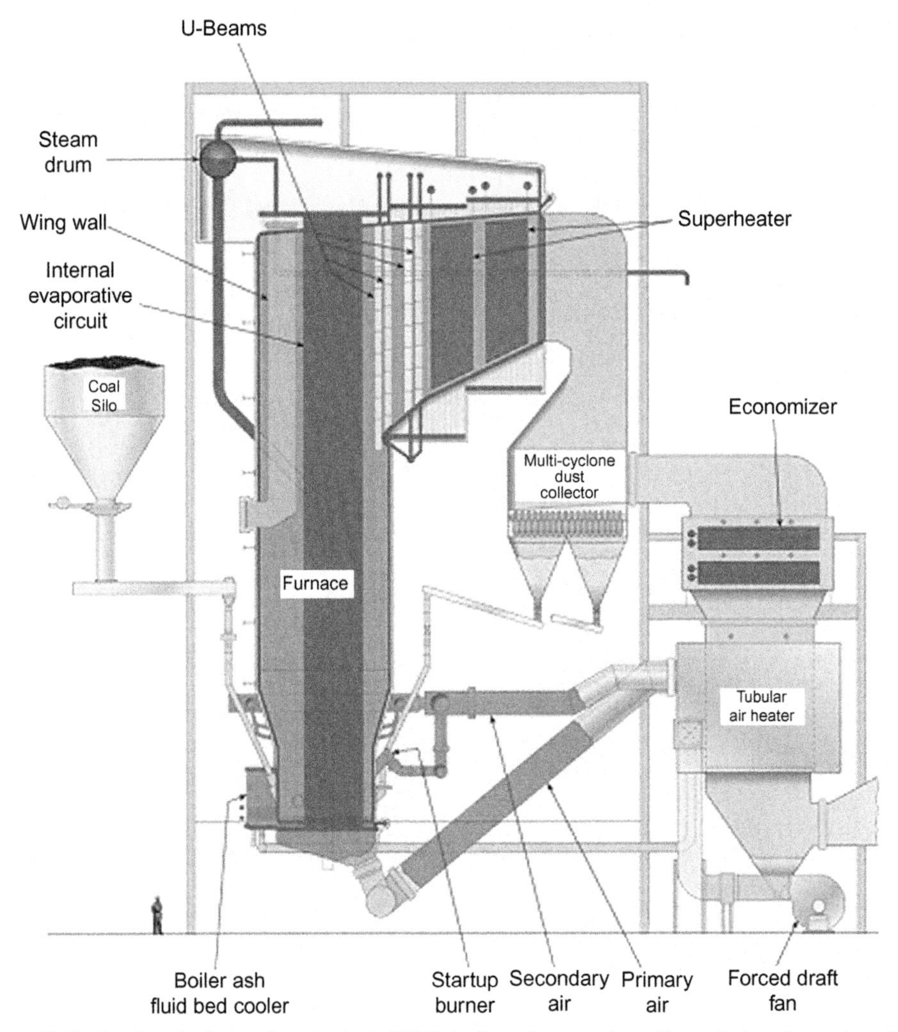

Figure 5.18 Sectional view of a typical CFBC boiler. *Source: http://www.babcock.com/products/Pages/Circulating-Fluidized-Bed-Boiler.aspx.*

k. No region like the free-board region of BFBC, where there can be substantial temperature gradients
l. Less tendency to show particle segregation and agglomeration
m. Easier to have staged processes
n. Because of superior radial mixing, fewer solid feed-points are needed
o. Higher solid flux through the reactor

Some of the disadvantages of a CFBC boiler over a BFBC boiler are:

a. Uneconomical for smaller size unit
b. Higher capital cost

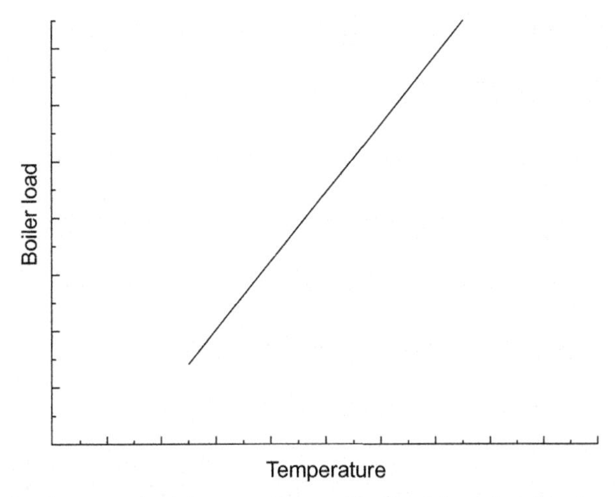

Figure 5.19 Boiler load versus bed temperature with fixed bed height. *Source: Zhengshun Wu, Hanping Chen, Dechang Liu, Jie Wang, Chuangzhi Wu and Haitao Huang. A STUDY OF OPERATION PERFORMANCE OF CIRCULATING FLUIDIZED BED COMBUSTION BOILER. Proceedings of the 4th National Workshop & Conference on CFB Technology and Revamping of Boilers in India, 2003.*

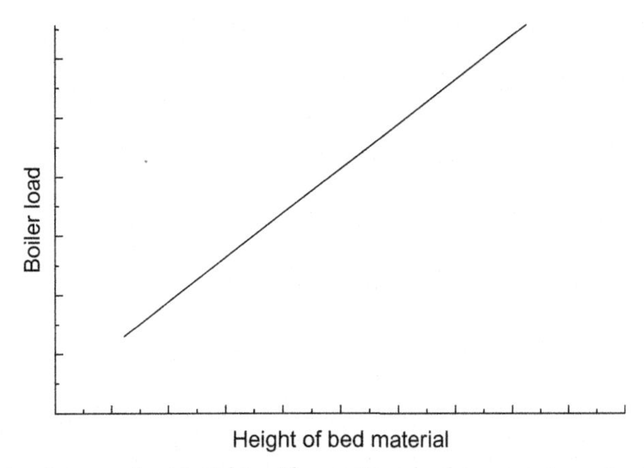

Figure 5.20 Boiler load versus bed height with constant bed temperature. *Source: Zhengshun Wu, Hanping Chen, Dechang Liu, Jie Wang, Chuangzhi Wu and Haitao Huang. A STUDY OF OPERATION PERFORMANCE OF CIRCULATING FLUIDIZED BED COMBUSTION BOILER. Proceedings of the 4th National Workshop & Conference on CFB Technology and Revamping of Boilers in India, 2003.*

c. Erosion of cyclone and superheater tubes
d. Increased overall reactor height
e. Decreased suspension-to-wall heat transfer coefficients for given particles
f. Added complexity in designing and operating re-circulating loop
g. Increased particle attrition

5.7 PRESSURIZED FLUIDIZED-BED COMBUSTION (PFBC) BOILERS

As discussed earlier by burning coal in a fluidized bed at low temperature it is possible to avoid sintering of coal ash and to minimize volatilization of alkali metals in the fuel. Thus, hot gas can be safely used in a gas turbine since the above properties ensure reduction of corrosion and erosion of gas turbine blades. To harness these benefits the pressurized fluidized-bed combustion (PFBC) boiler was developed in the late 1980s and considerable effort was devoted to the development of PFBC during the 1990s. Today, a PFBC is capable of burning all types of coal, even low-grade coal, be it high moisture, high ash, high sulfur, low gross calorific value (GCV) etc. [11].

This type of boiler is used with a combined-cycle system, incorporating both steam and gas turbines [11]. A PFBC boiler generates steam to drive a steam turbine, while gas from the boiler drives a gas turbine. Such combination of Rankine cycle and Brayton cycle along with FBC provides higher cycle efficiency than the efficiency of conventional coal-firing systems, lower emissions, and harmless waste products. The steam turbine produces 80–90% of the generated power, and the gas turbine 10–20%. These boilers operate at pressures of 1.0–1.5 MPa with combustion temperatures of 1073–1173 K and can be used in compact units. Typical size of PFBC boilers is only 1/5 size of atmospheric boilers [11]. The combustion air is pressurized in the compressor section of the gas turbine.

Combustion at high pressure yields the following benefits:

1. More uniform bed temperature distribution
2. Reduced size of boiler island components
3. Lower bed-surface area
4. Reduced maintenance
5. Prevents formation of free lime.

The benefits of PFBC relative to AFBC are:

i. Specific power output of PFBC is higher, and hence the capital cost is less.
ii. Efficiency of the combined cycle is higher than that of the conventional cycle.
iii. Practically all the combustion takes place within the bed and free-board combustion is negligible, facilitating gas turbine operation with ease.

Along with these benefits PFBC also shares the inherent benefits of FBC, i.e., low SO_2 emissions and less formation of NO_X than in a pulverized coal-fired boiler. PFBC units are built to give an efficiency value of over 40% and low emissions. Systems using more advanced cycles should achieve efficiencies of over 45%. As with AFBC, two configurations, e.g., bubbling-bed and circulating-bed, are also possible in PFBC. Currently commercial-scale operating units all use bubbling beds.

The combustor and hot gas cyclones in a PFBC are enclosed in a pressure vessel. Both coal and sorbent, which reduce SO_2 emissions, are fed across the pressure boundary.

The bed material contains 90–95% of coal ash. The pressurized coal combustion system heats steam, in conventional heat transfer tubing, and produces a hot flue gas. Particulates are then removed from the hot flue gases prior to expanding the gases through a gas turbine. The exhaust gases are cooled, generating more steam for power production.

Figures 5.21 and 5.22 show a typical PFBC co-generation plant comprised of a steam generator with its steam turbine, gas turbine, fuel-handling, and ash-handling systems. The fuel is a mixture of crushed coal, sorbent, and water supplied as a paste with the help of piston pumps.

The world's largest PFBC with a capacity of 360 MW in Karita, Japan was commissioned in 2000 and began commercial operations in July 2001 [17]. Global application of PFBC, however, is limited due to following requirement/restrictions:

- **i.** Feeding coal has to be completely dried, virtually free from surface moisture, to maintain the coal-feed rate to the combustor.
- **ii.** The need to pressurize the feed coal, limestone, and combustion air, and to depressurize the flue gases and the ash-removal system introduces significant operating complications.
- **iii.** Heat release with each bed area is much greater in pressurized systems, and bed depths of 3–4 m are required to accommodate the heat-exchange area necessary for the control of bed temperature.
- **iv.** At reduced load, bed material is extracted, so that part of the heat exchange surface is exposed.
- **v.** PFBC has a large thermal capacity. Thus the heating rate of PFBC is much lower than that of the gas turbine, requiring some external arrangement for start-up of gas turbine.
- **vi.** Hot gas from the combustor has to be sufficiently cleaned (from a dust concentration value of 10,000–40,000 ppm at the combustor outlet down to less than 300 ppm at the gas turbine inlet) to avoid erosion of gas turbine blades.
- **vii.** The response of the gas turbine to load change is fast, while that of PFBC is sluggish, and as a result control becomes complicated.
- **viii.** A PFBC power plant has huge auxiliary equipment that needs to be controlled along with many final control elements. Thus, the control system required for such a plant is essentially large in scale.

5.8 START-UP AND LOADING OF PFBC BOILERS

Prior to starting (black start-up) a PFBC boiler the gas turbine generator is started as a synchronous motor taking its power supply from the grid. This facilitates starting of the gas turbine compressor, which supplies combustion air to the furnace. Then gas or oil-fired bed pre-heaters are lit. As the fuel is fed into the furnace both bed

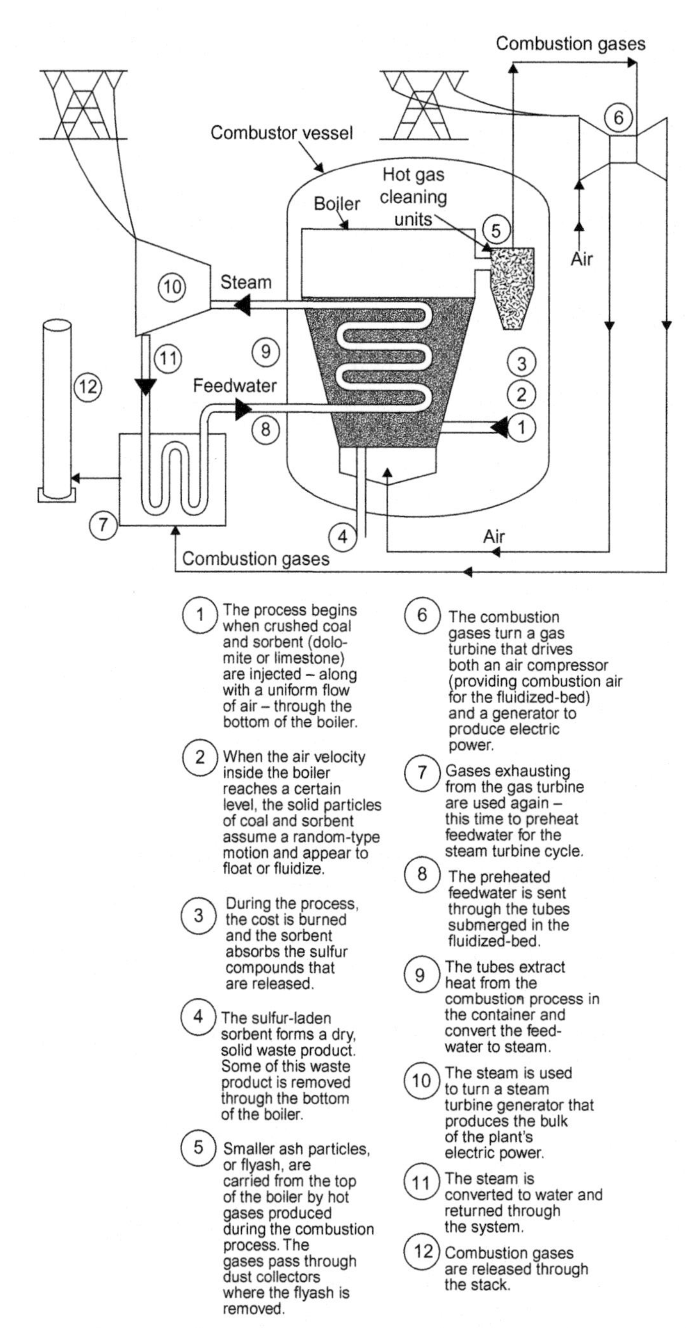

Figure 5.21 PFBC combustor with associated equipment. *Source: FIG. 18, P17-14, CH17. Fluidized-Bed Combustion. STEAM ITS GENERATION AND USE. 41st Edition. Babcock & Wilcox Company 2005.*

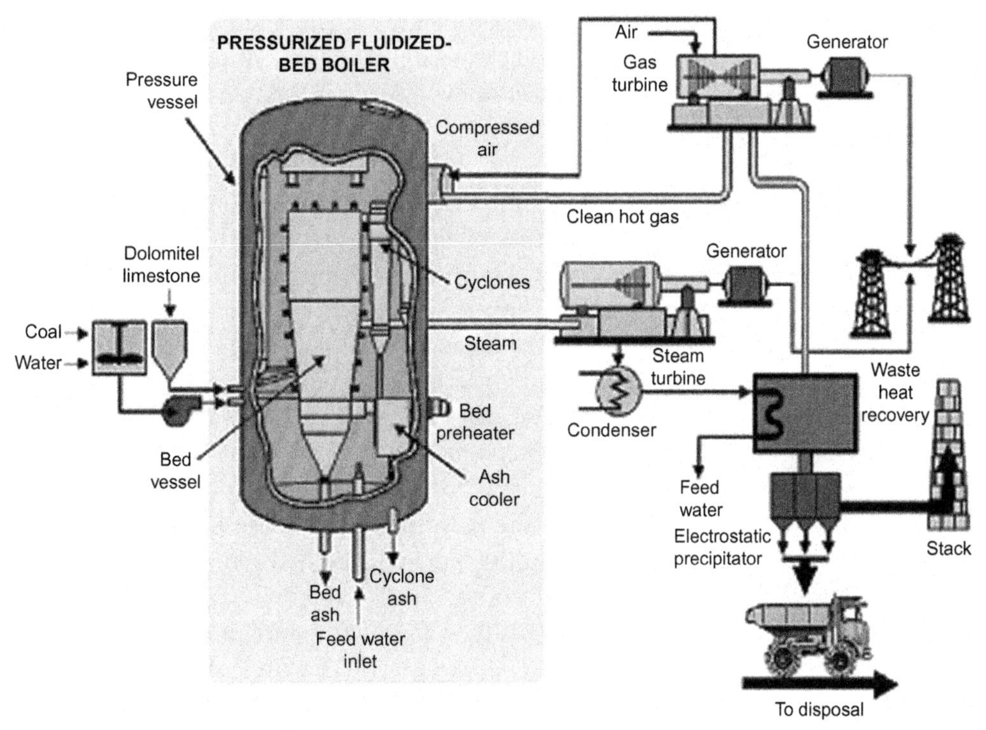

Figure 5.22 PFBC combustor with associated equipment. *Source: http://dc222.4shared.com/doc/ckLfFp5e/preview.html.*

temperature and bed height start rising. When the bed temperature exceeds the ignition temperature of coal (the average temperature value is 923 K), continuous coal feeding to the bed takes place, and the coal will burn rapidly. When the bed temperature becomes uniform the supplementary firing system is gradually withdrawn. The boiler is now in a position to generate enough steam to run the steam turbine. The hot gas from the burning coal after proper treatment is passed through the gas turbine. As the load is increased the gas turbine generator switches from the synchronous motor to a synchronous generator. Both the bed temperature and main steam temperature are controlled by controlling the coal-feed rate and the bed height simultaneously. (Detailed treatment on 'Start-up and Shut Down of Steam Generator' is addressed in Chapter 12.)

5.9 PROBLEMS

5.1 Determine the minimum fluidization velocity of 300 μm sand, if the density of sand is $2500\ kg/m^3$, density of gas is $0.316\ kg/m^3$, and viscosity of gas is $4.49 \times 10^{-5}\ N.s/m^2$.

(**Ans.:** $v_m = 0.0368\ m/s$)

5.2 The mean bed particle size of a fluidized bed is 350 μm and the average density of solid particles is 1900 kg/m^3. Find the terminal velocity of particles, if the gas density is 1.16 kg/m^3 and the gas viscosity is 1.84×10^{-5} $N.s/m^2$.

(**Ans.:** $v_t = 2.305$ m/s)

5.3 Find the Archimedes number, minimum fluidization velocity, and terminal velocity of particles in a fluidized-bed combustion chamber if the mean bed particle size is 400 μm and the average density of the solid particles is 2500 kg/m^3. The maximum and minimum bed temperature is maintained at 1173 K and 298 K, respectively. Other given conditions are:

Gas density, kg/m^3: 0.237 (at 1173 K) and 1.162 (at 298 K)

Gas viscosity, $N.s/m^2$: 4.493×10^{-5} (at 1173 K) and 1.833×10^{-5} (at 298 K)

(**Ans.:** At 1173 K: $Ar = 184.26$, $v_s = 0.065$ m/s, $v_t = 4.000$ m/s At 298 K: $Ar = 5425.84$, $v_s = 0.150$ m/s, $v_t = 3.164$ m/s)

5.4 A fluidized bed combustion chamber is fed with crushed coal of size between 6 mm and 20 mm. Within the bed the mean bed particle size is 300 μm and the average density of solid particles is 1300 kg/m^3. The bed temperature is maintained at 1123 K and the solid height of the bed is 1200 mm. Calculate the minimum gas velocity that fluidizes the coal bed and the pressure drop across the bed. (Assume the coefficient of drag is 0.6, average void fraction of collapsed bed is 0.25, and density of gas is 1.32 kg/Nm^3.)

(**Ans.:** $v_s = 5.15$ m/s, $\Delta P = 11.47$ kPa)

BIBLIOGRAPHY

[1] El-Wakil MM. Power Plant Technology, McGraw-Hill Book Company; 1984.
[2] Black and Veatch. Power Plant Engineering, CBS Publishers and Distributors; 2001.
[3] Rajaram S. Next generation CFBC, *Bharat Heavy Electricals Limited, Tiruchirapalli, India.* Chem Eng Sci 1999;54:5565–71.
[4] Lagisza Power Plant, Poland. <http://www.power-technology.com/projects/lagisza/>.
[5] Ilkka Venäläinen *Foster Wheeler Energia Oy Finland.* Rafa Psik *Foster Wheeler Energia Polska Sp. z o.o. Poland.* 460 MWe Supercritical CFB Boiler Design For Lagisza Power Plant.
[6] Patrick Laffont *ALSTOM Power Plant Segment (France),* Jacques Barthelemy *ALSTOM Power Boilers & Environment Segment (France),* Brendon Scarlin *ALSTOM Power Turbomachinery Segment (Switzerland),* Christian Kervenec *ALSTOM Power I&C Plant Segment (France).* A Clean and Efficient Supercritical Circulating Fluidized Bed Power Plant.
[7] Stultz SC, Kitto JB, editors. Steam its Generation and Use, 41st ed. The Babcock and Wilcox Company; 2005.
[8] Goodall PM, (General ed.), The Efficient Use of Steam. Westbury House; 1980.
[9] National Fire Protection Association, U.S.A. NFPA 85_2004 Boiler and Combustion Systems Hazards Code.
[10] FBC Boilers. Bureau of Energy Efficiency, India; 2010.
[11] PFBC Clean Coal Technology. PFBC Environmental Energy Technology, Inc.
[12] Bozzuto C, editor. Clean Combustion Technologies, fifth ed. Alstom; 2009.
[13] Singer JG, editor. Combustion–Fossil Power, Combustion Engineering Inc.; 1991.

[14] Reznikov MI, Lipov YuM. Steam Boilers of Thermal Power Stations, Mir Publishers, Moscow; 1985.
[15] Saha SK, Gupta KKD, Datta D, Chowdhuri AK, editors. Proceedings of the 4th National Workshop & Conference on CFB Technology and Revamping of Boilers in India. Mechanical Engineering Department – Bengal Engineering College (A Deemed University); 2003.
[16] Karita PFBC Power Plant, Japan. <http://www.power-technology.com/projects/karita/>.
[17] Technical Study Report: Biomass Fired FBC Boiler for Co-generation. UNEP-DTIE Energy Branch. <http://www.unep.fr/energy>; 2007.

CHAPTER 6

Steam Turbines

6.1 INTRODUCTION

The objective of all heat engines is to convert heat energy into mechanical work. In all heat engines some working fluid is used to which heat is applied, causing it to expand and to perform work overcoming external resistances. After completion of the expansion process heat of the working fluid is rejected to a sink. Thereafter, the fluid receives fresh supply of heat from the source.

The working fluid of a steam power plant is "water." The plant is comprised of a *steam generator*, in which water receives heat from the heat of combustion of fossil fuels and gets converted to high-energy steam, which then expands through a *steam turbine* and undergoes changes in pressure, temperature, and heat content. During this process of expansion the high-energy steam also performs mechanical work. The plant also consists of a *condenser*, where steam rejects heat in cooling water and returns to the original state (Figure 6.1). In addition to fossil fuels, energy from gas turbine exhausts, nuclear reactors, solar, biomass, etc., may also be used to generate steam.

Of all stationary prime movers that are used for electric power generation, the steam turbine bears the burden of handling the maximum power demand, which is evident from the status that as of 2012 fossil fuel power plants contributed 67.9% of global electricity generation as against 16.2% by hydropower, 10.9% by nuclear and 5.0% by renewable energy [1]. Compared with other prime movers like equivalent capacity reciprocating engines, the advantages of steam turbines are:

- **i.** Require less floor space
- **ii.** Lighter foundations
- **iii.** Less operator attention
- **iv.** Lower lubricating oil consumption
- **v.** Oil-free exhaust steam
- **vi.** Free from reciprocating vibrations
- **vii.** Great overload capability
- **viii.** High reliability
- **ix.** Low maintenance cost
- **x.** Excellent regulation
- **xi.** Capable of operating with very high steam temperature
- **xii.** Capable of expanding to lower exhaust pressure

Thermal Power Plant
DOI: http://dx.doi.org/10.1016/B978-0-12-801575-9.00006-8

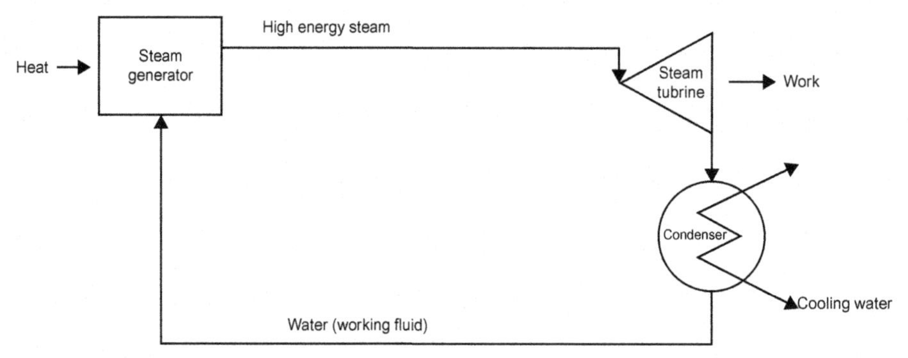

Figure 6.1 Schematic diagram of a thermal plant.

Figure 6.2 Hero's reaction turbine.

xiii. Single unit can be built with greater capacity
xiv. Cost per unit power of large turbines is much less

Historically, it has been believed that an early form of steam turbine utilizing the expansive force of vapor was in existence as far back as 150 B.C. However, the first rotary engine based on the reaction principle was invented by Hero of Alexandria in A.D. 50 (Figure 6.2), and the first impulse turbine was developed by Giovanni Branca in the year

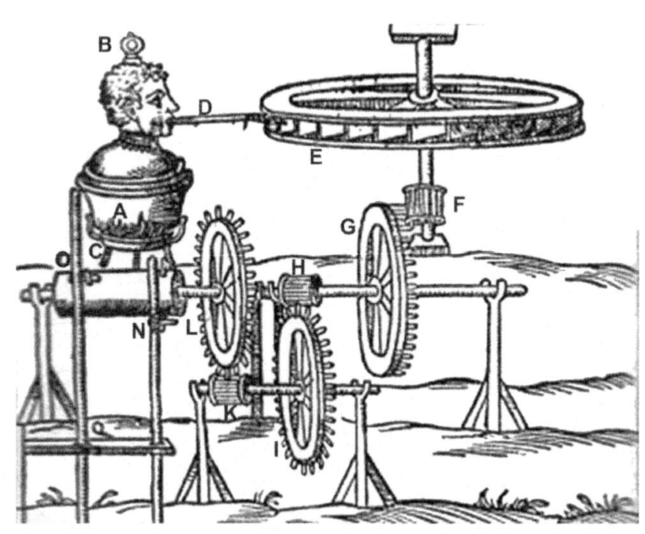

Figure 6.3 Giovanni Branca's impulse turbine. *Source: Fig. 211 P410, CHXIV THE THEORY OF THE STEAM TURBINE [2].*

1629 (Figure 6.3). But it took more than 250 additional years to develop the early form of the present-day impulse steam turbine [2].

In addition to the above inventors, credit also goes to British engineer Charles A. Parsons (1854–1931) for inventing the reaction turbine and to Swedish engineer Gustav De Laval (1845–1913) for inventing the impulse turbine; both worked independently between 1884 and 1889. Modern steam turbines, however, frequently employ both reaction and impulse in the same unit.

In the basic design of a reaction turbine the change in momentum of the steam coming out of the nozzle causes a reaction on the radial tube that made it rotate (Figure 6.4) [3]. The simple impulse turbine, developed by De Laval, revolved at an exceedingly high speed of 30,000 rpm with a steam velocity of about 1 km/s (Figure 6.5) [2]. The change in momentum of the steam from the nozzle creates an impulse on the wheel. Figures 6.6 and 6.7 depict the typical layout and sectional view of a large steam turbine.

A steam turbine is a *constant volume machine,* i.e., the volume of the steam flowing at any point within the machine remains unchanged at all loads and is never affected by a change in load. The change in load causes the governing system of a turbine to change the mass flow of steam through the turbine according to the "load demand." From experimental results it has been established that the relationship between load and steam consumption (i.e., throttle flow of steam) of a turbine is linear in nature and can be represented by the Willans line (Figure 6.8) [3]. The line can be predicted by joining two special points, e.g., throttle flow at the rated-load point and throttle flow at the no-load point. The steam consumption of a condensing type steam turbine at no-load varies from 0.10–0.14 times the steam consumption at full-load.

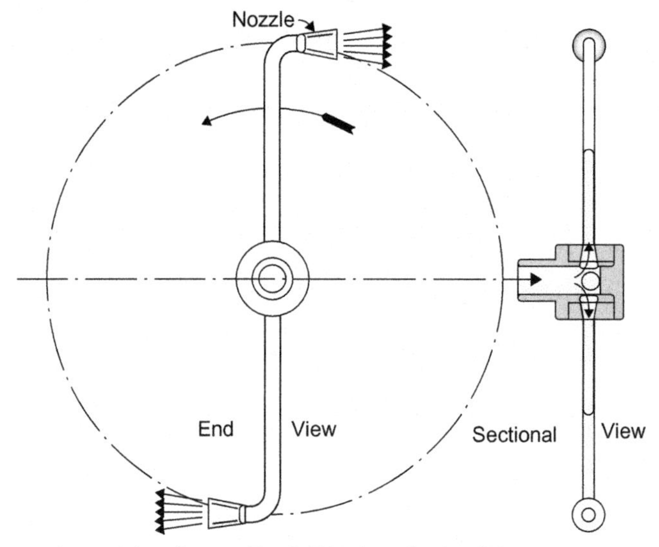

Figure 6.4 Pure reaction turbine. *Source: Fig. 5, P11, Introduction [3].*

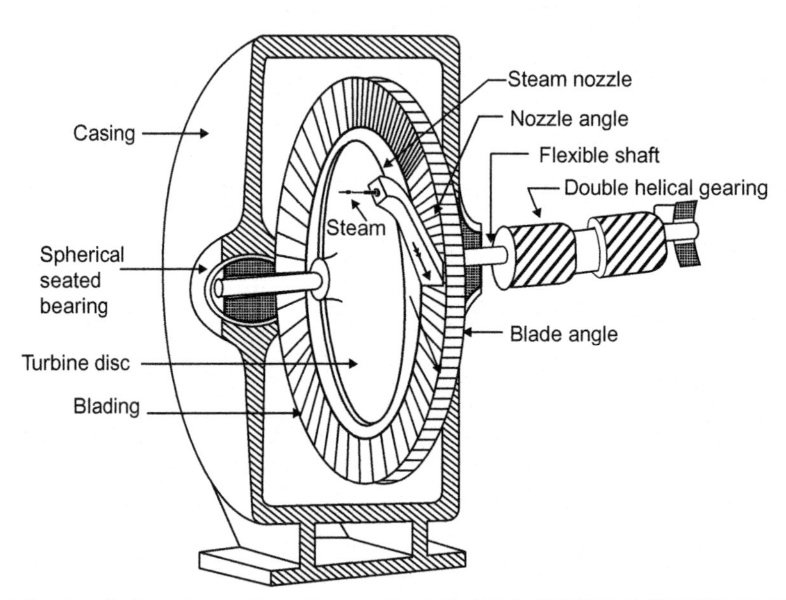

Figure 6.5 De Laval's impulse turbine. *Source: Fig. 213, P412, CHXIV THE THEORY OF THE STEAM TURBINE [2].*

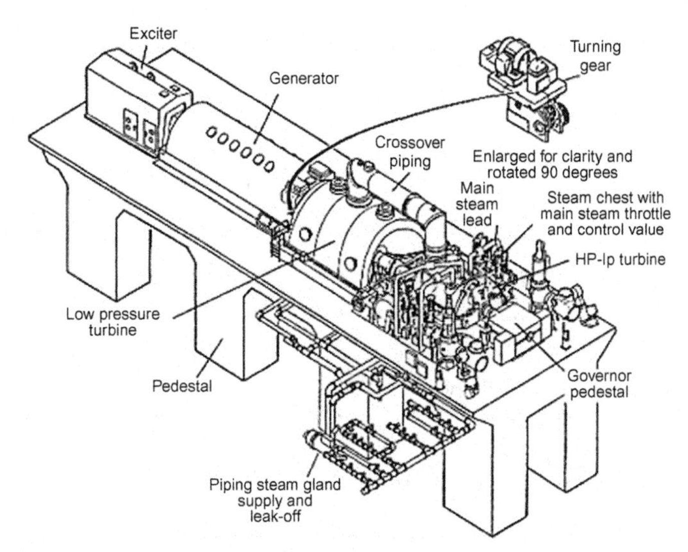

Figure 6.6 Layout of a typical large steam turbine. *Source: Steam Turbine – http://www.emt-india.net.*

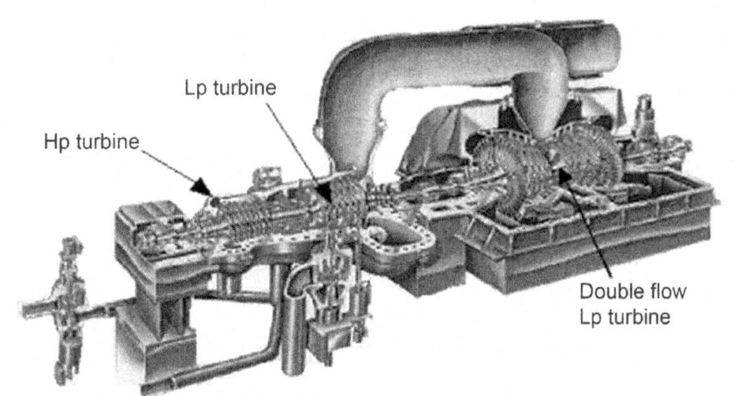

Figure 6.7 Sectional view of a typical large steam turbine. *Source: http://www.emt-india.net.*

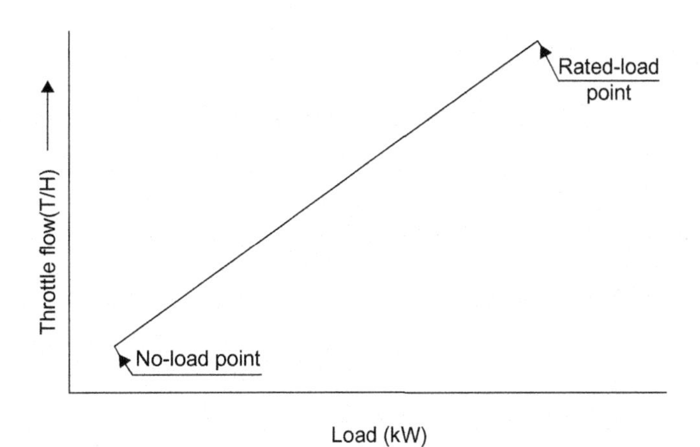

Figure 6.8 Throttle flow versus load.

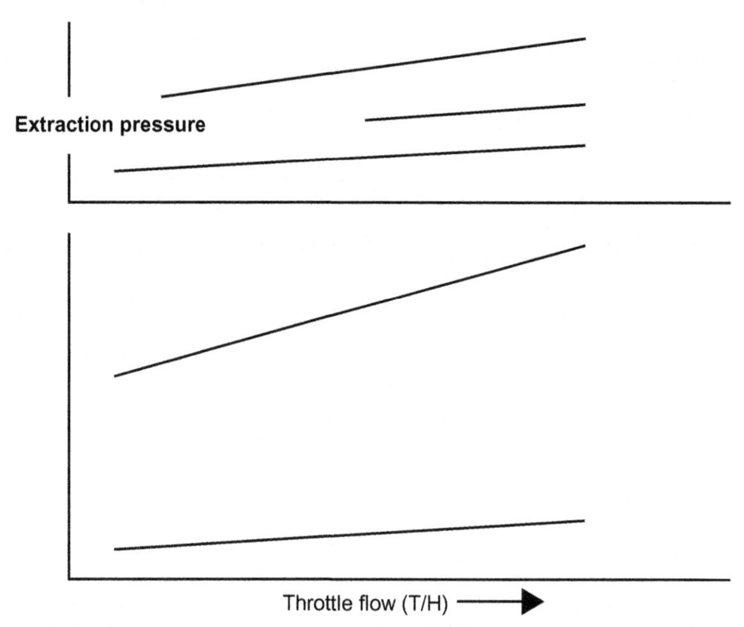

Figure 6.9 Nozzle box/extraction pressure versus throttle flow.

As a constant volume flow machine, the steam pressure changes within a turbine are unique. Various experimental results reveal that steam flow through a turbine is directly proportional to both the nozzle box pressure and stage pressures, but the steam flow is not proportional to the pressures at the last 2 or 3 stages of the low-pressure turbine [3]. This proportionality between the steam flow and steam pressure is linear in nature. Combining these two linear proportionalities it can be affirmed that the relationship between the nozzle-box pressure and stage pressures vs. load is also linear in nature and can be represented by a straight line (Figure 6.9).

6.2 TYPE OF STEAM TURBINE

In a steam turbine the inlet and exhaust pressures and the inlet temperature of the steam determine the theoretical energy available at the turbine inlet. To release the energy inside the turbine, steam flows over an alternate series of fixed and moving blades and as a result expands from high pressure to low pressure. A turbine stage consists of a stationary set of blades or nozzles, followed by a moving set called buckets or rotor blades. These two different blade types together cause steam to do mechanical work on the rotor. The stationary nozzles turn and accelerate steam converting the potential energy of steam (pressure) into kinetic energy (velocity). The buckets convert the kinetic energy of steam leaving the nozzles into both an impulse force (due

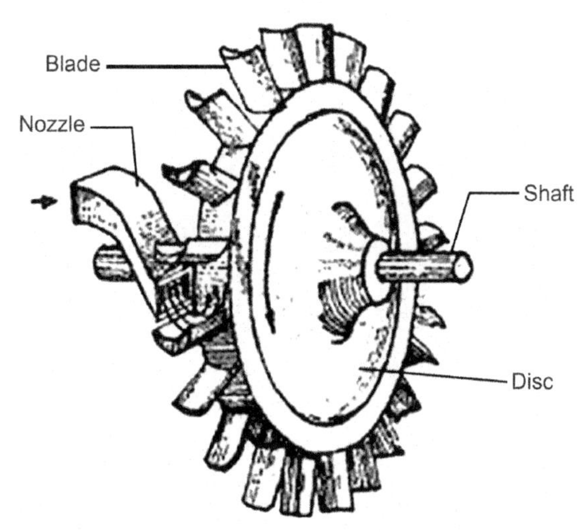

Figure 6.10 Single disc steam turbine.

to a change in direction of the steam) and a reaction force (due to a pressure drop across rotating blades), causing the shaft to rotate and generate power.

In the simplest single-disc steam turbine the expansion of steam is achieved from its initial pressure to its final one in a single nozzle located in front of the blades of the rotating disc (Figure 6.10). With the drop in steam pressure in the nozzle its heat content reduces, raising the velocity of the steam coming from the nozzle. Thus, the resulting kinetic energy of the jet of steam exerts an impulsive force on the blades performing mechanical work on the rotor.

6.2.1 Impulse and reaction

In the impulse design the complete process of expansion of steam takes place primarily in stationary nozzles, and the kinetic energy is transformed into mechanical work on the turbine blades without any further expansion. An impulse turbine has fixed nozzles that orient steam flow into high-speed jets. Moving rotor blades, shaped like buckets, absorb the kinetic energy of these jets and convert it to mechanical work, resulting in shaft rotation as the steam jet changes direction. A pressure drop occurs only across stationary blades, with a net increase in steam velocity across the stage (Figure 6.11).

As steam flows through the nozzle its pressure falls from steam chest pressure to condenser pressure (or any other pressure as conceived by the designer). Due to this relatively higher ratio of expansion of steam in the nozzle, steam leaves the nozzle with a very high velocity. The velocity of steam leaving the moving blades is a large portion of the maximum velocity of steam when leaving the nozzle.

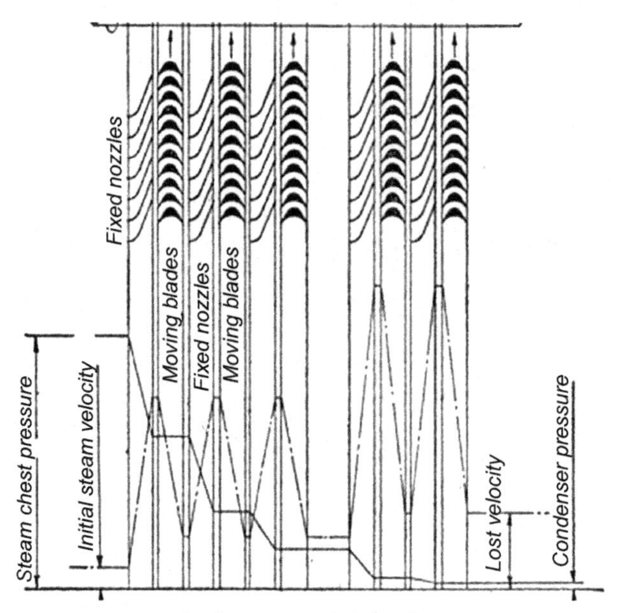

Figure 6.11 Change in pressure and velocity in impulse turbine. *Source: Fig. 2, P5, Introduction [3].*

The loss of energy due to this higher exit velocity is commonly called the "carry over velocity" or "leaving loss" [3].

In contrast, the expansion of steam in the reaction turbine occurs both in stationary or guide blades and in rotating or moving blades such that the pressure drop is about equally divided between stationary and rotating blades. In this type of turbine the expansion and decrease in the heat content of the steam per unit takes place continuously in both guide and moving blade passages. Nozzles and buckets in a reaction turbine are fixed and are designed so that relatively little expansion or pressure drop occurs across them. Upon expanding across the moving blades steam comes out as a high-velocity steam jet that does the mechanical work by the reaction of expanding steam as it leaves the blade wheel.

In the reaction turbine rotor blades themselves are arranged to form convergent nozzles. This type of turbine makes use of the reaction force produced as steam accelerates through the nozzles formed by the rotor. Steam is directed on to the rotor by the fixed blades of the stator. It leaves the stator as a jet that fills the entire circumference of the rotor. The steam then changes direction and increases its speed relative to the speed of the blades. A pressure drop occurs across both stationary blades and moving blades, with steam accelerating through stationary blades and decelerating through moving blades, with no net change in steam velocity across the stage but with a decrease in both pressure and temperature, reflecting the work performed in driving the rotor (Figure 6.12).

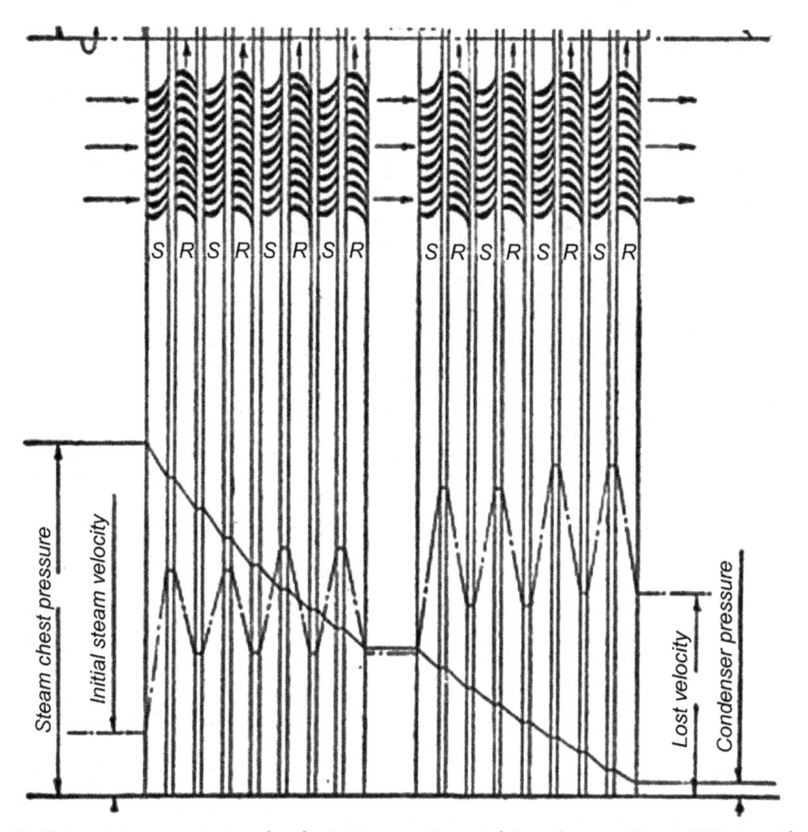

Figure 6.12 Change in pressure and velocity in reaction turbine. *Source: Fig. 6, P12, Introduction [3].*

For identical speed and efficiency, the impulse design requires only about half the number of stages of the reaction design. The fewer stages in impulse design allow for more space per stage and a more rugged construction. With the impulse design, since the majority of the pressure drop is taken across the nozzle and very little across the bucket the thrust is not as much of a problem as with the reaction design. The large pressure drop across buckets in the reaction design causes a large thrust loading on the rotor. To balance this high-thrust load one of the solutions is to use an opposed flow or double-flow design [4].

The impulse stage is best suited for the high-pressure region and for small steam quantities. These blades are short, since the specific volume of steam at high pressure is low and requires a much smaller flow area. From a thermal efficiency standpoint the small blade heights are undesirable. But the efficiency at low loads is improved with the help of nozzle governing. The reaction stage is advantageous at the lower pressure region, where a large volume of steam must be handled. At the low-pressure end of the turbine the specific volume increases so rapidly that the expansion is restricted to the reaction type. The supply of steam to the turbine is controlled by throttle governing.

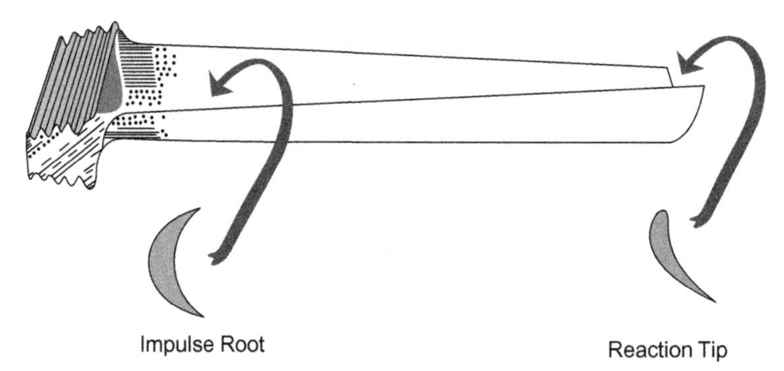

Figure 6.13 Combination of impulse-reaction blade. *Source: Fig. 6.8, P11, [5].*

From this discussion it may appear that there are only two basic types of turbines, but in practice most modern turbines are a blend of both principles. Nearly all turbines use the impulse principle for preliminary high-pressure stages to reduce losses across blade ends and to raise the steam volume. The low-pressure section consists only of reaction stages. A combined impulse-reaction turbine is simple to construct as well as less expensive to make. To overcome various shortcomings of both types of designs, blades are made twisted, a combination of part impulse and part reaction, etc. (Figure 6.13). Traditionally, General Electric (GE) and Associated Electrical Industries (AEI) turbines are impulse type and Westinghouse, Kraftwerk Union AG (KWU), and Parsons turbines are reaction type.

6.2.2 Impulse principle

As discussed earlier an impulse turbine is comprised of a stage of stationary nozzles followed by a stage of curved buckets called turbine blades. When a fluid jet impinges on a frictionless curved blade (Figure 6.14: simple impulse stage) the resulting velocity diagram will be as shown in Figure 6.15 [6], where:

C_1: Absolute velocity of steam leaving nozzle and entering the blade
C_2: Absolute velocity of steam leaving the blade
U: Velocity of the blade in the horizontal direction
V_1: Relative velocity of steam entering the blade
V_2: Relative velocity of steam leaving the blade
α_1: Nozzle/steam entrance angle
α_2: Steam exit angle
β_1: Blade entrance angle
β_2: Blade exit angle

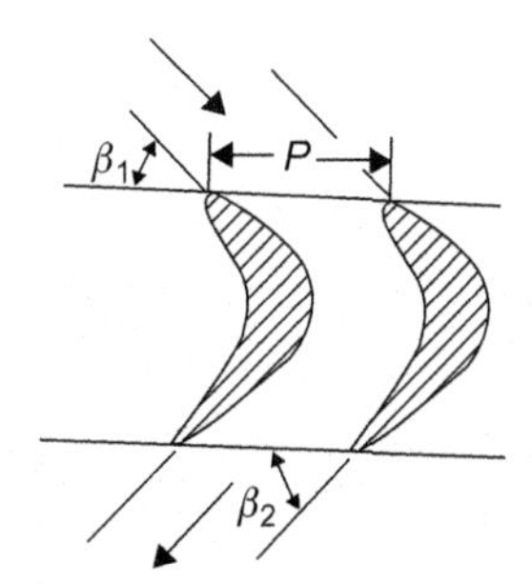

Figure 6.14 Simple impulse stage. *Source: Fig. 19.6, P432, CH 19, Rotary Expanders and compressors [6].*

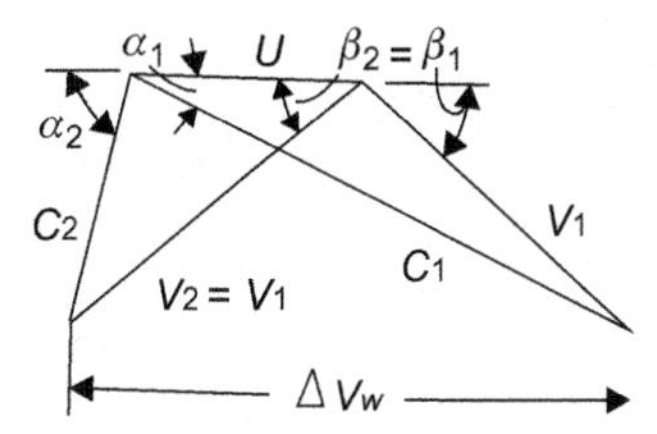

Figure 6.15 Velocity diagram of an impulse stage. *Source: Fig. 19.6, P432, CH 19, Rotary Expanders and compressors [6].*

Assuming neither expansion nor contraction of steam between blade entry and exit the change in the tangential component of relative velocity at the inlet and outlet is given by

$$\Delta V_w = V_1 \cos \beta_1 + V_2 \cos \beta_2 \tag{6.1}$$

If steam flow through the blade passage areas remains unchanged following the continuity equation, the areas at the inlet and outlet should be the same. Thus, V_1 must be equal to V_2 and $\beta_1 = \beta_2$. Eq. 6.1 then reduces to

$$\Delta V_w = 2V_1 \cos \beta_1 \tag{6.2}$$

or,

$$\Delta V_w = 2(C_1 \cos \alpha_1 - U) \tag{6.3}$$

Hence, work done by the stage

$$W = \frac{m}{g_c} U \Delta V_w = 2\frac{m}{g_c} U(C_1 \cos \alpha_1 - U) \tag{6.4}$$

where

m = mass-flow rate of steam, kg/s
g_c = conversion factor, 1 kg.m/N.s^2

The energy available to the rotor blades of an impulse stage is

$$E = \frac{mC_1^2}{2g_c} \tag{6.5}$$

Hence, the diagram efficiency of an impulse stage is as follows:

$$\eta_d = \frac{W}{E} = \frac{2mU(C_1 \cos \alpha_1 - U)/g_c}{mC_1^2/2g_c} = 4\frac{U}{C_1}\left(\cos \alpha_1 - \frac{U}{C_1}\right) \tag{6.6}$$

The term U/C_1 is known as the *blade speed ratio*. To find the value of blade speed ratio that will yield the maximum diagram efficiency, the first-order differential of Eq. 6.6 has to be zero. Hence,

$$\frac{d(\eta_d)}{d(U/C_1)} = 4\cos \alpha_1 - 8\frac{U}{C_1} = 0 \tag{6.7}$$

or,

$$\frac{U}{C_1} = \frac{\cos \alpha_1}{2} \tag{6.8}$$

Putting the value of Eq. 6.8 in Eqs. 6.4 and 6.6

$$W_{\text{max}} = 2\frac{m}{g_c}U^2 \tag{6.9}$$

$$(\eta_d)_{\text{max}} = \cos^2 \alpha_1 \tag{6.10}$$

6.2.3 Reaction principle

A reaction turbine is constructed of rows of fixed blades and rows of moving blades. The fixed blades act as nozzles. The moving blades move as a result of the impulse of steam received (caused by a change in momentum) and also as a result of expansion and acceleration of the steam relative to them. In other words, they also act as nozzles (Figure 6.16).

The velocity diagram of a reaction stage is symmetrical as is evident from Figure 6.17. Therefore, $C_1 = V_2$ and $C_2 = V_1$, as discussed for the impulse principle [6].

For a 50% reaction design, the similarity between various angles is $\alpha_1 = \beta_2$ and $\alpha_2 = \beta_1$.

Hence,

$$\Delta V_w = 2C_1 \cos \alpha_1 - U \tag{6.11}$$

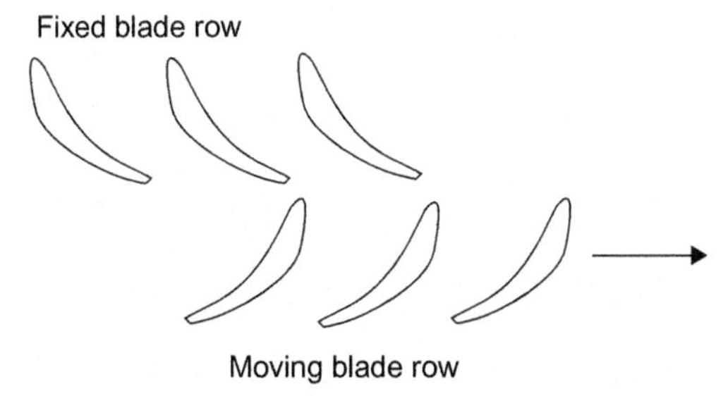

Figure 6.16 Simple reaction stage.

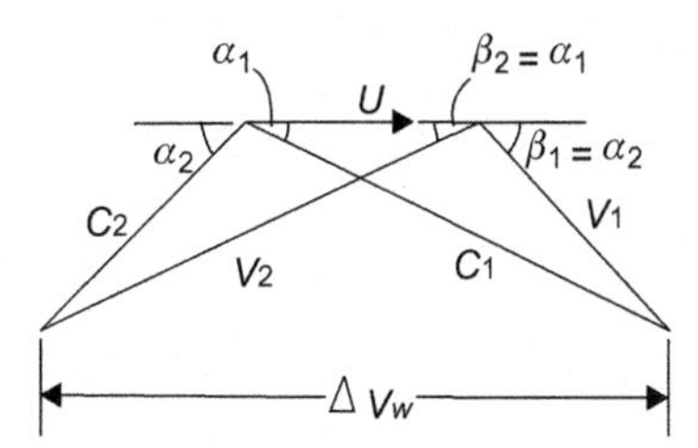

Figure 6.17 Velocity diagram of a reaction stage. *Source: Fig. 19.10, P438, CH 19, Rotary Expanders and compressors [6].*

Work done by the stage is

$$W = \frac{m}{g_c} U \Delta V_w = \frac{m}{g_c} U(2C_1 \cos \alpha_1 - U) \tag{6.12}$$

In a reaction stage there is an enthalpy drop in the rotor blade, so the energy available to the rotor blade is given by

$$E = \frac{m}{g_c}\left\{\frac{C_1^2}{2} + \frac{V_2^2 - V_1^2}{2}\right\} = \frac{m}{g_c}\left(C_1^2 - \frac{V_1^2}{2}\right) \tag{6.13}$$

(since $V_2 = C_1$).

Again, $V_1^2 = C_1^2 + U^2 - 2C_1 U \cos \alpha_1$, and Eq. 6.13 changes to

$$E = \frac{m}{2g_c}(C_1^2 - U^2 + 2C_1 U \cos \alpha_1) \tag{6.14}$$

From Eq. 6.12 and Eq. 6.14 it follows that the diagram efficiency of a reaction stage is

$$\eta_d = \frac{2U(2C_1 \cos \alpha_1 - U)}{(C_1^2 - U^2 + 2C_1 U \cos \alpha_1)} = \frac{2\frac{U}{C_1}\left(2 \cos \alpha_1 - \frac{U}{C_1}\right)}{1 - \left(\frac{U}{C_1}\right)^2 + 2\frac{U}{C_1} \cos \alpha_1} \tag{6.15}$$

The first-order differential of Eq. 6.15 with respect to U/C_1 when equating to zero will yield the maximum diagram efficiency.

$$\frac{U}{C_1} = \cos \alpha_1 \tag{6.16}$$

Putting the value of Eq. 6.16 into 6.12 and 6.15, we get

$$W_{\max} = \frac{m}{g_c} U^2 \tag{6.17}$$

$$(\eta_d)_{\max} = \frac{2 \cos^2 \alpha_1}{1 + \cos^2 \alpha_1} \tag{6.18}$$

Comparing Eqs. 6.9 and 6.17, it may be concluded that the maximum power (work done per unit time) of an impulse stage is twice that of a 50% reaction stage.

6.3 CLASSIFICATION OF STEAM TURBINES

Steam turbines may be classified into different categories depending on their construction, process by which heat drop is achieved, initial and final conditions of steam used, and their industrial use as discussed in the following. These turbines may be either coupled directly or through a reduction gear to the driven machine. They may also be classified as condensing, non-condensing, automatic extraction, controlled extraction, regenerative extraction, mixed-pressure, reheat, etc. This chapter focuses on the stationary type of steam turbines with constant speed of rotation. Figure 6.18 presents some of these turbines and shows the relationships between the common types of turbines in terms of pressure levels and division of steam flow.

Turbines are classified descriptively in various ways as follows [7]:

According to the direction of steam flow

a. Axial turbines in which steam flows in a direction parallel to the axis of the turbine

b. Radial turbines in which steam flows in a direction perpendicular to the axis of the turbine. One or more low-pressure stages in such turbines are made axial

According to the number of pressure stages

a. Single-stage turbines with one or more velocity stages usually of small power capacities. These turbines are mostly used for driving centrifugal compressors, blowers, and other similar machinery

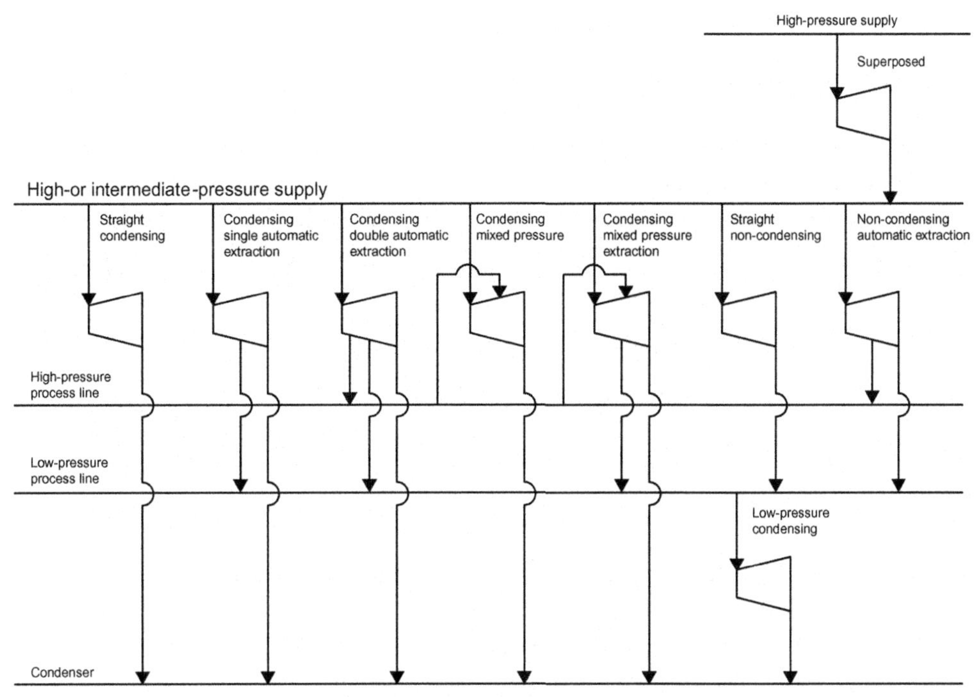

Figure 6.18 Typical classification of steam turbines in regard to steam pressure.

b. Multi-stage impulse and reaction turbines; they are made in a wide range of power capacities varying from small to large

According to the number of exhaust stages

a. Double-flow single LP turbine
b. Double-flow twin LP turbine

According to the number of cylinders

a. Single-cylinder turbines
b. Double-cylinder turbines
c. Three-cylinder turbines
d. Four-cylinder turbines

According to the shaft arrangement

a. Multi-cylinder turbines, which have their rotors mounted on one and the same shaft and coupled to a single generator, are known as single-shaft tandem compound turbines
b. Multi-cylinder turbines with separate rotor shafts for each cylinder placed parallel to each other are known as multi-axial cross-compound turbines

According to the method of governing

a. Turbines with throttle governing in which fresh steam enters through one or more (depending on the power developed) simultaneously operated throttle valves
b. Turbines with nozzle governing in which fresh steam enters through two or more consecutively opening regulators
c. Turbines with bypass governing in which steam in addition to being fed to the first stage is also directly led to one, two, or even three intermediate stages of the turbine

According to the principle of the action of steam (as determined in stage design)

a. Impulse turbines in which the potential energy of steam is converted into kinetic energy in nozzles or passages formed by adjoining stationary blades. The kinetic energy of steam is then converted into mechanical energy in the moving blades
b. Axial reaction turbines in which expansion of steam between blade passages, both guide and moving blades of each stage, takes place nearly to the same extent
c. Radial reaction turbines without any stationary guide blade
d. Radial reaction turbines having stationary guide blades

According to the heat drop process (by exhaust conditions)

a. Condensing turbines with regenerators in which steam at a pressure less than atmospheric is directed to the condenser. Steam is also extracted from intermediate stages for feedwater heating. The latent heat of exhaust steam during the process of condensation is completely lost in these turbines
b. Condensing turbines with intermediate stage extractions at specific pressures for industrial and heating purposes
c. Back-pressure turbines, the exhaust steam from which is utilized for industrial and heating purposes
d. Topping turbines are also the back-pressure type with the difference that the exhaust steam from these turbines is further utilized in medium- and low-pressure condensing turbines. These turbines operate at high initial conditions of steam pressure and temperature
e. Back-pressure turbines, with intermediate stage extractions at specific pressures, are meant for supplying steam at various pressure and temperature conditions
f. Low/exhaust pressure turbines in which the exhaust steam from the reciprocating steam engines, power hammers, presses, etc., is utilized for power generation
g. Mixed-pressure turbines with two or three pressure stages with supply of steam to its intermediate stages

According to the steam conditions at turbine inlet (by steam supply conditions)

a. Low-pressure turbines, using steam at pressures of 120—200 kPa
b. Medium-pressure turbines, using steam at pressures up to 4 MPa
c. High-pressure turbines, utilizing steam at pressures above 4 MPa
d. Very high-pressure turbines, using steam at pressures at or above 16.5 MPa and temperatures at or above 813 K
e. Supercritical pressures using steam at or above 22.12 MPa and temperature at or above 647.14 K

According to the usage in industry (by type of driven equipment)

a. Stationary steam turbines with constant speed of rotation primarily used for driving alternators
b. Stationary steam turbines with variable speed of rotation meant for driving turbo blowers, air circulators, pumps, etc.
c. Non-stationary steam turbines with variable speed of rotation usually employed in steamers, ships, etc.

In practice, however, a specific turbine may be described as a combination of two or more of these classifications, e.g., single casing straight condensing unit, single casing back pressure unit, tandem compound multi-casing reheat regenerative unit, etc.

6.4 COMPONENTS OF STEAM TURBINES

The following are the main components of a steam turbine:

1. Rotor
2. Casing
3. Bearings
4. Turning gear

6.4.1 Rotor (Figures 6.19 and 6.20)

The turbine rotor is comprised of a number of discs mounted on the shaft. When the working temperature is at or below 673 K discs are shrunk on the shaft. Rotors operating at temperatures much higher than 673 K and having relatively small diameters are made of a single-piece solid forging with the discs directly machined on it. In some designs the first few high-pressure stages are made in one piece with the shaft and on low-pressure stages discs are shrunk on the shaft. Small diameter rotors are made as a solid forging in one piece. The control stage disc is usually large in diameter and is either machined directly on the solid forgings or made separately and shrunk

Figure 6.19 A complete rotor assembly placed in position. *Source: Fig. 6, P207, Designing High Performance Steam Turbines with Rotordynamics as a Prime Consideration by Stephen L. Edney and George M. Lucas http://turbolab.tamu.edu/proc/turboproc/T29/t29pg205.pdf.*

on the shaft. In large diameter rotors discs are welded at the periphery. The transverse rigidity of the rotor is very high and the critical speed is usually well above the operating speed. In a tandem compound, the rotors of various sections of the turbine are mutually connected by means of rigid couplings.

6.4.2 Casing

The casing of a turbine could be either the barrel type or the horizontally split flanged type. The barrel type casing is designed without an axial joint (Figure 6.21). This casing permits rapid start-up and does not require external heating to arrest relative expansions of the casing and the rotor. Since the casing is symmetric around its axis, the shape of the barrel type casing also remains unchanged and leak-proof during start-up, shut down, changes in load, and even under high pressure when the temperature changes

Figure 6.20 A turbine rotor fixed with a diaphragms and blades. *Source: STEAM TURBINE. Wikipedia the Free Encyclopedia.*

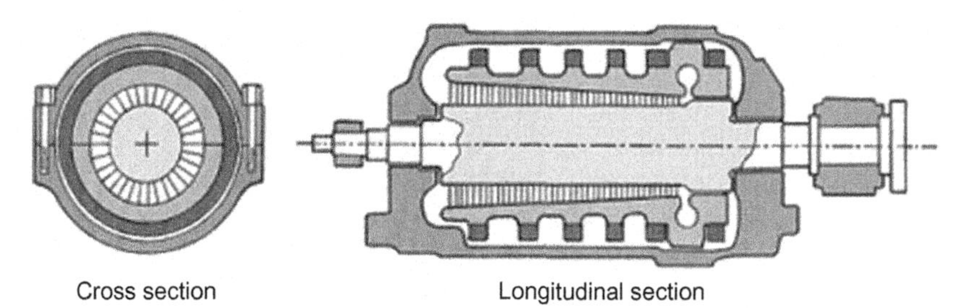

Figure 6.21 Typical barrel type casing. *Source: FIG. 2, [8].*

rapidly. This facilitates quick start-up and loading of a turbine with barrel type casing. In a typical design an axially split guide blade carrier is arranged in the barrel type casing. The flanges of horizontally split casing are designed to heat up by steam during start-up to avoid excessive relative expansions of the casing and the rotor. The split casings may be designed with single-, double-, or triple-shell construction.

Figure 6.22 shows a double-shell split casing where an inner casing is supported in the outer casing. Steam is supplied to the turbine but only the inner casing is subjected to high-temperature steam, while the outer casing is subjected to low-pressure,

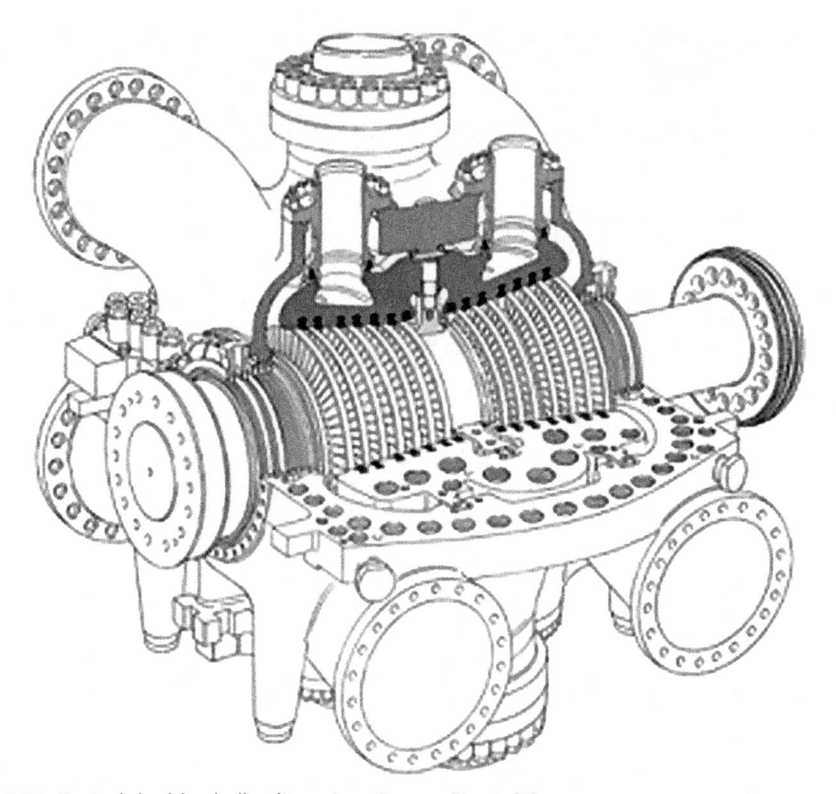

Figure 6.22 Typical double shell split casing. *Source: Fig. 4, [8].*

low-temperature steam from the inner casing exhaust. Thus, reducing deformation of casing due to non-uniform temperature rise during start-up and shut down. A triple-shell split casing is shown in Figure 6.23. The casings of different sections of the turbine are interconnected by a system of guide keys through bearing pedestals. As a result, the alignment of all casings is ensured along with free thermal expansion of the casings.

The high pressure (HP) turbine casing houses the HP control valves and may house the HP emergency stop valves (ESVs). High-pressure, high-temperature steam is supplied to the HP turbine from the superheater outlet of the steam generator. In a specific design steam extractions from the HP turbine casing to the HP feedwater heaters may also be provided.

Similar to HP turbine casing the intermediate pressure (IP) turbine casing has provisions for IP control valves and may house IP interceptor valves (IVs). The IP turbine receives medium-pressure, high-temperature steam from the reheater outlet of the steam generator. Contrary to HP turbine casing, all turbine designs keep provisions for steam extractions from the IP turbine casing to the HP feedwater heaters.

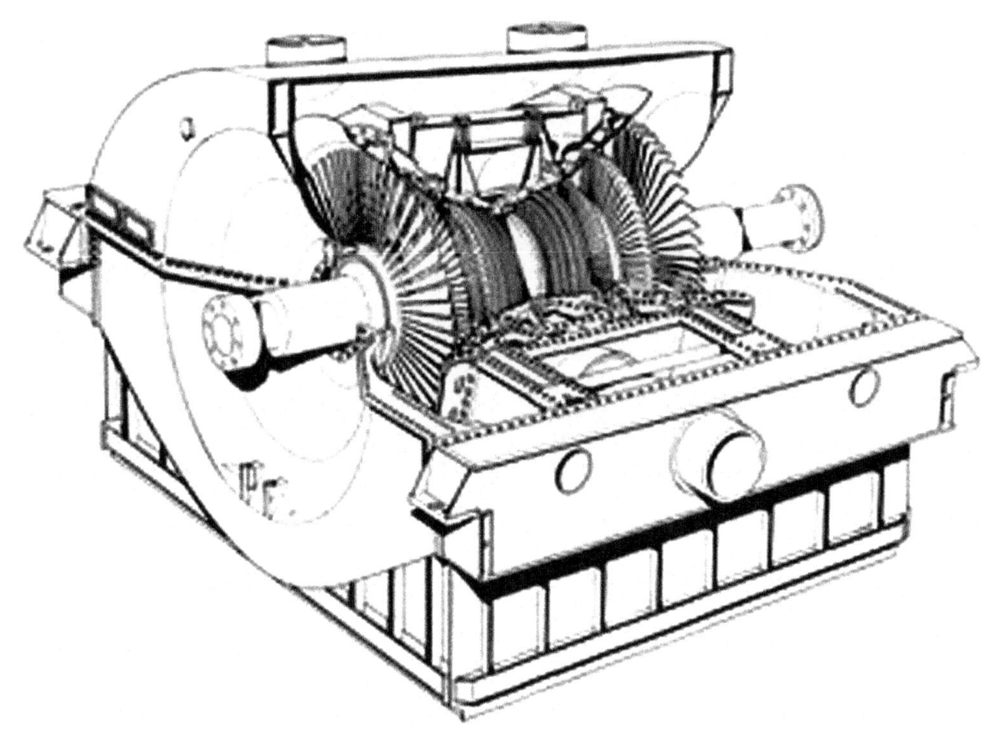

Figure 6.23 Typical shell casing. *Source: Fig. 5, [8].*

The low pressure (LP) turbine receives low-pressure, medium-temperature steam from the IP turbine exhaust. The LP turbine casing is connected to the condenser at the LP turbine exhaust area. This casing is always designed keeping provisions for steam extractions to LP feedwater heaters.

6.4.3 Bearings

The function of bearings is to carry the full weight of the turbine rotor and allow it to rotate with the least friction. Metal-to-metal contact will ruin the lining of a bearing and hence an oil film is established between the stationary surface and the rotor to separate them. There are two types of turbine bearings: journal bearings and thrust bearings.

6.4.3.1 Journal bearings (Figure 6.24)

The function of a journal bearing is to support the turbine rotor. The journal bearing is comprised of upper and lower shells with a babbit face. The bearings are supplied with lubricating oil from both sides of the shells. Clearance is provided between the shaft and the bearing surface to accommodate the lubricating oil film. A sustained oil

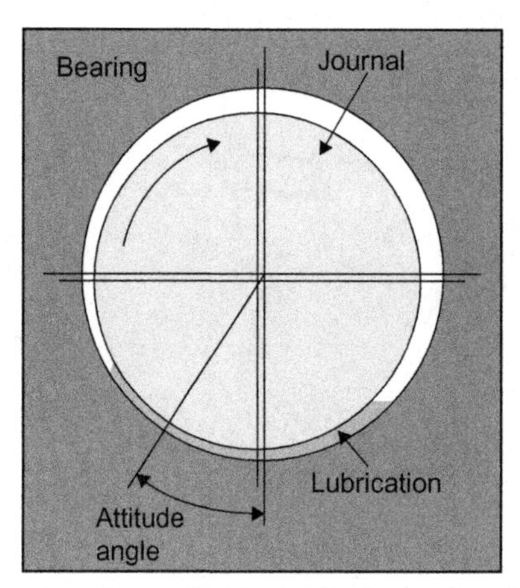

Figure 6.24 Typical journal bearing. *Source: PLAIN BEARING – Wikipedia the Free Encyclopedia.*

film between the shaft and the bearing surface is developed only when the oil pressure is able to balance the weight of the shaft. The shaft center at rest is situated eccentrically relative to the bearing center. With increasing shaft speed, the shaft center tends to reach the bearing center. At infinite speed of rotation both centers will coincide. High-pressure oil, called jacking oil, is supplied under the journals at low speed during start-up and shut-down of the turbine rotor. As a result, dry friction is prevented and breakaway torque during start-up with turning gear is reduced.

6.4.3.2 Thrust bearings (Figure 6.25 and Figure 6.26):

Axial thrust on the turbine rotor is caused by pressure and velocity differences across the rotor blades, pressure differences from one side to the other on wheels or rotor bodies, and pressure differences across the shaft labyrinths. The net thrust is equal to the algebraic sum of all these effects and is usually in the direction of the steam flow. The function of a thrust bearing is to fix its rotor position as well as to carry the residual axial thrust present in a combined turbine generator shaft system that is unable to be compensated for by the balance piston. The magnitude and direction of the axial thrust to be overcome by the thrust bearing is dependent on the load conditions of the turbine.

The thrust bearing consists of a collar, turned on the rotor shaft, thrust pads found on the collar surface, and a self-aligning pad held in position by a support that is rigidly fixed on the bearing pedestal (Figure 6.25). The collar and thrust pads permit equal loading of the thrust bearing in either direction. The thrust pads are the tilting type and can adjust themselves to the requisite shaft sag by finding their proper position on the spherical seating of the support (Figure 6.26). The lubricating oil is supplied to the bearing under pressure at the shaft surface. The thrust bearing is closed from both ends to prevent the escape of lube oil.

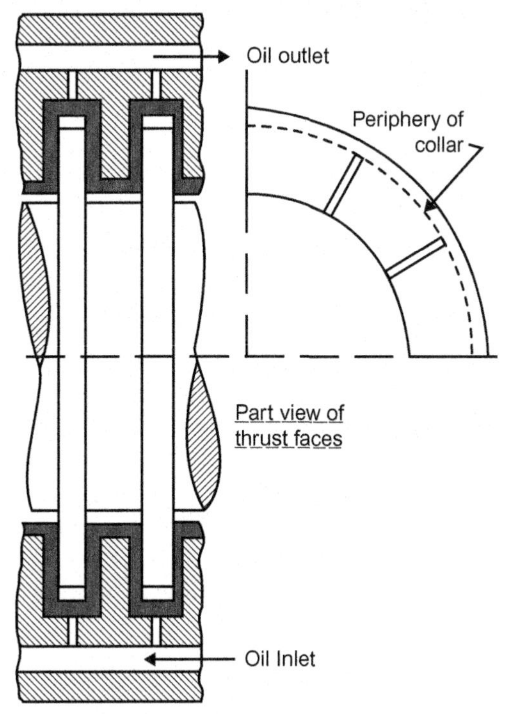

Figure 6.25 Typical thrust bearing. *Source: Fig. 389, P598, CHXXV BEARINGS AND LUBRICATION [3].*

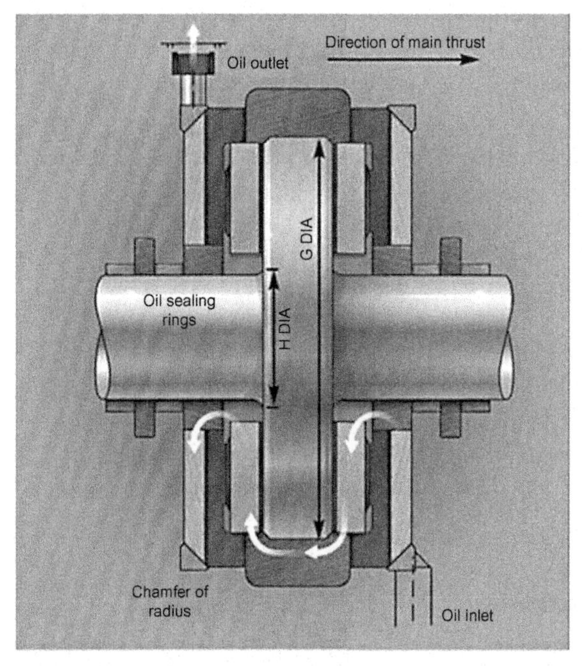

Figure 6.26 Thrust bearing with tilting pads. *Source: Fig. 1-Flooded Lubrication. TILTING PAD THRUST BEARING by Heinz P. Blotch. HYPERLINK "http://www.machinerylubrication.com/Magazine/Issue/Machinery%20Lubrication/7/2006"Machinery Lubrication (7/2006).*

6.4.4 Turning gear

Turning-gear systems are used with large turbomachines, such as large steam turbine-generators, to cause slow rotation of turbine rotor whenever the turbine is being started up, shut-down, or at other times as it becomes necessary to jog the rotor into a different position. The slow rotation during start-up and shut-down is carried out to ensure that the rotor remains free from undesirable temporary or permanent deformation or sag between long bearing spans due to different temperatures in the rotor sections perpendicular to the axis of the rotor, if left for too long in one position.

The turning gear may be located on the bearing pedestal between the LP turbine and either the IP turbine or the generator and is driven either hydraulically or electrically. In a typical design the turning gear drives the turbine rotor through a bull gear attached to the rotor shaft. When steam is allowed to enter into the turbine and when the turbine rotor speed exceeds the turning gear speed, the turning gear or the drive pinion is automatically thrown out of engagement or decoupled from the rotor bull gear by centrifugal force. During shut-down of the turbine as the rotor coasts down when its speed falls below the turning gear speed a low-speed signal re-engages the turning gear. The turning gear may also be engaged manually.

Before starting the turning gear, its lubricating oil supply must be ensured. The jacking oil, if required by the design to overcome the initial break-away torque and to prevent dry friction, should also be supplied to lift the rotor before putting the turning gear into operation. In a hydraulically operated turning gear, oil from the jacking oil system should also be supplied by engaging the drive pinion. The turning gear must not be started until the drive pinion is fully engaged. In the event of failure of the normal turning gear, the combined rotor shaft may be rotated manually with the help of a mechanical turning/barring gear.

6.5 METHODS OF TURBINE GOVERNING

Speed and power output of the turbine are controlled by varying the steam input to the turbine using throttle valves placed between the steam generator and the turbine. An emergency stop valve precedes the throttle valves and completely shuts off steam supply to the turbine as and when required during normal shut-down and also during emergencies. There are three basic methods of steam admission to the turbine, discussed as follows.

6.5.1 Throttle governing

In throttle governing the supply of steam to the turbine is controlled through single batch of nozzles either by a single, two, or multiple control/throttle valves operating in parallel (Figure 6.27). These valves are located in the HP cylinder steam chest.

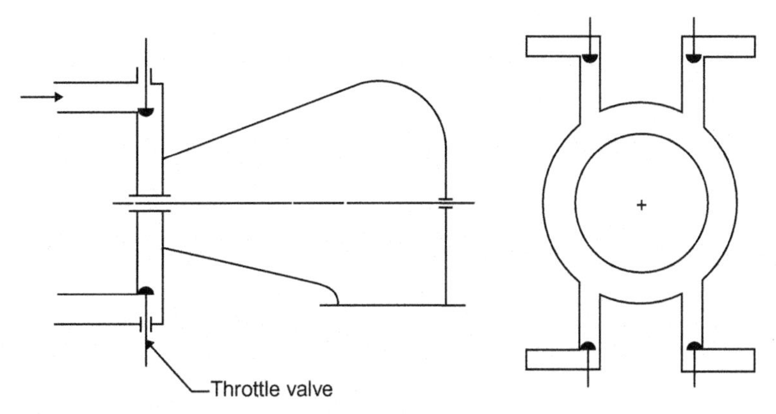

Figure 6.27 Simple throttle governing with common inlet annulus.

The steam flow through the turbine is controlled by the throttle valves. The upstream steam pressure at the throttle valves will be constant, irrespective of the turbine loading, but the pressure after the throttle valves will reduce as the valve opening reduces. This will cause the available energy per unit of steam due to expansion to be reduced, as well as the mass flow of steam. This is shown for a non-reheat turbine in Figure 6.28. At the turbine stop valve the steam conditions are 16.8 MPa and 838 K. At full-load, the loss of steam pressure due to the throttling effect of the turbine stop and the throttle valve is about 5%. Hence, the steam pressure after the throttle valve is about 16 MPa (point A). The steam does work in the turbine by expansion up to 9.6 kPa back pressure. At half-load, the steam pressure after the throttle valve will be half of that at full-load, as shown at point C and at 25% load pressure will be one-fourth of the full-load pressure (point E). Although the specific enthalpy of the steam at the turbine inlet remains unchanged, the available energy gets reduced as the loading gets reduced (from AB to CD to EF). This reduction of heat drop and steam flow causes the turbine output to be reduced as the throttle valves close. Nevertheless, efficiency of the expansion remains the same for all loads as is evident from the slope of expansion lines [9].

The main effect of throttling, apart from reducing steam flow, is to reduce adiabatic heat drop. A secondary effect of throttling is to increase initial superheat and to delay the dew point. It should also be noted that throttling incurs losses as the steam pressure and temperature are untowardly reduced. It is not suitable for part-load operation or for fluctuating load demand.

6.5.2 Nozzle governing

In nozzle governing, control is necessarily restricted to the first stage of the turbine, and the nozzle areas in the other stages remain unchanged. The first-stage nozzles are

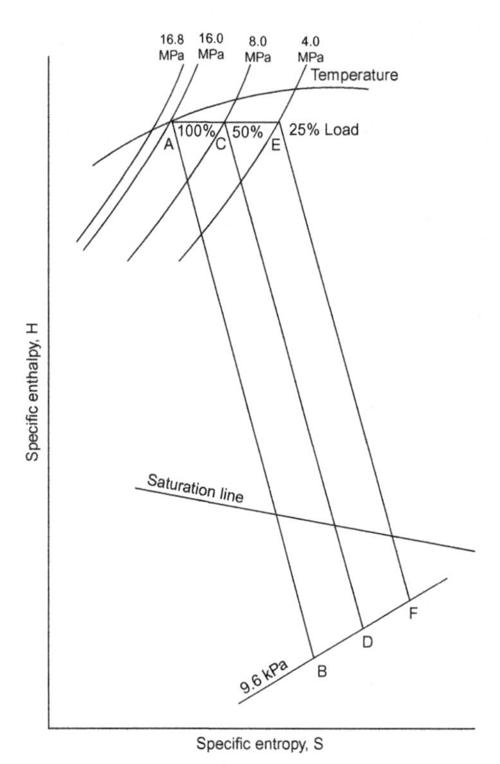

Figure 6.28 Enthalpy-entropy diagram of throttle governing. *Source: Fig. 7.65. P509. Article 5.10.2.: Throttle Governing. CH7: Plant Performance and Performance Monitoring [9].*

divided into a number of groups, from three in a simple system, to six or higher in a more elaborate arrangement. The steam supply to each group of nozzles is controlled by a valve, and the number of valves opened in turn varies according to the load on the turbine. For example, if six valves are opened to meet any given load demand then five of them will remain fully open and the actual control will be taken care of by the sixth valve.

Figure 6.29 shows a four-valve arrangement. At standstill all four valves remain closed. Between no-load and 25% load only the first valve starts opening, which opens full at 25% load. The first valve remains fully open and the second valve comes into service between 25% and 50% load, followed by opening of the third valve at 50% to 75% load. Finally, in a fully open condition of the three valves the only valve operating in throttling mode between 75% and 100% load is the last valve, which permits maximum loading of the turbine when opened fully. Thus, the losses due to throttling in nozzle governing are much less than the losses due to throttling in throttle governing type, so operation is more efficient.

It can be seen in Figure 6.30 that at full-load, the expansion is from A to B, the same as the throttle governing type. However, at 50% load the required pressure is

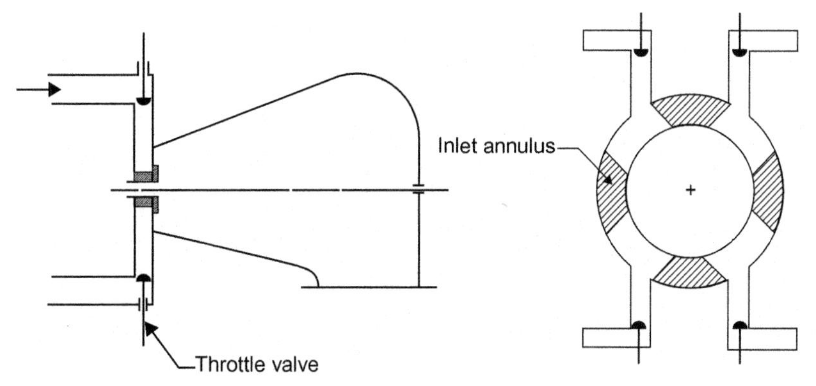

Figure 6.29 Nozzle governing.

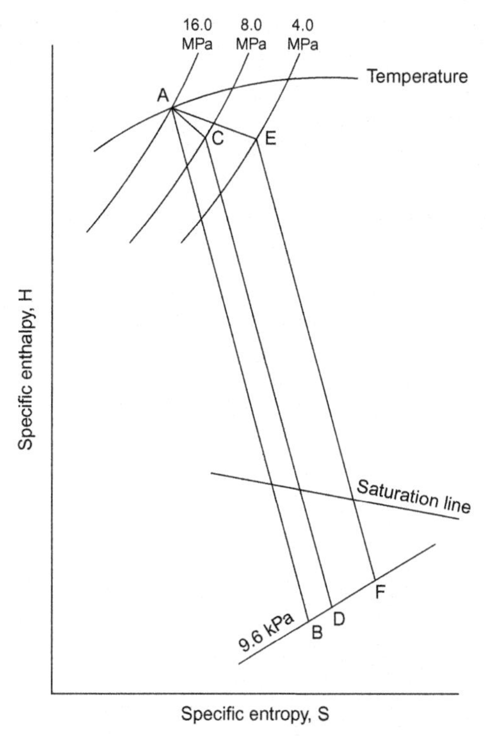

Figure 6.30 Enthalpy-entropy diagram of nozzle governing. *Source: Fig. 7.67. P510. Article 5.10.2.: Throttle Governing. CH7: Plant Performance and Performance Monitoring [9].*

again 8 MPa. Therefore, with two-nozzle operation this load is achieved more efficiently than before (line AC and expansion CD). Thereafter, for one nozzle operation with 4 MPa required pressure, throttling is from A to E, and the expansion is from E to F.

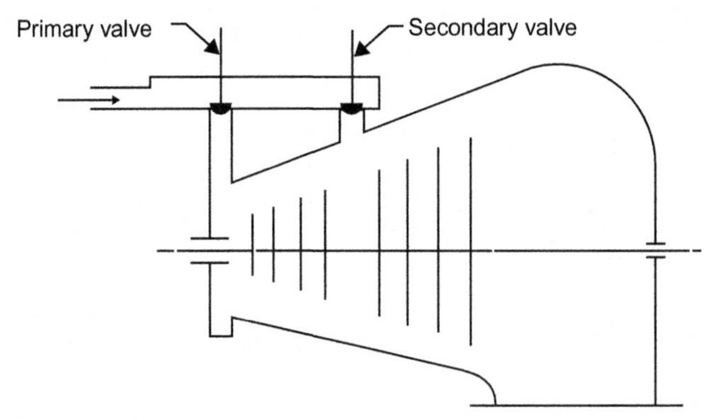

Figure 6.31 Bypass governing.

The chief advantage of nozzle governing is higher turbine efficiency at part loads because of reduced throttling losses. Therefore, nozzle-governed machines are preferable where the loading regime involves prolonged operation under part-load conditions.

6.5.3 Bypass governing

In bypass governing steam is normally supplied through a primary valve and is adequate to meet the bulk portion of the maximum load, which is called the economic load. At lower loads the regulation is done by throttling steam through this valve. When the load demand exceeds the economic load the primary valve remains fully open and a secondary valve is opened (Figure 6.31). This secondary valve bypasses the first stage and some high-pressure stages and supplies throttled steam downstream of these stages. This high-enthalpy steam increases the energy of the partially spent steam admitted through the primary valve, developing additional blade torque to meet the increased load.

6.6 TURBINE GOVERNING SYSTEM

The development of a simple but effective turbine-governing system is essential for:

i. Increasing electric power system capacities
ii. Increasing unit sizes
iii. Meeting system load variations without large deviations in the nominal electrical frequency

The system has to be such that the response of the turbine to load variations is quick and smooth. The governing system also is comprised of devices that protect the turbine from abnormal conditions that may arise during normal operation. Figure 6.32 presents a block diagram of a typical governing system.

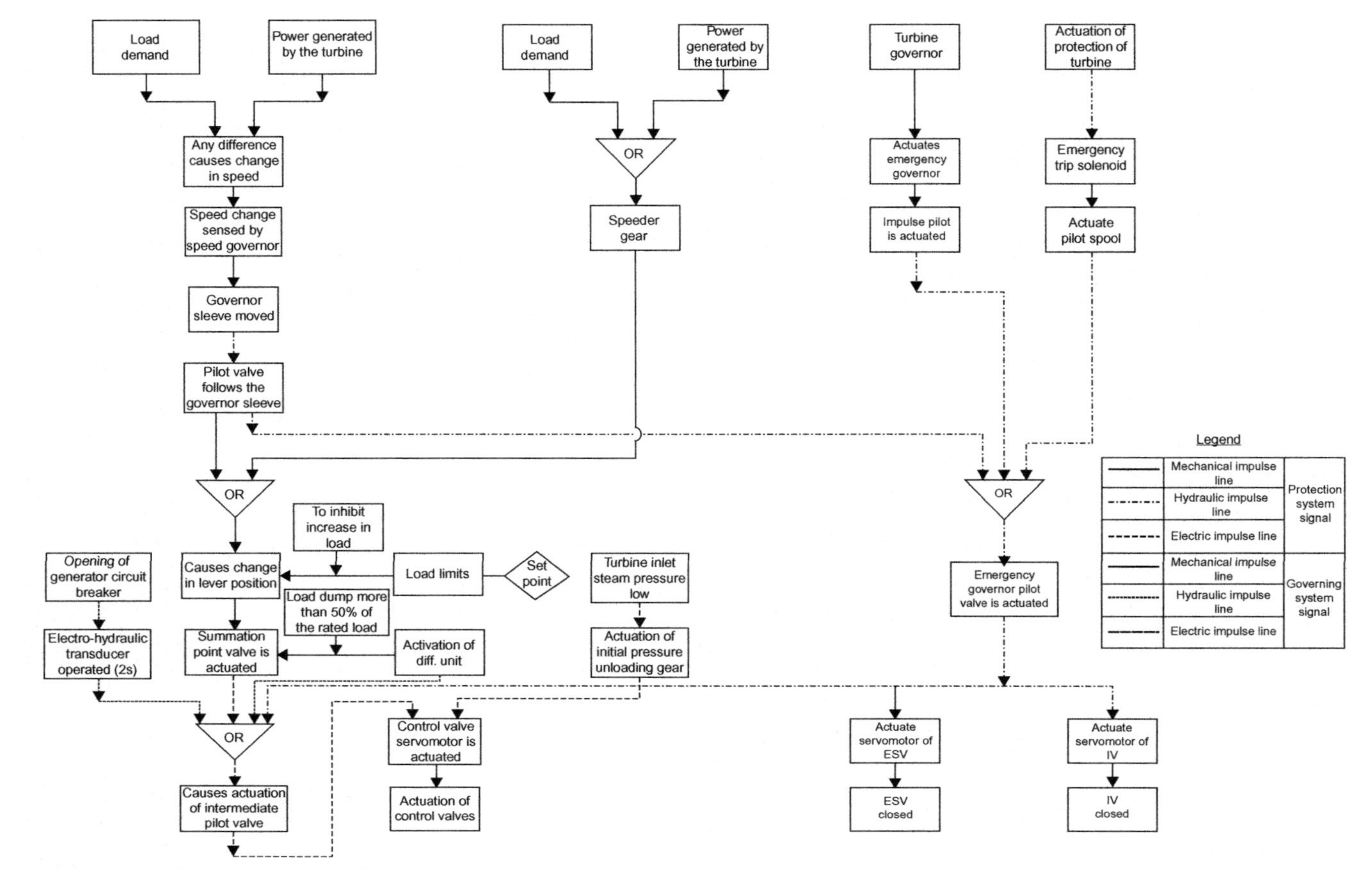

Figure 6.32 Block diagram of a typical governing system.

In power-station turbines a load increase will cause the shaft speed to slow down. The governor will respond according to the increased load until the rated speed is attained. The speed-governing system is arranged to ensure this by automatically regulating the steam input. The speed governor is a speed-sensing device that senses the change in the magnitude of speed within a preset range and produces a corresponding displacement that can be used as a corrective signal to control the steam flow. However, without some form of *droop*, the turbine speed regulation will always be unstable. Droop is a straight-line function that has a specific speed reference for every load condition. It is normally around 3–5% of the reference speed over the full range of load. Thus, a 5% droop governor with a reference speed of 3000 rpm at 100% load will have a reference speed of 3150 rpm at no-load (50 Hz at full-load and 52.5 Hz at no-load).

Utility turbines, connected to the grid, share loads with other running turbines operating in parallel. The grid determines the frequency of the connected alternator of the turbine. In the event the governor speed falls below the grid frequency, the power will flow from the grid and the alternator will act as a motor. Should the governor speed be higher than the grid frequency, the governor will go to full-load to make an attempt to raise the grid frequency. To prevent these problems the droop acts as a savior and causes the governor speed reference to decrease as the load increases. Normally hydraulic governors are provided with adjustable droop, but in most mechanical governors adjustment of droop is difficult.

6.6.1 Types of speed governors

6.6.1.1 Mechanical

The earliest known automatic turbine speed governor was a centrifugal fly ball type mechanical governor, which was driven by the main turbine shaft directly or through gearing. The centrifugal force acting on two revolving weights is opposed by the elastic force of a spring, so that the weights take up different radius for each speed and produce a proportional displacement of the sleeve linked to the fly balls through hinges (Figure 6.33).

6.6.1.2 Electrical

An electrical governor uses robust servomechanism and circuit components. An AC generator, driven by the turbine shaft, provides an electrical signal of a frequency proportional to the speed. A frequency-sensitive circuit produces a voltage proportional to this frequency that after amplification is fed to a torque motor that in turn produces a proportional displacement. (Additional details given in section 6.8.)

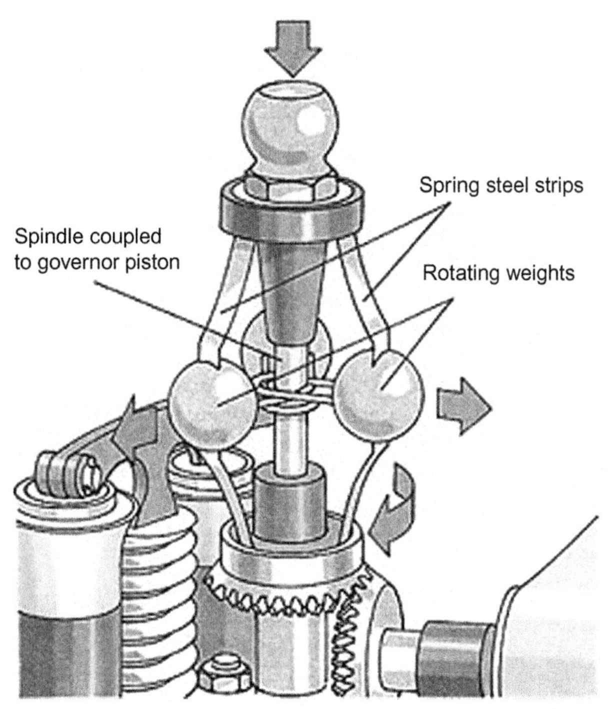

Figure 6.33 Flyball governor. *Source: Images for Flyball Governor. Google Web Page.*

6.6.1.3 Hydraulic

Put simply, a hydraulic governor for a turbine consists of a centrifugal pump driven from the turbine main shaft. The pressurized oil from it is fed into a cylinder containing a spring-loaded piston. The pressure is proportional to the square of the speed so that the position of the piston is a function of the speed.

6.6.1.4 Electro-hydraulic

An electro-hydraulic governor is a more recent innovation. The electrical part of an electro-hydraulic-governing system is comprised of equipment for detecting measured values (i.e., control valve lift, speed, load, and initial pressure) and processing of signals. The processed electrical signal is introduced at a suitable point in the hydraulic circuit. The hydraulic controls provide continuous control of large positioning forces for control valves. The electro-hydraulic converter is used as the connecting link between the electronic modules and the valve actuators. A change from electro-hydraulic governing to hydraulic governing during operation is possible. When one system fails, the other one comes into operation automatically. The displacement of the piston in a hydraulic governor, of the torque motor in the electrical governor, or of the sleeve of the flyball governor, which is a function of the shaft speed, is used to control the throttle valve of the turbine through hydraulic relay action or mechanical linkages or a combination of both.

The governing system in addition to ensuring the falling load-speed characteristics of the turbine, as mentioned above, also ensures the following functions:

- **i.** Run up of the turbine from rest to rated speed and synchronizing with the grid
- **ii.** Meets system load variations in a predetermined manner, when running in parallel with other machines
- **iii.** Protects the machine by reducing the load or shutting off completely in abnormal and emergency situations

6.6.1.5 Devices in governing system

1. Governing Devices:
   - **a.** Speed governor with pilot valve
   - **b.** Speeder gear or load-speed changer
   - **c.** Load limiting gear
2. Protection Devices:
   - **a.** Emergency trip valve
   - **b.** Over speed governor
   - **c.** Acceleration governor
   - **d.** Over-speed limiting gear
   - **e.** Pre-emergency governor
   - **f.** Low vacuum run back/unloading unit
   - **g.** Initial pressure regulator/low initial pressure unloading gear

6.6.2 Brief description of governing and protection system and devices

The stop valves and control valves in the steam lines to the turbine are actuated by hydraulic servomotors. The servomotors consist of a cylinder and a spring-loaded piston that is held in the open position by admission of high-pressure oil against the spring force. This spring force ensures positive closing of valves when the oil is drained out. The high-pressure oil supplied by an oil pump to the governing system is fed to the servomotors through their pilot valves. The position of the pilot valve determines the opening or closing of servomotor. The high-pressure oil that actuates the servomotor is usually called *power oil or control oil.*

The pilot valves of stop valve servomotors are positioned in the open position by yet another type of oil called *protection oil/trip oil*, either directly or through hydraulic relays. This protection oil is the same high-pressure oil but supplied through an *emergency trip valve.* In the 'reset' position the emergency trip valve admits oil through

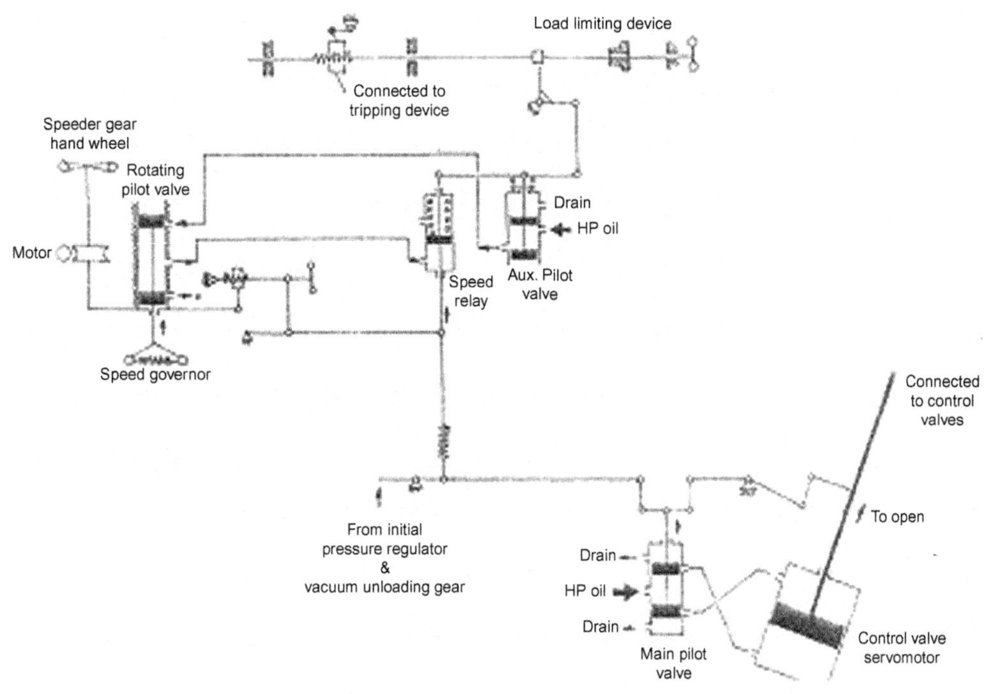

Figure 6.34 Speed governor pilot valve.

it to various pilot valves of servomotors, thereby enabling opening of emergency stop and control valves. In the trip position it shuts off the oil supply and drains the oil from the lines downstream of it, thereby ensuring quick closure of emergency stop and control valves.

The protection oil being fed to the HP control valve servomotor pilot valve-actuating device is regulated through the *speed governor pilot valve* (Figure 6.34). The change in speed that causes a corresponding change in governor pilot valve varies the oil pressure, which in turn regulates the position of the control valve through the servomotor and its pilot valve. There are variations in this arrangement. The other governing and protection devices, such as *load limiting gear, low vacuum unloading gear,* and *initial pressure regulator,* are hooked up into this control valve governing system through hydraulic relays and linkages.

In reheat turbines in cases of partial or full load throw off even after the HP control valves are fully closed, the entrained steam in reheaters and hot reheat steam line is more than enough to speed up the turbine above over speed limits. Hence, it is necessary to provide interceptor valves (stop valves) and control valves on the hot reheat steam inlet line to the IP turbine. While the IP interceptor valve is operated similar to the HP emergency stop valve, the IP control valve is controlled similar to the HP control valve but at a higher speed range by a secondary or pre-emergency governor as it is called. In specific designs the valve remains fully open at rated speed and starts closing at about 3% over speed and fully closed at about 5% over speed.

6.6.3 Governing devices

6.6.3.1 Speed governor pilot valve

The pilot valve consists of a movable sleeve with ports for oil inlet, outlet, and drain inside which the double bobbin valve moves, actuated by centrifugal flyball governor. The high-pressure oil or protection oil is admitted through the inlet ports of cylinder and sleeve. The pressure of the outlet oil to the control valve servomotor pilot valve-actuating relay is regulated by relative displacement/position of the piston and sleeve. Any change in shaft speed produces a corresponding change in pilot valve position, thereby causing a corresponding change in the control valve opening.

6.6.3.2 Speeder gear

A speeder gear is essential in the governing system of turbines in electricity generating stations for synchronizing the machine with the grid and to vary the load when operating in parallel. A speeder gear is needed to match the speed of the turbine to that of the grid while synchronizing. After synchronizing (the speed being determined by the grid frequency) the speeder gear is used to raise or lower load on the machine. The speeder gear is either operated manually locally or by a small motor remotely.

6.6.3.3 Load-limiting gear

This device is incorporated in the governing system to limit the maximum opening of HP control valves to the desired upper limit, in the event of loss of an auxiliary, e.g., boiler-feed pump, ID fan, FD fan, etc. This can be done mechanically by stopping the movement of linkages connected with relays in the control system or by limiting the sensitive oil pressure in the hydraulic system, thereby restricting the movement of speed relay by shutting off or draining the oil. A small motor is usually provided for remote operation of the gear.

6.6.4 Protection devices

6.6.4.1 Emergency trip valve

The function of the emergency trip valve has already been explained. Figure 6.35 shows a typical trip valve in the 'reset' position with a provision for manual and oil injection tripping and resetting. Remote tripping with an emergency push button can also be incorporated to trip the trip valve through a solenoid.

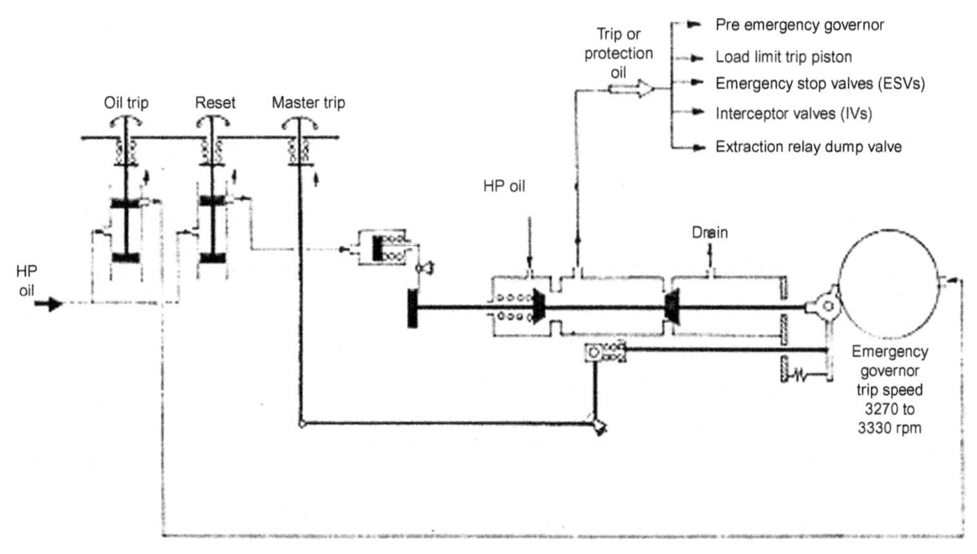

Figure 6.35 Emergency trip valve.

6.6.4.2 Over-speed limiting gear

The purpose of an over-speed limiting gear is to limit the over speed that will occur if a sudden loss of load takes place. The control operates on each of the emergency stop valves. On reheat turbines, it operates on both the HP emergency stop valves and the IP interceptor valves. The gear is comprised of an additional solenoid-operated pilot valve that releases oil from the steam stop valve power cylinder. When the solenoid is energized the valve closes rapidly under the action of spring, and the solenoid will de-energize and open the valve when the steam pressure drops.

6.6.4.3 Acceleration-sensing device

To deal with the rapid rise in speed that may occur in the event of a sudden loss of load, an acceleration-sensitive governor is fitted. On sudden loss of load this device causes the rapid closure of both the HP and IP control valves. When the acceleration of the turbine ceases the speed will have reduced to such a value that the main governor will keep the acceleration-sensitive governor closed. If for any reason the speed continues to rise, the over-speed governor will come into action and trip the turbine causing closure of the HP emergency stop valves and the IP interceptor valves.

6.6.4.4 Pre-emergency governor

The IP control valves/interceptor valves, either independently or along with the HP control valves, are operated either by another centrifugal governor or the same main speed governor by an acceleration-sensing differentiator and pilot valve. The operation

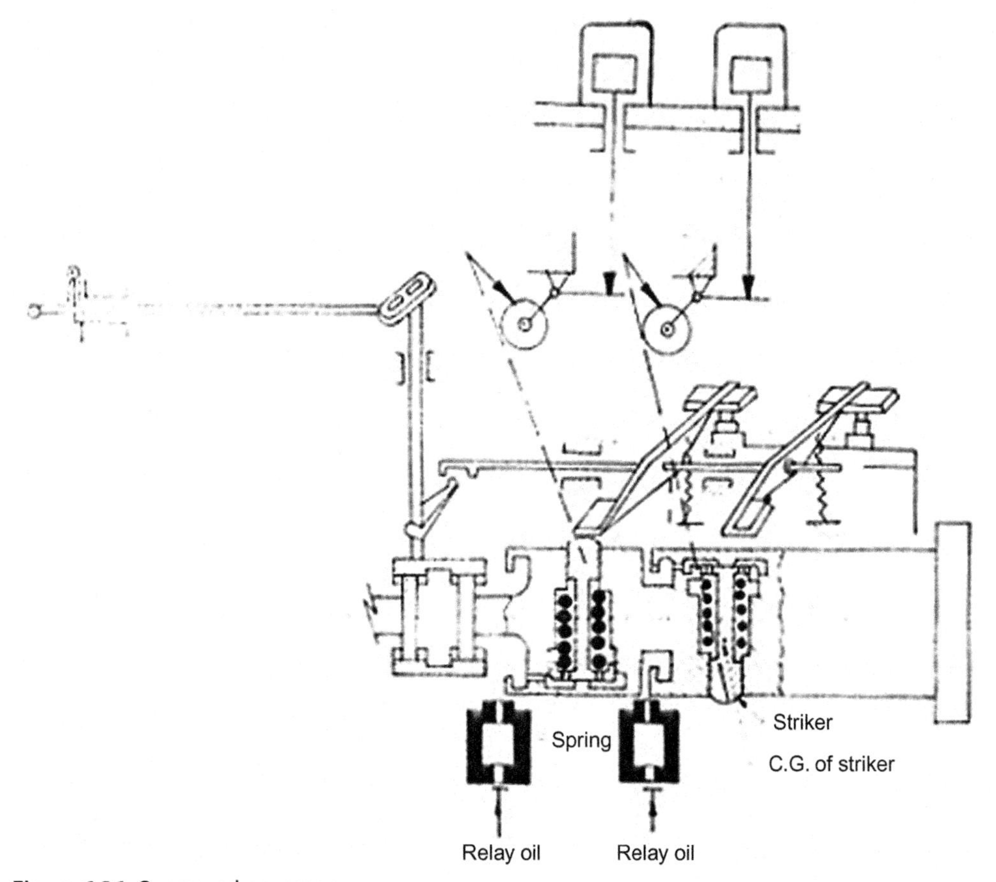

Figure 6.36 Overspeed governor.

is similar to an HP control valve. The IP control valves are normally fully open at a rated speed. In a typical case, they will begin to close at a speed above 3% and completely close at 5% above normal. The final speed rise, due to a time delay in the relay operation and the effect of the steam in loop pipes, is about 7–8% above normal. The pre-emergency governor does not function during normal operation.

6.6.5 Over-speed governor/emergency governor (Figure 6.36)

In the unlikely event of speed increasing to 109–111%, emergency or over-speed governors are provided to trip the machine. The emergency governor is normally mounted on the front end of the HP rotor and consists of two strikers for reliability. The center of gravity of the striker is away from the center line of rotation and is held in place by springs. As soon as the speed rises to 109–111% of the norm, the centrifugal force on the striker overcomes the force of the springs and the strikers fly out; once the strikers begin to move out, the distance between the center of gravity of strikers and the center line of rotation increases, causing further increase in centrifugal force. Thus strikers, once dislodged from the stable position, continue to move out until checked by the stop of the body. One or more of the strikers flying out will hit the trip levers placed close to the tip, which upon being hit will instantaneously trip the emergency trip valve causing complete shut off of all steam input to the machine (both HP and IP).

6.6.5.1 Testing of strikers by Oil Injection

It is necessary to periodically test the strikers to ensure their free movement. This is done by oil injection, without actually raising the speed of the machine. An actual over-speed test can be done in some turbines during start-up or shut-down by tripping the turbine. Provisions are also made in some turbines to test them during operation where the emergency trip valve is kept-gagged in the 'reset' position or the over-speed trip level is made out of action by the operator during the test period. A tell-tale indication, e.g. striker will remain in strike-out position till manually reset, is normally provided to show that the striker has flown out.

6.6.5.2 Low-vacuum unloading gear (Figure 6.37 and Figure 6.38)

In addition to normal control of the turbine a low-vacuum unloading gear is also provided to gradually close the turbine control valves when the pressure in the condenser rises above a preset value. The device works as follows: Displacement of a bellow connected to the condenser vacuum space is used to reduce the control oil pressure directly or by changing the position of the pilot valve, thus reducing the control valve opening. A provision is made on the device by means of which automatic reloading of the turbine with restoration of the condenser vacuum is prevented and reloading is manual. The stroke of the unloading gear is limited so that it can only close the valves down to a no-load position, to prevent the machine from motoring.

6.6.5.3 Initial pressure regulator/initial steam pressure unloading gear (Figure 6.39 and Figure 6.40)

The device is similar to low-vacuum unloading gear. If the throttle pressure falls, this device comes into operation and reduces the turbine load by closing the control valves until a balance is reached. It is provided with a "stop" to limit the stroke of the unloading gear so that the control valves close up to their 'no-load' position to avoid motoring and is also equipped with a cut-out device to completely block the regulator to permit start-up with lower boiler pressure. In the event of operation of this device, the reason for the pressure drop should be determined immediately and corrective measures should be taken to restore the pressure.

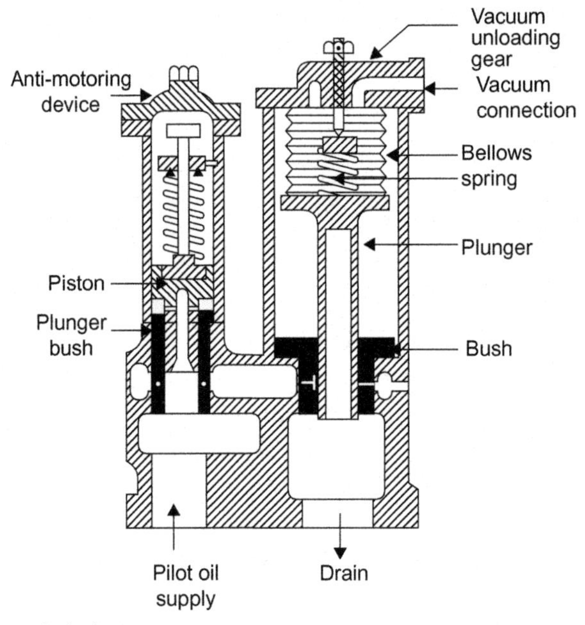

Figure 6.37 Vacuum unlodading gear.

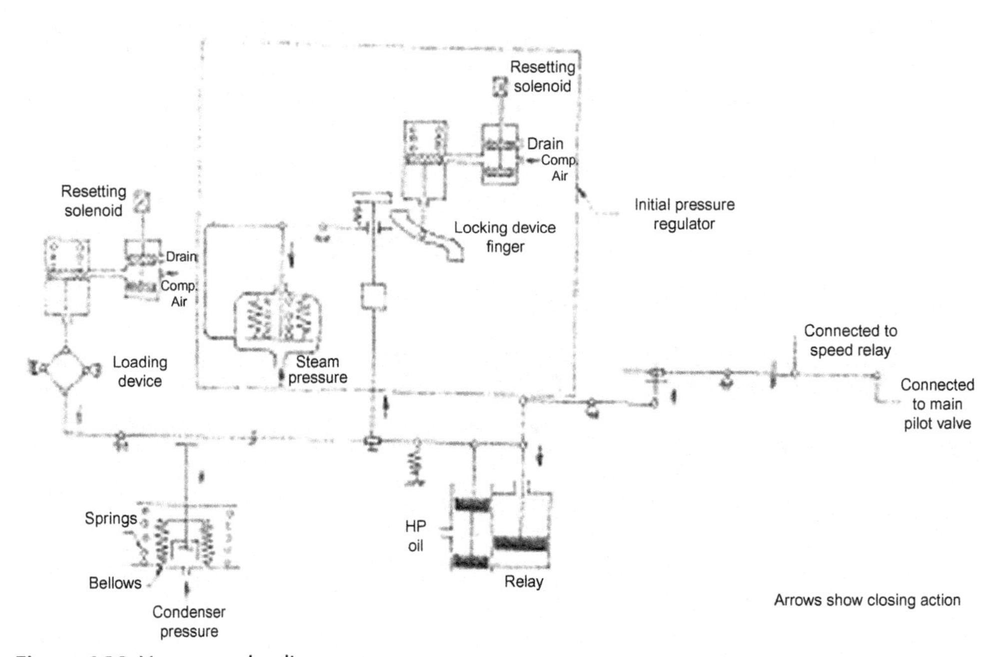

Figure 6.38 Vacuum unloading gear.

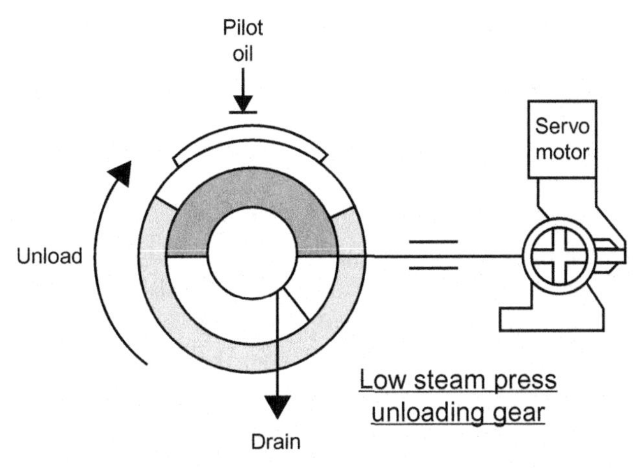

Figure 6.39 Typical initial pressure unloading gear.

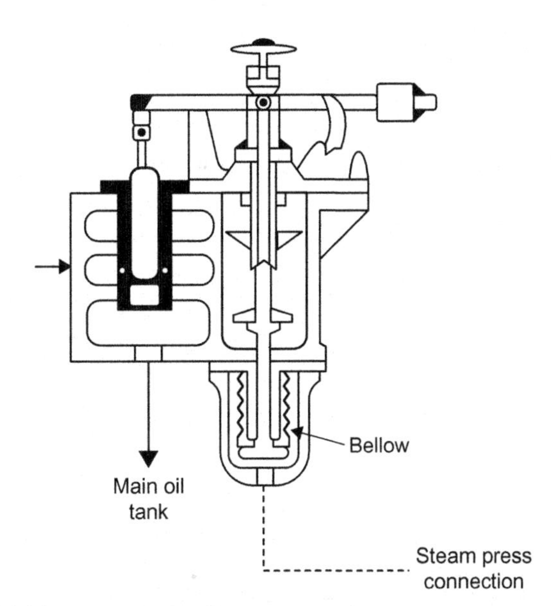

Figure 6.40 Typical initial pressure unloading gear.

6.6.5.4 On-load testing devices

The risk of serious damage to a large turbine due to sticking of one of the steam valves in the event of sudden loss of load is so great that provision is normally made for testing the freeness of movement of each valve stem while the turbine is running on load. In general, this is done with a device that releases the power oil from beneath the servomotor/relay piston of the valve being tested. Provision is made such that only one set of valves can be tested at a time with the machine load limited to the capability of the other set.

6.7 TURBINE LUBRICATION SYSTEM

Lubrication of the steam turbine is required to minimize rotating friction in journal bearings, thrust bearings, and reduction gears as well as to cool the journal and other bearing surfaces. Heat from hot parts of the turbine is also transferred from the shaft to bearings. Heat thus transmitted is carried away by lube oil, thereby maintaining bearing temperature at a safe level. Depending on the size and type of the steam turbine and also on their age the lubrication systems of steam turbines vary widely and may fall into either one or a combination of a ring oiling and circulating (pressure or gravity) system.

6.7.1 Ring oiling

In small turbines, the oil quantity required for lubrication is small, and the supply system is relatively simple. A ring rides loosely on the journal, and its lower part dips into the oil reservoir. As the shaft rotates, the ring revolves and correspondingly oil from the reservoir clings to the ring and is carried to the top of journal to which it transfers by direct contact. Grooves in the bearing permit the oil to travel in axial directions and provide proper lubrication at bearing areas along the length of the journal. Excess oil drains back to the reservoir beneath the bearing and aids in cooling the journal and bearing. In small turbines, the surface areas of the reservoir are large enough to radiate this accumulated heat and maintain the oil at a safe temperature. For some turbines, it is necessary to cool the oil by passing the circulating water through a chamber adjacent to the reservoir or through a coil immersed in the oil.

6.7.2 Pressure circulating systems

In larger units, the need for greater oil flow and larger heat dissipation dictate the use of the oil-circulating system (Figure 6.41). During normal operation, the main oil pump, mounted on and driven by the turbine shaft, draws oil from the main oil tank and discharges it at a high pressure to the bearing lubricating oil system. The lubricating oil pressure is usually maintained constant at about 100–150 kPa, depending on the particular turbine and bearing design. During the starting and stopping operations of the turbine, when the rotor and main oil pump are not up to the rated speed, an externally driven auxiliary oil pump is needed to supply the oil.

The main oil tank is equipped with vapor extractors to keep the oil tank atmosphere free of oil vapor and strainers to arrest suspended impurities. Oil used for the bearing lubricating oil system is cooled in the turbine oil coolers to control the

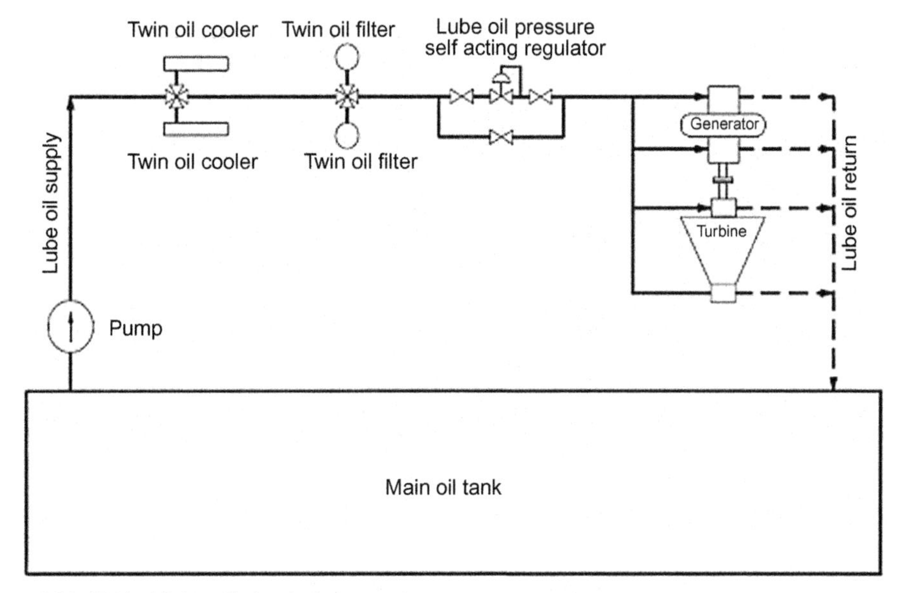

Figure 6.41 Typical lube oil circulating system.

bearing oil/metal temperature. Thereafter, oil passes through the oil filter to supply dirt-free lube oil to bearings. The oil that has passed through the filter is then led to bearings via a lube oil pressure-reducing regulator. From the bearings the lube oil drains back, by gravity, to the main oil tank, where any heavy impurities that may have been picked up tend to settle out.

6.7.3 Oil coolers

Some form of oil cooling is found in all lubrication systems. While direct radiation, water jackets, and simple coils are common in small turbines, separate oil coolers are usually used for circulation systems. They serve to remove the heat developed in bearings that in turn maintain low lube oil temperature and retard oxidation of oil. Excessive cooling of the oil, however, is detrimental as it causes precipitation of soluble products of oxidation on oil cooler tubes. The water pressure is maintained less than the oil pressure so that, in the case of a leak, no water will enter the oil.

6.7.4 Oil pumps

Circulation systems are designed with the main oil pump for bearing oil supply that is driven from the main turbine shaft. Auxiliary oil pumps for lubrication, when the turbine is not up to speed, are usually the motor-driven type. The auxiliary oil pumps are particularly useful during starting and stopping of the turbine.

6.7.5 Care and supervision of a turbine lubrication system

A careful check of the operating conditions of the lubrication systems and proper treatment of oil are important for prolonging oil life and maintaining adequate lubrication. In general, the following basic procedures are recommended:

1. Practice a regular method of purifying oil in service using an oil-purification or treatment system. Maintain treatment equipment in good order, e.g., by cleaning or changing filters and cleaning centrifuges and precipitation tanks.
2. Clean oil strainers regularly, usually daily.
3. Drain the bottom of the oil reservoir every day to remove water and other contaminants.
4. Check on water leakage by inspecting the turbine steam and water glands and oil cooler.
5. Check for oil leakage by inspecting bearing seals and oil piping.
6. Follow the manufacturer's recommendations for periodic cleaning and overhaul, as the frequency varies widely in different plants with service conditions. Yearly cleaning is generally good practice for most turbines.
7. Check oil level daily, and keep a record of the type and quantity of make-up oil added. Reservoirs of ring-oiled bearings should be drained monthly, and the oil replaced by new or clean oil. In circulation systems, users of inhibited oils should secure a brand of oil for make-up similar to that in the system.
8. When starting, follow the manufacturer's recommendations, but check whether the main oil pump is functioning and whether the correct oil pressure is developed.
9. Keep a daily log of bearing oil pressure and oil pressure to the governor mechanism.
10. Keep a daily log of water temperatures entering the cooler and leaving the cooler. A decrease in the normal temperature indicates sludge or water scale in the cooler.
11. Keep a daily log of oil temperatures entering and leaving the bearings, entering and leaving the cooler, and entering and leaving the gearbox.
12. Keep a log of all maintenance items such as overhauls, oil changes, oil additions, filter cleanings, centrifuge cleanings, etc.

6.8 SPEED CONTROL SYSTEM

The speed and power output of a turbine are controlled by the throttle valves. While there are various ways of achieving such control, two basic ways, in a very simplified form, are discussed in the following.

6.8.1 Hydro-mechanical speed control

A typical hydro-mechanical speed control system is presented in Figure 6.42, where it can be seen that the position of the pointer sets the desired speed of the turbine.

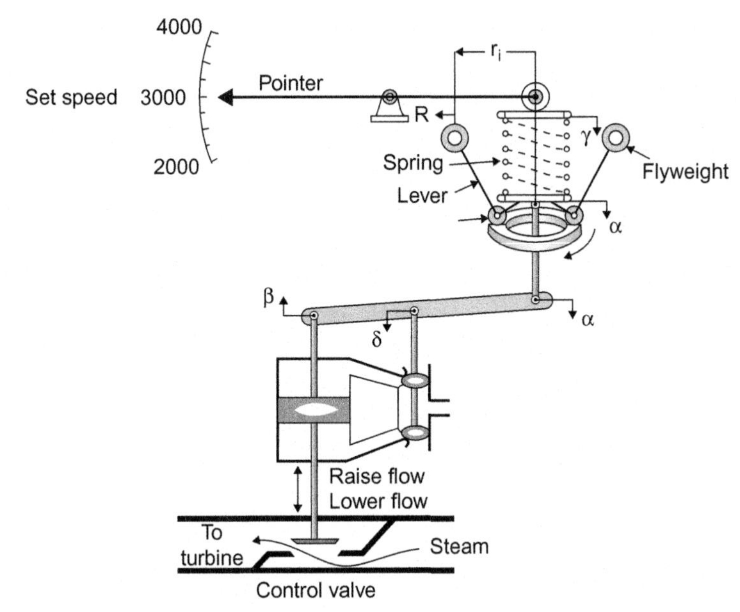

Figure 6.42 Hydro-mechanical speed control.

The flyweights are located so that their center of gravity lies at a distance of r_i from the center of their rotation. These flyweights are connected directly to the output shaft, so their angular speed is proportional to the output speed of the shaft.

A lever rotates with the flyweights as a unit and transmits the centrifugal force from the flyweights to the bottom of a lower spring seat. In the event that the speed of the turbine drops below its reference value the centrifugal force of the flyweight decreases, which in turn causes the force exerted on the bottom of the spring to decrease. The combined effect of such a reduction is to move both α and δ downward. This downward movement of δ causes the control fluid to flow to the bottom of a big piston raising β upward and opening the flow control valve wider. By supplying more steam/fuel, the speed of the turbine will increase until equilibrium is again reached.

If the speed of the turbine exceeds the rated speed reverse mode of actions as above will follow.

6.8.2 Electrical speed control

Figure 6.43 shows a typical electrical speed control system. The potentiometer in this control system provides a reference input voltage e_m, which is proportional to the desired speed n_m. The output of the speed pick-up in this control system is voltage e_d, which is again proportional to the controlled output speed n_d. The error $(e_m - e_d)$ is then amplified by the amplifier, whose output e_F when applied to the field induces a

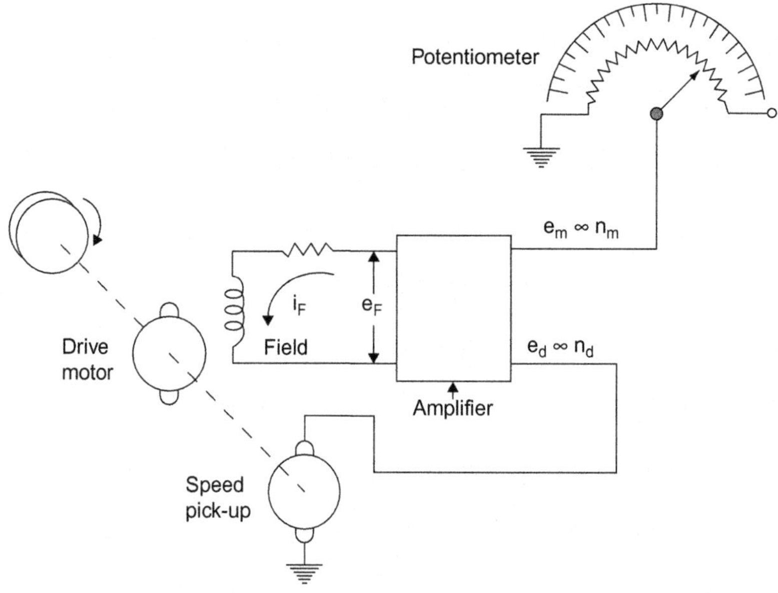

Figure 6.43 Electric speed control.

field current i_F, which in turn is proportional to the torque exerted on the rotor by the drive motor. Thereby, the speed of the turbine is restored even though the changed load condition is maintained.

6.9 TURBINE LOSSES [3,4]

Like the operation of all roto-dynamic equipment, the operation of steam turbines also suffers from various losses. First, loss occurs at the control valves, where a throttling loss of about 5% of the main steam stop valve upstream pressure is imminent for speed regulation. Bearings and thrust blocks involve frictional losses. In addition, there is radiation loss as well as more serious losses, as discussed below. The objective of each design is to keep these losses to a minimum.

6.9.1 Fluid friction

Friction and eddies cause small pressure drops in the high-pressure nozzle box. Further, loss in the nozzle itself is quite substantial, caused by a change in the direction of the steam flow path while passing through an angle. The high-pressure turbine disc runs in a dense medium, resulting in a considerable fluid resistance called disc friction. In the case of partial admission of steam in the first stage, blades that are not influenced by the steam jet idly churn the steam in the casing causing windage loss. In addition, the high speed of the rotor creates a centrifugal force on the steam, causing a portion of the steam to move radially and get dragged by the moving blades. The cumulative effect of all above losses is called fluid friction, the cumulative value of this loss is the highest among all turbine losses.

6.9.2 Leakage

Steam leak through turbine gland seals does not perform any useful work, hence results in losses. Leakage of steam also takes place from bearing ends. While flowing through the turbine steam leaks through the clearance between the turbine casing and the tip of the moving blades. Energy contained in leaking steam cannot be utilized for doing any work and is accounted as a loss of available energy.

6.9.3 Leaving loss

The exhaust steam from a low-pressure turbine enters the condenser with substantial velocity and the kinetic energy corresponding to this velocity represents a loss of energy.

6.9.4 Supersaturation

When superheated vapor expands to point 1 across the saturated vapor line (Figure 6.44) a change of phase should begin to occur. However, if this expansion

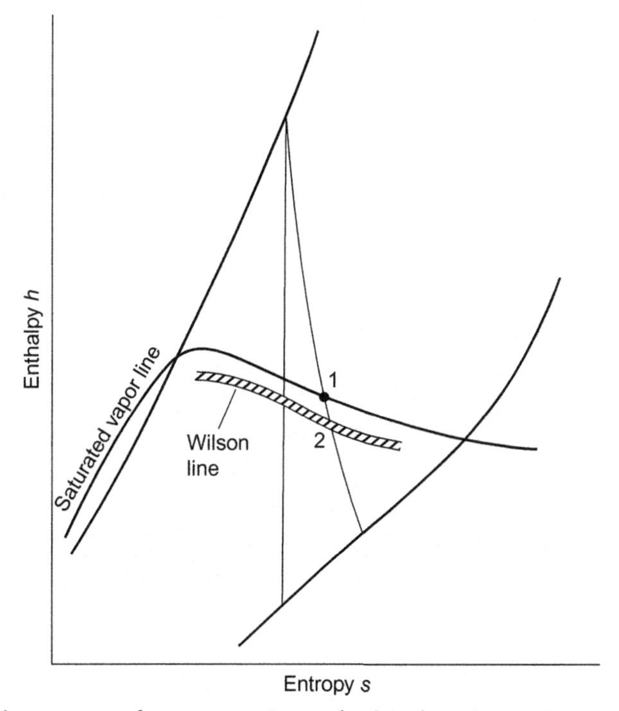

Figure 6.44 Enthalpy-entropy of supersaturation and wilson line. *Source: Figure 5-19, P198 [4].*

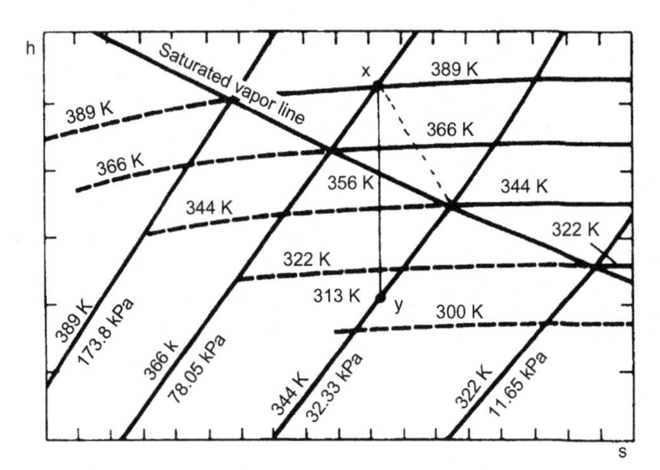

Figure 6.45 Enthalpy-entropy diagram of supersaturation.

takes place rapidly, as in the case of flow through nozzles and turbines, the achievement of equilibrium between the liquid and vapor phases is delayed, and the vapor continues to expand in a dry state beyond point 1. Condensation of vapor starts at some point 2 in the wet region, such that states of the fluid between point 1 and point 2 are not states of stable equilibrium, but of *transient equilibrium*. At point 2 condensation of vapor *suddenly* takes place and the condition of the vapor once again comes back to *thermodynamic equilibrium*. Between region 1 and 2 the vapor is said to be *supersaturated* or *undercooled* [3]. The latter term is adopted to denote that at any pressure between points 1 and 2 the temperature of the vapor is always less than the saturation temperature corresponding to that pressure. For example, at any given pressure a supersaturated fluid at point y (Figure 6.45) has a lower temperature, about 313 K, than if it were in thermodynamic equilibrium (356 K). The locus of point 2 at various pressures is called the Wilson line.

6.9.5 Moisture loss

Large steam turbines are designed so that dryness fraction of exhaust steam will not fall below about 88%. In the event that the dryness fraction falls below 88% or the moisture content of the exhaust steam exceeds 12%, the liquid droplets will strike the moving blades at different angles, thereby reducing the mechanical work of the rotor. Some liquid droplets will also get carried away with high-velocity steam and thus with the exchange of momentum reduce the energy content of steam. Higher size droplets will also cause erosion of blade materials. From an operational point of view, contaminants in the steam, such as oxygen causes corrosion, sodium, and chlorine from a water treatment plant, can cause stress-corrosion cracking and erosion.

6.9.6 Heat-transfer losses

All modes of heat transfer, e.g., conduction, convection, and radiation, are responsible for causing energy losses from a turbine. While heat loss through conduction in a turbine takes place both internally among metal parts, and externally between turbine casing and the foundation, convection and radiation losses take place from the high-temperature zone of the turbine casing to its surroundings.

6.9.7 Mechanical and electrical losses

Mechanical losses, e.g., frictional losses in bearings, governor mechanism, etc., are usually constant and independent of load. The turbine must also supply mechanical work to accessories such as oil pumps, etc. In general, mechanical losses are fairly small, amounting to 1% or less of the total available energy. The efficiency of modern large electric generators is around 98–99%, and the remaining portion is considered loss.

6.10 OPERATING REGIME

In any steam power plant the operating regime of the steam turbine should be selected to ensure that it will perform the following:

- **i.** Provide the minimum thermal stresses in turbine material
- **ii.** Ensure the most economical generation of power
- **iii.** Require the least demand on operating personnel

Normally the steam temperature entering a a turbine is kept as high as permitted by the metallurgical limit of internal materials used in high-pressure and intermediate-pressure turbines. Throughout the operation of a turbine the steam temperature is also maintained constant to harness the benefit of maximum attainable efficiency. Hence, to meet these duties the following three operating regimes are commonly used [9]:

1. Constant pressure operation
2. True sliding pressure operation
3. Modified sliding pressure operation

6.10.1 Constant pressure operation

The start-up of a steam turbine from any state, be it cold, warm, or hot, begins from a steady state of reduced steam conditions. During start-up initial (throttle) steam pressure and initial (throttle) steam temperature are selected following the manufacturer's guidelines. Once the turbine is synchronized and starts raising loads both pressure and temperature are gradually raised to reach their rated conditions. During normal running of the unit the changes in power output are carried out at rated initial steam conditions. However, while reducing load, although throttle steam pressure remains unchanged, the throttle steam temperature is kept unaltered down to a load of about 60% in a typical pulverized coal-fired boiler. Below this load it is imperative to reduce the steam temperature for stability of boiler operation. Nozzle control is normally used for this mode.

6.10.2 True sliding pressure operation

When the turbine is started, the initial steam pressure is increased gradually to its rated value beginning at the reduced initial value. During normal operation the main steam flow through the turbine is controlled by sliding the throttle steam pressure progressively from the initial value to the rated value in proportion to the generation output, while keeping the turbine control valves fully open. In true sliding pressure operation, changes in both pressure and load take place linearly, attaining 100% throttle pressure at 100% load.

At part load the thermal efficiency of the steam turbine operating with true sliding pressure operation will be better than the efficiency of the turbine if it operates with constant pressure operation, due to following reasons:

i. Smaller governing valve loss enables improvement of high-pressure turbine internal efficiency

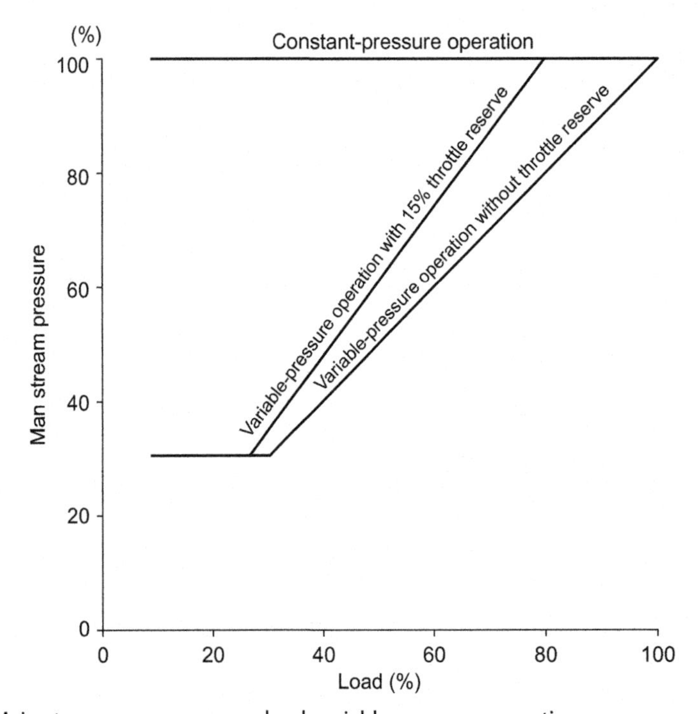

Figure 6.46 Main steam pressure versus load-variable pressure operation.

ii. Boiler feed pump throughput gets reduced

iii. Boiler reheat steam temperature can be higher because of higher temperatures in the high-pressure turbine exhaust steam

Another benefit of adopting variable pressure operation is this mode results in extremely fast turbine start-up time.

6.10.3 Modified sliding pressure operation

In the case of sudden load demand the turbine control valves can react almost instantaneously to take corrective action, but the boiler fuel control cannot react with the same speed because of its sluggish response. To address this problem, a modified sliding pressure operation with some throttle reserve has become attractive to system designers. With this mode, while throttle steam pressure is raised progressively from an initial value the rated throttle steam pressure is attained at 80–85% control valve opening, instead of 100%. Therefore, whenever there is sudden load demand, the turbine control valves open immediately, allowing more steam to flow through the turbine according to the demand, but with a sacrifice of throttle steam pressure. Boiler fuel control then follows the turbine control to raise the throttle steam pressure to its rated value.

Throttle governing is the most favorable method of control in the above two modes. Figure 6.46 shows the "main steam pressure vs. load" characteristic of "constant pressure", "true sliding pressure", "modified sliding pressure" operations.

BIBLIOGRAPHY

[1] Key World Energy Statistics – International Energy Agency, <http://www.iea.org/publications/freepublications/publication/keyworld2014.pdf>.
[2] Wrangham DA. Theory and practice of heat engine. The English Language Book Society; 1961.
[3] Kearton WJ. Steam turbine theory and practice. The English Language Book Society; 1962.
[4] El-Wakil MM. Power plant technology. Singapore: McGraw-Hill Book Company; 1984.
[5] The Natural Edge Project 2008. Lesson 6—Electricity from Steam. Sustainable Living Challenge.
[6] Rogers GFC, Mayhew YR. Engineering thermodynamics work and heat transfer. The English Language Book Society; 1967.
[7] Baumeister T, Marks LS, editors. Mechanical engineers handbook. 6th ed. McGraw-Hill Book Company, Inc.; 1958.
[8] Chaplin RA. Thermal power plants—Vol III. UNESCO – Encyclopedia of Life Support Systems (EOLSS).
[9] British Electricity Institute. Modern power station practice volume G station operation and maintenance. London: Pergamon Press; 1991.

CHAPTER 7

Gas Turbine and Heat Recovery Steam Generator

7.1 INTRODUCTION

Back in 1500 Leonardo da Vinci (1452–1519) sketched a machine, called the "Chimney Jack," that extracted mechanical energy from a gas stream. Hot air from a fire rose through a series of fans that connected and turned the roasting spit. Giovanni Branca developed a machine in 1629 in which jets of steam rotated a turbine that then rotated the driven machinery of a stamping mill. In 1678 Ferdinand Verbeist built a model carriage relying on a steam jet for power. In 1791 a basic turbine engine was patented with all the same elements as today's modern gas turbines to power a horseless carriage. However, until the nineteenth century no practical implementation of such machines was considered. That is, until the American engineer George Brayton (1830–1892) proposed a cycle consisting of a gas compressor, mixing chamber, and expander that used a combustion chamber exhausting to the atmosphere.

In the original nineteenth-century Brayton engine, ambient air is drawn into a piston compressor, where it is compressed isentropically. The compressed air then runs through a mixing chamber where fuel is added at constant pressure (isobaric process). The heated (by compression) pressurized air and fuel mixture is then ignited in an expansion cylinder and energy is released, causing the heated air and combustion products to expand isentropically through a piston/cylinder. Some of the work extracted by the piston/cylinder is used to drive the compressor through a crankshaft arrangement. In 1872 German engineer F. Stolze patented a machine that predicted many of the features of a modern gas turbine engine, with a gas compressor, burner (or combustion chamber), and an expansion turbine, but the engine never ran under its own power.

The first gas turbine was built in 1903 by a Norwegian named Ægidius Elling that was able to produce more power than it needed to run its own components. At that time, knowledge of aerodynamics was limited, and Elling's invention was considered a remarkable achievement. Using rotary compressors and turbines it produced an equivalent amount of about 8 kW power (substantially high at that time in respect of prevailing scenario). Elling's work was later used by Sir Frank Whittle. The first application for a gas turbine engine was filed in 1914 by Charles Curtis. General Electric, one of the leading gas turbine manufacturers of today, started their gas

Thermal Power Plant
DOI: http://dx.doi.org/10.1016/B978-0-12-801575-9.00007-X

turbine division in the year 1918. In 1920 Dr. A. A. Griffith applied the experience of gas flow past air foils into gas flow through passages of a turbine. Sir Frank Whittle patented the design for a gas turbine for jet propulsion in the year 1930. His work on gas propulsion relied on all works carried out earlier in the same field. Whittle once stated that his invention was greatly influenced by the work of Ægidius Elling in the absence of which it would have been difficult for Whittle to achieve his goal. The first successful use of his engine was in April 1937. In the year 1936, Hans von Ohain and Max Hahn of Germany developed their own patented engine design at the same time that Sir Frank Whittle was developing his design in England.

All of the aforementioned chronological development led to the first modern gas turbine plant being installed in 1950.

The modern-day *gas turbine*, also called a *combustion turbine*, is a rotary engine that extracts energy from a flow of combustion gas. The gas is normally air, or products of combustion of fuel and air. The essential features of a gas turbine are [1]:

i. Compressor that raises the air pressure
ii. Combustion chamber/s in which fuel is sprayed into the pressurized air from compressor for combustion
iii. Gas turbine through which high-temperature products of combustion from the combustor pass and are expanded

The upstream compressor is directly coupled to downstream turbine, and the combustor or combustion chambers are located in-between (Figure 7.1). The air is compressed in the compressor, after which it enters a combustion chamber where energy is released when air is mixed with fuel and ignited in the combustor. As a consequence of the combustion at constant pressure the temperature of air is increased. After leaving the combustion chamber the resulting high-temperature gases are

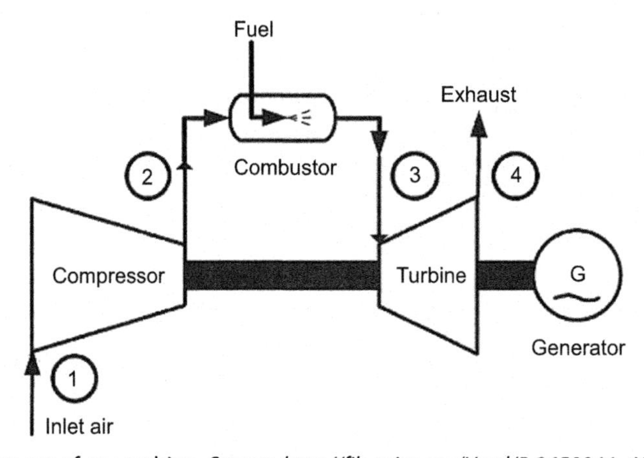

Figure 7.1 Equipment of gas turbine. *Source: http://file.scirp.org/Html/5-2650044_40880.htm.*

directed over the blades of the gas turbine, spinning the turbine, which in turn powers the compressor. Finally, the gases are passed through additional turbine blades generating more thrust by accelerating the hot exhaust gases by expansion back to atmospheric pressure thereby performing useful work. Energy is extracted in the form of shaft power, compressed air, and thrust, in any combination of these, and used to power aircraft, trains, ships, electrical generators, and even tanks.

Gas turbines for industrial and utility applications have many advantages. Compared with conventional steam power plants with large steam generators and bulky condensers, gas turbines, along with associated systems, are small in size and mass. Thus, the transport of gas turbine units is simpler.

Gas turbines offer flexibility in using a range of liquid and gaseous fuels, such as crude oil, heavy fuel oil, natural gas, methane, and distillate "jet fuel," gas produced by gasification processes using coal, municipal waste, and biomass, as well as gas produced as a by-product of oil refining.

Gas turbines are ideal for electricity generation in periods of peak electricity demand, since they can be started and stopped, quickly enabling them to be brought into service as required to meet energy demand peaks. They are smooth running, and their completion time to full operation is the fastest compared to other power generating plants. Gas turbines also have the lowest capital and maintenance costs. They are also subject to fewer environmental restrictions than other prime movers [2].

Their main disadvantage, however, is the high heat rate, which restricts its use for electricity generation. This restriction, nevertheless, could be overcome by adopting a higher turbine inlet temperature by using high-grade material capable of withstanding high metallurgical limit. For every 55 K increase in temperature, the work output of a gas turbine unit can be enhanced by about 10% and the heat rate reduced by approximately 1.5% [3]. Another disadvantage of the gas turbine is its incompatibility with solid fuels [2].

There are two basic types of gas turbines: aero derivative and industrial. Aero-derivative units were primarily designed as aircraft jet engines, modified subsequently to drive electrical generators. They are lightweight and thermally efficient, but usually more expensive than gas turbines built exclusively for stationary applications. Currently, the capacity of a single-unit stationary type of aero-derivative gas turbine is maximum 50 MW, with compression ratios in the range of 30:1, using a high-pressure external fuel gas compressor.

Industrial gas turbines are exclusively used for stationary power generation. The capacity of a single unit could be as high as 450 MW [4]. Industrial gas turbines have a lower capital cost per kilowatt installed than aero-derivative units. Other benefits of industrial gas turbines over aero-derivative gas turbines are they are more rugged, can operate longer between overhauls, and are more suitable for continuous base-load operation with longer inspection and maintenance intervals, and their compression ratio is low (up to 16:1), requiring no external fuel gas compressor. The disadvantages of industrial gas turbines are they are less efficient and much heavier.

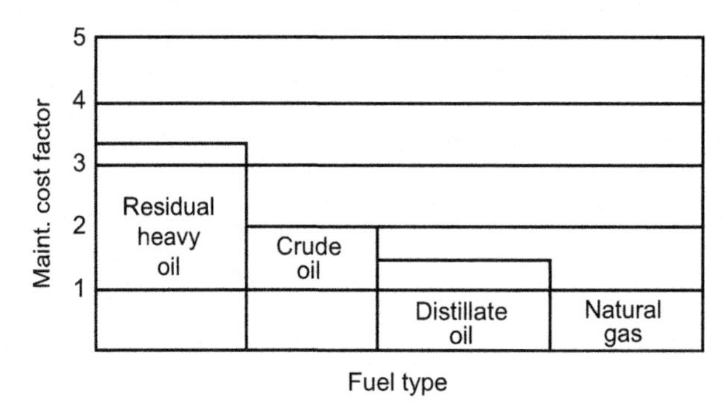

Figure 7.2 Effect of fuel on maintenance cost.

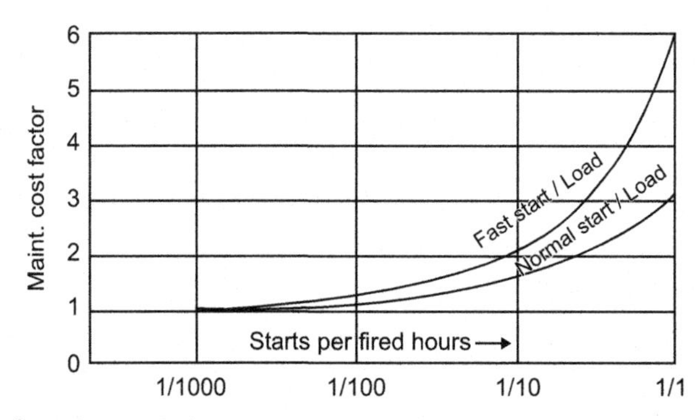

Figure 7.3 Effect of starting on maintenance cost.

The fuel to be used in a gas turbine must be free of chemical impurities and solids as these either stick to the blades of the turbine or damage components in the turbine that operate at high temperature. When natural gas is used, the power output and thermal efficiency of gas turbines are higher than when using most liquid fuels. In contrast, about 6% of liquid fuel is wasted during combustion as compared to 11% for natural gas [5]. The relative maintenance cost is the lowest while firing natural gas in a gas turbine (Figure 7.2). The maintenance costs of a gas turbine are greatly influenced by **i.** the type of starting (normal/fast) (Figure 7.3), **ii.** the frequency of starting in a specified period (Figure 7.3), as well as **iii.** the loading pattern (Figure 7.4).

Gas turbines can be used in a variety of configurations as follows:

1. Simple cycle operation, i.e., a single gas turbine producing power only.
2. Combined heat and power (CHP) operation, which is a simple cycle gas turbine with a heat recovery heat exchanger that recovers the heat from gas turbine exhaust to generate low-pressure steam or hot water.
3. Combined cycle operation in which high-pressure steam generated from recovered exhaust heat is passed through a steam turbine to produce additional power.

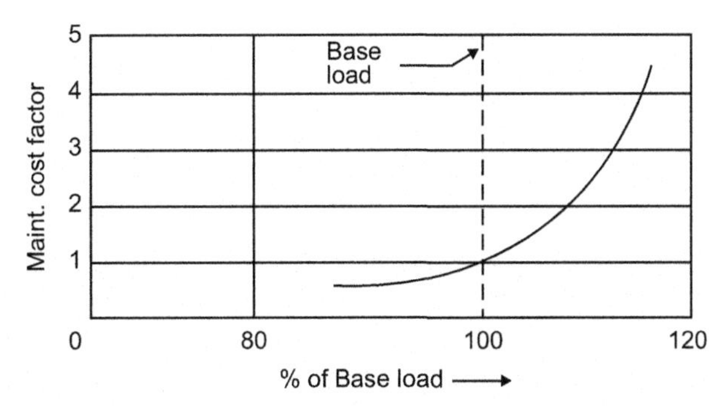

Figure 7.4 Effect of loading on maintenance cost.

When used in a simple cycle with no heat recovery, the heat left over in exhaust gases from the gas turbine set is substantial, to the extent of 60% or more of heat input to the set, and is lost to the atmosphere, resulting in poor thermal efficiency of a gas turbine. The gas exhausted from the turbine is not only plentiful and hot (673–823 K), but it also contains substantial amount of oxygen (normally a gas turbine operates with 300–400% excess air and its exhaust contains about 16–18% oxygen). A 50-MW gas turbine might discharge approximately 280 kg/s of exhaust gases at about 753 K, with an equivalent value of about $580*10^6$ kJ/hr of energy or approximately 75% of initial fuel input. These factors point to using the hot, oxygen-rich gas in a steam-generating plant, whose steam output will drive a steam turbine. The use of such wasted heat in a *heat-recovery steam generator* (HRSG) is the basis of a "*combined cycle gas turbine* (CCGT)" plant (Figure 7.5).

A gas bypass stack and silencers are frequently installed downstream of the gas turbine (Figure 7.6) so that it can be operated independently of the steam cycle. Silencers are also installed at air inlet to compressor.

The approximate site performance of a gas turbine may be predicted by applying the following rule of thumb correction factors based on Original Equipment Manufacturer (OEM) recommended International Standards Organization (ISO) performance parameters [5]:

i. For each 0.56 K (1°F) rise in ambient air temperature above 288 K there would be about 0.3–0.5% drop in power output, with a proportionate increase in heat rate.
ii. For every 305 m (1000 ft) increase in site elevation above sea level, gas turbine power output would reduce by about 3.3%.
iii. For every 25-mmwc inlet pressure drop, power output is expected to be reduced by 0.5% with 0.1% increase in heat rate.
iv. For every 25-mmwc outlet pressure drop, the power output is expected to be reduced by 0.15% with 0.1% increase in heat rate.
v. There would be about 2–3% less power output with 1–2% higher heat rate when operating on distillate fuel as compared to natural gas.

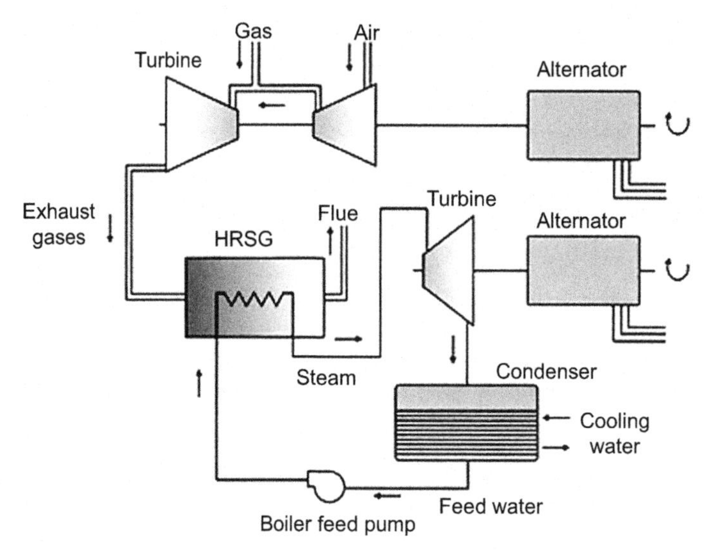

Figure 7.5 Combined cycle gas turbine flow diagram. *Source: http://electrical-engineering-portal.com/an-overview-of-combined-cycle-power-plant.*

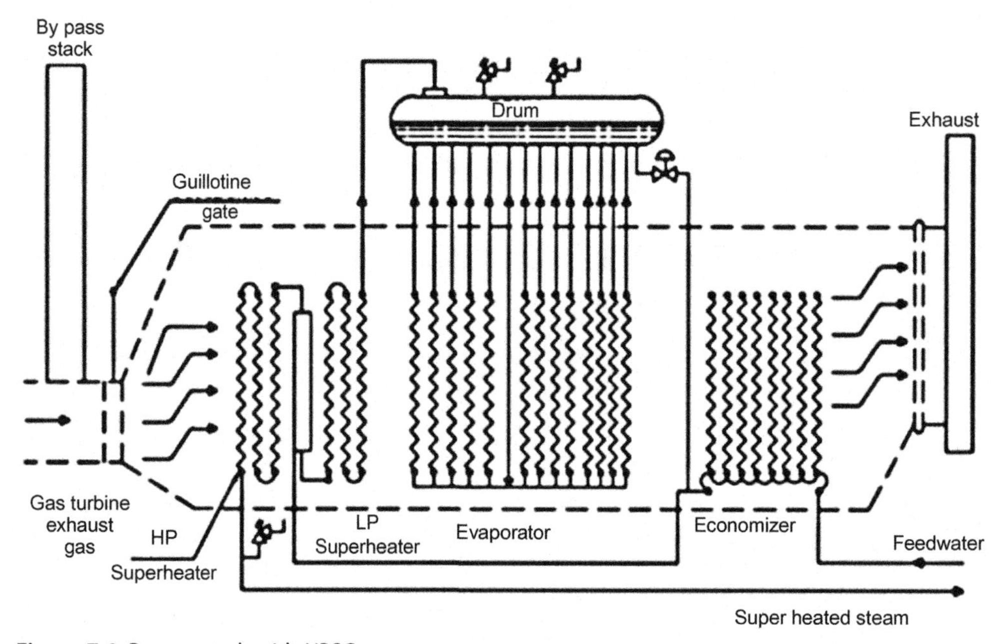

Figure 7.6 Bypass stack with HRSG.

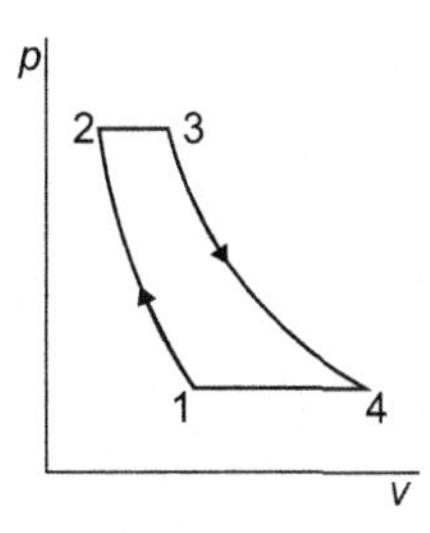

Figure 7.7 P-V diagram of ideal brayton cycle.

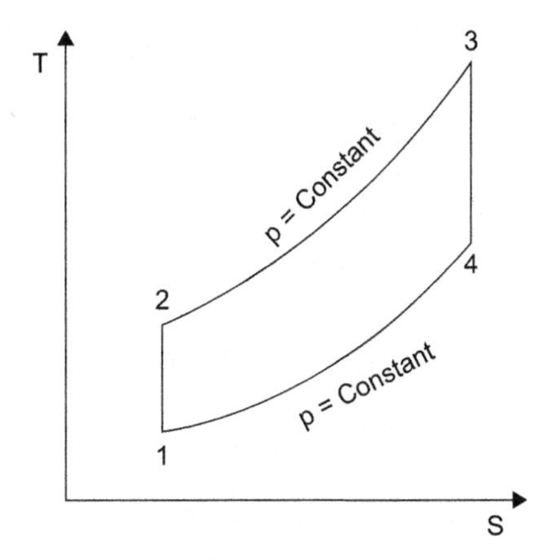

Figure 7.8 T-S diagram of ideal brayton cycle.

7.2 IDEAL BRAYTON CYCLE

The P-V and T-S diagrams of an ideal Brayton cycle, also known as the *Joule cycle*, are shown in Figures 7.7 and 7.8 for a perfect gas-like air with constant specific heats. The cycle is comprised of two isentropic and two isobaric processes. The compressor draws in ambient air and compresses it isentropically from position 1 to position 2 at pressure P_2. Then the compressed air is discharged into a combustion chamber, where fuel is burned, heating that air at constant-pressure from position 2 to position 3. The heated, compressed product of combustion (gas) then creates energy, as it expands through a turbine isentropically from position 3 to position 4. The gas is then cooled from position 4 to position 1 at constant pressure P_1. In a closed cycle, cooling of gas takes place within a heat exchanger, but in an open cycle, gas is discharged to the atmosphere.

Figures 7.7 and 7.8 reveal that the steady-flow constant pressure processes during which heat is transferred are no longer constant temperature processes. Hence, the ideal efficiency of the Brayton cycle has to be much less than the efficiency of the

Carnot cycle based on the maximum and minimum temperatures of the cycle. Also, in comparison with the positive expansion work, $C_P\ (T_3-T_4)$, the negative compressor work, $C_P\ (T_1-T_2)$, is in substantially high proportion, so the net work output is quite low. Some of the work extracted by the turbine is used to drive the compressor. Thus, the efficiency of the Brayton cycle is much poorer than that of the Rankine cycle.

The net work output and cycle efficiency of an ideal Brayton cycle may be calculated by using the ideal gas law $PV = nRT$ and $\gamma = C_P/C_V$. Since the adiabatic condition $PV^{\gamma} =$ constant, the following relations hold:

$$\frac{V_3}{V_2} = \frac{V_4}{V_1} = \frac{T_3}{T_2} = \frac{T_4}{T_1} \tag{7.1}$$

$$\frac{T_2}{T_1} = \frac{T_3}{T_4} = \left(\frac{V_1}{V_2}\right)^{\gamma-1} = \left(\frac{V_4}{V_3}\right)^{\gamma-1} = \left(\frac{P_2}{P_1}\right)^{\frac{\gamma-1}{\gamma}} \tag{7.2}$$

Assuming constant specific heat of air/gas and relatively negligible change in the kinetic energy of gas at inlet and outlet of each component, work and heat transfers may be expressed as

Work input to compressor

$$W_{12} = C_P(T_1 - T_2) \tag{7.3}$$

Work output from turbine

$$W_{34} = C_P(T_3 - T_4) \tag{7.4}$$

The heat supplied to the cycle

$$Q_{23} = C_P(T_3 - T_2) \tag{7.5}$$

Thus, the cycle efficiency is

$$\eta = \frac{W_{12} + W_{34}}{Q_{23}} = \frac{(T_3 - T_4) - (T_2 - T_1)}{(T_3 - T_2)} \tag{7.6}$$

(for the ideal case there are no pressure losses in the cycle). Hence, for isentropic compression and expansion the pressure ratio across the compressor and the turbine is

$$r_p = \frac{P_2}{P_1} = \frac{P_3}{P_4} \tag{7.7}$$

Putting the value of Eq. 7.7 in Eq. 7.2

$$T_4 = \frac{T_3}{r_p^{\frac{\gamma-1}{\gamma}}} \tag{7.8}$$

$$T_1 = \frac{T_2}{r_p^{\frac{\gamma-1}{\gamma}}} \tag{7.9}$$

Putting values of T_4 and T_1 from Eq. 7.8 and Eq. 7.9 in Eq. 7.6, the ideal air-standard efficiency of Brayton cycle becomes

$$\eta = \frac{T_3\left(1 - \frac{1}{r_p^{\frac{\gamma-1}{\gamma}}}\right) - T_2\left(1 - \frac{1}{r_p^{\frac{\gamma-1}{\gamma}}}\right)}{T_3 - T_2} \tag{7.10}$$

or,

$$\eta = 1 - \frac{1}{r_p^{\frac{\gamma-1}{\gamma}}} \tag{7.11}$$

From Eq. 7.11 it is evident that for any ideal gas, since γ remains unchanged, the thermal efficiency of the cycle is a function of r_p alone and increases with it. The efficiency is also independent of the minimum and maximum cycle temperatures T_1 and T_3. Figure 7.9 depicts the same phenomenon, but also shows that by using the high-pressure ratio the gain in efficiency is not very large.

From Eq. 7.3 and Eq. 7.4 the net work output of the cycle is

$$W = W_{12} + W_{34} = C_P(T_1 - T_2) + C_P(T_3 - T_4) \tag{7.12}$$

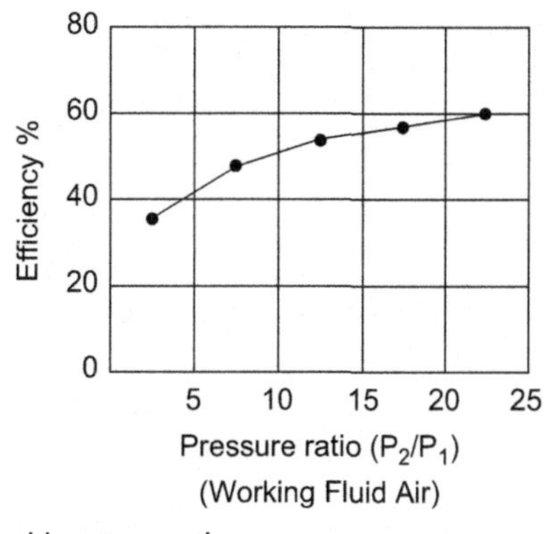

Figure 7.9 Efficiency of an ideal brayton cycle.

Putting the values of T_4 and T_2 from Eq. 7.8 and Eq. 7.9 into Eq. 7.12 and assuming a uniform pressure ratio for expansion and compression, the net work output is found to be

$$W = C_P\left\{T_1\left(1 - r_p^{\frac{\gamma-1}{\gamma}}\right) + T_3\left(1 - \frac{1}{r_p^{\frac{\gamma-1}{\gamma}}}\right)\right\} \tag{7.13}$$

Scrutinizing Eq. 7.11 and Eq. 7.13 it is noted while η increases indefinitely with r_p, the net work output does not increase indefinitely. At $r_p = 1$, W is zero. Again, at $r_p = \left(\frac{T_3}{T_1}\right)^{\frac{\gamma}{\gamma-1}}$, the value of W is zero. These phenomena are reflected in Figure 7.10, which shows three ideal cycles operating between the same temperatures T_3 and T_1 and have the same inlet and exhaust pressures but different values of r_p [6]. The net work output in each case is represented by the enclosed area of the cycle.

To find out the optimum value of r_p ($=r_{po}$) that would yield maximum net work output, let us differentiate W of Eq. 7.13 with respect to r_p and equate dW/dr_p to zero:

$$\frac{dW}{dr_p} = -C_P T_1 \frac{\gamma-1}{\gamma} r_{po}^{-\frac{1}{\gamma}} + C_P T_3 \frac{\gamma-1}{\gamma} r_{po}^{\frac{1-2\gamma}{\gamma}} = 0 \tag{7.14}$$

or,

$$T_1 \frac{\gamma-1}{\gamma} r_{po}^{\frac{1}{\gamma}} = T_3 \frac{\gamma-1}{\gamma} r_{po}^{\frac{1-2\gamma}{\gamma}} \tag{7.15}$$

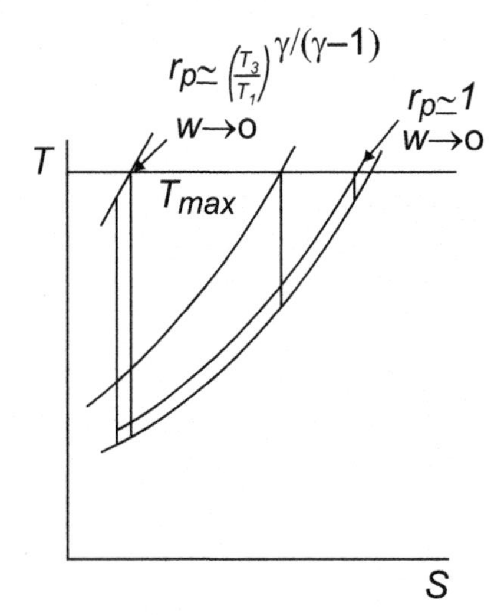

Figure 7.10 Three ideal cycles operating between temperatures T_3 & T_1. *Source: Fig. 12.4, Page 236, Gas Power Cycles, [6].*

Thus,

$$r_{po} = \left(\frac{T_3}{T_1}\right)^{\frac{\gamma}{2(\gamma-1)}} \tag{7.16}$$

The optimum pressure ratio for the ideal cycle, r_{po}, would be maximized if T_3 could be raised as high as possible and T_1 could be lowered as low as possible. In practice, T_3 is restricted to a metallurgical limit of about 1500 K for the alloy usually used in present-day large gas turbines and T_1 is limited to 288 K atmospheric temperature at sea level. Hence the optimum pressure ratio-under these temperature limits is about 18, which would yield maximum net work output. Note that as γ increases the quantity $\gamma/2(\gamma-1)$ decreases and hence also r_{po} decreases.

Referring to Eq. 7.13 it could be concluded that

i. For constant T_1, T_3, r_p, and γ, the net work output of the gas is directly proportional to C_P

ii. For constant T_1, T_3, r_p, and C_P gases with higher values of γ, i.e., higher $(\gamma-1)/\gamma$, produce more net work output of gas than gas with lower values of γ.

iii. For constant T_1, T_3, γ, and C_P increasing the compression ratio r_{po}, up to the optimum limit, is the most direct way to increase the power output of a Brayton cycle.

Example 7.1

For an ideal air-standard Brayton cycle operating between temperatures 298 K and 1000 K, find the value of r_{po} and the corresponding cycle efficiency. (Assume $\gamma = 1.4$.)

Solution: From Eq. 7.16,

$$r_{po} = \left(\frac{T_3}{T_1}\right)^{\frac{\gamma}{2(\gamma-1)}} = \left(\frac{1000}{298}\right)^{\frac{1.4}{0.8}} = 8.32$$

Applying the above value in Eq. 7.11, we get

$$\eta = \left(1 - \frac{1}{8.32^{\frac{0.4}{1.4}}}\right) \text{x} 100 = 45.41\%$$

7.3 REAL BRAYTON CYCLE

The real Brayton cycle includes the effect of irreversibility i.e., friction in the compressor and the turbine. Hence, neither compression nor expansion is truly isentropic The Brayton cycle with fluid friction is represented on the T-S diagram of Figure 7.11, where the isentropic end points are denoted with the superscript "prime." Both compression process with fluid friction 1–2 and expansion process with fluid friction 3–4 show an increase in entropy as compared with the corresponding ideal processes "1–2" and "3–4." Drops in pressure during heat addition (process

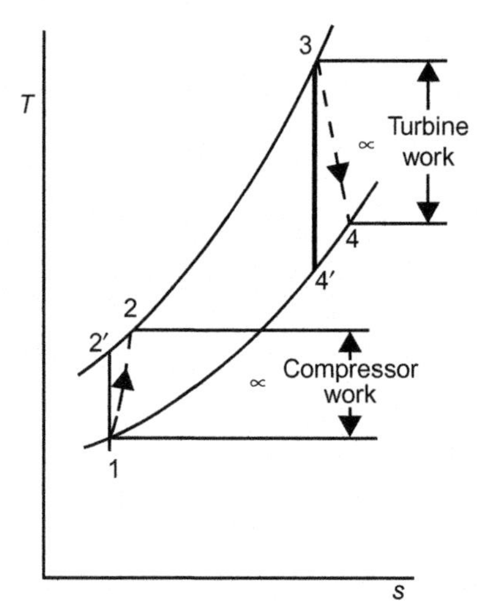

Figure 7.11 Efficiency of an ideal brayton cycle. *Source: Fig. 12.6, Page 237, Gas Power Cycles, [6].*

2–3) and heat rejection (process 4–1) are neglected in this analysis, so the turbine pressure ratio equals the compressor pressure ratio as before.

The isentropic efficiency of compression and expansion processes with fluid friction can be expressed as follows:

Isentropic efficiency of compressor:

$$\eta_c = \frac{h_2' - h_1}{h_2 - h_1} \tag{7.17}$$

Isentropic efficiency of turbine

$$\eta_t = \frac{h_3 - h_4}{h_3 - h_4'} \tag{7.18}$$

where h is the specific enthalpy of flowing gas.

For constant specific heat Eq. 7.17 and Eq. 7.18 change to

$$\eta_c = \frac{T_2' - T_1}{T_2 - T_1} \tag{7.19}$$

$$\eta_t = \frac{T_3 - T_4}{T_3 - T_4'} \tag{7.20}$$

The net work output of the real cycle (= output of turbine–output of compressor) is

$$W = C_P\{(T_3 - T_4) - (T_2 - T_1)\} \tag{7.21}$$

or,

$$W = C_P\left\{\eta_t(T_3 - T_4') - \frac{T_2' - T_1}{\eta_c}\right\} \tag{7.22}$$

or,

$$W = C_P T_1 \left\{ \eta_t \frac{T_3}{T_1} - \eta_t \frac{T_4'}{T_1} - \frac{1}{\eta_c} \frac{T_2'}{T_1} + \frac{1}{\eta_c} \right\} \tag{7.23}$$

As before for isentropic compression and expansion the pressure ratio across the compressor and the turbine is r_p.

Putting $T_4' = \frac{T_3}{r_p^{\frac{\gamma-1}{\gamma}}}$ and $\frac{T_2'}{T_1} = r_p^{\frac{\gamma-1}{\gamma}}$, Eq. 7.23 can be expressed as

$$W = C_P T_1 \left\{ \eta_t \frac{T_3}{T_1} \left(1 - \frac{1}{r_p^{\frac{\gamma-1}{\gamma}}} \right) - \frac{r_p^{\frac{\gamma-1}{\gamma}}}{\eta_c} \left(1 - \frac{1}{r_p^{\frac{\gamma-1}{\gamma}}} \right) \right\} \tag{7.24}$$

$$W = C_P T_1 \left(\eta_t \frac{T_3}{T_1} - \frac{r_p^{\frac{\gamma-1}{\gamma}}}{\eta_c} \right) \left(1 - \frac{1}{r_p^{\frac{\gamma-1}{\gamma}}} \right) \tag{7.25}$$

For the same fluid operating with the same pressure ratio the second item in the parentheses of Eq. 7.25, i.e., $\left(1 - \frac{1}{r_p^{\frac{\gamma-1}{\gamma}}} \right)$, is the same as the efficiency of the corresponding ideal cycle, as explained earlier in section 7.2. As in the case of an ideal cycle, the net work output of the real cycle attains a maximum value at some optimum pressure ratio and is a direct function of the specific heat of the gas used.

For maximizing net work output let us differentiate W of Eq. 7.24 with respect to r_P and equate dW/dr_P to zero:

$$\frac{dW}{dr_p} = C_P T_3 \eta_t \frac{\gamma - 1}{\gamma} r_{po}^{\frac{1-2\gamma}{\gamma}} - C_P T_1 \frac{1}{\eta_c} \frac{\gamma - 1}{\gamma} r_{po}^{-\frac{1}{\gamma}} = 0 \tag{7.26}$$

or,

$$\frac{T_1}{\eta_c} \frac{\gamma - 1}{\gamma} r_{po}^{-\frac{1}{\gamma}} = T_3 \eta_t \frac{\gamma - 1}{\gamma} r_{po}^{\frac{1-2\gamma}{\gamma}} \tag{7.27}$$

Thus,

$$r_{po} = \left(\frac{T_3}{T_1} \eta_t \eta_c \right)^{\frac{\gamma}{2(\gamma-1)}} \tag{7.28}$$

Note from Eq. 7.28 that due to the presence of terms η_t and η_c, r_{po} of a real cycle would be much less than that of an ideal cycle.

The heat added to the cycle, Q_{23}, is given by

$$Q_{23} = C_P(T_3 - T_2) = C_P\left(T_3 - \frac{T_2' - T_1}{\eta_c} - T_1\right) = C_P\left\{T_3 - \frac{T_1 r_p^{\frac{\gamma-1}{\gamma}} - T_1}{\eta_c} - T_1\right\}$$

$$= C_P\left\{T_3 - T_1\left(\frac{r_p^{\frac{\gamma-1}{\gamma}} - 1}{\eta_c} + 1\right)\right\}$$

or,

$$Q_{23} = C_P T_1\left\{\frac{T_3}{T_1} - \left(1 + \frac{r_p^{\frac{\gamma-1}{\gamma}} - 1}{\eta_c}\right)\right\} \tag{7.29}$$

Thus, the efficiency of the real cycle can be obtained by dividing Eq. 7.25 by Eq. 7.29. Unlike with the ideal cycle, the efficiency of the real cycle is very much a function of the initial temperature, T_1, and the maximum temperature T_3 and assumes a maximum value at an optimum pressure ratio for each set of temperatures T_1 and T_3.

In a real cycle, the fluid friction in heat exchangers, piping, etc., results in pressure drop between end points 2 and 3 and also between end points 4 and 1. As a result, the pressure at point 2 would be higher than that at point 3 and the pressure at point 1 would be less than that at point 4. Hence, the pressure ratio across the compressor (r_{pc}) would be greater than the pressure ratio across the turbine (r_{pt}).

Further, in a real cycle there are mechanical losses in bearing friction and auxiliaries as well as heat losses from combustion chambers, thus these are also responsible for lower efficiency of a real cycle than that of an ideal cycle.

Example 7.2

A simple air-standard gas turbine plant is operating on a real Brayton cycle with the maximum and the minimum temperatures at 1000 K and 298 K, respectively. Determine the pressure ratio, the net work output, and the cycle efficiency of this gas turbine if the isentropic efficiencies of both the compressor and turbine are 90%. (Assume the specific heat of the air at constant pressure is 1.005 kJ/kg.K and γ is 1.4.)

Solution: From Eq. 7.28 we get

$$r_{po} = \left(\frac{T_3}{T_1}\eta_t\eta_c\right)^{\frac{\gamma}{2(\gamma-1)}} = \left(\frac{1000}{298}0.9 * 0.9\right)^{\frac{1.4}{0.8}} = 5.75$$

Temperature of the fluid after isentropic compression is

$$T_2' = T_1 r_{po}^{\frac{\gamma-1}{\gamma}} = 298 * 5.75^{\frac{0.4}{1.4}} = 491.21 \text{ K}$$

Hence, the actual temperature after compression is

$$T_2 = T_1 + \frac{T_2' - T_1}{\eta_c} = 298.00 + \frac{491.21 - 298.00}{0.9} = 512.68 \text{ K}$$

Temperature of the fluid after isentropic expansion is

$$T_4' = \frac{T_3}{r_{po}^{\frac{\gamma-1}{\gamma}}} = \frac{1000}{5.75^{\frac{0.4}{1.4}}} = 606.67 \text{ K}$$

The actual temperature after expansion is

$$T_4 = T_3 - \eta_t(T_3 - T_4') = 1000 - 0.9(1000 - 606.67) = 646.00 \text{ K}$$

Hence, the net work output of the cycle is

$$W = C_P\{(T_3 - T_4) - (T_2 - T_1)\} = 1.005\{(1000.00 - 646.00) - (512.68 - 298.00)\} = 140.03 \text{ kJ/kg}$$

The heat added to the cycle is

$$Q_{23} = C_P(T_3 - T_2) = 1.005(1000 - 512.68) = 489.76 \text{ kJ/kg}$$

Thus, the efficiency of the cycle is

$$\eta = \frac{W}{Q_{23}}100 = \frac{140.03}{489.76}100 = 28.59\ \%$$

(Comparing the results of Example 7.1 and Example 7.2 note that the pressure ratio and the efficiency of a real cycle are far below the pressure ratio and the efficiency of an ideal cycle, due to the irreversibility that crept in during the process of compression and expansion.)

7.4 IMPROVEMENT OF PERFORMANCE

Considering the inferior result of a real Brayton cycle its performance may be substantially improved by adopting any one or combination of following modifications.

7.4.1 Regeneration

This is the internal exchange of heat within the Brayton cycle. In a cycle operating with normal pressure ratio and turbine inlet temperature, the turbine outlet temperature is always higher than the compressor outlet temperature (Figures 7.12 and 7.13), i.e., T_d is higher than T_b and heat addition is from "b" to "c." The performance of this cycle may be improved by adding a surface-type heat exchanger called the *regenerator*, where heat from exhaust gases leaving the turbine at "d" is transferred to the compressed gas at "b" before it enters the combustion chamber.

In an ideal process, it is theoretically possible to raise the temperature of the compressed air entering the combustion chamber from T_b to $T_{b''} = T_d$, while gas leaving the turbine is cooled from T_d to $T_{d''} = T_b$. The effect of the regenerator is to reduce the amount of heat to be supplied from $C_P\ (T_c - T_b)$ to $C_P\ (T_c - T_{b''})$.

Thus, while the net work output of the cycle remains unchanged at $W = C_P\{(T_c - T_d) - (T_b - T_a)\}$, the corresponding cycle efficiency increases because of reduced heat supply. The addition of a regenerator in a cycle leads to lower fuel consumption and less power loss as waste heat. In reality, however, it is

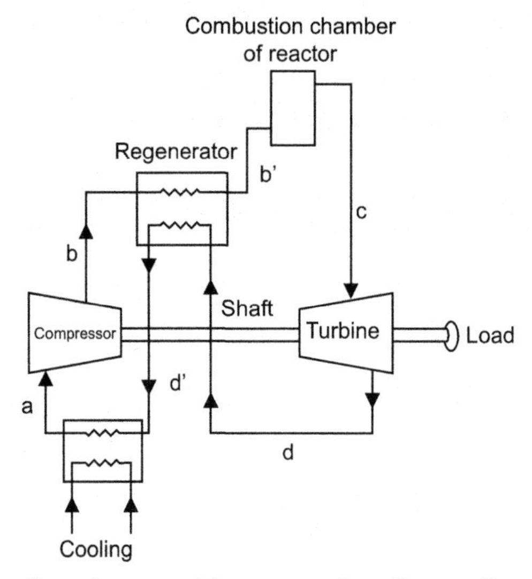

Figure 7.12 Gas turbine flow diagram with regeneration. *Source: Figure 8-13a, Page 324, Gas-Turbine and Combined Cycles, [2].*

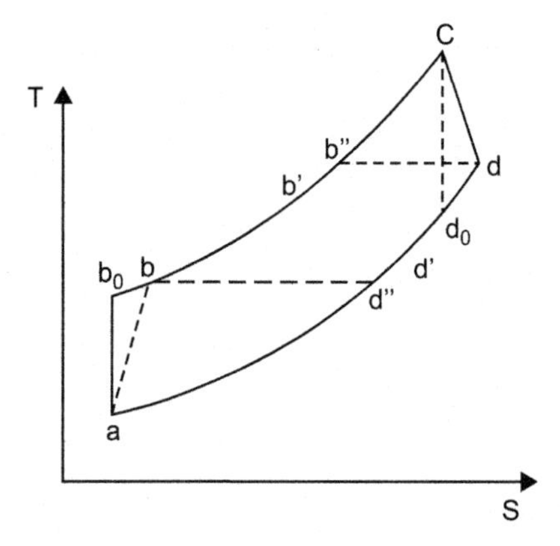

Figure 7.13 T-S diagram with regeneration. *Source: Figure 8-13b, Page 324, Gas-Turbine and Combined Cycles, [2].*

difficult to get a regenerator with 100% effectiveness, and the actual temperature of the compressed gases inlet to the combustion chamber is always lower than T_d, somewhere between $T_{b''}$ and T_b, at a temperature of say $T_{b'}$ The corresponding effectiveness of the regenerator may be expressed as [2]

$$R_{eff} = \frac{T_{b'} - T_b}{T_d - T_{d''}} = \frac{T_{b'} - T_b}{T_d - T_b} \tag{7.30}$$

Example 7.3

Referring to Figure 7.13 express the efficiency of an ideal cycle.

Solution: Since the net work output of the cycle remains unchanged, from Eq. 7.13

$$W = C_P\left\{T_a\left(1 - r_p^{\frac{\gamma-1}{\gamma}}\right) + T_c\left(1 - \frac{1}{r_p^{\frac{\gamma-1}{\gamma}}}\right)\right\} = C_P\left\{\left(r_p^{\frac{\gamma-1}{\gamma}} - 1\right)\left(\frac{T_c}{r_p^{\frac{\gamma-1}{\gamma}}} - T_a\right)\right\}$$

Heat added to the cycle is

$$Q_{cb''} = C_P(T_c - T_{b''}) = C_P T_c\left(1 - \frac{1}{r_p^{\frac{\gamma-1}{\gamma}}}\right) = C_P\left(r_p^{\frac{\gamma-1}{\gamma}} - 1\right)\frac{T_c}{r_p^{\frac{\gamma-1}{\gamma}}}$$

Hence, the efficiency is

$$\eta = \frac{W}{Q_{cb''}} = 1 - \frac{T_a}{T_c} r_p^{\frac{\gamma-1}{\gamma}} \tag{7.31}$$

From Eq. 7.31 it may be inferred that the lower the pressure ratio the higher the efficiency of an ideal regenerative cycle, and it becomes Carnot efficiency when the pressure ratio is unity. However, at this stage both the heat added and the net work output are also zero [6]. Thus, the benefit of regeneration can only be ripped if the value of r_p adopted is less than one. Note that although the addition of a regenerator increases cycle efficiency it does not enhance the net work output of the cycle. To increase the net work output from a gas turbine either the compressor work should be reduced or the turbine work has to be increased.

7.4.2 Compressor intercooling (Figure 7.14 and Figure 7.15)

First, consider the compressor work. For a perfect gas working with a constant pressure ratio the work is directly proportional to temperature, from which (as well as by

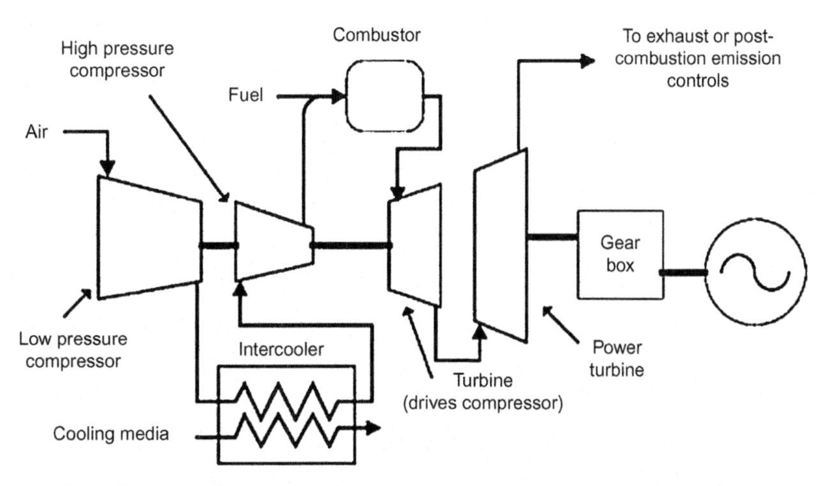

Figure 7.14 Flow diagram of a compressor intercooling. *Source: FIG. 4-5. Article 4.3 Gas Turbines. © Copyright Energy Solutions Center, DG Consortium 2004.*

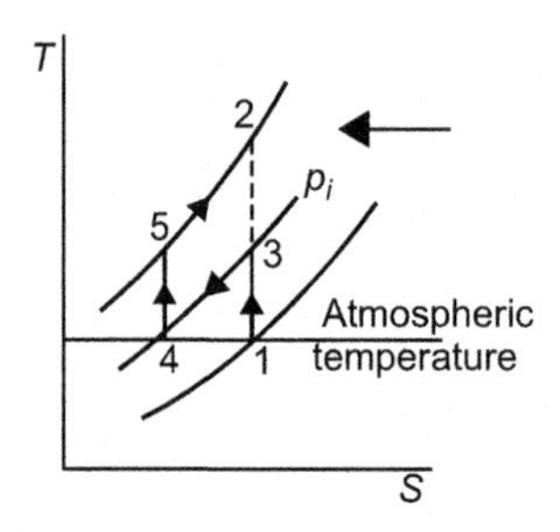

Figure 7.15 T-S diagram of compressor intercooling. *Source: Fig. 12.10, Page 240, Gas Power Cycles, [6].*

looking at Section 7.2) it may be inferred that the compressor work, $C_P(T_1-T_2)$, can be decreased by keeping the gas temperature (T_2) in the compressor low.

Figure 7.15 shows that a compressor working between points 1 and 2 would expend more and more work as the gas approaches point 2. If the compression takes place further to the left of 1–2 the work required to drive the compressor would reduce. Since compressor work reduces the net cycle work output, it is advantageous to keep the compressor outlet temperature low while reaching the desired pressure P_2. This can theoretically be done by continuous cooling of the compressed gas to keep T_1 at the atmospheric temperature, as shown by the lower horizontal line of Figure 7.15. However, this is not physically possible, so cooling is done in stages.

Figure 7.15 shows two stages of compression, 1–3 and 4–5, and one stage of intercooling where gas is partially compressed from point 1 to point 3, cooled back to point 4 at constant pressure (ideally) p_i between the stages, then finally compressed to point 5. Ideally $T_1 = T_4$ and $T_3 = T_5$. It is clear from the figure that the sum of the temperature rises, $(T_3-T_1) + (T_5-T_4)$, is less than the temperature rise (T_2-T_1).

Hence, the total compressor work with isentropic compression and complete intercooling as shown below is less than the compressor work $C_P(T_2-T_1)$.

$$W_c = C_P\{(T_3 - T_1) + (T_5 - T_4)\} \tag{7.32}$$

$$W_c = C_P T_1 \left[\left\{ 1 - \left(\frac{p_i}{p_1} \right)^{\frac{\gamma-1}{\gamma}} \right\} + \left\{ 1 - \left(\frac{p_2}{p_i} \right)^{\frac{\gamma-1}{\gamma}} \right\} \right] \tag{7.33}$$

To find the value of p_i that would yield minimum compressor work with isentropic compression, let us differentiate W_c with respect to p_i and equate dW_c/dp_i to zero and putting $T_1 = T_4$.

$$\frac{dW_c}{dp_i} = C_P T_1 \frac{\gamma - 1}{\gamma} \left\{ \left(\frac{p_i}{p_1} \right)^{-\frac{1}{\gamma}} - \left(\frac{p_2}{p_i} \right)^{-\frac{1}{\gamma}} \right\} = 0 \tag{7.34}$$

or,

$$\frac{p_i}{p_1} = \frac{p_2}{p_i} = r_{pi} = \sqrt{\frac{p_2}{p_1}} = \sqrt{r_p} \tag{7.35}$$

and

$$p_i = \sqrt{p_1 p_2} \tag{7.36}$$

From Eq. 7.35 it is found that the condition for minimum compressor work is that both the compression ratio and work input for all stages must be equal. If the goal is to reduce the compression work further then number of stages of intercooling must be increased. This would, however, call for additional expenditure. Note from the T-s diagram that while the work required to drive the compressor is reduced by the area 3-4-5-2-3, the heat added has increased by $C_P\,(T_2 - T_5)$. Since the work required by the compressor has reduced while work output from the turbine remains unchanged, net result is increase in work ratio, which in turn increases specific work output [6].

7.4.3 Turbine reheat

Section 7.4.2 concluded that the purpose of compressor intercooling is to minimize the compressor work input to enhance the work ratio. This section will argue that the work ratio may also be increased by reheating the gas after partial expansion in the gas turbine as far as permissible for a given compression ratio without exceeding the metallurgical limit, thereby increasing the turbine work output and decreasing the effect of component losses.

From Eq. 7.13 it is clear that the turbine work output can be enhanced theoretically by keeping the gas temperature (T_3) in the turbine high complemented with continuous heating of the gas as it expands through the turbine. Since continuous heating is not practical, reheat is done in steps or stages. Figure 7.16 shows that the portion of the cycle with bearing on the turbine reheat comprises two stages of the turbine expansions and one stage of the reheating to the metallurgical limit of $T_9 = T_6$.

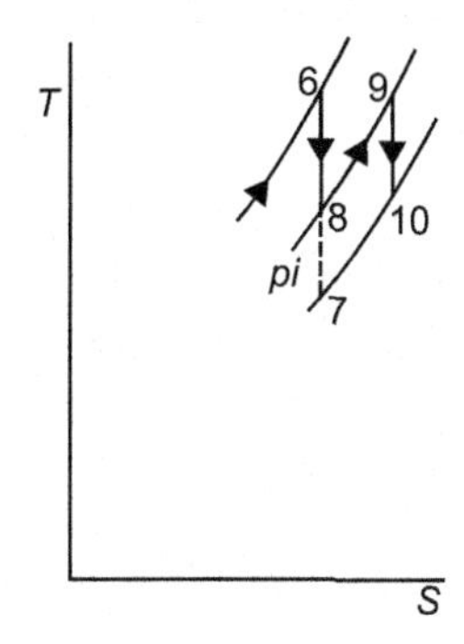

Figure 7.16 T-S diagram of turbine reheat. *Source: Fig. 12.11, Page 241, Gas Power Cycles, [6].*

The gas expands in the high-pressure section of the turbine from point 6 to point 8, it is reheated then at constant pressure (ideally) to point 9, and finally expands in the low-pressure section of the turbine to point 10. The consequent output from the turbine has increased from W_{67} to

$$W_t = C_P\{(T_6 - T_8) + (T_9 - T_{10})\} \tag{7.37}$$

or,

$$W_t = C_P T_6 \left[\left\{ 1 - \left(\frac{p_6}{p_i} \right)^{\frac{\gamma-1}{\gamma}} \right\} + \left\{ 1 - \left(\frac{p_i}{p_7} \right)^{\frac{\gamma-1}{\gamma}} \right\} \right] \tag{7.37}$$

Differentiating W_t with respect to p_i and equating dW_t/dp_i to zero the value of p_i that would yield the maximum turbine work output with isentropic expansion is given by

$$p_i = \sqrt{p_6 p_7} \tag{7.38}$$

and

$$\frac{p_6}{p_i} = \frac{p_i}{p_7} = r_{pi} = \sqrt{\frac{p_6}{p_7}} = \sqrt{r_p} \tag{7.39}$$

In the T-S diagram (Figure 7.16) the increase in cycle work is shown by the area 8-9-10-7-8, whereas the heat added is increased by C_P ($T_9 - T_8$). The net effect is an increase in both work ratio and specific work output.

Figure 7.17 reveals that with continuous isothermal cooling and continuous isothermal heating we may arrive at a lower horizontal line and upper horizontal line, respectively, in the T-S diagram. If the rest of the cycle is ideal, we may arrive at an ideal Ericsson cycle, which has the same efficiency as a Carnot cycle operating between two extreme temperature limits [7].

It should be noted that by adopting "compressor intercooling" and/or "turbine reheating" although net work output gets increased, the ideal cycle efficiency may decrease, since heat supplied is also increased. The full benefit can only be harnessed if regeneration is also added with compressor intercooling and turbine reheat as shown in Figure 7.18 [6]. The additional heat required for the colder air leaving the compressor at point 4 is obtained from the hotter exhaust gases from turbine at point 9. The hotter air then enters the combustion chamber at point 5, resulting in a gain in ideal cycle efficiency along with increased work output.

On the T-S diagram (Figure 7.19) for ideal intercooling

$$T_1 = T_3 \text{ and } T_2 = T_4$$

In an ideal regenerator $T_5 = T_9$

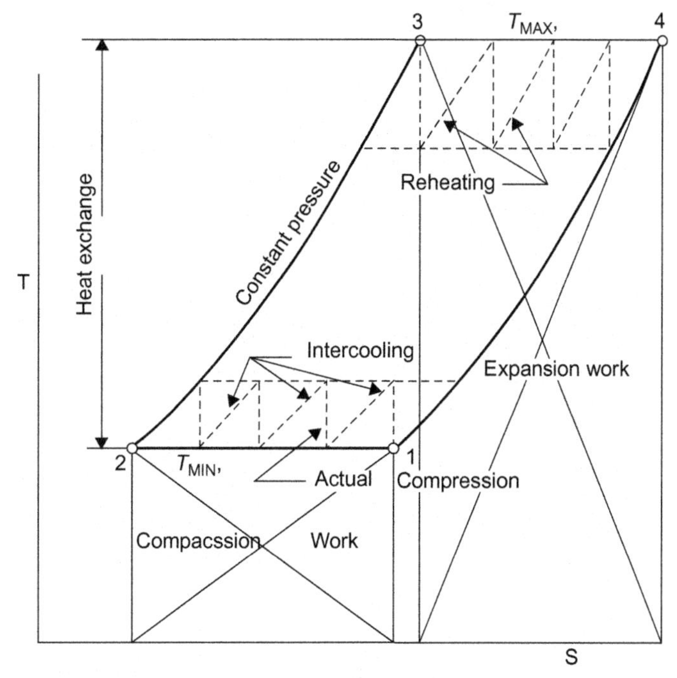

Figure 7.17 T-S diagram of compressor intercooling & turbine reheat. *Source: Fig. 404, Page 791, Jet Propulsion and the Gas Turbine, [7].*

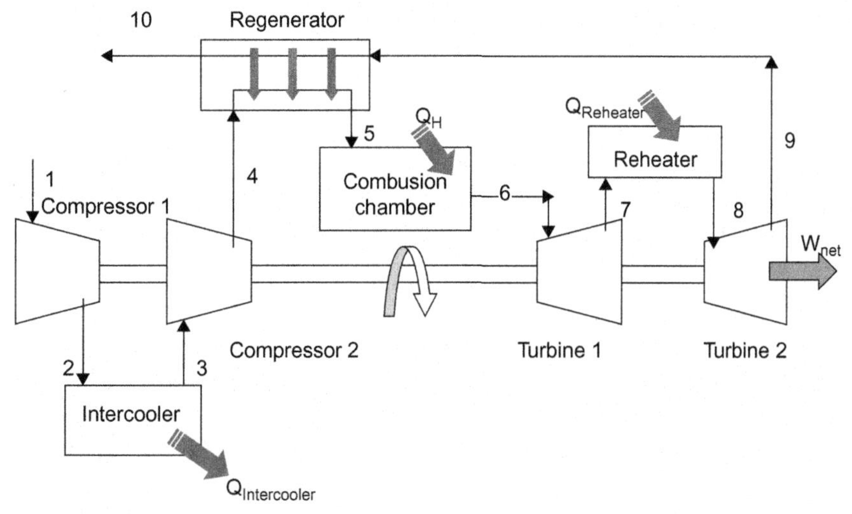

Figure 7.18 A gas turbine with intercooling, reheating & regeneration. *Source: Figure 9, Page 9, Section 8.9-10 The Brayton Cycle With Regeneration, Intercooling And Reheating by Denise Lane.*

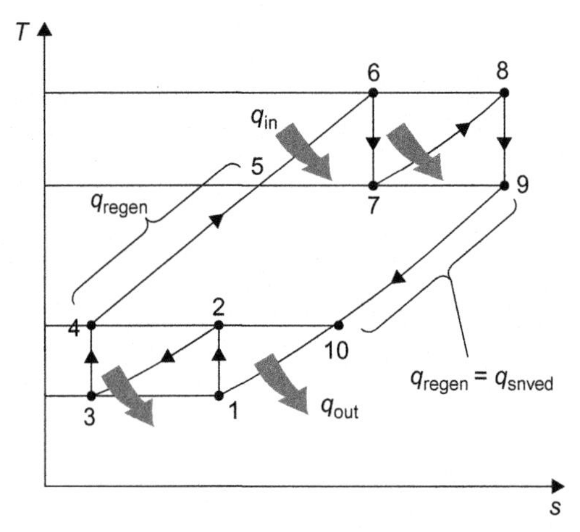

Figure 7.19 T-S diagram for an ideal brayton cycle with intercooling, reheating and regeneration. *Source: Figure 10, Page 9, Section 8.9-10 The Brayton Cycle With Regeneration, Intercooling And Reheating by Denise Lane.*

In ideal reheating

$$T_8 = T_6 \text{ and } T_7 = T_9$$

As explained earlier the net work input to compressor is minimized when

$$\frac{P_2}{P_1} = \frac{P_4}{P_3}$$

And net work output from turbine is maximized when

$$\frac{P_6}{P_7} = \frac{P_8}{P_9}$$

(Note: A gas turbine plant incorporating intercooling, reheating, and regeneration would operate at lower pressure ratios than a plant without intercooling, reheating, and regeneration.)

7.4.4 Water injection

Water injection downstream of the compressor or between two stages of the compressor has the same effect as intercooling. As the air temperature rises due to compression water is injected in the air, and the heat of vaporization of water then reduces the air temperature. As a result compressor work gets reduced. By adopting this method the higher power output of a gas-turbine cycle may be obtained with marginal gain in efficiency. Figure 7.20 shows a schematic arrangement of water injection between compressor and regenerator.

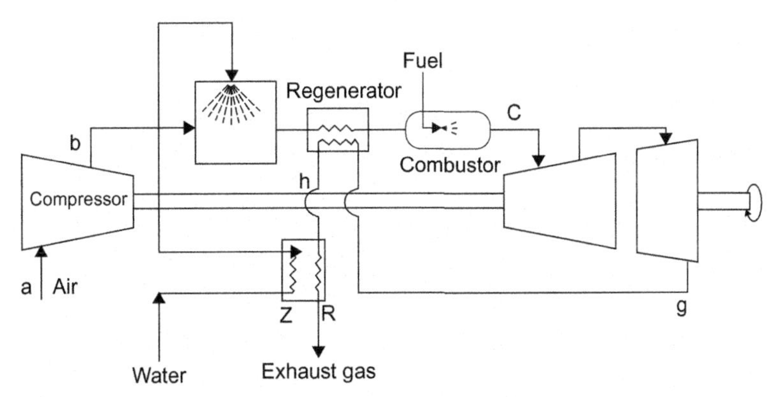

Figure 7.20 water injection between compressor & regenerator. *Source: Figure 8-15, Page 328, Gas Turbine and Combined Cycles, [2].*

Hot compressed air coming out of compressor is cooled at nearly constant pressure by the evaporating water. The cooled air enters the regenerator for preheating. Heat added in the regenerator is obtained from turbine exhaust gases that circulate through the regenerator. With water injection there is an increase in turbine work due to the increased mass-flow rate of air and water vapour without a corresponding increase in compressor work.

7.4.5 Co-generation or (Combined Heat and Power CHP) [8,9]

A co-generation plant combines a conventional steam power plant and a process steam plant to provide simultaneous generation of both electricity and useful heat from a single power plant, thereby increasing the overall efficiency of the plant. In separate production of electricity some energy must be rejected as waste heat, whereas in separate production of heat the potential for production of high quality energy (electricity or work) is lost. The co-generation plant uses the available energy in more conservative way than that used in either a steam power plant or a process steam plant. Co-generation, therefore, is thermodynamically the most efficient use of fuel. This plant captures the byproduct heat for domestic heating or industrial heating, as in chemical industries, paper mills, etc. Heat remaining in the exhaust gases from gas turbine plants is used as the source of energy to steam generators of co-generation plants for generating steam, which is subsequently used in steam turbines. Steam turbines in a co-generation plant may be of two configurations, i.e., a back-pressure turbine and a pass out turbine.

The back-pressure turbine is usually employed if the power generated by the expanding steam from the initial steam pressure down to the required process steam pressure is more than the power required by the process. Power and heat demand must be fairly steady and well matched. Figure 7.21 shows a typical back-pressure turbine co-generation plant, wherein steam generated in the boiler is admitted to the

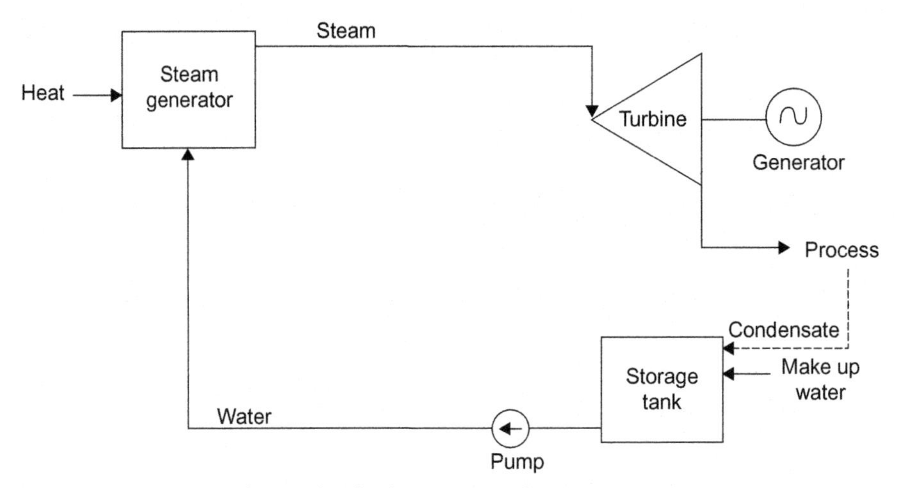

Figure 7.21 Co-generation plant with a back pressure turbine.

turbine. The exhaust steam of the turbine is supplied to the process and gets entirely condensed. Depending on the quality of the condensate, it may or may not be returned to the steam generator.

This plant, however, suffers from the following limitations:

i. If there is a demand for power supply there should be an equivalent quantity of process steam demand and vice-versa.
ii. If the power demand reduces, the process steam supply has to be reduced proportionately and vice-versa.
iii. In the event process steam is shut-down, an additional dump condenser will be required for handling the exhaust steam.

These limitations may be overcome if a pass-out turbine is used instead. A co-generation plant employing a pass-out turbine is depicted in Figure 7.22. This turbine may be considered as a combination of two turbines, i.e., a high-pressure section that drops the steam pressure at the turbine inlet to the steam pressure required by the process, while a low-pressure section handles the steam flow to develop the required power by the process. However, this steam flow is not required by the process and expands to the condenser pressure.

7.4.6 Combined cycle (Figure 7.5 and 7.23)

When the ratio of power generation to heat demand is high, the combined cycle plant becomes attractive. In a combined-cycle power plant both gas and steam turbines supply power to the net work. The heat recovery steam generator (HRSG) lies within the gas turbine and the steam turbine. The heat associated with the gas turbine

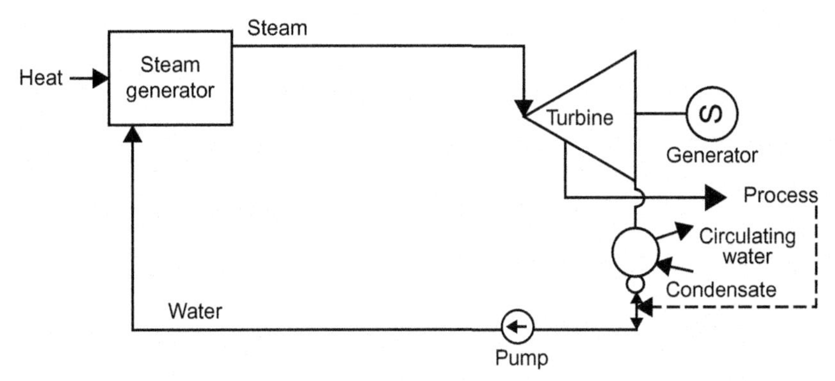

Figure 7.22 Co-generation plant with a pass-out turbine.

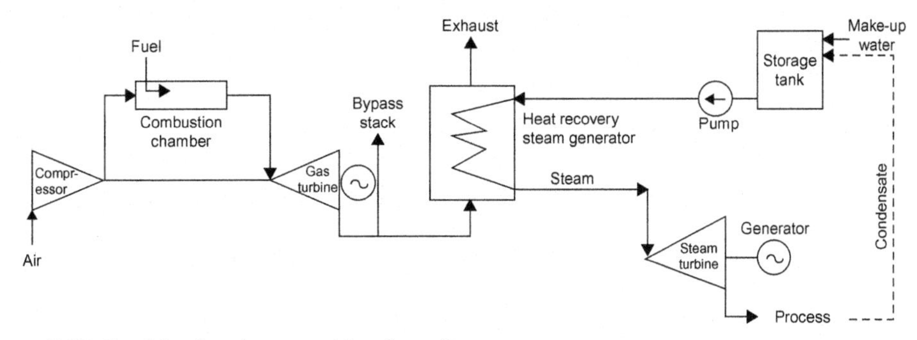

Figure 7.23 Combined cycle gas turbine flow diagram.

exhaust is used in the HRSG to produce steam that drives a steam turbine and generates additional electric power. When heat is used to generate steam this way, the whole plant becomes a binary unit. This joint operation of the gas turbine and the steam turbine is called a combined-cycle power plant. Thermodynamically, this implies joining a high temperature (1373–1923 K) *Brayton* (gas turbine) *cycle* and a moderate and low temperature (813–923 K) Rankine cycle, employing features of both the cycles to achieve an overall thermal efficiency of about 50% that is simply not possible with either cycle on its own. The combination most widely accepted for commercial power generation is "a gas topping cycle" with "a water-steam bottoming cycle." In most applications, the steam turbine (ST) will produce approximately 30–35% of the total plant output, with the remaining 65–70% being supplied by the gas turbine (GT).

This cycle is characterized by flexibility, quick part-load starting, suitability for both base-load and cyclic operation, and a high efficiency over a wide range of loads. The only disadvantage is in their complexity in combining two technologies in one plant. Combined cycle power plants also have low investment costs and short construction times compared to large coal-fired power stations. The other benefits of combined cycles are high efficiency and low environmental impact. Non-greenhouse gas emissions such as SO_2, NO_x and particulate matter are also relatively low. The start-up time of an HRSG is also low. An HRSG boiler system can usually be brought from a cold state to full load steam generation in about 60 minutes. The efficiency of advanced combined cycle plants at full load is close to 60% (LHV basis), while full-load efficiency of a supercritical pulverized coal-fired power plant is about 46% (LHV basis) [10].

7.4.7 Cheng cycle (Figures 7.24 and 7.25)

The output of a simple cycle gas turbine may be enhanced at a lower cost by adopting the Cheng cycle. As in a combined cycle, the Cheng cycle is also based on recovering and utilizing waste heat energy in the gas turbine exhaust. However, unlike the

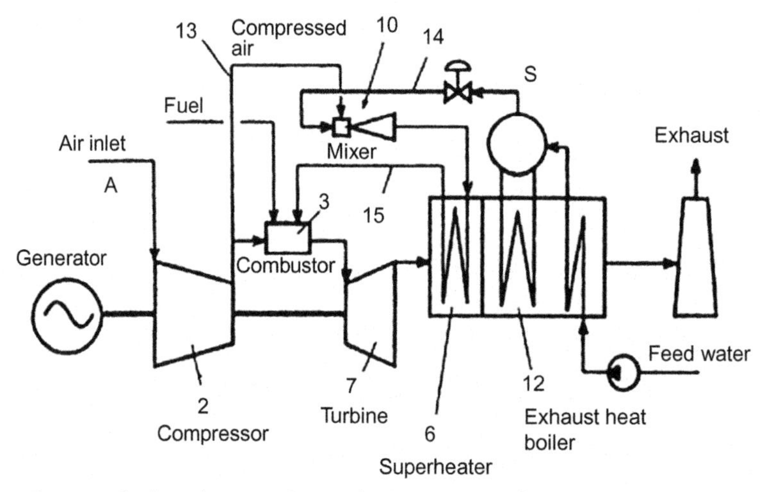

Figure 7.24 Cheng cycle flow diagram. *Source: http://www.google.com/patents/EP1069282B1?cl=en.*

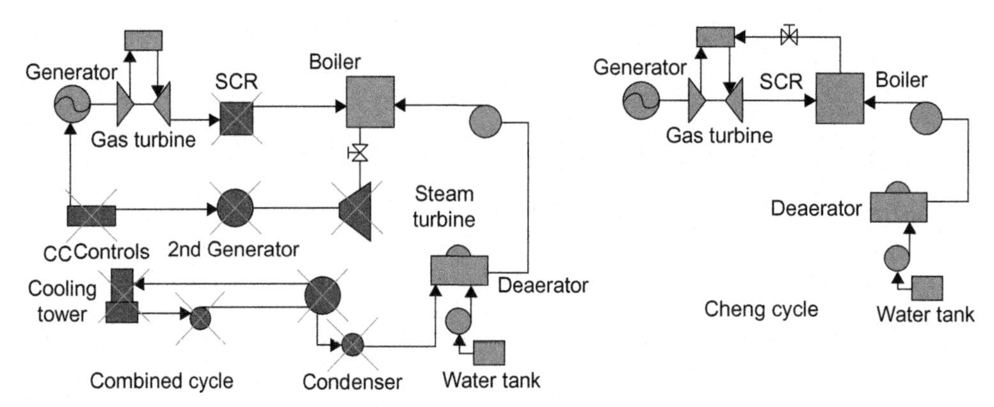

Figure 7.25 Cheng cycle versus combined cycle. *Source: [11]*

combined cycle, in the Cheng cycle the steam generated by the waste heat recovery of energy in the gas turbine exhaust is injected directly into the combustion chamber of the gas turbine, instead of passing the steam through a separate steam turbine generator. The injected steam is heated by additional fuel, reaches the working temperature of the gas turbine, and mixes with the air as additional working fluid then passes through the gas turbine to generate much enhanced power [11]. Thus, electrical power output from the gas turbine generator increases as more superheated steam is injected into the gas turbine due to the increased gas mass flow and more specific heat.

The Cheng cycle is simpler, more compact, and less expensive than a combined cycle plant due to the absence of steam turbine, associated generator, condenser, and cooling tower/condenser cooling water system. Another added advantage of eliminating the steam turbine is fast start-up and quick ramp rate to full load similar to those of a simple cycle gas turbine. The Cheng cycle has the potential to increase the output of a simple cycle gas turbine by 70% with decrease in heat rate by about 40% based on manufacturer's design parameters [11].

7.5 GAS TURBINE SYSTEMS AND EQUIPMENT

Depending on its size, location of installation, operating environment, and the proprietary design a gas turbine unit may include some or all of following systems and equipment:

- **i.** Air intake system
- **ii.** Anti-icing system
- **iii.** Evaporative cooling system
- **iv.** Exhaust system
- **v.** Starting system
- **vi.** Lubrication and power oil system
- **vii.** Hydraulic and pneumatic control system
- **viii.** Fuel gas system
- **ix.** Fuel oil system
- **x.** Water injection system
- **xi.** Turbine governing system
- **xii.** Fire fighting system
- **xiii.** Compressor unit
- **xiv.** Combustion components
- **xv.** Annular combustor
- **xvi.** Spark plugs and igniter
- **xvii.** Flame monitors
- **xviii.** Turbine section

7.5.1 Air intake system

The air intake system usually consists of an integrated air cleaning filter housing, silencer, expansion joint, inlet manifold, etc., leading to the compressor suction of the turbine. The filter elements prevent compressor fouling, erosion, and physical damage to compressor blades that may take place if the amount of the airborne dirt and contaminants is more than 25–30 microns. The dirt that accumulates on the filter elements is cleaned by the jets of compressed air in the counter direction of the main flow.

7.5.2 Anti-icing system

Anti-icing systems prevent formation of ice in the air intake system and on compressor blades when the ambient air temperature falls below a low temperature, say +280 K, while the relative humidity is at or above 70%. This system warms the compressor intake air by circulating hot air from the compressor exhaust and then mixing hot air with the intake air.

7.5.3 Evaporative cooling system

The evaporative cooling system is used to raise both power output and efficiency of gas turbines. The power output and efficiency rating of gas turbines are usually based on ISO conditions (inlet air temperature 288 K and 65% relative humidity). If the inlet air is hotter and drier than ISO conditions, the power of the gas turbine decreases. The purpose of the evaporative system is to cool the intake air by introducing water in the air system, where evaporation energy of water reduces enthalpy of air, resulting in increase in compressor inlet mass flow.

7.5.4 Exhaust system

Hot exhaust gases coming from the gas turbine are evacuated to the atmosphere through a stack. The gases pass through a silencer to attenuate the sound of the exhaust gases before being released. In the exhaust system an expansion joint is provided to compensate for thermal expansion in the exhaust ducting.

7.5.5 Starting system

Before admitting fuel into the combustor a gas turbine must be started by a start-up device. This start-up device may be either an AC motor, a diesel engine, or a "static starting device." With the help of a start-up device the gas turbine is accelerated to its ignition speed when the turbine burners are ignited. Once the flame is stabilized in the combustor the turbine accelerates further, exceeding its self-sustaining speed, resulting in the start-up device automatically disconnecting and the turbine governor taking over and accelerating the set to synchronous speed.

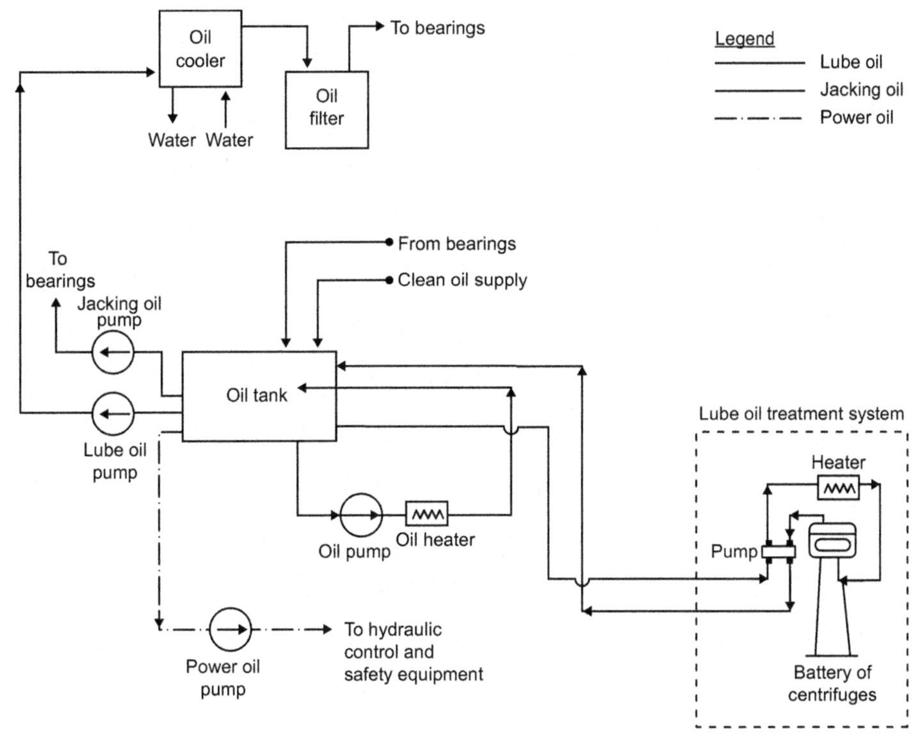

Figure 7.26 Lube oil system.

A static starting device is a variable frequency device that gains its starting power from the grid (HV system). Once the power is established the gas turbine is started by using its generator as a synchronous motor.

7.5.6 Lubrication and power oil system (Figure 7.26)

The lubrication and power oil system is provided to meet the lubrication needs of gas turbine bearings, compressor bearings, generator bearings, thrust bearings, etc., and depending on equipment design a portion of the fluid may be diverted for use by gas turbine hydraulic control devices.

The lube oil is stored in a tank provided with a heater that maintains the lube oil at the minimum temperature required for operation during prolonged periods of downtime. During operation, the lube oil pump takes the suction from the oil tank and circulates the lube oil through the oil filters to the bearings and other consumers. After lubricating the bearings the oil flows back through various drain lines to the oil tank. The lube oil is also circulated through lube oil coolers to maintain oil temperatures within the preset range.

The same oil is also supplied to bearings for lifting the turbine and generator rotors slightly during start-up or rotor barring operation. This prevents wear on the bearings and also reduces the starting torque required. The system is called as the jacking oil system. A separate pump is used to supply power oil to the gas turbine hydraulic control and safety equipment. The lube oil treatment system is used to clean the oil circulating in the gas turbine lubrication and power oil system. Oil is drawn by a pump from the oil tank, and on completion the desired treatment oil is returned to the oil tank.

7.5.7 Hydraulic and pneumatic control system

This system is the backbone of gas turbine control and protection system that ensures safe, efficient, and trouble-free operation. The hydraulic protection system protects the gas turbine from serious damage in the event of control system failure. The hydraulic control system operates the variable inlet guide vane mechanism and fuel regulation system. The pneumatic control system controls compressor blow-off valves.

7.5.8 Fuel gas system (Figure 7.27)

Prior to supplying fuel gas to the gas turbine skid the incoming gas is treated in a "knock-out drum" (KOD) followed by a "filter separator" unit to strip the gas of all solids and liquids. Any solid particle carried with the gas stream is separated first. Liquid separation takes place subsequently. Thereafter, the fuel gas may enter into a booster compressor for boosting the gas pressure to the pressure required by the gas turbine.

The compressor discharge is sent to the gas coolers to limit flue-gas temperature to comply with the requirement of the gas turbine. The cooled gas is then passed through another set of filters to remove any liquid condensed after cooling of the compressed gas before the fuel gas is conveyed to the fuel gas skid of gas turbine.

Fuel gas flows into the gas turbine burner zone when both the main and burner gas shut-off valves are opened. During start-up the pilot gas valves are opened for ignition. Once gas is ignited, the main gas valves gradually open and pilot control

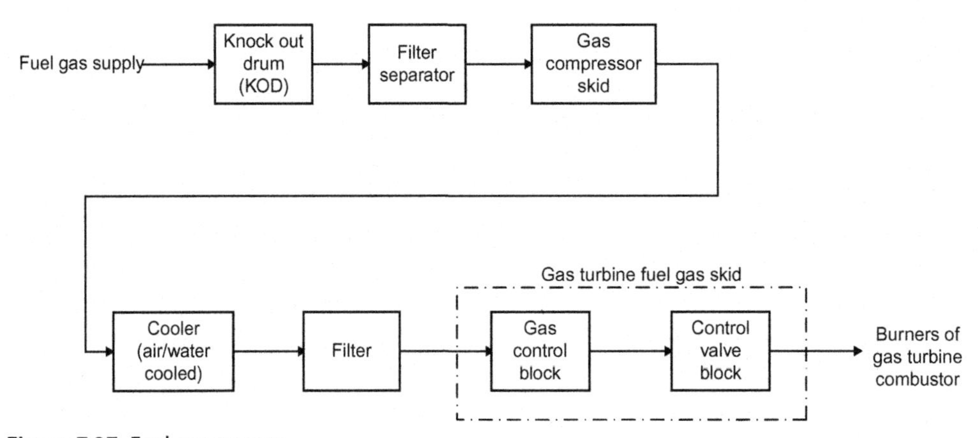

Figure 7.27 Fuel gas system.

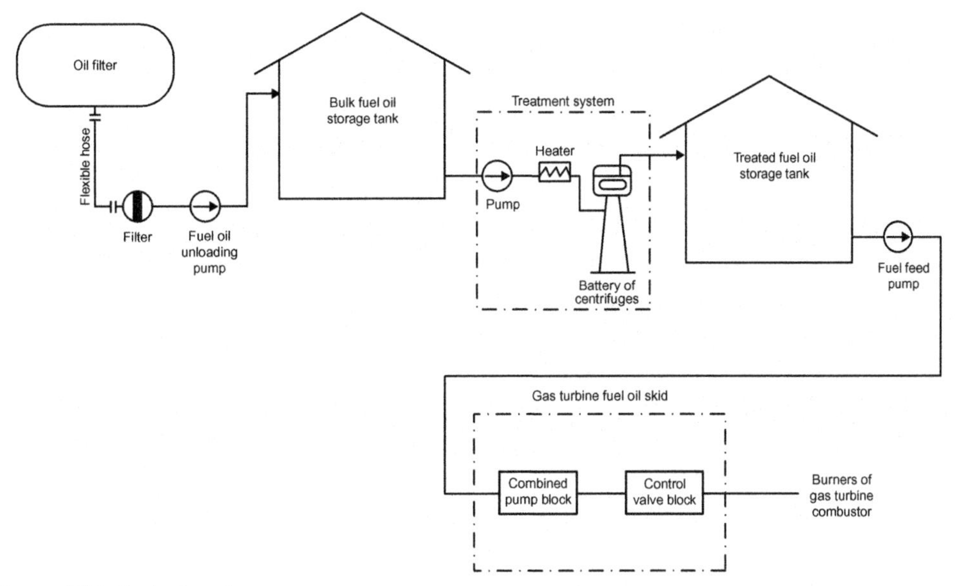

Figure 7.28 Fuel oil system.

valves closed, thereby the desired load on the machine is maintained. During shut-down of the turbo set both the main and pilot control valves close.

7.5.9 Fuel oil system (Figure 7.28)

Fuel oil received from an oil refinery is unloaded in fuel oil storage tanks. A fuel oil treatment system is provided near storage tanks to remove the water-soluble sodium ($Na+$) and potassium ($K+$) salts in the fuel oil to a level equal to or better than that acceptable to the gas turbine operation. The treated oil is then taken to the suction of fuel-oil forwarding pumps of the fuel oil skid of the gas turbine.

The fuel oil pump forwards the fuel oil to the burners at adequate pressure needed for atomization. The fuel oil is supplied to the burners only when the main oil shut-off valve is opened. The control valves of the burners are then gradually opened for ignition of oil and thereafter open further as per the load demand. During shut-down of the set, all control valves and shut-off valves are closed.

7.5.10 Water injection system

Water is mixed with fuel oil, primarily for the reduction of NO_X emission. The mixing also enhances the output power from the machine. (For details refer to Section 7.4.)

7.5.11 Gas turbine governing system

This system regulates fuel flow to the gas turbine to ensure smooth start-up and a high degree of fuel regulation from the minimum turbine running conditions to full-load operation within safe working parameters. The governing system serves the following basic control functions:

i. Start-up control
ii. Frequency/load control
iii. Firing temperature control
iv. Emergency shut-down control

7.5.11.1 Start-up control

The start-up control regulates fuel flow to the gas turbine during start-up of the turbo-set until the nominal speed is reached. It also controls the fuel-air ratio so the outlet gas temperature does not exceed the permissible limits at maximum acceleration. Once the turbine speed matches the grid frequency the unit is made ready to be synchronized with the grid.

7.5.11.2 Frequency/Load control

The frequency/load controller comes into action during idling, synchronization, and normal operation of the turbine. Once the turbine accelerates to operating speed the start-up control mode is switched over to the frequency/load control mode. This controller derives the set point for the turbine load from the output of the frequency controller. The fuel flow to the turbine is then controlled to generate the desired power output.

7.5.11.3 Firing temperature control

The temperature controller regulates the rate of the fuel addition during the acceleration stage of the turbine so that the turbine is not inflicted with over temperature condition. The controller monitors both the inlet and outlet temperatures of the turbine for the control of fuel flow. It also maintains the minimum fuel flow to the combustor so the flame does not blow out, leading to trip gas turbine.

7.5.11.4 Emergency shut-down control

The emergency shut-down control protects the gas turbine from serious damage whenever any control system fails.

7.5.12 Firefighting system

At any place around the plant a fire may start accidentally due to shortcircuiting or ignition of fuel and lubricants or of common combustible materials, e.g., paper, cotton waste, etc. A number of portable/semi-portable soda-acid type fire extinguishers are provided at various places to fight minor fires as soon as they are detected.

The fuel oil system is protected from fire by a foam and spray water system. The generated foam provides a blanket over the oil surface and cuts off the oxygen supply, which is essential for supporting combustion. As an alternative means, sand stored in the bins near the dyke walls of the fuel oil storage tanks can also be used to prevent outbreak of minor fire. Sand is sprinkled evenly on the spillage surfaces by using a shovel or bucket. Small fires can also be handled successfully using sand. Ordinary water is not generally suitable for fighting oil fire, as oil has a tendency to float and spread further.

Carbon dioxide fire extinguishers are used for fighting electrical fire. Wherever it is practicable, the electric supply to the equipment on fire should be switched off before using the extinguisher. In the event of a big fire, the portable/semi-portable extinguishers may prove ineffective. In such cases, the central firefighting team should be called in as soon as possible.

7.5.13 Compressor unit

The air compressor used in a gas turbine is made up of several rows of blades. The compressor develops a high-compression ratio and ensures an adequate air supply to combustors. The compressed air may be withdrawn from the compressor and directed to the parts in the hot gas path in the turbine zone to cool that zone.

Axial flow compressors are used in all larger gas turbine units because of their high efficiency and capacity. Centrifugal compressors are more stable than axial flow ones but of much low capacity and not as efficient. Axial flow compressors experience surge problem during start-up, low-load, and low-speed operations. Axial flow compressors used in gas turbine power plants must be designed to avoid operation in such instability region, which is damaging to the equipment. To avoid compressor instability under these conditions any or a combination of the following methods are used, inter-stage bleed, discharge bleed, variable compressor inlet vanes, turbine variable vanes, and bypass to ensure fast and smooth starting and low-load operation of the gas turbine.

7.5.14 Combustion components

The combustion system consists of a combustor, burners, igniter, and flame monitors. The *combustor* must bring the gas to a controlled uniform temperature with minimum impurities and minimum loss of pressure. The major problems of combustor design, in addition to clean combustion and proper mixing of gases, are flame stabilization; elimination of pulsation and noise; reduction of pressure loss; and maintenance of steady, closely controlled outlet temperature. The gas turbine power plants are arranged with single or multiple combustors. The single combustor design is easier to control but is larger and less compact than the multiple types.

The combustor is placed between the turbine and the compressor within turbine casing. It is a direct-fired heater in which fuel is burnt to supply heat energy to the gas turbine. The air flow through the combustor has three functions: to oxidize the fuel, to cool the metal parts, and to adjust the extremely hot combustion products to the desired turbine inlet temperature. The temperature of the gas in the combustors and entering the turbine can reach up to 1623 K or even more. Usually low NO_X burners are used to reduce the concentration of NO_X in the exhaust gas to less than 25 ppm at full load. Sometimes water or steam can be injected into the combustors to reduce the concentration of NO_X in the exhaust gas.

The primary zone of the combustor accommodates *burners*. Gaseous fuel is admitted directly into each burner and mixes with air. When liquid fuel is used it is atomized in the nozzle using high-pressure air. The atomized fuel/air mixture is then sprayed into the combustor. During oil firing, water is mixed into the oil to ensure low NO_X emission level.

Igniters are used for initial ignition of the fuel-air mixture during start-up of the gas turbine. Once energized, the spark plugs of the igniters start to spark and ignite the mixture of ignition gas and air. The main fuel then flows into the combustor through burners and is ignited there by the ignition flame. The flame then spreads from burner to burner without further sparking once these burners are supplied with fuel. After ignition is accomplished, the spark plug is switched off.

Flame monitors are installed in the combustor to indicate the presence or absence of flame. The flame monitor is sensitive to the presence of radiation, typically emitted by a hydrocarbon flame.

7.5.15 Turbine section

Power required to drive the compressor, various auxiliaries, and the load package is supplied by the turbine rotor. Turbines in gas turbine plants are almost all the axial flow type, except for a few smaller sizes, which are radial flow. Gas turbines have been built using both air and liquid cooling, to permit higher operating temperatures or the use of less critical materials. The gas turbine is capable of rapid start and loading and has no stand-by losses. The gas turbine is normally capable of firing both fuel gas and fuel oil.

When the turbine is at a standstill the variable inlet guide vanes remain in the closed position. During start-up, the variable inlet guide vanes are opened to their predefined starting position. When the turbine starting system is actuated ambient air is drawn through the air inlet assembly, filtered and compressed in the compressor. To reach the optimum efficiency with the lowest emissions, the air flow is controlled by variable inlet guide vanes.

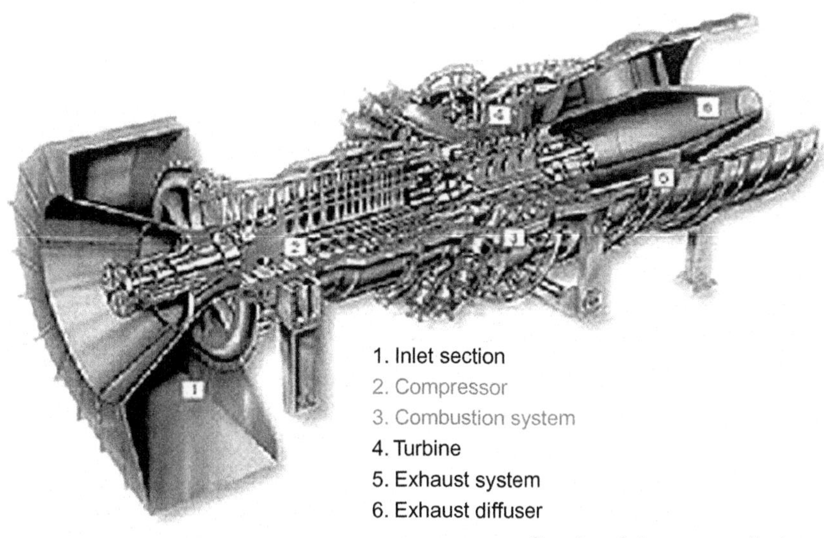

Figure 7.29 Sectional view of typical gas turbine.

The main components of a typical gas turbine unit are as follows:

1. Rotor
2. Journal and thrust bearings
3. Turbine blades and vanes
4. Air-cooling system for hot gas path components
5. Turbine and compressor casing
6. Compressor blades and vanes
7. Compressor inlet variable guide vanes
8. Air intake manifold
9. Compressor intake housing
10. Exhaust diffuser
11. Burners

Figure 7.29 shows a sectional view of a typical gas turbine design.

7.6 HEAT RECOVERY STEAM GENERATOR (HRSG)

A *heat recovery steam generator* (HRSG) or a *waste heat recovery boiler* (WHRB) is a heat exchanger that recovers heat from a hot gas stream. It produces steam that can be used in a process or used to drive a steam turbine. A common application for HRSG is in a combined cycle power plant, where hot exhaust from a gas turbine is fed to a HRSG to generate steam, which in turn drives a steam turbine (Figure 7.23). This combination produces electricity more efficiently than either a gas turbine or steam turbine alone, as explained earlier. HRSG is also an important component in co-generation plants. Where the HRSG supplies at least part of the steam to a process, the application is referred to as *co-generation*.

The exhaust from a gas turbine is directed to the HRSG, which is designed to accept the maximum exhaust temperature and gas flow of the connected gas turbine. It may be of either vertical or horizontal design for natural or controlled circulation.

HRSG can be designed with one to four separate operating pressure circuits to optimize heat recovery and cycle efficiency. Low-pressure steam may be used in deaeration and feedwater heating, thus avoiding the steam extraction regenerative feedwater heating used in conventional power cycles. However, conventional power cycle may also be followed depending on cycle configuration.

Gas turbine exhaust concurrently supplies sensible heat and oxygen to the furnace of a steam generator. A high level of oxygen remaining in the gas turbine exhaust (16–18% oxygen) may be used as an oxygen source to support further combustion. Supplementary (or duct) firing systems can be installed upstream of the HRSG. Duct burners provide additional energy to the HRSG, which produces more steam. This permits greater operating flexibility, improved steam temperature control, and higher overall power output. Generally, duct firing provides electrical output at lower capital cost but inferior efficiency compared with combined cycle generation. It is therefore often used for peaking.

The thermal efficiency of a typical unfired combined cycle plant may be found by using the following formula:

$$\eta = \frac{(W_{out} - W_{in})_{GT} + (W_{out} - W_{in})_{ST}}{Q_{TOTAL}} \tag{7.40}$$

where

$(W_{out} - W_{in})_{GT}$: net work output of gas turbine
$(W_{out} - W_{in})_{ST}$: net work output of steam turbine
Q_{TOTAL}: total heat supplied to the unit

The "steam production capability" and associated "heat rejection through stack" of a typical HRSG are shown in Figure 7.30 and Figure 7.31, respectively, which were drawn based on the following inputs:

i. The gas turbine was firing 8.8 kg/s of fuel with 425.0 kg/s of air
ii. The HRSG was receiving exhaust gas from above gas turbine at a temperature of 862 K

7.6.1 Design

Temperatures available in the furnace of conventional steam generators by combustion of fuel are very high (1200–1500 K, depending on the type of fuel), while gas turbine exhaust temperatures upstream of HRSG are very low (673–823 K). The design of

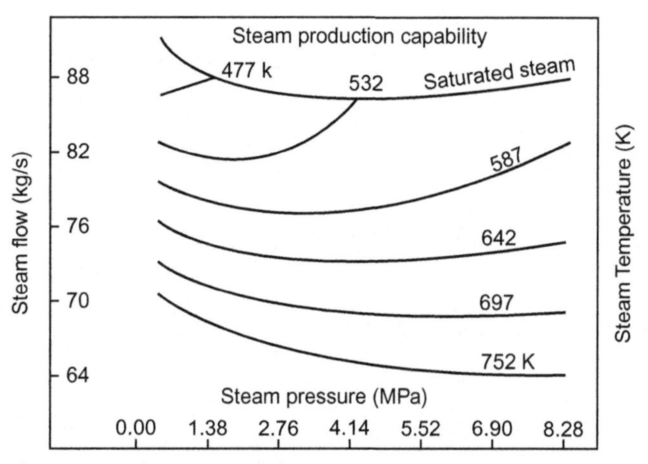

Figure 7.30 Typical steam production capability.

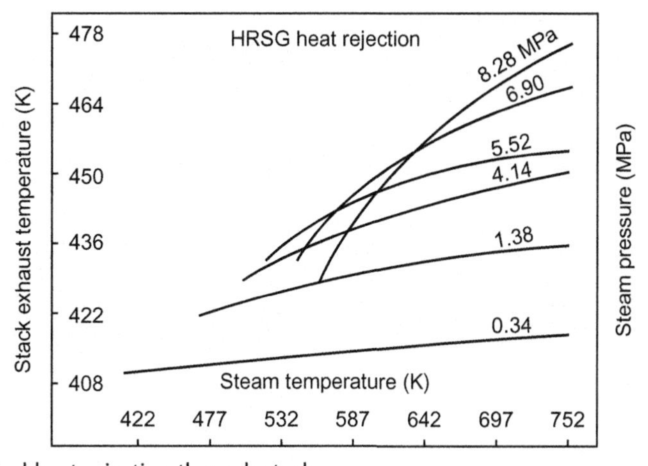

Figure 7.31 Typical heat rejection through stack.

HRSG thus becomes critical since fall in its inlet gas temperature will affect the energy absorbed in various heat transfer surfaces, especially the economizer, and eventually the exit gas temperature from the HRSG. Hence, to determine the steam generation and to predict gas-steam profiles, an economic design of HRSG is ensured considering the following five key parameters:

1. Allowable back-pressure
2. Steam pressure and temperature
3. Pinch point
4. Superheater and economizer approach temperatures
5. Stack outlet temperature

The *back-pressure* is significantly influenced by the HRSG cross-sectional flow area. Higher back pressures reduce HRSG cost but also reduce gas turbine efficiency. Thus, from a techno-economic standpoint back-pressures that are typically maintained at HRSG inlet are 2.50−3.75 kPa.

The best place to locate the superheater would be in the hottest area of the gas path. Since, at this location, it would take the least amount of surface to exchange the maximum heat to generate *the hottest steam at the highest pressure.*

The superheat attemperator to desuperheat steam may be placed at the outlet of the superheater, or at an intermediate position in superheat heat exchange path. The intermediate position, however, has an added advantage of preventing accidental entry of water that may damage downstream equipment.

HRSG is generally designed for natural, "thermosiphon" circulation. A vertical-tube HRSG generating very high steam pressure, however, may be designed with forced circulation system. The water is distributed within an inlet header to various parallel circuits and the steam and water mixture is collected in an outlet header and returned to the drum for separation of steam and water.

An HRSG without supplementary firing is generally designed with a circulation ratio of 5:1. However, suffice it to say that one with supplementary firing may be designed with a lower circulation ratio of 3:1. Figure 7.32 depicts a simple, single-pressure HRSG with the usual disposition of superheater, evaporator, and economizer sections.

A detailed flow scheme of a dual-pressure cycle with deaerator is shown in Figure 7.33 that is comprised of an HRSG with two steam circuits, one high-pressure (HP) and one low-pressure (LP) circuit. While the HP steam feeds the HP turbine, the latter feeds the LP turbine.

Each of the circuits has its own economizer, steam drum, evaporator, and super-heater. The exhaust gas leaving the gas turbine enters the HRSG. On passing through the HRSG gas exchanges its heat at various heat recovery sections, thereby getting

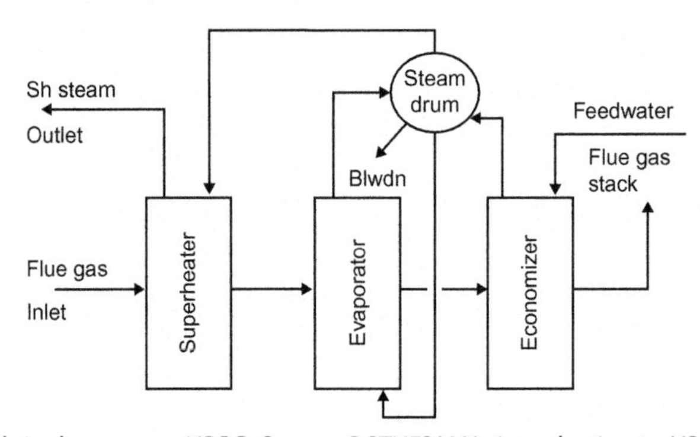

Figure 7.32 Typical single pressure HRSG. *Source: PGTHERMAL. Introduction to HRSG Design.*

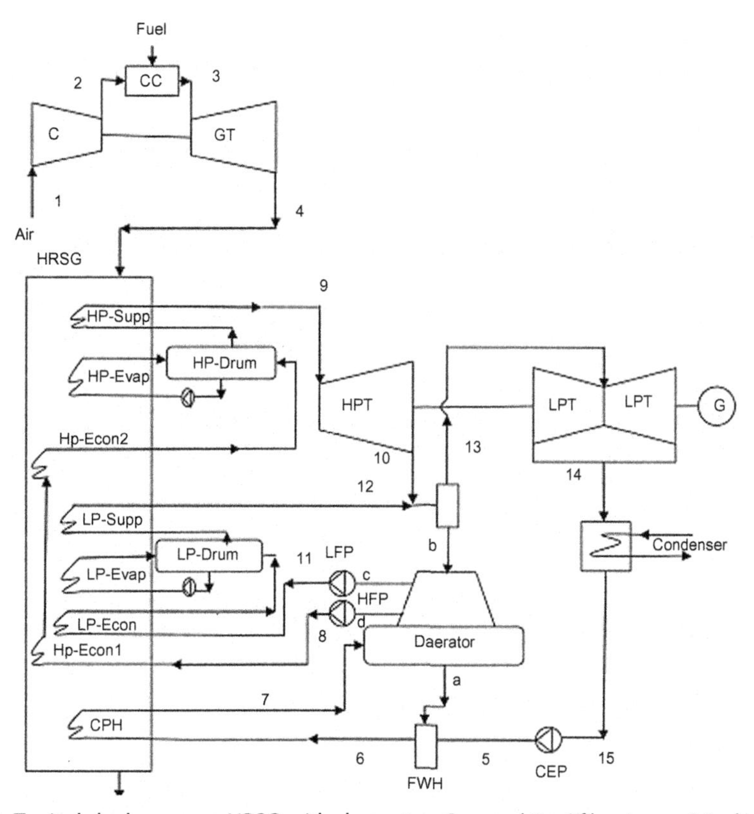

Figure 7.33 Typical dual pressure HRSG with deaerator. *Source: http://file.scirp.org/Html/.*

cooled prior to leaving through the stack. The condensate extraction pump draws condensate from the condenser hotwell and then circulates the condensate through two closed-type feedwater heaters (FWH & CPH) and deaerator. From the deaerator feedwater storage tank one high pressure boiler feed pump (HFP) and one low pressure boiler feed pump (LFP) take their suctions. Feedwater line from LFP discharge after passing through a low pressure economizer enters the low pressure drum, followed by low pressure evaporator and low pressure superheater, wherefrom superheated steam is transported through the low-pressure steam turbine to deliver power. Similarly feedwater line from HFP discharge passes through the high-pressure economizer, steam drum, evaporator and superheater in that sequence. This superheated steam enters the high-pressure stage of the same steam turbine to boost power generation.

The *pinch point temperature* and *approach temperatures* have a significant impact on the overall unit size. While small pinch point and superheater approach temperatures result in larger heat transfer surfaces and higher capital costs, the economizer approach temperature is typically set to avoid the economizer steaming at the design point.

Pinch and approach points are selected in unfired mode at design gas flow and inlet gas temperature conditions. Once they are assumed, the surface areas of the HRSG evaporator, superheater, and economizer are indirectly fixed or considered selected. Once selected, pinch and approach will vary with other conditions of gas flow and steam parameters.

The evaporator pinch is defined as the temperature difference between the flue gas leaving the evaporator and the saturation temperature of the fluid flowing through evaporator. The evaporator pinch limits the amount of heat that can be recovered in most HRSG designs. A pinch point of 28 K is generally considered for economic design of HRSG used in refineries and chemical plants. In a combined cycle or co-generation plants the pinch point is generally considered as 11 K, but could be as low as 5 K.

The economizer approach is the difference between the water temperature leaving the economizer and the saturation temperature of water. The "approach" of the economizer significantly affects the design and should be considered judiciously. This approach provides sufficient safety for load swings. Too low an approach may result in steaming in the economizer, which may upset the operation at certain operating conditions. Nevertheless, just because the economizer is steaming, does not necessarily mean it is a problem.

During the design stage itself the effect of steaming in the economizer could easily be overcome by taking the appropriate steps and factors. Most designs, however, are conservative and avoid this condition. In general, an average approach to the economizer is considered to be 11 K at the design stage, which is expected to ensure smooth, trouble-free operation. There are, however, many units in operation where the approach to the economizer is almost nil.

The higher the approach in superheaters, the less surface it will take to exchange heat. As a result, most of the flow in these sections is counter current to the gas flow, which provides a higher approach. For a lower approach steam flow through the superheater is usually cross flow to flue-gas flow. The absolute value of the superheater approach, however, does not affect the overall design of HRSG much.

The following ranges of pinch point and approach provide an economical and technically satisfactory design of HRSG, although a specific design may consider different values based on economical considerations:

- Evaporator Pinch Point: 8 to 28 K
- Superheater Approach: 22 to 33 K
- Economizer Approach: 6 to 17 K

The minimum *flue-gas exit or stack temperature* needs to be controlled to avoid corrosion due to acid condensation. In the event the exit temperature reaches or falls below the dew point of the flue gases resulting in formation of moisture film in the cold end, then such moisture formation may cause corrosion and fouling in that zone. These adverse conditions would exacerbate further if sulfur-bearing fuels are fired in gas turbines.

Example 7.4

A single-pressure HRSG, similar to Figure 7.32, receives 200 kg/s of flue gas from the exhaust of a gas turbine at a temperature of 800 K. The steam pressure and temperature at the HRSG superheater outlet are 4.25 MPa and 672 K, respectively. The feedwater temperature at the HRSG economizer inlet is 381 K. Determine:

a. Steam generation capability of the HRSG
b. Flue-gas temperature downstream of superheater
c. Evaporator duty
d. Economizer duty
e. Flue-gas temperature downstream of economizer

Given:

i. Economizer pressure drop is 0.07 MPa
ii. Economizer approach is 11 K
iii. Evaporator pinch is 28 K
iv. Superheater pressure drop is 0.10 mpa
v. Heat loss to atmosphere is 2% of heat available at inlet to each heat receiving section i.e., Superheater, evaporator, economizer, etc.
vi. Boiler drum blow-down is 2% of HRSG outlet steam flow
vii. Specific heat of flue gas at HRSG inlet is 1.10 kJ/kg/K
viii. Specific heat of flue gas at evaporator outlet is 1.07 kJ/kg/K
(Neglect pipe line losses.)

Solution: Let:

1. Economizer pressure drop be ΔP_1 ($= 0.07$ MPa)
2. Superheater pressure drop be ΔP_2 ($= 0.10$ MPa)
3. Economizer approach be ΔT_1 ($= 11$ K)
4. Evaporator pinch be ΔT_2 ($= 28$ K)
5. HRSG outlet steam flow be M_{SH}

Based on these inputs the following table may be prepared:

Parameter	Unit	Steam at superheater outlet	Feedwater at economizer inlet	Steam-water at boiler drum	Feedwater at economizer outlet	Flue gas at evaporator outlet
Pressure	MPa	4.25	$4.25 + \Delta P_1 + \Delta P_2 = 4.42$	$4.25 + \Delta P_2 = 4.35$		
Temperature	K	672	381	528.50 (Sat. temp.)	$528.50 - \Delta T_1 = 517.50$	$528.50 + \Delta T_2 = 556.50$
Enthalpy	kJ/kg	3208.7	455.36	$h_s = 2798.6$ $h_{wd} = 1112.0$	$h_{we} = 1058.5$	

Heat rejected by flue gas between HRSG inlet and evaporator outlet:

$$H_{SG} = 200 \times 1.10 \times (800.00 - 556.50) = 53570.00 \text{ kJ/s}$$

Net heat available to superheater and evaporator considering 2% heat loss:

$$H_{Net} = 53570.00 \times (1.00 - 0.02) = 52498.60 \text{ kJ/s}$$

Considering 2% blow-down, the heat required by HRSG between the superheater and evaporator to generate steam,

$$H_{Req} = M_{SH} \times \{(3208.7 - 1112.0) + (1.00 + 0.02) \times (1112.0 - 1058.5)\}$$
$$= M_{SH} \times 2151.27 \text{ kJ/s}$$

Since H_{Req} has to be equal to H_{Net},

$$\mathbf{M_{SH}} = 52498.60/2151.27 = \mathbf{24.40\ kg/s}$$

Heat required at superheater:

$$H_{SH} = 24.40 \times (3208.70 - 2798.60) = 10006.44 \text{ kJ/s}$$

Considering 2% heat loss, flue-gas temperature downstream of superheater, T_{SH}, is:

$$200 \times 1.10 \times (800 - T_{SH}) = 10006.44/(1.00 - 0.02)$$
$$\mathbf{T_{SH} = 753.59\ K}$$

The evaporator duty considering 2% blow-down:

$$\mathbf{H_{EV}} = 24.40 \times \{(2798.6 - 1112.0) + (1.00 + 0.02) \times (1112.0 - 1058.5)\}$$
$$= \mathbf{42484.55\ kJ/s}$$

Thus, the steam generated in the evaporator:
$\mathbf{M_{EV}} = 42484.55/(2798.6 - 1112.0) = \mathbf{25.19\ kg/s}$
The economizer duty considering 2% blow-down:

$$\mathbf{H_{ECO}} = 25.19 \times (1.00 + 0.02) \times (1058.50 - 455.36)$$
$$= \mathbf{15496.96\ kJ/s}$$

Considering 2% heat loss, the flue-gas temperature downstream of the economizer, T_{ECO}, is:

$$200 \times 1.07 \times (556.50 - T_{ECO}) = 15496.96/(1.00 - 0.02)$$
$$\mathbf{T_{ECO} = 482.61}$$

7.7 PROBLEMS

7.1 An ideal simple cycle gas turbine takes in air at 300 K temperature and 101.3 kPa pressure. The combustor raises the gas temperature to 1100 K. Both the compression pressure ratio and the expansion pressure ratio are 4. Determine:

a. Compression work
b. Expansion work
c. Thermal efficiency
d. Energy lost in the exhaust gas
e. Inlet gas pressure to turbine

(Assume specific heat of air at constant pressure is 1.01 kJ/kg.K and γ is 1.4.)

(**Ans.:** (a) 147.26 kJ/kg, (b) 363.35 kJ/kg, (c) 26.74%, (d) 444.65 kJ/kg, (e) 405.2 kPa)

7.2 A simple cycle gas turbine operates between temperatures 1255 K and 293 K using helium as its working fluid. The isentropic efficiencies of the compressor and the turbine are 80% and 85%, respectively. The specific heat of helium is 5.2 kJ/ kg/ K and γ of helium is 1.66. Assuming both compression and expansion pressure ratios are 6, determine

a. temperatures around the cycle
b. Efficiency of the cycle
c. Total heat to be supplied to generate 50 MW of power

(**Ans.:** $T_2 = 673.48$ K, $T_4 = 711.46$ K, $\eta = 28.04\%$, $Q_{Total} = 178316.69$ kJ/s)

7.3 A real Brayton cycle uses carbon dioxide as the working medium with polytropic efficiencies of the compressor and the turbine as 80% and 90%, respectively. A regenerator with 85% effectiveness is used in the cycle to improve the efficiency. The inlet and the maximum temperatures of the cycle are 313 K and 1323 K, respectively. Find the efficiency of this cycle considering the maximum output. Given: the specific heat of carbon dioxide is 0.87 kJ/kg/K and γ of carbon dioxide is 1.28.

(**Ans.:** $\eta = 39.98\%$)

7.4 A single-pressure HRSG receives 425 kg/s of flue gas from the exhaust of a gas turbine at a temperature of 862 K. The steam pressure and temperature at the HRSG superheater outlet are 5.5 MPa and 790 K, respectively, and the feedwater temperature at the HRSG economizer inlet is 400 K. While the fluid is flowing it suffers pressure drop across the superheater and economizer as 0.11 MPa and 0.09 MPa, respectively. If the economizer approach is 10 K and the evaporator pinch is 25 K, estimate the following neglecting any loss in the pipeline. (Assume the specific heat of flue gas at the HRSG inlet is 1.13 kJ/kg/K, and at evaporator outlet is 1.08 kJ/kg/K.)

a. Steam-generation capability of the HRSG
b. Flue-gas temperature downstream of superheater
c. Evaporator duty
d. steam generated in the evaporator
e. flue-gas temperature downstream of economizer

(**Ans.:** $M_{SH} = 60.42$ kg/s, $T_{SH} = 776.68$ K, $H_{EV} = 99578.20$ kJ/s, $M_{EV} =$ 62.33 kg/s

$T_{ECO} = 487.31$ K)

7.5 A dual-pressure HRSG, similar to the flow scheme in Figure 7.34 but with an integral deaerator, receives 315 kg/s of exhaust gas from a gas turbine at a temperature of 713 K. The high-pressure (HP) steam of 25 kg/s is generated from

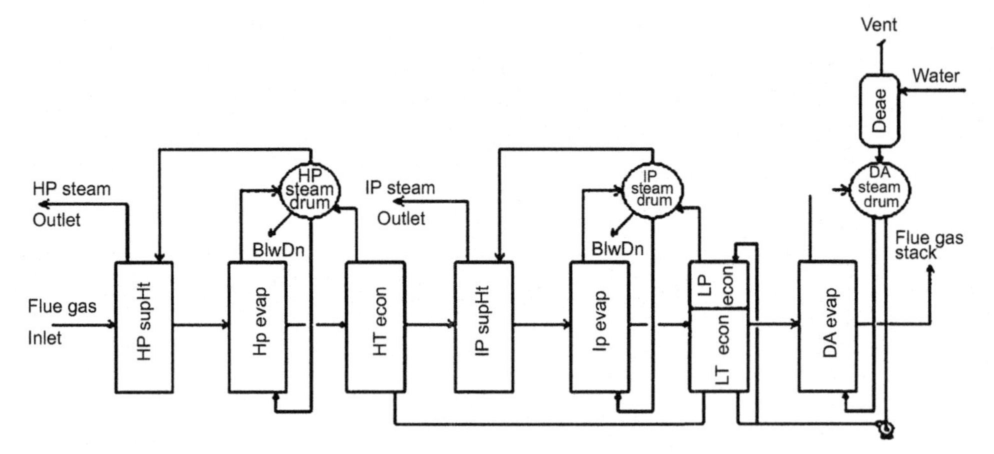

Figure 7.34 Flow scheme of dual pressure HRSG. *Source: https://upload.wikimedia.org/wikipedia/commons/9/9f/HRSG_cycle.jpg.*

the HRSG at a pressure and temperature of 5 MPa and 693 K, respectively, while the intermediate pressure (IP) steam of 10 kg/s is generated at pressure and temperature of 0.5 MPa and 573 K, respectively. Each of the HP and IP circuits uses one superheater, one evaporator, and one economizer. Condensate is returned to the deaerator at a pressure and temperature of 0.6 M Pa and 327 K, respectively. To avoid corrosion at the cold end of the HRSG the exit flue-gas temperature is maintained no less than 405 K. Pressure drops of the working fluid at various sections of HRSG are as follows:

i. HP superheater 0.40 MPa
ii. HP economizer 0.30 MPa
iii. IP superheater 0.05 MPa
iv. IP economizer 0.03 MPa

The "pinch points" at the HP, IP, and LP (deaerator) evaporators are 14, 10, and 12 K, respectively, while for the "economizer approach" it is 10 K at the HP circuit and 6 K at the IP circuit. Consider the average specific heat of gas is 1.07 kJ/kg/K. Neglecting any loss of working fluid from the cycle, calculate the temperature of the gas leaving each heat transfer surfaces. Neglect also temperature drop between:

1. HP superheater and HP evaporator
2. HP evaporator and HP economizer
3. HP economizer and IP superheater

4. IP superheater and IP evaporator
5. IP evaporator and IP economizer
6. IP economizer and LP evaporator

(**Ans.:** (a) 679.19 K, (b) 555.91 K, (c) 510.85 K, (d) 501.57 K, (e) 438.62 K, (f) 433.82 K.)

BIBLIOGRAPHY

[1] Baumeister T, Marks LS, editors. Mechanical Engineers" Handbook, sixth ed., McGraw-Hill Book Company, Inc.; 1958.
[2] El-Wakil MM. Power Plant Technology. McGraw-Hill, Inc.; 1984.
[3] Boyce MP. Gas Turbine Engineering Handbook. Gulf Publishing Company Book Division; 1982.
[4] Gas Turbine World 2013 GTW Handbook. Pequot Publishing Inc.
[5] Gas Turbine World January–February 2013.
[6] Rogers GFC, Mayhew YR. Engineering Thermodynamics-Work and Heat Transfer. English Language Book Society and Longmans, Green and Co. Ltd.; 1967.
[7] Wrangham DA. Theory and Practice of Heat Engine. The English Language Book Society; 1962.
[8] COMBUTION March 1981. A Comparison of Power Plants for Co-generation of Heat and Electricity.
[9] COMBUTION September 1978. What is the True Cost of Electric Power from a Co-generation Plant.
[10] Seebregts AJ. IEA ETSAP (Energy Technology Systems Analysis Programme). Technology Brief E02–April 2010.
[11] Gas Turbine World March–April 2013.

CHAPTER 8

Diesel Power Plant

8.1 INTRODUCTION

In 1885, the English engineer Herbert Akroyd Stuart invented an engine using paraffin oil as the fuel. This was the first internal combustion engine to use a pressurized fuel injection system. This engine adopted a low-compression ratio, so the compressed air did not have adequate temperature in the combustion chamber to initiate ignition of fuel. Hence, the combustion of oil used to take place in a separate chamber called the "vaporizer" mounted on the cylinder head, which was heated externally by an oil lamp to raise the temperature of the chamber sufficient enough to ignite the oil. Compressed air from the cylinder is passed through a constricted passage into the vaporizer, where oil is then sprayed and ignited by the temperature of the vaporizer chamber for combustion. In 1892 Thomas Henry Barton built a high-compression internal combustion engine, thus eliminating the use of a separate vaporizer.

Credit for the invention of the modern diesel engine, however, goes to French engineer Rudolf Diesel. In 1892 Diesel adopted a high compression ratio in his engine, where air could be compressed so much that the temperature attained by air far exceeded the ignition temperature of the fuel, thereby ensuring appropriate combustion of the fuel gradually introduced into the combustion chamber. Diesel's original 1897 designed engine is depicted in Figure 8.1.

A diesel power plant uses a diesel engine, similar to one developed by Rudolf Diesel, as the prime mover for the generation of electrical energy. This engine, also called a compression-ignition engine, is a reciprocating internal combustion engine. Unlike a spark-ignition engine, a diesel engine uses the heat of compression of air to initiate ignition and burning of the fuel in a combustion chamber.

Of all type of standard prime movers of equivalent size, whether internal or external combustion types, the diesel engine has the highest thermal efficiency (almost 52%) due to its very high compression ratio. Low-speed diesel engines can have a thermal efficiency greater than 50% [1].

A diesel power plant is generally compact. The largest unit at the time of writing is about 85 MW, which was put into service aboard a container ship. The world's largest two-stroke diesel power-generating unit on land is of capacity 50 MW installed at GMR Vasavi power plant in Chennai, India.

Thermal Power Plant
DOI: http://dx.doi.org/10.1016/B978-0-12-801575-9.00008-1

Figure 8.1 Diesel's original 1987 engine. *Source: http://en.wikipedia.org/wiki/Diesel_engine.*

Based on its requirement a diesel power plant may serve any one of the following three purposes [2]:

i. Base load station running continuously as sole source of power
ii. Standby station sometimes running continuously at rated load
iii. Emergency unit running for short time at rated load

The thermal efficiency of a diesel engine is inherently better than that of a steam or gas turbine. The efficiency of the engine could further be improved (to the extent of about 60% [1]) by recovering waste heat in exhaust gas in a heat recovery steam generator for generation of steam either to be used in a steam turbine to generate additional power or be used as process steam. In spite of its high efficiency a diesel power plant is generally used as emergency supply station because of high cost of diesel oil in most countries. As a smaller version a diesel power plant is a portable type that may be located *in situ* as and when there is an emergency demand. Diesel engine generators are attractive due to their long life. Today's diesel engines emit less noise and generally do not require much maintenance.

Diesel engines are classified as two-stroke and four-stroke engines. While most diesel engines, particularly of domestic type, generally use the four-stroke one, larger engines are usually two-stroke type. Domestic diesel generator sets require single-phase power supply, whereas industrial or commercial application generators usually require three-phase power.

Most diesel engines have large pistons and therefore draw more air and fuel, which results in bigger and more powerful combustion. In internal-combustion engines, it is presumed that the combustion process occurs at either constant pressure or at constant volume or by a combination of these two processes. The constant-pressure process is observed in the slow-speed compression-ignition (C.I.) or Diesel cycle; the constant volume process is noted in the spark-ignition (S.I.) or Otto cycle; combination of these two processes, however, is called mixed or limited-pressure cycle and appears in high-speed compression-ignition engines.

Compression ratio (CR) of an internal combustion engine is defined as the ratio of "the volume between the piston and cylinder head before and after a compression stroke." In the Diesel cycle mixing of air and fuel takes place near the end of compression, while in the Otto cycle mixing takes place before compression. In the diesel engine the heat of compression is used to initiate ignition to burn the fuel, while in the Otto cycle a spark is required to begin combustion of fuel. The Diesel cycle can operate with a higher compression ratio than the Otto cycle because only air is compressed in the Diesel cycle while a charge of fuel-air mixture is compressed in the Otto cycle. As a result there is no risk of auto-ignition of the fuel in the Diesel cycle.

In compression-ignition engines no separate source of ignition energy, like a spark plug, is required for igniting fuel. These engines use liquid fuels of low volatility. In the diesel engine the most common working fluid is air, which is introduced into the combustion chamber to burn fuel. The air is compressed with a compression ratio, typically between 11.5:1 and 22:1, resulting in a compression pressure ranging between 2.8 MPa and 4.8 MPa [3]. This high compression heats the air to 823 K, then fuel is injected directly into the compressed air in the combustion chamber. The fuel injector ensures that the fuel is broken down into small droplets, and is distributed as evenly as possible. The vapor is then ignited by the heat from the compressed air in the combustion chamber, where the droplets continue to vaporize from their surfaces and burn, getting smaller, until all the fuel in the droplets has been burnt. The start of vaporization causes a delay during ignition, resulting in an abrupt increase in pressure above the piston. The rapid expansion of combustion gases then drives the piston downward, supplying power to the crankshaft. The load and speed in this engine are controlled by varying the injected fuel flow.

Spark-ignition engines use high volatile liquids and gases as fuel and have compression ratios between 4:1 and 12:1. The compression pressure ranges from below 0.7 MPa to 2.0 MPa. S.I. engines use carburetors and fuel injection systems. Load and speed are usually controlled by throttling the charge [3].

The combustion process in an internal combustion engine is governed by the air-fuel ratio in the cylinder. In a typical compression-ignition engine burning hydrocarbon fuels the stoichiometric air-fuel ratio is 14.9. The limiting relative air-fuel ratio for combustion at full load ranges between 1.2 and 1.6 [4].

8.1.1 Advantages of diesel power plant over steam power plant

1. Simple design and layout of plant
2. Occupies less space and is compact
3. Can be started quickly and picks up load in a short time
4. Requires less water for cooling
5. Thermal efficiency (about 50% or more) is better than that of a steam power plant of same size
6. Overall cost is less than that of steam power plant of same size
7. Requires low operating staff
8. No stand-by losses

8.1.2 Disadvantages of diesel power plant over steam power plant

1. High running charges due to costly price of diesel
2. Plant does not work efficiently under prolonged overload conditions
3. Generates a comparatively small amount of power
4. Specific lubricating oil consumption and its costs are very high
5. Maintenance charges are generally high

8.1.3 Mean effective pressure (M.E.P.)

In a reciprocating engine it is difficult to isolate the positive and negative pressure in the cycle, since the processes are carried out in a single component. Hence, while comparing air-standard cycles of reciprocating engines a term *mean effective pressure* is usually referred to in lieu of the work ratio. Mean Effective Pressure, or in short M.E.P., is defined as the mean pressure that, if imposed on the pistons uniformly from the top to the bottom of each power stroke, would produce the net work of the cycle. A cycle operating with large M.E.P. will produce large work output per unit of swept volume. As a result size of an engine based on this cycle will be small for a given output.

For naturally aspirated four-stroke diesel engines, the maximum M.E.P. is in the range of 700 to 900 kPa. Two-stroke diesel engines have comparable values of M.E.P., while very large low speed diesel engines may have a M.E.P. up to 1900 kPa.

8.1.4 Four-stroke engine

A four-stroke engine comprises of the following:

i. *Induction stroke*: Air is drawn into the cylinder.
ii. *Compression stroke*: The air is compressed and just before the piston reaches the top dead center liquid fuel is sprayed into the cylinder. The temperature of the air at the end of compression is significantly high to vaporize the liquid droplets and ignite them.

iii. *Expansion or working stroke*: Combustion is almost completed at the start of the expansion stroke, where the product expands and pressure falls steadily. As the exhaust valve opens blow-down occurs and the pressure in the cylinder becomes slightly more than the atmospheric pressure.
iv. *Exhaust stroke*: The gases remaining in the cylinder at the end of the expansion stroke are displaced by the piston during the exhaust stroke.

8.1.5 Two-stroke engine

In a two-stroke engine the induction and exhaust strokes are eliminated and part of the compression and expansion strokes are used for exhaust and induction. At the end of the compression stroke the inlet port opens and air is admitted to the cylinder, while at the end of the expansion stroke the exhaust port is opened to allow blow-down of gases. Eventually the piston stroke of this engine is longer than that of a four-stroke engine of the same capacity.

The relative advantages/disadvantages of four-stroke and two-stroke engines are as follows:

8.1.6 Advantages of four-stroke engines

i. Better scavenging, hence higher M.E.P.
ii. More flexibility
iii. Low fuel and lubricating oil consumption
iv. Less noisy exhaust
v. Compression better maintained, since the engine runs cooler, and the average load on the piston rings is less
vi. Cylinder wears less than in some two strokes

8.1.7 Advantages of two-stroke engine

i. Larger power per cylinder
ii. Shorter crankshaft and more even torque reduces the risk of torsional oscillation
iii. Engine easily reversible
iv. Mechanically operated valves are eliminated
v. Few spares required
vi. Inferior fuel may be used
vii. Higher fuel consumption so greater output

Two-stroke engines have one major disadvantage – leakage of scavenge air through the exhaust ports accounts for a loss of air almost equal to 50% of the swept volume of the cylinder.

8.2 THERMODYNAMIC CYCLE

As discussed earlier a reciprocating internal-combustion engine may operate either on Diesel cycle or on Otto cycle. The following sections discuss these cycles in detail.

8.2.1 Diesel cycle

Figure 8.2 and Figure 8.3 depict the P-v and T-s diagrams for an ideal Diesel cycle, where P is the pressure and v is the specific volume. The ideal Diesel cycle follows four distinct processes: compression, combustion, expansion, and cooling. The cycle follows the numbers 1–4 in the clockwise direction.

- Process 1 to 2 is a isentropic compression process. The air is compressed isentropically through a volume ratio v_1/v_2. During this process work (W_{in}) is done by the piston compressing the working fluid.
- Process 2 to 3 is a reversible constant-pressure heat addition process. Heat (Q_{in}) is supplied by combustion of the fuel while the air expands at constant pressure to volume v_3 during this process.
- Process 3 to 4 is an isentropic expansion process to the original volume v_1. Work (W_{out}) is done by the working fluid expanding on to the piston, which produces usable torque.
- Process 4 to 1 is a reversible constant volume heat rejection (cooling) process. During this process heat (Q_{out}) is rejected by exhausting air until the cycle is completed.

The stages of these four processes are discussed in the following. Note that theoretically each of the strokes of the cycle complete at either top dead center (TDC) or

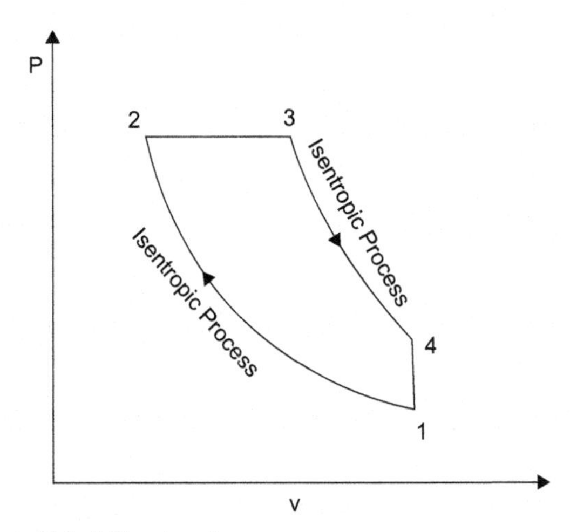

Figure 8.2 P-v Diagram of ideal Diesel cycle.

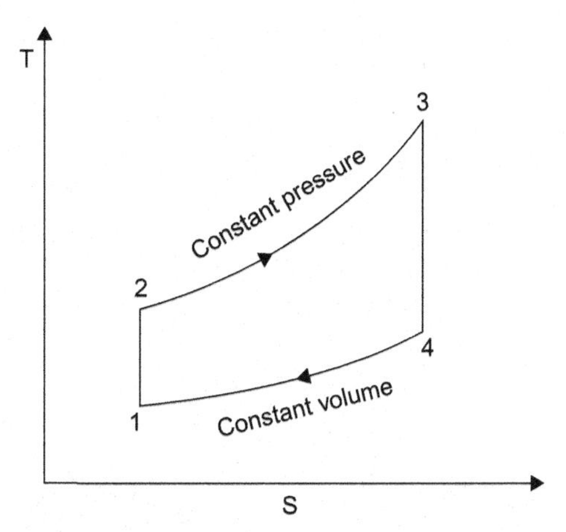

Figure 8.3 T-s Diagram of ideal Diesel cycle.

bottom dead center (BDC). However, due to delay in the opening and closing of the inlet and exhaust valves, the inertia of inlet air and exhaust gas, each of the strokes invariably begin and end beyond the TDC and BDC.

8.2.1.1 Compression

The cycle is started by drawing air at ambient conditions into the engine (Figure 8.4). This air is then compressed adiabatically by moving the piston upward in the cylinder from BDC to TDC (Figure 8.5). This compression raises the temperature of the air to a level where the fuel mixture, which is formed by injecting fuel once the air is compressed, spontaneously ignites. It is in this part of the cycle that work is applied *to* the air. This compression is considered to be isentropic.

The ratio of the volume of the working fluid inducted into the cylinder before the compression process (v_1) to its volume after the compression (v_2) is defined as the *volumetric compression ratio*, r_c $(=v_1/v_2)$.

8.2.1.2 Combustion (Figure 8.6)

Next, the fuel is injected into the combustion chamber during the final stage of compression, thereby adding heat to the air by combustion of fuel. This process begins just as the piston leaves its TDC position. Because the piston is moving during this part of the cycle, the heat addition process is called isochoric (constant volume). The combustion process, however, takes place at constant pressure.

8.2.1.3 Expansion (Figure 8.7)

In the Diesel cycle, fuel is burned to heat compressed air and as the rapidly burning mixture attempts to expand within the cylinder, it generates a high pressure that forces the piston downward from the TDC position to the BDC position in the cylinder.

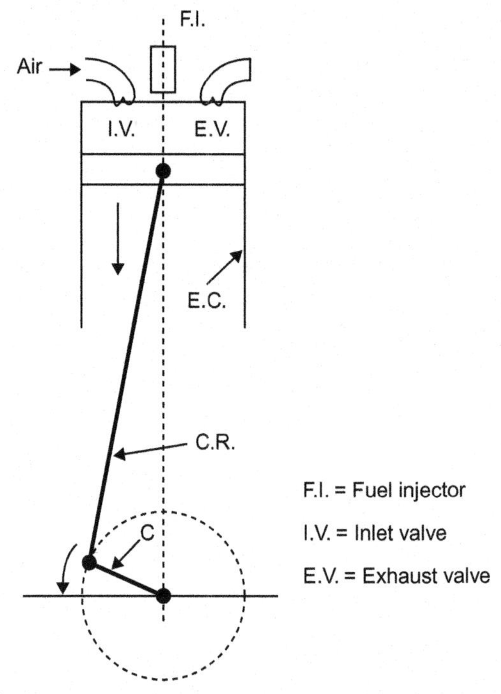

Figure 8.4 Induction of air into the cylinder in Diesel cycle. *Source: http://image.slidesharecdn.com/asintroductionicengine.*

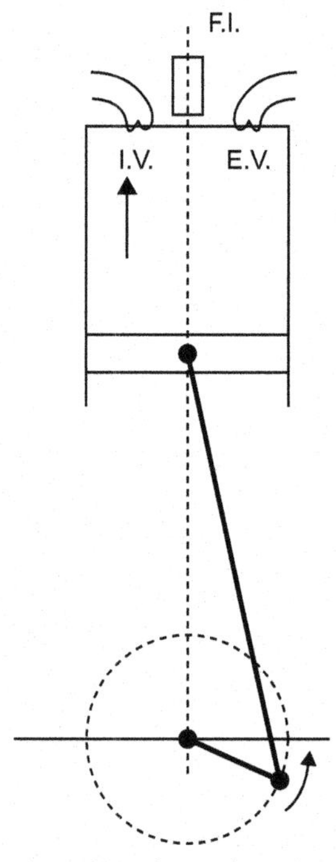

Figure 8.5 Compression of air in Diesel cycle. *Source: http://image.slidesharecdn.com/asintroductionicengine.*

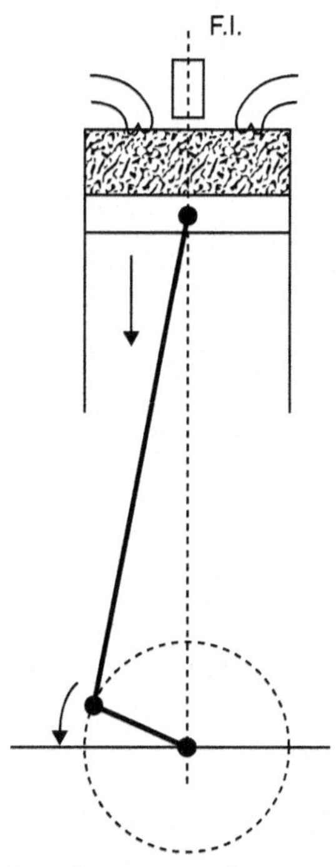

Figure 8.6 Combustionin in Diesel cycle. *Source: http://image.slidesharecdn.com/asintroduction-icengine.*

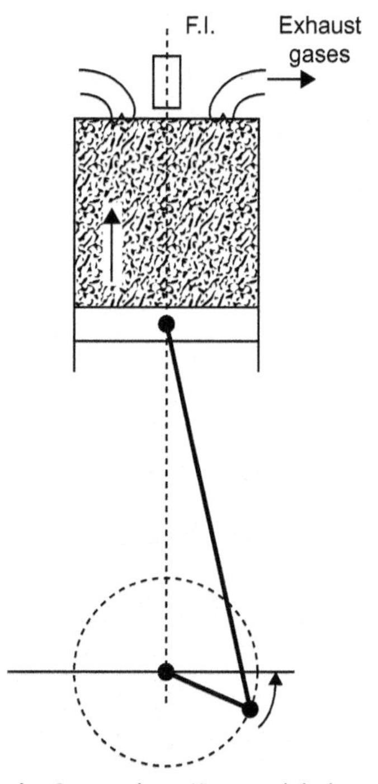

Figure 8.7 Expansion in Diesel cycle. *Source: http://image.slidesharecdn.com/asintroductionicengine.*

It is in this phase that the cycle contributes its useful work, rotating the engine crankshaft. This stage of an ideal Diesel cycle is considered as isentropic.

The ratio of the volume of the working fluid completing the expansion process ($v_4 = v_1$) to its volume at the beginning of expansion (v_3) is called the *volumetric expansion ratio*, r_e $(=v_1/v_3)$.

The expanded mixture or gas is then ejected from the cylinder to the environment, which is replaced with a fresh supply of cold air from ambient conditions. Since this happens when the piston is at the TDC position in the cycle and is not moving, this process is isochoric (no change in volume). During actual operation there is a point in the cycle when both inlet and exhaust valves remain open simultaneously.

8.2.1.4 Thermal efficiency [5]

As explained above the compression and power strokes of an ideal Diesel cycle are adiabatic, hence the thermal efficiency of this ideal cycle can be calculated from the constant-pressure and constant-volume processes. Accordingly, the heat input (Q_{in}), heat output (Q_{out}), and thermal efficiency (η_{th}) can be calculated if the temperatures and specific heats of various end states are known. Assuming constant specific heats for the air and considering unit mass flow of the fluid, using Figure 8.3 we can write:

$$Q_{in} = C_p(T_3 - T_2) \tag{8.1}$$

$$Q_{out} = C_v(T_1 - T_4) \tag{8.2}$$

$$\eta_{th} = \frac{Q_{in} + Q_{out}}{Q_{in}} \tag{8.3}$$

or,

$$\eta_{th} = 1 + \frac{Q_{out}}{Q_{in}} = 1 + \frac{C_v(T_1 - T_4)}{C_p(T_3 - T_2)} \tag{8.4}$$

where

C_p = Specific heat at constant pressure
C_v = Specific heat at constant volume
T = Temperature at various end states of the cycle

From Figure 8.2 it can also be written that the compression ratio $r_c = v_1/v_2$ and the expansion ratio $r_e = v_1/v_3$. Now, using the ideal gas law $Pv = nRT$ and $\gamma = C_p/C_v$ and for adiabatic condition Pv^γ = constant, Eq. 8.4 can be written as

$$\eta_{th} = 1 - \frac{1}{\gamma}\frac{(T_4 - T_1)}{(T_3 - T_2)} \tag{8.5}$$

For adiabatic compression

$$\frac{T_2}{T_1} = r_c^{(\gamma-1)} \tag{8.6}$$

therefore

$$T_1 = \frac{T_2}{r_c^{(\gamma-1)}} \tag{8.7}$$

For constant-pressure process

$$\frac{T_3}{T_2} = \frac{v_3}{v_2} = \frac{v_3}{v_1} * \frac{v_1}{v_2} = \frac{r_c}{r_e} \tag{8.8}$$

hence,

$$T_3 = T_2 \frac{r_c}{r_e} = T_1 \frac{r_c^{\gamma}}{r_e} \tag{8.9}$$

Again, for adiabatic expansion,

$$\frac{T_3}{T_4} = r_e^{(\gamma-1)} \tag{8.10}$$

or,

$$T_4 = \frac{T_3}{r_e^{(\gamma-1)}} = T_2 \frac{r_c}{r_e} \frac{1}{r_e^{(\gamma-1)}} = T_2 \frac{r_c}{r_e^{\gamma}} = T_1 \left(\frac{r_c}{r_e}\right)^{\gamma} \tag{8.11}$$

Substituting the values of T_1, T_3 and T_4 from Eq. 8.7, Eq. 8.9, and Eq. 8.11, respectively, into Eq. 8.5 we arrive at

$$\eta_{th} = 1 - \frac{1}{\gamma} \frac{\left(T_2 \frac{r_c}{r_e^{\gamma}} - \frac{T_2}{r_c^{(\gamma-1)}}\right)}{\left(T_2 \frac{r_c}{r_e} - T_2\right)} = 1 - \frac{1}{\gamma} \frac{\left(\frac{r_c}{r_e^{\gamma}} - \frac{1}{r_c^{(\gamma-1)}}\right)}{\left(\frac{r_c}{r_e} - 1\right)} = 1 - \frac{1}{\gamma} \frac{\left(\frac{1}{r_e^{\gamma}} - \frac{1}{r_c^{\gamma}}\right)}{\left(\frac{1}{r_e} - \frac{1}{r_c}\right)}$$

or,

$$\eta_{th} = 1 - \frac{1}{\gamma} \frac{\left(r_e^{-\gamma} - r_c^{-\gamma}\right)}{\left(r_e^{-1} - r_c^{-1}\right)} \tag{8.12}$$

Sometimes a "cut-off" ratio, as follows, is used to carry out various calculations:

$$r_v = \frac{v_3}{v_2} = \frac{v_1}{v_2} * \frac{v_3}{v_1} = \frac{r_c}{r_e} \tag{8.13}$$

Inputting the value of Eq. 8.13 into Eq. 8.12

$$\eta_{th} = 1 - \frac{1}{\gamma} \frac{r_c^{-\gamma}\left\{\left(\frac{r_e}{r_c}\right)^{-\gamma} - 1\right\}}{r_c^{-1}\left\{\left(\frac{r_e}{r_c}\right)^{-1} - 1\right\}} = 1 - \frac{r_c^{(1-\gamma)}}{\gamma} \frac{\left\{\left(\frac{1}{r_v}\right)^{-\gamma} - 1\right\}}{\left(\frac{1}{r_v}\right)^{-1} - 1} = 1 - \frac{r_c^{(1-\gamma)}}{\gamma} \frac{(r_v^{\gamma} - 1)}{(r_v - 1)} \quad (8.14)$$

Equations 8.12 and 8.14 give the ideal thermal efficiency of the Diesel cycle, which depends on the compression ratio as well as either the expansion ratio or the cut-off ratio, but not on the peak temperature T_3. However, the thermal efficiency of actual Diesel cycle will be much lower than the ideal efficiency due to heat and friction losses, power absorbed by auxiliaries, e.g., lubricating oil pumps, cooling water pumps, fuel pumps, radiator fans, etc.

8.2.1.5 Mean effective pressure (M.E.P.)

As explained in Section 8.1.3 that M.E.P. is defined as the mean pressure, which, if imposed on the pistons uniformly from the top to the bottom of each power stroke, would produce the net work of the cycle.

In a cycle net work done = heat added − heat rejected.

Therefore,

$$W = C_p(T_3 - T_2) - C_v(T_4 - T_1) \quad (8.15)$$

Again, a change in volume,

$$v_1 - v_2 = v_1\left(1 - \frac{v_2}{v_1}\right) = \frac{RT_1}{P_1}\left(1 - \frac{1}{r_c}\right) = \frac{(C_p - C_v)T_1}{P_1}\left(\frac{r_c - 1}{r_c}\right) = \frac{C_v(\gamma - 1)T_1}{P_1}\left(\frac{r_c - 1}{r_c}\right) \quad (8.16)$$

Using the values from Eq. 8.15 and Eq. 8.16,

$$M.E.P. = \frac{W}{v_1 - v_2} = \frac{C_p(T_3 - T_2) - C_v(T_4 - T_1)}{\frac{C_v(\gamma - 1)T_1}{P_1}\left(\frac{r_c - 1}{r_c}\right)} = \frac{P_1 r_c}{(r_c - 1)} \frac{1}{(\gamma - 1)}\left\{\gamma\frac{(T_3 - T_2)}{T_1} - \frac{(T_4 - T_1)}{T_1}\right\} \quad (8.17)$$

Inputting the values from Eq. 8.7, Eq. 8.9, and Eq. 8.11 into Eq. 8.17

$$M.E.P. = \frac{P_1 r_c}{(r_c - 1)}\frac{1}{(\gamma - 1)}\left\{\gamma\left(\frac{r_c^\gamma}{r_e} - r_c^{(\gamma-1)}\right) - \left(\frac{r_c}{r_e}\right)^\gamma + 1\right\}$$

or,

$$M.E.P. = \frac{P_1 r_c}{(r_c - 1)(\gamma - 1)}\{\gamma r_c^{(\gamma-1)}(r_v - 1) - (r_v^\gamma - 1)\} \qquad (8.18)$$

Example 8.1

1 kg of air at 298 K temperature and 101.3 kPa pressure is taken through the ideal Diesel cycle of Figure 8.1. The compression ratio, r_c, of the cycle is 15 and the heat added is 1900 kJ. Find the ideal thermal efficiency and M.E.P. of this cycle.

(Assume $R = 0.287$ kJ/(kg.K), $C_p = 1.005$ kJ/kg/K, and $\gamma = 1.4$ for air and conversion; factors: 1 kJ = 102 m.kg and 1 kPa = 102 kg/m^2.)

Solution: For the unit mass flow the initial volume of the air is

$$V_1 = \frac{RT_1}{P_1} = \frac{0.287 * 298}{101.3} = 0.844\ m^3$$

After compression

$$T_2 = T_1 * r_c^{(\gamma-1)} = 298 * 15^{0.4} = 880.34\ K$$

and

$$V_2 = \frac{0.844}{15} = 0.056\ m^3$$

With heat added

$$Q_{in} = C_p(T_3 - T_2)$$

or,

$$1900 = 1.005(T_3 - 880.34)$$

hence,

$$T_3 = 2770.89\ K$$

The expansion ratio is

$$r_e = \frac{V_4}{V_3} = \frac{V_1}{V_3} = \frac{V_1}{V_2} * \frac{V_2}{V_3} = \frac{V_1}{V_2} * \frac{T_2}{T_3} = 15 * \frac{880.34}{2770.89} = 4.766$$

hence,

$$T_4 = T_3\frac{1}{r_e^{(\gamma-1)}} = 2770.89\frac{1}{4.766^{0.4}} = 1483.74\ K$$

With heat rejected

$$Q_{out} = C_v(T_1 - T_4) = \frac{1.005}{1.4}(298.00 - 1483.74) = -851.2\ kJ/kg$$

Net work

$$W = Q_{in} + Q_{out} = 1900 - 851.2 = 1048.8\ kJ/kg$$

Therefore, the ideal thermal efficiency is

$$\eta_{th} = \frac{1048.8}{1900.0} * 100 = 55.2\%$$

and

$$M.E.P. = \frac{1048.8}{(0.844 - 0.056)} = 1330.96\ kPa = 1.331\ MPa$$

An alternative solution of η_{th} using Eq. 8.12:

$$\eta_{th} = 1 - \frac{1}{1.4}\frac{(4.766^{-1.4} - 15^{-1.4})}{(4.766^{-1} - 15^{-1})} = 1 - \frac{1}{1.4}\frac{(0.1124 - 0.0226)}{(0.2098 - 0.0667)} = 1 - \frac{0.0898}{0.2003} = 0.5518,\ i.e.\ 55.18\%$$

Example 8.2

Considering the unit mass flow of the fluid determine the thermal efficiency, the final temperature of fluid after expansion, and the compression ratio of an air-standard Diesel cycle, in which the inlet pressure and temperature are 100 kPa and 288 K, respectively. The heat addition to the cycle is 1045.9 kJ and the final temperature after heat addition is 2073 K. (Given $C_p = 1.005$ kJ/kg/K, $C_v = 0.718$ kJ/kg/K, and $\gamma = 1.4$ for air.)

Solution: From Eq. 8.1, $1045.9 = 1.005(2073 - T_2)$
Hence, $T_2 = 1032.3$ K
From Eq. 8.6, $r_c^{0.4} = 1032.3/288$
Wherefrom, $r_c = 24.32$
From Eq. 8.8 and Eq. 8.13, the cut-off ratio is

$$r_v = 2073/1032.3 = 2.008$$

From Eq. 8.11

$$T_4 = 288 * 2.008^{1.4} = 764.3\ K$$

Thus, from Eq. 8.14, the thermal efficiency of the air-standard Diesel cycle is

$$\eta_{th} = \left\{1 - \frac{24.32^{-0.4}}{1.4}\left(\frac{2.008^{1.4} - 1}{2.008 - 1}\right)\right\} * 100 = \left(1 - \frac{1}{5.018}\frac{1.654}{1.008}\right) * 100 = 67.3\%$$

8.2.2 Otto cycle

The Otto cycle was developed by German engineer Nikolaus Otto in 1876. This cycle is very similar to the Diesel cycle, except that the Otto cycle forms the basis of

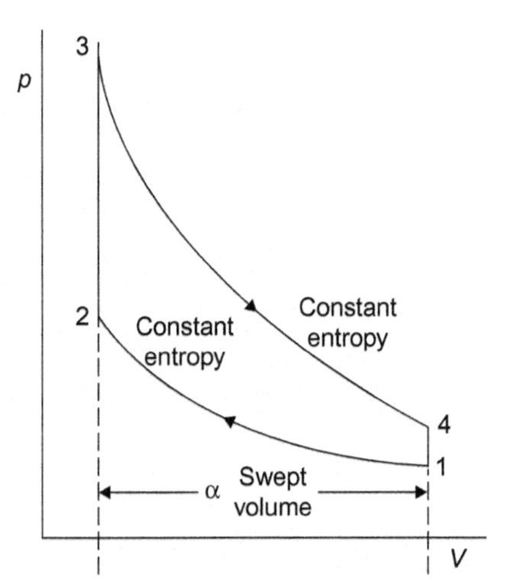

Figure 8.8 P-v Diagram of Otto cycle. *Source: Fig. 12.14, Page 245, Gas Power Cycles, [6].*

a *spark-ignition* (S.I.) engine, which requires a spark to begin combustion of fuel. Figure 8.8 represents a P-V diagram of an Otto cycle.

As in an ideal Diesel cycle four distinct processes are also followed in the Otto cycle: reversible adiabatic compression, addition of heat at constant volume, reversible adiabatic expansion, and rejection of heat at constant volume as follows:

i. Air enters the cycle at point 1 when compression starts and it ends at point 2. The air is compressed in an isentropic process through a volume ratio v_1/v_2, known as the *compression ratio* r_c (the ratio of the volume of the working fluid before the compression process to its volume after).

ii. Heat addition starts at point 2 and ends at point 3. A quantity of heat Q_{in} is added at constant volume, and the working fluid temperature rises.

iii. The isentropic expansion process starts at point 3 and it ends at point 4 to the original volume.

iv. The heat Q_{out} is rejected at constant volume from point 4 to point 1. Air gets cooled and the working fluid temperature goes down at a constant volume until the cycle is completed.

Stages of above four processes are described below:

8.2.2.1 Compression

In this part of the cycle as the piston moves from the TDC to the BDC, the inlet valve is open and the exhaust valve is closed. A fresh charge of fuel-air mixture is drawn into the cylinder (Figure 8.9). As the inlet valve closes the piston moves from the BDC upward to the TDC, and in the process compresses the fuel-air mixture (Figure 8.10). In this stroke work is contributed to the air. In an ideal Otto cycle, this compression is considered to be isentropic.

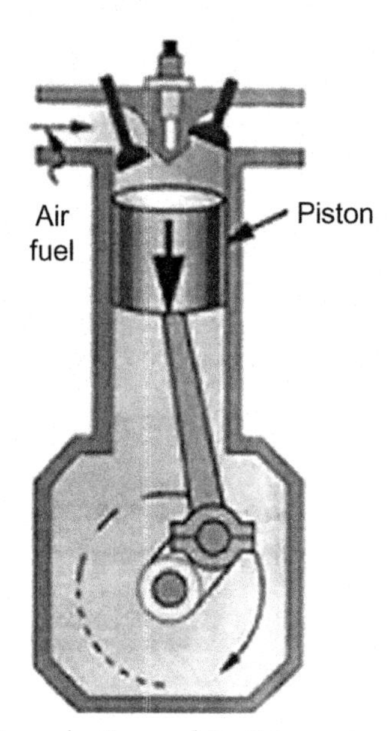

Figure 8.9 Induction of air in Otto cycle. *Source: http://enx.org/resources/beb.*

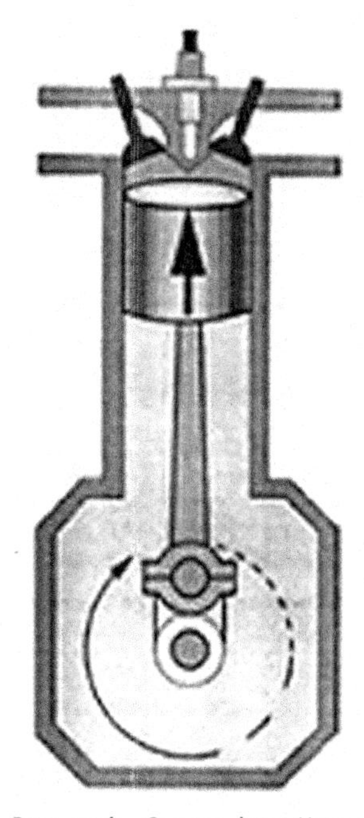

Figure 8.10 Compression of air in Otto cycle. *Source: http://enx.org/resources/beb.*

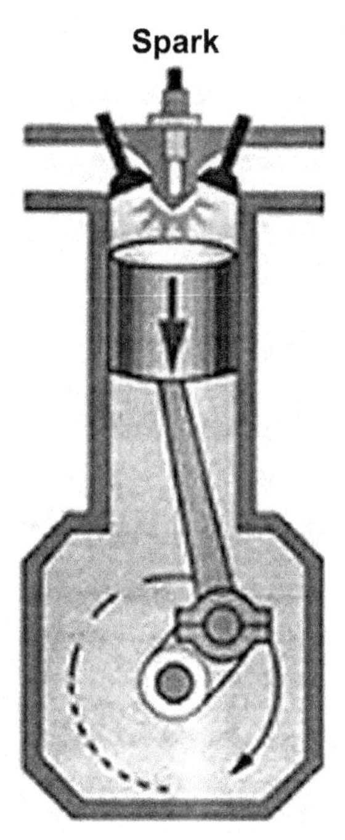

Figure 8.11 Combustion on Otto cycle. *Source: http://enx.org/resources/beb.*

8.2.2.2 Combustion (Figure 8.11)

In this stage heat is added to the fuel-air mixture by combustion of fuel when the piston is at the TDC position. Combustion is initiated by a spark with the help of a spark plug, which is the most important and fundamental difference between the Otto cycle and the Diesel cycle. The piston during heat addition remains at a standstill and hence the process of heat addition is known as isochoric (constant volume).

8.2.2.3 Expansion (Figure 8.12)

While tending to expand, the rapidly burning fuel-air mixture generates high pressure thus forcing the piston to travel downward from the TDC to the BDC in the cylinder. It is in this stage that the cycle contributes its useful work, rotating the engine crankshaft. In an ideal Otto cycle this process is isentropic.

In an engine, the cooling process involves exhausting the gas from the engine to the environment and replacing it with fresh air. Since this happens when the piston is at the TDC position in the cycle and remains immobile, this process is also isochoric (no change in volume).

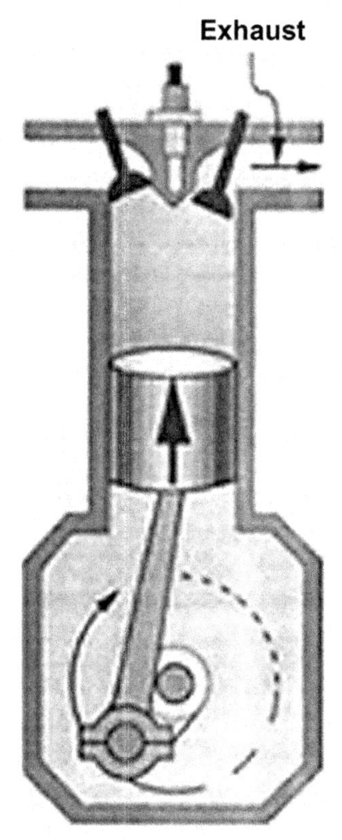

Figure 8.12 Expansion in Otto cycle. *Source: http://enx.org/resources/beb.*

8.2.2.4 Thermal efficiency [5]

The thermal efficiency of the Otto cycle can be expressed by dividing external work by heat input to the working fluid as follows:

$$\eta_{th} = \frac{Q_{in} + Q_{out}}{Q_{in}} = 1 + \frac{Q_{out}}{Q_{in}} \tag{8.19}$$

Hence, considering the unit mass flow of the fluid and assuming constant specific heats for the air, for constant volume heating and cooling we can use

$$\eta_{th} = 1 + \frac{C_v(T_1 - T_4)}{C_v(T_3 - T_2)} = 1 - \frac{T_4 - T_1}{T_3 - T_2} \tag{8.20}$$

We know that for ideal gas $Pv = nRT$ and $\gamma = C_p/C_v$ and for adiabatic condition $Pv^{\gamma} = \text{constant}$. Using Figure 8.8 $v_3 = v_2$ and $v_4 = v_1$. Hence,

$$\frac{T_2}{T_1} = \left(\frac{P_2}{P_1}\right)^{\frac{\gamma-1}{\gamma}} = \left(\frac{v_1}{v_2}\right)^{(\gamma-1)} \tag{8.21}$$

or,

$$T_1 = \frac{T_2}{\left(\frac{v_1}{v_2}\right)^{(\gamma-1)}} \tag{8.22}$$

Again,

$$\frac{T_4}{T_3} = \left(\frac{P_4}{P_3}\right)^{\frac{\gamma-1}{\gamma}} = \left(\frac{v_3}{v_4}\right)^{(\gamma-1)} \tag{8.23}$$

or,

$$T_4 = T_3\left(\frac{v_2}{v_1}\right)^{(\gamma-1)} = \frac{T_3}{\left(\frac{v_1}{v_2}\right)^{(\gamma-1)}} \tag{8.24}$$

Using the values of Eq. 8.22 and Eq. 8.24 in Eq. 8.20 and expressing the compression ratio as $r_c = v_1/v_2$ the thermal efficiency of the Otto cycle changes to

$$\eta_{th} = 1 - \frac{1}{\left(\frac{v_1}{v_2}\right)^{(\gamma-1)}} = 1 - \frac{1}{r_c^{(\gamma-1)}} \tag{8.25}$$

where

η_{th} = Thermal efficiency
Q_{in} = Heat added to the working fluid
Q_{out} = Heat rejected from the working fluid
C_p = Specific heat at constant pressure
C_v = Specific heat at constant volume

Equation 8.25 shows that the thermal efficiency of the Otto cycle depends only on the compression ratio and not on the peak temperature T_3.

Comparing Eqs. 8.12/8.14 and Eq. 8.25 it is noted that the formula for the thermal efficiency of the Diesel cycle is more complex than that of the Otto cycle. This is due to the fact that the heat addition in the Diesel cycle is at constant pressure and the heat rejection is at constant volume, while both heat addition and heat rejection in the Otto cycle take place at constant volume. For the same compression ratio the thermal efficiency of the Otto cycle is always greater than that of the Diesel cycle. In practice, however, the Diesel cycle yields higher thermal efficiency because it adopts higher compression ratios than those adopted by the Otto cycle, which is evident from Example 8.3.

Example 8.3

Determine the thermal efficiency, the final temperature after expansion, and the compression ratio of an air-standard Otto cycle for the same problem as given for Example 8.2.

Solution: From Eq. 8.19 and Eq. 8.20, $1045.9 = 0.718(2073 - T_2)$

Hence, $T_2 = 616.3$ K

From Eq. 8.22, $r_c^{0.4} = 616.3/288 = 2.14$

Therefore, $r_c = 6.7$

From Eq. 8.24, $T_4 = 2073/6.7^{0.4} = 968.7$ K

From Eq. 8.25 $\eta_{th} = \left(1 - \frac{1}{6.7^{0.4}}\right) * 100 = 53.3\%$

Comparing the results of Example 8.2 and Example 8.3 it is clear that the thermal efficiency of the air-standard Diesel cycle (67.3%) is much higher than that of the Otto cycle. This is because the compression ratio of the Diesel cycle (24.3) greatly exceeds the compression ratio of the Otto-cycle, which is 6.7.

In practice many reciprocating engines adopt dual or mixed cycles as shown in Figure 8.13, where heat is added in part in the constant volume process and the rest in the constant-pressure process.

To find out the thermal efficiency of the dual cycle, as before, let us assume constant specific heats of air and consider unit mass flow of the fluid. If the compression ratio between point 2 and point 3 is expressed as $r_p = P_3/P_2$ and as before the compression ratio between point 1 and point 2 as $r_c = v_1/v_2$ and the cut-off ratio between point 4 and point 3 as $r_v = v_4/v_3$, then the efficiency of this cycle could be expressed as follows:

$$\eta_{th} = 1 + \frac{Q_{out}}{Q_{in}} = 1 + \frac{C_v(T_1 - T_5)}{C_p(T_4 - T_3) + C_v(T_3 - T_2)} = 1 + \frac{T_1 - T_5}{\gamma(T_4 - T_3) + (T_3 - T_2)} \tag{8.26}$$

Converting Eq. 8.26 from temperature to compression-ratio, pressure-ratio, and cut-off ratio

$$\eta_{th} = 1 - \frac{1}{r_c^{(\gamma-1)}} \frac{r_p r_v^{\gamma} - 1}{(r_p - 1) + \gamma r_p (r_v - 1)} \tag{8.27}$$

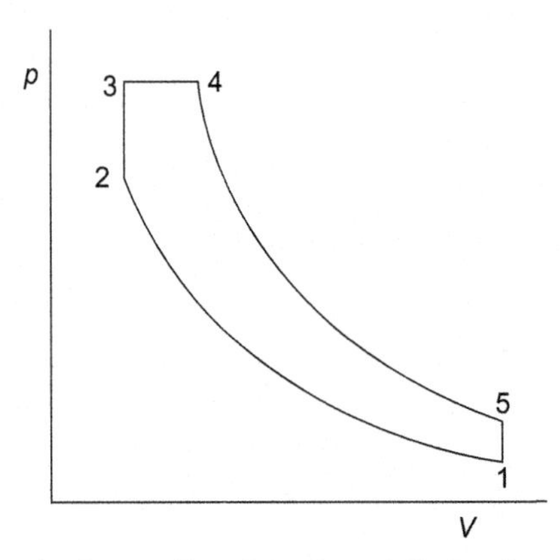

Figure 8.13 Dual or mixed cycle. *Source: Fig. 12.17, Page 247, Gas Power Cycles, [6].*

8.3 DESIGN [2,4]

The fundamental decision regarding the design of a diesel-generating station starts with the desired power output. Once the power output is finalized the next step is to decide whether to select a two-stroke-cycle or a four-stroke-cycle diesel engine. Large, double-acting low-speed (80–200 rpm) power engines are the two-stroke type. A four-stroke-cycle should be preferred for lightweight, high-speed (1000–2500 rpm) engines. Medium-speed (200–1000 rpm) engines may operate on both cycles.

The individual parts of a low-speed, two-stroke engines are comparatively heavy, while the parts of four-stroke engines are much lighter. Two-stroke engines are equipped with fewer cylinders essentially there are less moving parts. Thus, maintenance cost of two-stroke engines is usually lower than that of medium or high speed engines. For the same output height of engine room of a four-stroke engine is lower than the height of engine room required by a two-stroke engine.

The lubrication of the piston rings in four-stroke engines is better than that in two-stroke engines. The specific lubricating oil consumption of a two-stroke engine instead is lower than that of a four-stroke engine.

Low-speed engines are generally built with horizontal cylinders. For higher speeds vertical cylinders are preferred.

The rotational speed of the engine must always match the frequency of the grid for the power-generating station in all respects. Hence, the speed of the engine shall be calculated as follows:

$$n = 60f/p \tag{8.28}$$

where

n = Rotational speed, rpm
f = Frequency of the grid, Hz
p = Number of pair of poles

In a compression ignition engine the pressure of compression must be high enough, typically between 11.5:1 and 22:1, to ensure ignition at any conditions. The approximate compression pressure ranges from 2.8 MPa to 4.8 MPa. A higher compression ratio yields higher thermal efficiency or higher power output, but results in higher stress on working parts.

Another factor that influences the design is how many cylinders of the engine need to be considered. Once the number is known, the bore (d) and stroke (l) of the engine can be determined. The ratio of the stroke to the bore (l/d) generally varies between 0.9 and 1.9. Higher values of l/d up to as high as 2.5 or so may be chosen for opposed piston two-stroke engines.

Knowing all the data as explained above the expected power output of the designed unit may be determined using Eq. 8.29:

$$W = plan/333.5 \tag{8.29}$$

where

W = Power output, kW
p = Compression pressure, MPa

l = Stroke, m
a = Inner cross-sectional area of the cylinder, cm^2
n = Rotational speed, rpm

Thereafter, it should be determined whether the expected output conforms to the desired output. (Note: 1 MPa = 10.2 kg/cm^2, 1 hp = 0.746 kW = 76.0 kg.m/s.)

8.4 EQUIPMENT AND ACCESSORIES

Depending on its size, the engine of a Diesel-generating station comprises some or all of the following equipment and accessories:

i. Fuel injection equipment
ii. Piston
iii. Governor
iv. Starting system
v. Intake silencer
vi. Air filters
vii. Exhaust mufflers
viii. Supercharger

8.4.1 Fuel injection equipment

Correct functioning of the fuel injection equipment begets efficient combustion in a Diesel engine. Fuel is injected through nozzles at adequate pressure for proper atomization of fuel. The nozzle area should be small enough to allow fuel intake for complete vaporization without remaining in suspension as droplets. Depending on the size of the engine the atomizing pressure developed by the fuel injection pump/s may vary from 40 MPa to 100 MPa. The atomized fuel thus obtained mixes with pressurized air in the correct proportion in the combustion chamber so that complete combustion is ensured. This equipment ensures supply of metered quantity of fuel. The fuel injection equipment also controls fuel-air mixing rate according to the required power output. This equipment needs to be designed with high standards to limit gaseous emissions, noise, and smoke in compliance with the local environmental regulations.

8.4.2 Piston

In all internal combustion engines the design of a piston is very critical since it is the most vulnerable member. It is subjected to two prone attacks: one from the mechanical loading resulting from the maximum cylinder pressure and another thermal loading due to the heat of compression. Gas loads are transmitted to the crank shaft with the help of the piston. The design of a piston is influenced by the following parameters:

i. Engine rating
ii. Duty cycle

iii. Quality of fuel
iv. Maximum allowable weight, etc.

The sealing of gas and oil in a cylinder is provided by piston rings.

8.4.3 Governor

Like any rotary power-generating device every engine of a Diesel-generating plant is invariably provided with a governor to maintain the speed of the engine automatically by regulating the flow of fuel to meet the desired power output. A governor may be a hydro-mechanical or electronic type.

8.4.4 Starting system

To start a Diesel engine from standstill sufficient torque to the crankshaft has to be provided from the outside to rotate the engine at the desired speed, which is essentially done by a starting gear. The most common starting gear is the *manual* type, e.g., hand cranking or a rope and pulley system. It is simple, does not require an external source to break the starting torque, and its costs are low. The use of manual starting gear, however, is restricted to low-speed, small engines. The most commonly used starting gears for medium-speed and high-speed engines are *electric.* They consist of DC motors and battery packs to supply electric energy. In medium-speed and high-speed engines another starting gear that is widely used is pneumatic. An air motor or a power cylinder is supplied with pressurized air from an air compressor, which facilitates breaking the starting torque.

8.4.5 Intake silencer

When intake air from atmosphere rushes through the intake valve with high velocity it produces a hissing noise, which is intolerable to human ears. This noise is reduced by providing a silencer at the intake manifold.

8.4.6 Air filters

A filter is installed at the air intake to prevent wear of pistons, piston rings, cylinder, valves, etc., by airborne suspended impurities. An air filter also assists in the attenuation of intake noise.

8.4.7 Exhaust mufflers

When leaving the exhaust pipe exhaust gases produce an unbearable howling noise, which is much louder sound than intake noise. Hence, it is imperative to provide different means, e.g., expansion of exhaust gas, or change of direction of flow of gas, or cooling of gas with water, etc., at the exhaust system – for abatement of howling noise.

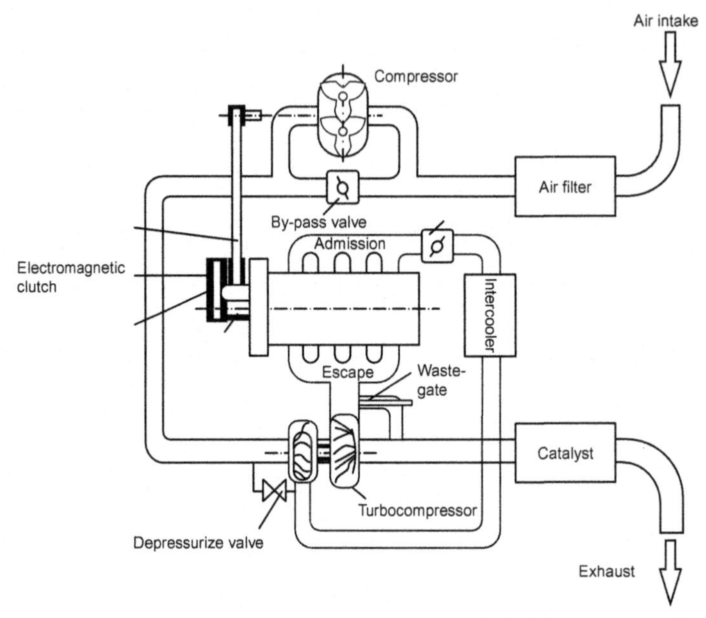

Figure 8.14 Turbocharging of an engine. *Source: TURBOCHARGER, http://www.mechanicalengineeringblog.com/tag/turbo-lag/.*

8.4.8 Supercharger

Supercharging of an engine can be defined as the artificial raising of the intake pressure of air above atmospheric pressure. The pressure is raised by the use of a compressor. If the compressor is driven by a turbine that operates on the exhaust gases from the engine it is called a turbocharger (Figure 8.14).

The objectives of supercharging are to:

1. Increase the power output of the engine
2. Maintain the power output regardless of the altitude
3. Increase the air-fuel ratio, and thereby improve the thermal efficiency of the engine

8.5 DIESEL ENGINE SYSTEMS

For smooth trouble-free operation a Diesel engine is equipped with some or all of following systems. The necessity of each system is determined by the engine's size, location of installation, and operating environment:

i. Fuel Oil System
ii. Lubricating Oil System
iii. Cooling Water System
iv. Compressed Air System
v. Firefighting System

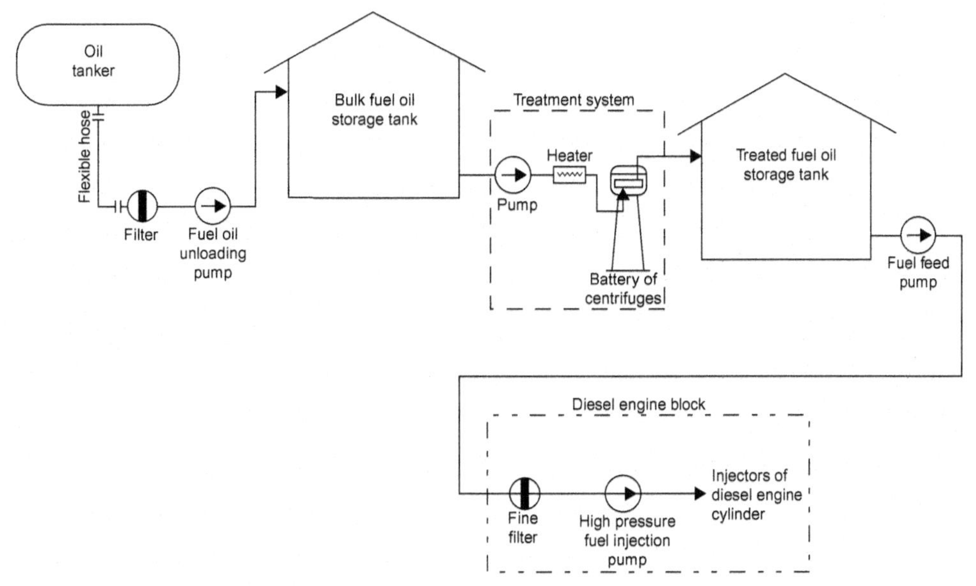

Figure 8.15 Fuel oil system.

8.5.1 Fuel oil system (Figure 8.15)

The fuel oil used in a Diesel engine should be completely free from water and mechanical impurities. Solid and liquid contaminants are cleaned before use to prevent damage to fuel pumps and engine components. A diesel engine is capable of burning a wide range of low-quality liquid. It should be ensured that heating of fuel to the required viscosity is carried out without developing thermal cracking. For start-up of a diesel engine either high-speed diesel (HSD) or light diesel oil (LDO) is used. For normal running, a heavier grade fuel oil, e.g., heavy fuel oil (HFO) or low sulphur heavy stock (LSHS), is generally used.

The fuel oil system deals with oil receipt, storage, treatment, and forwarding of oil to the Diesel engine. Fuel oil may be delivered to a plant by road tankers or railway wagons or if the plant is located near a port through the pipeline from the port. Oil delivered by road tankers or railway wagons is unloaded with the help of unloading pumps equipped with suction strainers. Connection is made with the tankers or wagons through screwed hoses. Pressurized fuel oil from unloading pumps is supplied through the pipeline to bulk fuel oil storage tanks for storing. These storage tanks are usually vertical, cylindrical, cone-roof steel tanks. While storing heavy-grade fuel oil, the minimum temperature in bulk oil storage tanks should be maintained at around 353 K by electric or steam heating to ensure the fluidity of the oil. However, for ensuring the required viscosity of heavy-grade fuel oil at the engine end the minimum temperature should be maintained in the treated oil storage tank at around 393 K.

The fuel system of large diesel engines generally includes the following:

i. Fuel feed or lift pump
ii. Fine filters
iii. High-pressure fuel injection pumps
iv. Injectors
v. High-pressure pipes connecting the injection pumps to the injectors
vi. Low-pressure pipes

The fuel feed pump supplies fuel from the treated oil storage tank/s and delivers it through a fine filter to the high-pressure fuel injection pumps. These pumps transfer oil to the injectors located on each cylinder.

8.5.2 Lubricating oil system (Figure 8.16)

The lubricating oil system of a Diesel engine typically includes an *external lube oil circuit.* The primary function of this circuit is to filter and cool the circulating oil. It also serves to heat, prime, and scavenge the oil when preparing the diesel engine for starting. An *internal lube oil circuit* supplies oil to individual units and parts of the diesel engine. The pipelines of internal circuit are inside the engine.

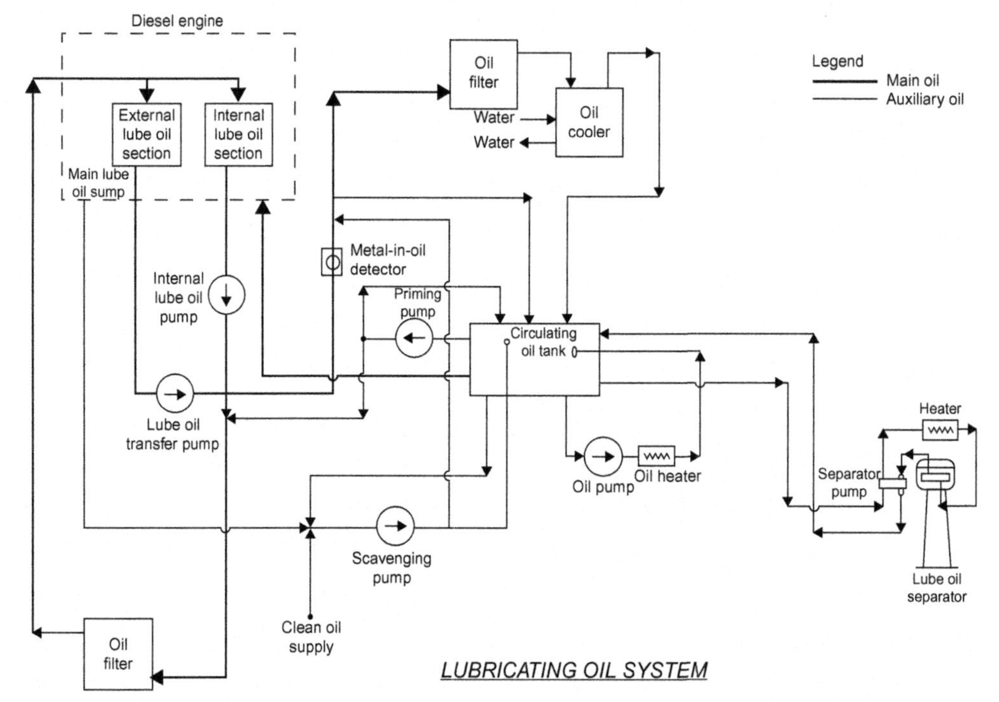

Figure 8.16 Lubricating oil system.

The external lube oil circuit in turn is typically divided into two groups, i.e., the main lube oil circuit and the auxiliary lube oil circuit. The main lube oil is either delivered by road tankers with the help of lube oil unloading pumps or downloaded from drums containing lube oil to the Diesel engine main lube oil sump. A *lube oil transfer pump* delivers lube oil from the lube oil sump through a metal-in-oil detector, oil filter, and oil cooler into a *circulating oil tank*. When the pressure in the system exceeds the maximum limit a by-pass valve diverts part of the lube oil directly to the circulating tank. The temperature of the lube oil leaving the oil cooler is controlled by adjusting the flow of cooling water through the oil cooler automatically by a thermo-regulator.

The temperature of the lube oil in the circulating oil tank is maintained with the help of an oil pump provided with oil heater, which cuts-in automatically in the event that the oil temperature drops down to 313 K and cuts-out when the temperature rises above 323 K.

A lube oil separator is used to clean the oil circulating in the lubrication system. Oil is drawn by the *separator pump*, either from the system or from the tank, and after separation it is returned to the circulating tank. A temperature of 328 K or more, as recommended, is maintained in this circuit, if required, with the help of an oil heater.

Oil in the *internal lube oil circuit* is circulated with the help of an engine-driven *internal lube oil pump*. This pump draws oil from the circulating tank and delivers it to the engine through an oil filter. Prior to starting the Diesel engine it is primed with oil by the *priming pump*. The internal circuit is exclusively used for "piston cooling and lubrication of crosshead and crankpin bearing" and "lubrication of camshaft main bearings." Oil is drained from the pistons and bearings into the Diesel engine sump, from which the *scavenge pump* takes its suction and delivers oil back to the circulating tank. Separate lube oil pipelines connect to various lubricating points of major and minor accessories, drives, and the turbo-charger.

8.5.3 Cooling water system (Figure 8.17)

The Diesel engine cooling system generally consists of two circuits: the *internal cooling circuit*, also known as jacket cooling system, and the *external cooling circuit*. The water circulating along the internal circuit cools the Diesel engine, while the water of the external circuit serves to cool the oil, the water of the internal cooling circuit and the air for supercharging. Distilled or de-mineralized water treated with a corrosion inhibitor is generally used in the inner circuit, while either raw water or filtered water is adopted in the external circuit.

The *inner circuit cooling water pump* takes water from the Diesel engine sump and circulates water through the engine jacket, jacket water cooler, and the turbo-compressor. From the supply manifolds, water also flows to cool the exhaust manifold sections and the working cylinders. The water from the working cylinders is drained

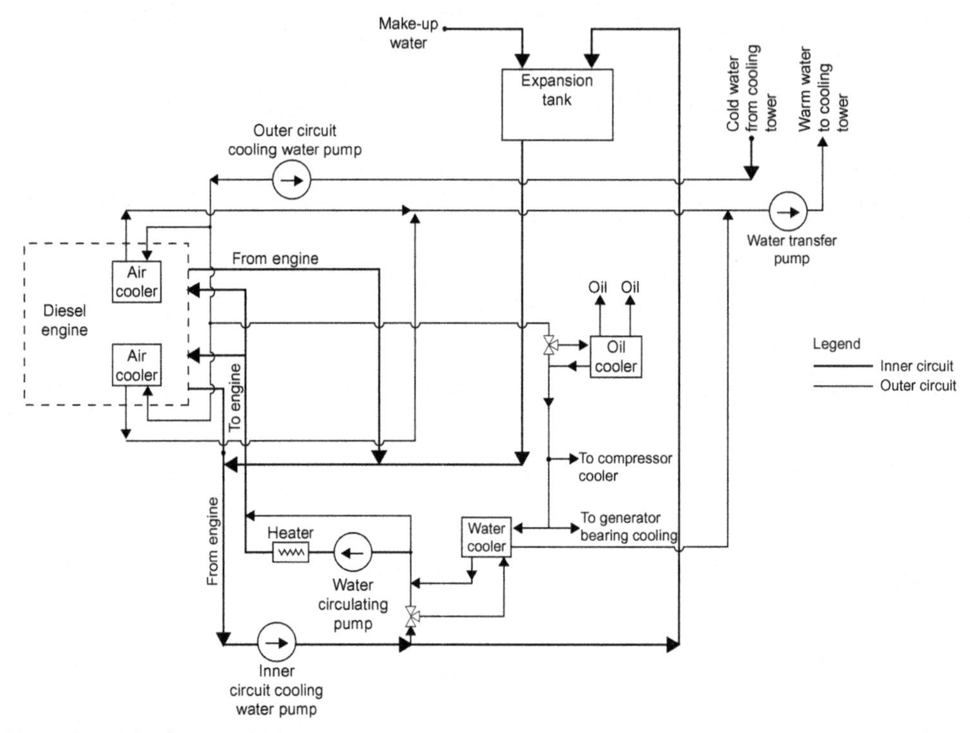

Figure 8.17 Cooling water system.

to return manifolds, from which it is directed to the pump suction line, completing the cycle. An expansion tank is connected to the suction side of the pump to ensure initial priming, as well as to ensure reliable functioning of the closed system when the water volume changes due to heating or leakage. An independent *water-circulating pump* is provided with an electric heater to warm up the engine with hot water, particularly when the engine remains idle. The heater automatically cuts-in when the engine water outlet temperature falls below about 303 K and cuts-out when the temperature exceeds 323 K.

The external cooling circuit is equipped with an *outer circuit cooling water pump.* This pump draws water from an evaporative mechanical-draft cooling tower sump and directs it through the lube oil cooler and internal cooling circuit water cooler and then returns to the cooling tower with the help of a *water transfer pump.* Cooling water from this circuit is supplied to air coolers, generator-bearing cooler, and compressor cooling system as well. The amount of water passing through the lube oil cooler is usually determined by a thermo-regulator control unit, to maintain the operating temperature of the oil flowing out of the cooler.

For small capacity engines, the inner circuit cooling water may be cooled by supplying air to a radiator with cooling air fans.

8.5.4 Compressed air system

The starting air of a diesel engine is delivered by the air compressor. The pressurized air is stored in vertically mounted air receivers. During start-up, pressurized air enters the Diesel engine main starting pipeline through the main starting valves to serve bank of cylinders. From the main starting valves air gets into the air distributors as well as into working cylinder-starting valves. Once the diesel engine starts operating on fuel, the starting valve is closed to cut the supply of air.

From the starting air receivers control air is led off through pressure-reducing valve to the oil and water thermo-regulators. This air then regulates the flow of water through the cooler or heat-exchanger by opening or closing the valves.

8.5.5 Firefighting system

In a Diesel power station fire may start accidentally due to short circuiting, ignition of fuel and lubricants or of common combustible materials, e.g., paper, cotton waste, etc. As soon as minor fires are detected portable/semi-portable fire extinguishers are used to extinguish it.

Oil fire is best tackled by a foam-type extinguisher along with a spray water system. The foam provides a blanket over the oil surface and cuts off the oxygen supply, which is essential for supporting combustion.

As an alternative means, sand stored in the bins near the dyke walls of fuel storage tanks can also be used to prevent outbreak of minor fire. Sand is sprinkled evenly on the spillage surfaces using shovels/buckets. Small fires can be handled successfully using sand. Ordinary water is not generally suitable for fighting oil fire, since oil has a tendency to float and spread further.

Electrical fire is fought with carbon dioxide fire extinguishers. Wherever practical, the electric supply to the equipment on fire should be switched off before using the extinguisher. In general, fire due to combustible materials such as cotton waste, paper, timber, etc., is tackled either by soda-acid type fire extinguishers or with water stored in fire buckets. In the event of a big fire, portable/semi-portable extinguishers may not extinguish the fire effectively. In such an event, the central firefighting team should be called as soon as possible.

8.6 PROBLEMS

8.1 In an ideal air standard Diesel engine, 1 kg of air is taken from atmosphere at a pressure of 101.3 kPa and temperature of 288 K. Find the ideal thermal efficiency and M.E.P. of this engine and final pressure and temperature after compression.

(Assume $\gamma = 1.4$, compression ratio $r_c = 15$ and expansion ratio $r_e = 5$.)

Ans.: $\eta = 55.1\%$, M.E.P. = 1.34 MPa, $P_2 = 4.49$ MPa, and $T_2 = 851$ K

8.2 Referring to Figure 8.6 determine the thermal efficiency of the ideal air standard Otto cycle, if 1 kg of air is taken from atmosphere at a pressure of 101.3 kPa and temperature 298 K. Also find the final pressure and temperature after compression.

(Assume compression ratio $r_c = 10$ and $\gamma = 1.4$.)

(**Ans.:** $\eta = 60.19\%$, $P_2 = 2.54$ MPa, and $T_2 = 749$ K.)

8.3 In an ideal S.I. engine, a charge enters at a pressure of 100 kPa and temperature of 330 K. If the charge is compressed isentropically through a compression ratio of 7:1, determine the temperature and pressure at the end of compression, assuming charge to be (i) pure air and (ii) a mixture of air and fuel.

(Assume $\gamma_{air} = 1.4$, $\gamma_{mixture} = 1.326$.)

(**Ans.:** $T_{2air} = 719$ K, $p_{2air} = 1.525$ MPa, $T_{2mixture} = 622$ K, $p_{2mixture} = 1.32$ MPa)

8.4 Find the rotational speed of a two-stroke Diesel generator, if the number of poles of the generator is 120.

(**Ans.:** 25 rpm)

8.5 A Diesel engine is equipped with a cylinder having a bore of 20 cm and a stroke of 35 cm. For a compression pressure of 4.8 MPa, estimate the power output of the engine. The speed of the engine is 50 rpm.

(**Ans.:** 7.92 kW)

8.6 The bore of a Diesel engine cylinder is 28 cm and its stroke is 50 cm. The pressure and temperature at the end of admission are 101.3 kPa and 363 K, respectively. For a compression curve that follows $PV^{1.34} =$ constant find the pressure and temperature at the end of compression, if the clearance volume, i.e., the volume at the end of compression, is 9800 cm^3.

(**Ans.:** 680.44 kPa and 588.56 K)

BIBLIOGRAPHY

[1] Wikipedia, Thermal Efficiency; Four-stroke Engine.
[2] Lilly LCR. Diesel engine reference book. Butterworth and Co. (Publishers) Ltd; 1984.
[3] Baumeister T, Marks LS. Mechanical engineers handbook. sixth ed McGraw-Hill Book Company, Inc; 1958.
[4] Maleev VL. Internal combustion engine theory and design. McGraw-Hill Book Company, Inc; 1945.
[5] Wrangham DA. Theory and Practice of Heat Engine. The English Language Book Society; 1962.
[6] Rogers GFC, Mayhew YR. Engineering thermodynamics-work and heat transfer. English Language Book Society and Longmans, Green and Co. Ltd; 1967.

CHAPTER 9

Steam Power Plant Systems

9.1 INTRODUCTION

A large pulverized coal-fired steam power plant encompasses numerous systems, both complex and simple, the details of which are difficult to present in one chapter. Nevertheless, systems that are complex in nature and constitute major area of steam power plant are addressed in Sections 9.2 and 9.3 as follows. The primary objective of this chapter is to illustrate just how vast and complex a large power plant is. Against this backdrop the materials presented in the following sections are aimed at helping power plant engineers execute their respective area of responsibilities, be it design or field engineering services. However, this chapter refrains from describing individual equipment, which is difficult to be covered in one book, let alone a single chapter.

In any project or industry, or even for a reasonably comfortable life, the first utility that is essential is "water" followed by "electrical power." In accordance with this philosophy this chapter first discussed the "intake/raw water" system followed by other mechanical systems. A general discussion on electrical power supply and distribution systems is presented in the concluding part of this chapter.

9.2 MECHANICAL SYSTEMS

In general, all mechanical systems consist of the following components:

i. Piping
ii. Hangers and/or supports
iii. Isolating valves, check valves, relief valves
iv. Pressure/temperature/level/flow measurement and control

In addition to the above, based on its complexity a system may comprise pump/fan/compressor, cooler, filter/strainer, storage tank/reservoir, control valve, flow elements, etc. All these components, however, are not individually discussed in following sections.

It is also to be noted that mechanical systems (Table 9.1) which have been discussed in some of the previous chapters are not repeated in this chapter.

9.2.1 Intake/raw water (RW) system (Figure 9.1 and Figure 9.2)

This system begins with the intake pump house, where the intake pumps are located. This pump house is situated on either a riverbank or canal-bank or sea coast. Water

Thermal Power Plant
DOI: http://dx.doi.org/10.1016/B978-0-12-801575-9.00009-3

Table 9.1 List of systems dicussed in previous chapters

Sl. No.	Name of the system	Section No./ Chapter No.	Name of the chapter
1.	Water and Steam Circuit	2.3/2	STEAM GENERATORS
2.	Furnace		
3.	Fuel Burning System		
4.	Draft System		
5.	Heat Recovery System		
6.	Coal Burners	4.4/4	PULVERIZED COAL-FIRED BOILERS
7.	Coal Preparation System	4.7/4	
8.	Coal Feeders	4.8/4	
9.	(Steam) Turbine Governing System	6.6/6	STEAM TURBINES
10.	(Steam) Turbine Lubrication System	6.7/6	
11.	Speed Control System	6.8/6	
12.	Air Intake System	7.5/7	GT and HRSG
13.	Anti-Icing System		
14.	Evaporative Cooling System		
15.	Exhaust System		
16.	Starting System		
17.	Lubrication and Power Oil System		
18.	Hydraulic and Pneumatic Control System		
19.	Fuel Gas System		
20.	Fuel Oil System		
21.	Water Injection System		
22.	Turbine Governing System		
23.	Firefighting System		
24.	Fuel Oil System	8.5/8	DIESEL
25.	Lubricating Oil System		POWER
26.	Cooling Water System		PLANTS
27.	Compressed Air System		
28.	Firefighting System		

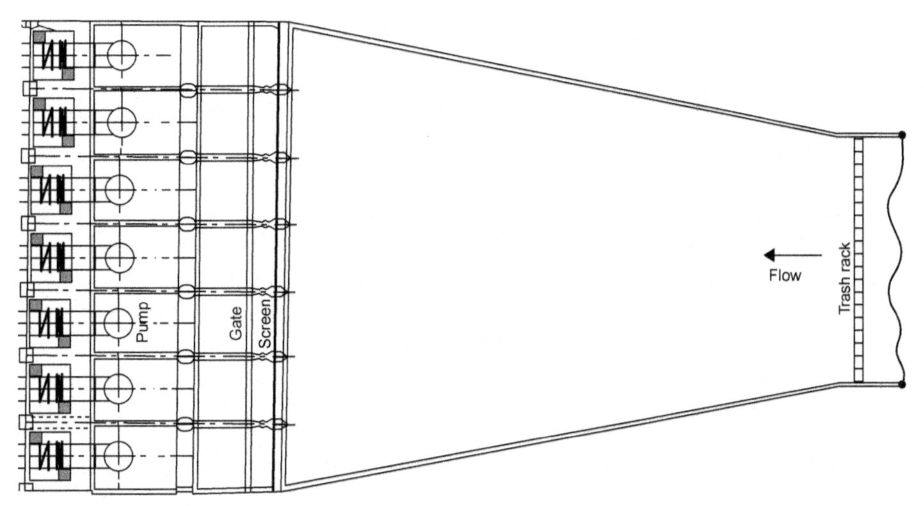

Figure 9.1 Intake pump house with forebay.

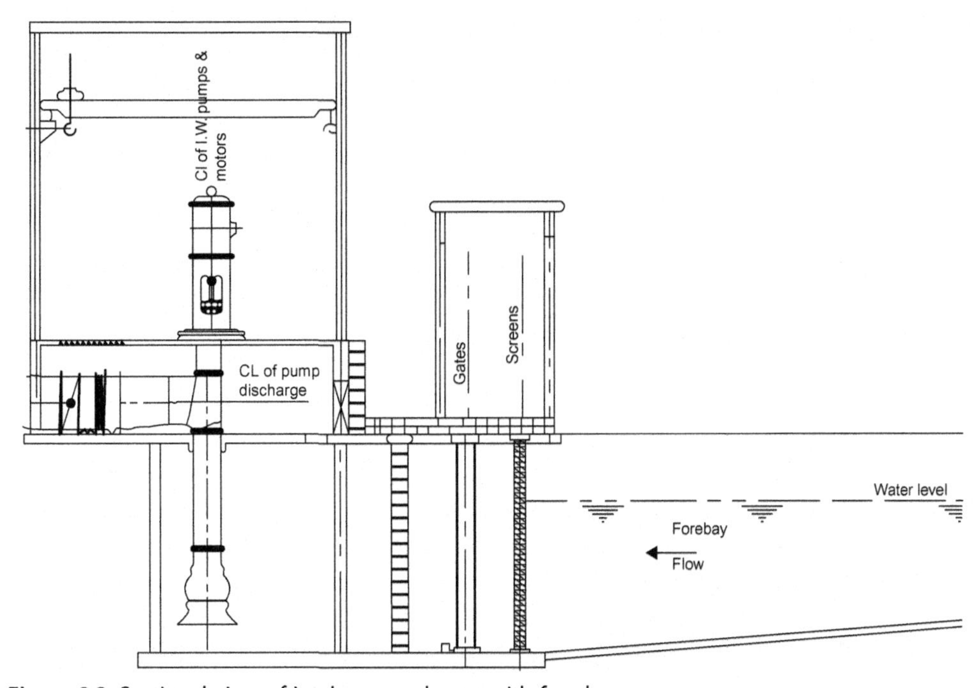

Figure 9.2 Sectional view of intake pump house with fore bay.

from the river/canal/sea flows into a forebay through a trash rack that catches large floating debris such as carcass, logs, framework of ships, etc. The forebay leads water to individual pump suction basin through traveling water screen and inlet gate. Floating and suspended debris from water that passed through the trash rack are caught in the traveling water screen. Clogged screens are then washed with high-pressure water jets, with the wash water supplied by screen wash pumps that also receive suction from the forebay. Each pump is provided with an isolation gate to facilitate maintenance of the pump. From the forebay each intake pump receives its suction and then discharges raw water to either raw water storage tank or raw water reservoir for storing.

Depending on the configuration of the plant water storage and supply system, raw water from this storage tank or reservoir may be distributed as service water to different users, i.e., fire-water storage tank, as make-up to cooling tower, or water pre-treatment plant for clarification and further treatment of water.

9.2.2 Water pre-treatment system

All natural water, commonly known in the industry as raw water, contains impurities. The quantum and type of impurities depend on the source of this natural water. If the source is well water the pre-treatment plant would likely constitute filtration of the water. However, if raw water is sourced from an open area, e.g., cooling pond, reservoir, lake, river, sea, etc., the pre-treatment system would need to be more complex.

Use of untreated raw water for industrial application may result in equipment/system fouling, unscheduled down time, and loss of production and waste of energy. Hence, prior to using this water it must be pre-treated, lest raw water will seriously affect the efficient and reliable operation of equipment cooling system and steam generating systems.

The water pre-treatment plant includes two sections: one section meets the clarified and filtered water requirements for the circulating (condenser cooling) water, service water, drinking water, etc., and another section serves as the clarified and filtered water input to the de-mineralized (DM) water plant.

In a clarified water reservoir sodium hypochlorite solution is dosed to prevent any micro-biological growth. The first step in the water pre-treatment process is "clarification," the principal mechanisms of which are coagulation, flocculation, and sedimentation.

Coagulation is the process of destabilizing suspended particles in raw water by neutralizing their charge. It is carried out in a rapid mixing tank to ensure complete and uniform dispersion of the coagulant throughout the entire mass of water. The retention time for this process may vary from 30–300 s.

To ensure the desired effect of the process of coagulation sometimes chemicals, such as alum (aluminum sulfate), ferrous sulfate, ferric sulfate, ferric chloride, activated silica, organic polyelectrolytes, etc., are dozed.

Flocculation is the process of agglomerating destabilized particles, and is carried out by gentle stirring of water in the flocculation tank. The detention time may vary from 600–3600 s.

Once the water is coagulated and flocculated the next step is the process of liquid-solid separation followed by settling of solid particles at the bottom of a tank. This

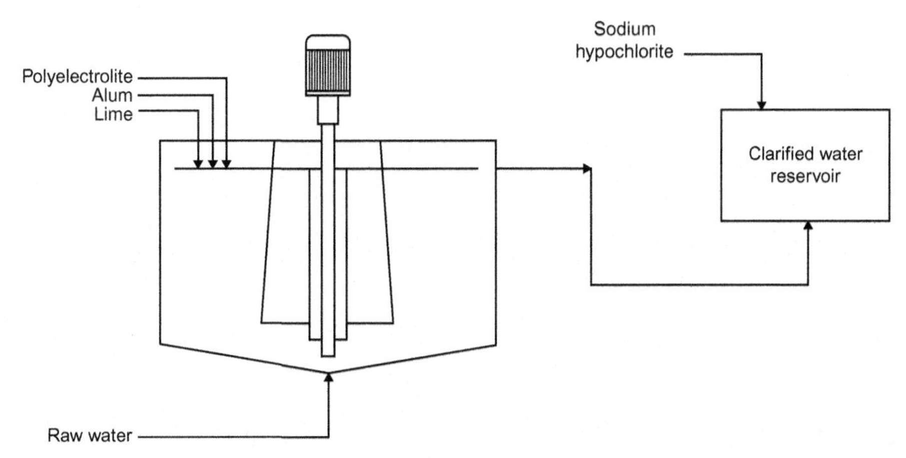

Figure 9.3 Clariflocculator with clarified water reservoir.

process is known as *sedimentation*. The tank is normally provided with mechanical sludge rakes/collector arms to scrap the sludge from the bottom of the tank.

Figure 9.3 shows how alum and lime are added to raw water in a rapid mixing tank, either a clariflocculator or a combination of flash mixer and inclined surface settler, to facilitate coagulation. While alum is added for settling of particles in the form of clusters, lime is added to remove the hardness (caused by the presence of calcium and magnesium salts) of raw water by the precipitation of calcium carbonate and magnesium hydroxide. A polyelectrolite solution may be dozed to raw water to reduce the level of colloidal silica, if there is any.

The clarified water is then supplied to the softening plant for abatement of residual hardness and alkalinity (resulting from mainly bicarbonates), which have a detrimental effect on steam generators, pipelines, heat exchangers, etc. To remove organic matter present in raw water, chlorine is dozed in that section of the clarified water that is dedicated for DM water service. Clarified water used for service water, dust suppression, etc., requires no further treatment.

9.2.3 Water softening system

If the hardness of the clarified water coming out of the pre-treatment plant is too excessive to use in condenser cooling water system, the clarified water will need to be passed through ion-exchange polishing units to reduce the hardness to a level not exceeding 5.0 ppm as $CaCO_3$. In an ion-exchange softener, sodium ions from solid phase are exchanged with calcium and magnesium ions from aqueous phase, thereby producing soft water.

Lime $\{Ca(OH)_2\}$ is dozed in the softening unit to react with calcium bicarbonate $\{Ca(HCO_3)_2\}$, magnesium bicarbonate $\{Mg(HCO_3)_2\}$, and magnesium sulfate

($MgSO_4$), while soda ash (Na_2CO_3) reacts with calcium sulfate ($CaSO_4$) as detailed hereunder:

$$Ca(OH)_2 + Ca(HCO_3)_2 = 2CaCO_3 + 2H_2O$$

$$2Ca(OH)_2 + Mg(HCO_3)_2 = 2CaCO_3 + Mg(OH)_2 + 2H_2O$$

$$Ca(OH)_2 + MgSO_4 = CaSO_4 + Mg(OH)_2$$

$$Na_2CO_3 + CaSO_4 = Na_2SO_4 + CaCO_3$$

Since the products of above chemical reactions that is calcium carbonate ($CaCO_3$) and magnesium hydroxide {$Mg(OH)_2$} are not soluble in water they settle at the bottom of the tank/chamber for removal time to time. The softened water is used for the condenser cooling and auxiliary cooling as well as for the air conditioning units.

9.2.4 DM water system

The DM (de-mineralized) water system primarily meets the heat-cycle make-up water demand to replenish losses from the blow-down of the boiler drum, leakages from valve glands and flange joints, steam venting, safety valve popping, etc. There are various methods of treating clarified water into DM water. The common feature of each method is some form of ion-exchange process, i.e., displacement of one ion by another. DM water coming out of the treatment plant is stored in DM water storage tank.

Raw water, irrespective of its source, contains many minerals, as cations (positively charged ions) and anions (negatively charged ions), in diverse concentrations, along with silica (SiO_2) and free carbon dioxide (CO_2). Cations in raw water are usually calcium, magnesium, barium, aluminum, iron, sodium, and potassium. Anions generally are chlorides, sulphates, nitrates, carbonates, and bicarbonates.

In an ion-exchange process, cations are replaced with hydrogen ions in a *cation-exchanger* and anions are replaced with hydroxide ions in an *anion-exchanger.* Resins generally used in an ion-exchange process are weak acid cation resin, strong acid cation resin, weak-base anion resin, and strong-base anion resin. The amount of resins needed in a plant depends on the quality of incoming water.

A weak acid cation unit exchanges cations of mainly alkaline salts (carbonates, bicarbonates of Na, K). If the cation content constituting bicarbonates is significant, the incoming clarified water is first passed through a weak acid cation unit. If the incoming clarified water contains cations of both neutral salts (chlorides, sulphates, nitrates of Ca, Mg, and Na) and alkaline salts they are exchanged with the hydrogen ions of the resin of a strong acid cation-exchanger. The effluent of strong acid cation-exchanger is usually called "decationized water." This water then enters the degasser tower where CO_2 is removed by scrubbing with an upward flow of air. The outgoing effluent from the degasser tower is called "decationized and decarbonated" water.

The use of a strong-base anion-unit exchanger is required in the event that incoming clarified water contains anions of both strong and weak acids, e.g., hydrochloric, sulphuric, nitric, carbonic, and silicic, to exchange them with free-to-move hydroxide ions of the resin. In plants where free mineral anionic content of clarified water constitutes a large amount of the total anionic content then the incoming decationized and decarbonated water is first passed through a weak-base anion-exchanger. This exchanger removes (by absorption only) the mineral acids that are strong acid molecules of hydrochloric, sulfuric, and nitric acids. Water coming out of a strong-base anion unit is DM water containing very minor dissolved compounds. These remaining impurities are then polished in a mixed bed unit where both cation and anion resins are mixed.

In general, various vessels, which contained chemicals and resins used for DM water production, comprise pressure sand filter (PSF), activated carbon filter (ACF), weak acid cation unit (WAC), strong acid cation unit (SAC), degasser, weak-base anion unit (WBA), strong-base anion unit (SBA), and mixed bed unit (MB). Figure 9.4 shows the arrangement of various vessels, without WAC and WBA units. Water enters each vessel at the top, flows downward through the filter media/resin bed, and then flows out of the bottom of each vessel.

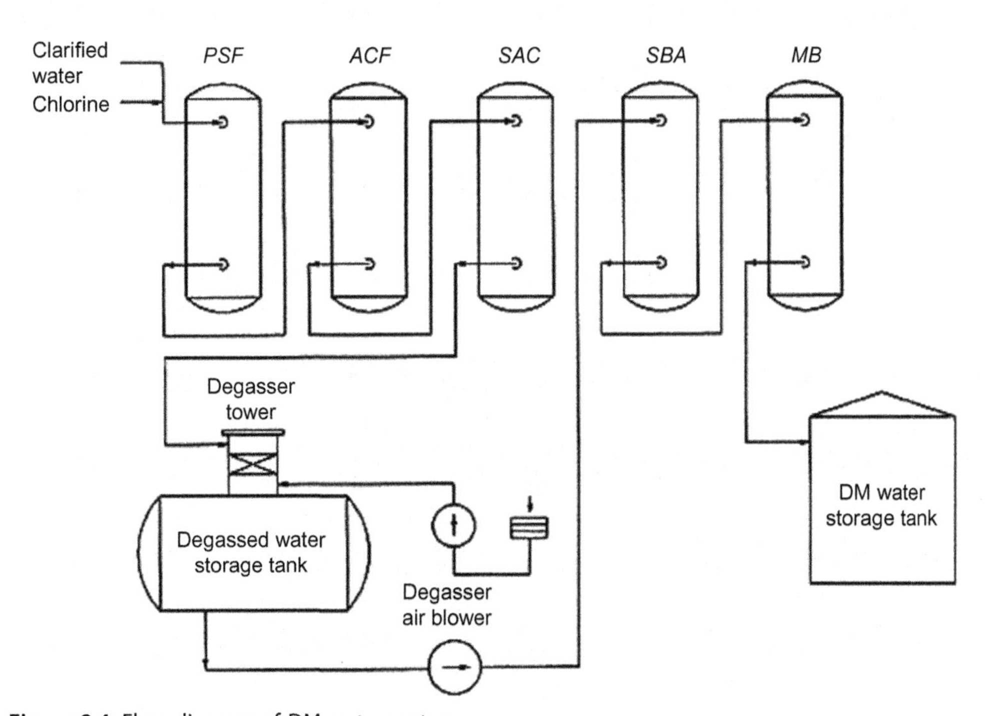

Figure 9.4 Flow diagram of DM water system.

In the pressure sand filter (PSF) turbidity of incoming water is reduced. PSF is followed by the activated carbon filter (ACF), where suspended impurities and organic matters are entrapped and "free chlorine" and "free iron" are removed. Flow of water is downward as explained above.

Once all available ions in an ion-exchanger become saturated and exhausted by repeated exchange of ions, the exchanger is regenerated. Regeneration of SAC (and also WAC, if provided) is carried out with HCl, whereas SBA (and also WBA, if provided) is regenerated with NaOH. The process of regeneration is classified as co-current when chemicals (HCl, NaOH) are dozed along with the flow of water or counter-current when chemicals are dozed in the reverse direction of flow of water. Figures 9.5 and 9.6 depict the condition of the resin bed of a typical strong acid cation-exchanger before and after co-current regeneration. The condition of the resin bed of a typical strong-base anion-exchanger before and after co-current regeneration is shown in Figures 9.7 and 9.8.

Quality of DM water downstream of MB usually is as follows:

Total Electrolyte (as $CaCO_3$)	: 0.1 ppm (max)
Total Hardness (as $CaCO_3$)	: NIL
Total Reactive Silica (as SiO_2)	: 0.01 ppm (max)
Iron (as Fe)	: NIL
Free CO_2 (as CO_2)	: NIL
pH at 298 K	: 6.8–7.2
Conductivity at 298 K	: $<0.1\ \mu S/cm$

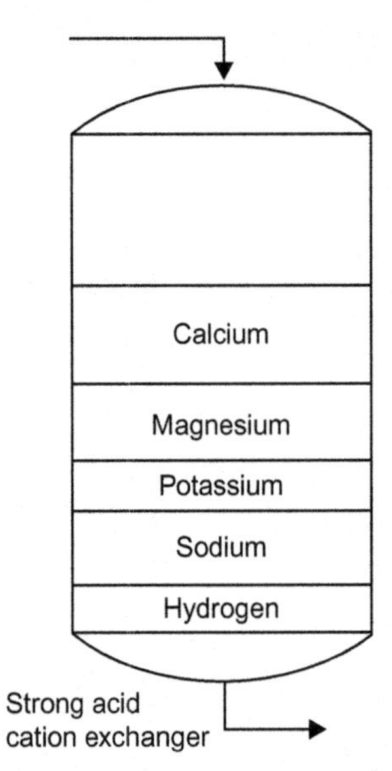

Figure 9.5 Resin bed of SAC before regeneration.

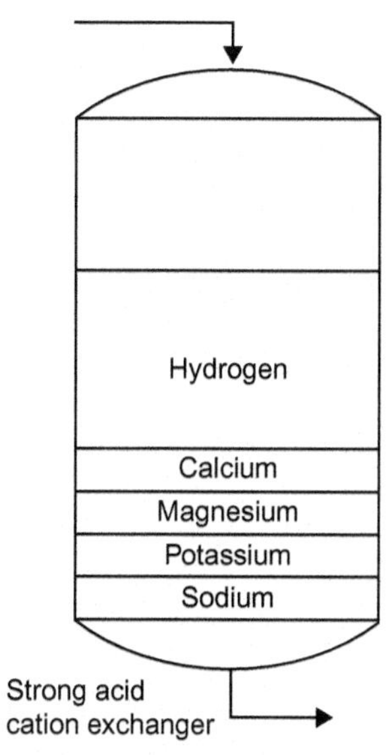

Figure 9.6 Resin bed of SAC after co-current regeneration.

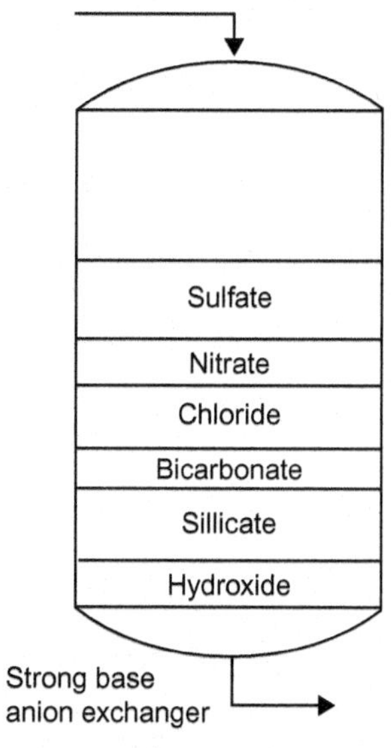

Figure 9.7 Resin bed on SBA before regeneration.

In plants where the requirement of DM water is not substantial, the treatment of clarified water uses an electrical de-ionization unit (EDI) based on membrane technology. The EDI process removes ionizable species from liquids using electrically active media, where an electrical potential influences ion transport. These units ensure continuous water purification without regeneration of chemicals.

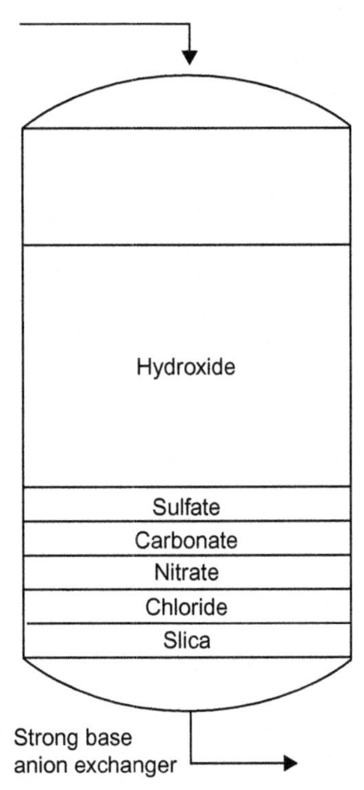

Figure 9.8 Resin bed of SBA after co-current regeneration.

EDI-based plants comprise cation-exchange membranes, anion exchange membranes, spacers, electrodes, and cation/anion exchange resin. The primary cation/anion exchange process is replaced with a reverse osmosis (RO) process. It is the initial demineralization step that removes ionic impurities, particles, and high molecular weight organics. EDI replaces the mixed bed ion-exchanger. It is the final demineralization step that removes residual ionic and ionizable impurities.

9.2.5 Auxiliary cooling water (ACW) system (Figure 9.9)

The auxiliary cooling water system uses clarified water in a common sump, catering to both auxiliary cooling water (ACW) and circulating (condenser cooling) water (CW) systems. Each of the ACW pumps is provided with a dedicated bay in this sump. These pumps receive suction from the common sump and supply cooling water to turbine lube oil coolers, heat exchangers of the closed cycle cooling water (CCCW) system, and as make-up to the ash water system, etc. Hot water returns from the coolers then flows to the CW system return line, which in turn flows to the hot water section of the cooling tower.

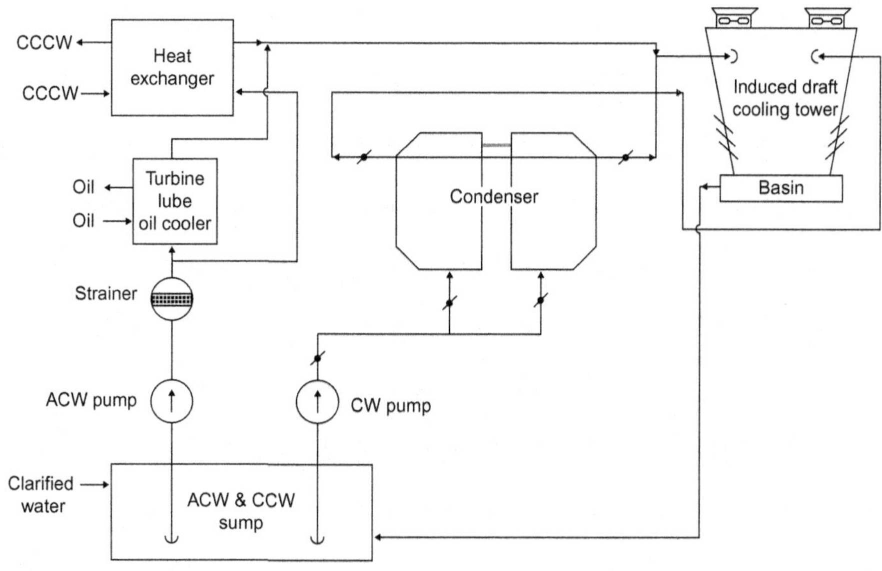

Figure 9.9 ACW system and recirculating type CW system.

9.2.6 Circulating (Condenser Cooling) water (CW) system

The condenser cooling water system, generally known in the industry as a circulating water (CW) system, provides a continuous supply of cooling water to the main condenser during plant operation in order to remove the heat rejected by the exhaust steam from the LP steam turbine at the condenser. The CW system could be either a "recirculating" type or a "once through" type.

In the "recirculating" type (Figure 9.9), the clarified water reservoir is the source of water. From this reservoir, water is pumped to the ACW and CW sump. This sump comprises a dedicated bay for each CW pump, which supplies cooling water to the condenser. Hot water from the condenser outlets is then carried to the hot water section of the cooling tower, where the downward flow of water is cooled by the upward flow of air, thereby rejecting heat to the atmosphere. The cooling tower may be the "induced draft" type or the "natural draft" type.

Cooled water is collected at the cold-water basin of the cooling tower. From the cold-water basin water travels to the ACW and CW sump. Since the cooling water travels through a closed cycle, only the make-up water is replenished from the clarified water reservoir to the ACW and CW sump.

The "once through" condenser cooling water system (Figure 9.10) can only be adopted in coastal areas. In this type of water system sea is the source of water. Circulating water (CW) pumps are installed near or on the coast, from which water is pumped through CW piping. The hot water from the condenser is returned to

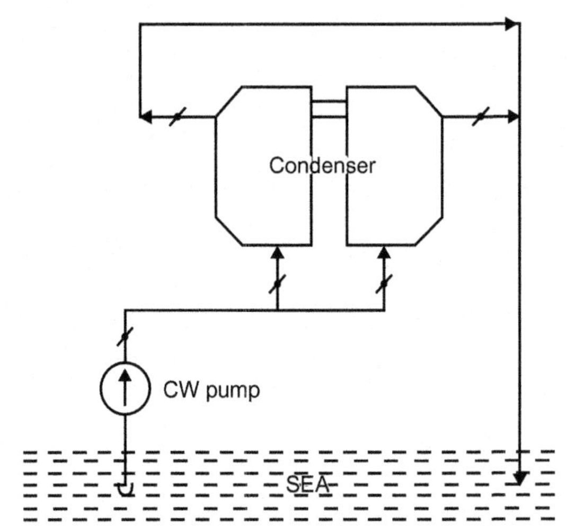

Figure 9.10 Once through type CW system.

the sea downstream of CW pump house, such that the maximum temperature difference between the supply water and the return water is restricted to 7 K, to protect marine life.

In both the "recirculating" type and "once through" type, cooling water enters the water box of the condenser at the bottom and subsequently flows through the tubes of the condenser. The cooling water absorbs the latent heat of the exhaust steam that passes over the outside of the tubes, causing steam to condense. A vacuum is thus created in the condenser by condensation of the exhaust steam. The condensed steam or the condensate is then directed to the condenser hotwell and warm circulating water exits through the condenser water box outlet at the top. With this configuration the CW flow through each shell of the condenser and the pressure (vacuum) in each shell would be identical, since the CW inlet temperature in each shell remains unchanged. This condenser is called a "single-pressure" condenser.

In addition to the above, there is another configuration where cooling water enters into one shell only, say shell 1 of the condenser, passes through the tubes of this shell and then exits through outlet water box. The outlet cooling water from shell 1, at a higher temperature than its inlet temperature, enters shell 2 as "cold water," then passes through the tubes of this shell and comes out through condenser outlets (Figure 9.11). With the new configuration of CW flow, the pressure in shell 1 would be less than the pressure in shell 2 because of dissimilar "cold water" inlet temperatures. Thus, the nomenclature given to this condenser is "multi-pressure" condenser.

The net result of the new configuration of CW flow is always a lower average back-pressure for the multi-pressure condenser compared to that obtained with the

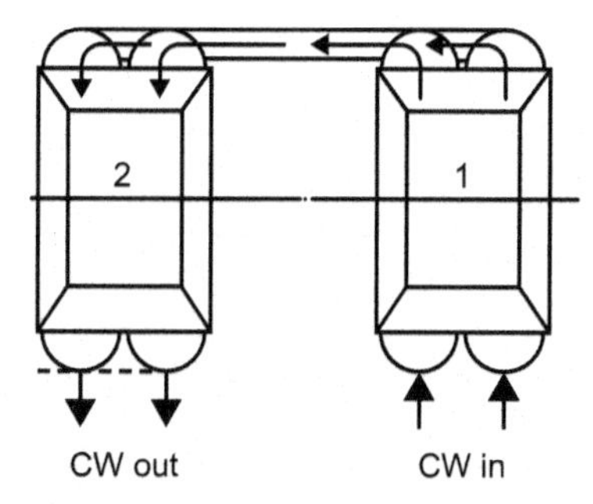

Figure 9.11 Flow scheme of multi-pressure condenser.

same water flow and surface in a conventional single-pressure condenser. A multi-pressure condenser would hence yield the following benefits:

i. With lower average back-pressure, the turbine heat rate will improve.
ii. If the average back-pressure of multi-pressure condenser is selected same as a single-pressure condenser, the turbine heat rate will remain unchanged, but the surface area and cooling water flow will be less. Thus, capital expenditure and running cost will be reduced.

9.2.7 Closed cycle cooling water (CCCW) system (Figure 9.12)

The source of water for a closed cycle cooling water (CCCW) system is DM water storage tank, from which water is transferred through a pump to an overhead CCCW make-up or expansion tank. This tank is atmospheric and is located at a height above the CCCW pump suction line to provide make-up to the system and also to act as a surge for expansion of the fluid in circulation due to temperature changes. The tank level is maintained by a supply of DM water.

CCCW system supplies DM cooling water through heat exchangers to various coolers of the steam generator (SG) package, turbo-generator (TG) package, compressors, ash-handling system, sample coolers, etc., with the help of CCCW pumps. Hot-water return from these coolers then flows back to the suction of pumps, the discharge from which flows through the heat exchangers. In these exchangers, water is cooled by rejecting heat to the auxiliary cooling water. Caustic soda (NaOH) or morpholine may be dozed in a CCCW pump suction header for maintaining a pH between 8 and 9. Oxygen, which may enter into closed circuit, is scavenged by intermittent dosing of hydrazine to the system.

9.2.8 Compressed air system (Figure 9.13)

Compressed air (CA) system comprises two sections; one section, i.e., the "service air" (SA), which provides compressed air for general house cleaning, pneumatic tools, and

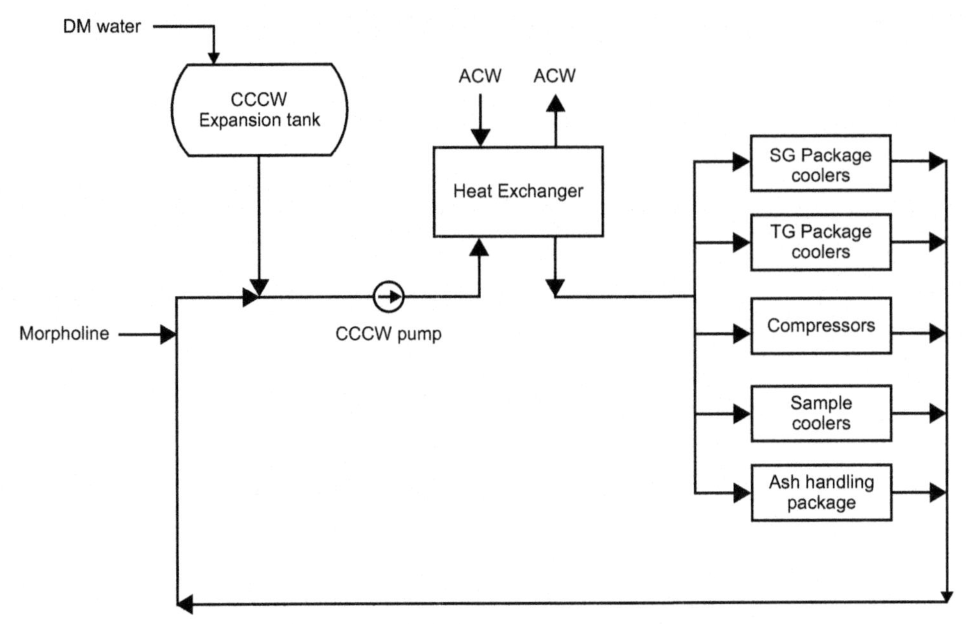

Figure 9.12 Closed Cycle cooling water (CCCW) system.

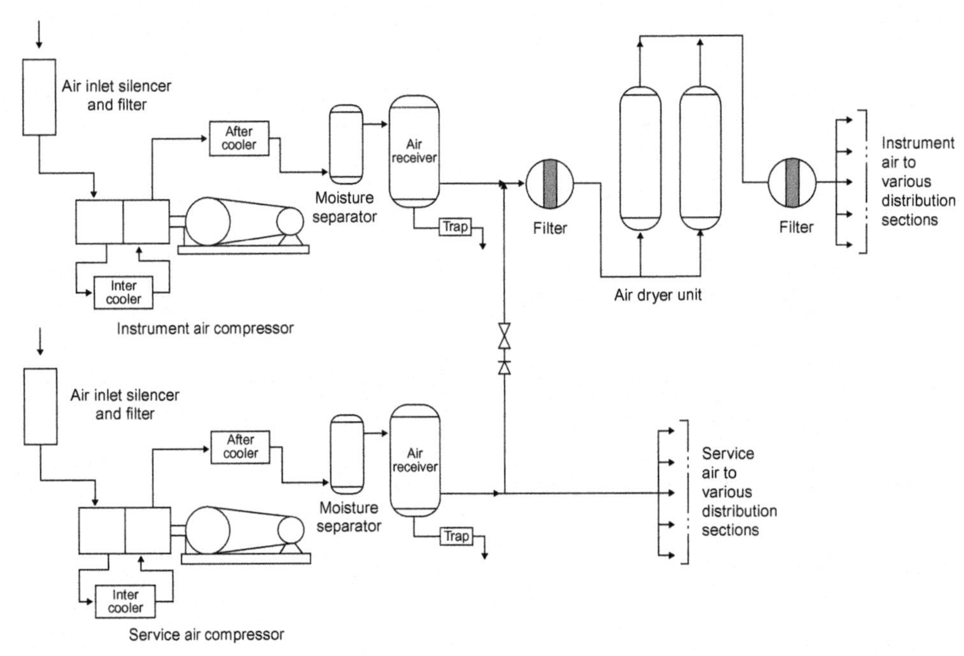

Figure 9.13 Compressed air system.

other miscellaneous purposes, and the other section, which meets the "instrument air" (IA) requirement for control of pneumatically operated instruments and drives. The compressed air system is capable of supplying all station air requirements, either intermittently or continuously, at a pressure of 700–800 kPa. The storage capacity of air receivers may be of 120–600 s operation of all air users at the maximum consumption rate.

The instrument air essentially has to be oil-free, dust-free, and dry to prevent condensation of moisture. The system uses compressors with associated suction filters, silencers, intercoolers, after-coolers, and moisture separators to convert ambient air into high-pressure instrument air. The discharge from compressors is sent to air receivers for storage and distribution to instrument air loads. The purpose of air receivers is to dampen pulsation in the airline and to meet sudden surging peak requirements beyond the capacity of the compressors. A drain trap is provided at the bottom of each air receiver to drain any accumulated moisture in the receiver. They also facilitate to removal of other impurities. Downstream of air receivers, an air-drying plant is provided to remove any entrained moisture. The air dryer unit comprises two chambers and is complete with pre- and after-filters to remove dust, dirt, etc. When in operation, one chamber remains in service and the other chamber is put into automatic regeneration. An air-drying plant could be the "desiccant air drying" type, the "heat of compression" type, or the "refrigerant" type.

The quality of instrument air must meet the following requirements:

i. The dew point value of instrument air at line pressure should be at least −7.8 K below the minimum temperature to which any part of the instrument air system is exposed at any season of the year, but in no case should exceed 275 K at line pressure.
ii. The maximum particulate size in the air stream should be 3 μm.
iii. The maximum total oil content should not exceed 1 ppm.
iv. The instrument air should be free of all corrosive contaminants and hazardous (inflammable or toxic) gases.

The service air system uses identical compressors as the instrument air system with the appropriate suction filters, silencers, intercoolers, after-coolers, and moisture separators. The discharge from compressors is sent to service air receivers for storage and distribution to service air loads at selected locations of the plant. At each location an outlet with a screwed cap and chain is provided to fit a hose.

The discharge header of the service air system is interconnected with the discharge header of instrument air system through an isolating valve and a non-return valve. In the event that there is exorbitant demand of instrument air, the branch-serving service air is automatically cut off and the service air is allowed to meet instrument air demand only.

9.2.9 Fire-water system (Figure 9.14)

The fire protection system uses firewater pumps, both electric motor driven and diesel engine driven, to provide raw water to hydrant system and spray systems. The hydrant

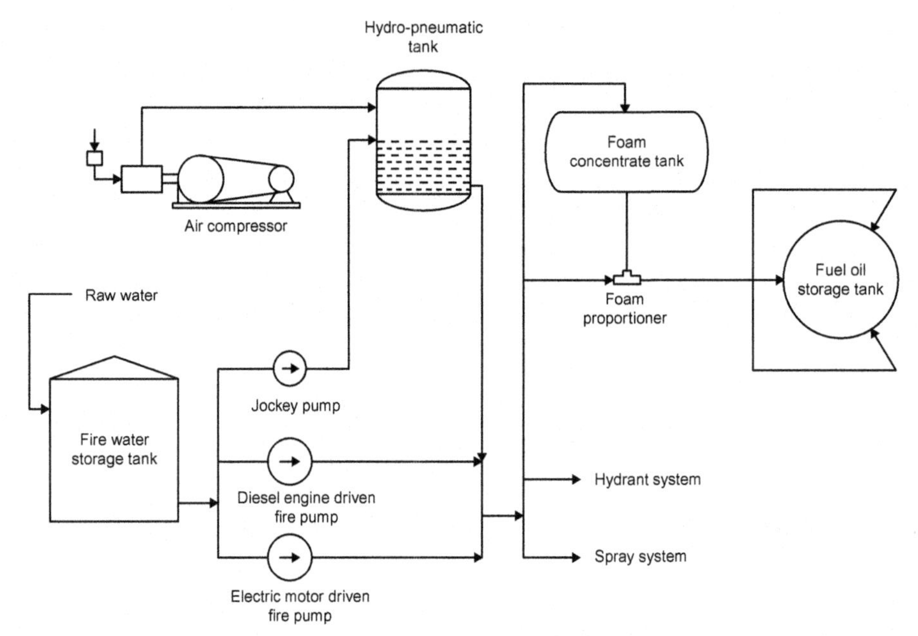

Figure 9.14 Fire water system.

system mains and the spray system mains are normally charged to 1.0 MPa pressure. The entire fire-water network is kept pressurized by one hydro-pneumatic tank, and pressure in this tank in turn is maintained by an air compressor. In the event that there is a fall in the water level in the hydro-pneumatic tank it is replenished with the help of a jockey pump.

The hydrant system essentially consists of a large network of piping, which feeds pressurized water to a number of hydrant valves located throughout the entire power station. In the event of a fire, fire hoses are coupled to hydrant valves, and jets of water are then directed toward the fire.

There are two sections in the spray system: the automatic spray system and the manually operated spray system. The automatic spray system detects and then controls and extinguishes any outbreak of fire automatically. The water supply to this system is controlled by pneumatically operated deluge valves. The water line forms a ring around the equipment to be protected. The air pipe also forms a ring around the equipment to be protected and frangible bulb type detectors are mounted on airlines at selected points. When the surrounding temperature exceeds the rated temperature of detectors, the frangible bulb collapses, releasing compressed air. The consequent fall in compressed air pressure opens the deluge valve and water starts coming out through nozzles toward the hot zone. This system protects generator transformers, station transformers, unit auxiliary transformers, CW system transformers, coal-handling plant transformers, etc.

In a manually operated spray system, fire is detected either visually (at the turbine lube oil tank area or near outdoor LT transformers) or by the use of heat detectors (around indoor LT transformers). Upon detection of fire, the operator would manually open the valve to put the spray system into service.

There are two types of detectors used in various areas of a plant: the "linear heat sensing cable" type used to detect fire in a coal-handling plant, mill reject-handling system, etc., and "smoke detectors" used to detect fire in cable vault, cable spreader room, etc.

The heavy fuel oil (HFO)/light diesel oil (LDO) tank areas are equipped with a foam supply system. In the event of a fire in a tank, foam is injected into the tank to extinguish it.

9.2.10 Main steam, cold reheat, and hot reheat system (Figure 9.15)

As discussed in Chapter 2, *Steam Generators*, low-temperature water receives heat from the combustion of fuel in a steam generator to become high-pressure, high-temperature steam. Before leaving the steam generator steam is passed through superheaters to reach the superheat state and through attemperators, where the main steam temperature is maintained by spray water. Steam thus generated is known as the *main steam* and is sent from a steam generator superheater outlet to the high-pressure (HP) turbine inlet. The steam then expands through the HP turbine to carry out mechanical work.

The exhaust steam from HP turbine is known as *cold reheat steam*, which is sent back to the steam generator in reheater section, where it is again superheated. At inlet

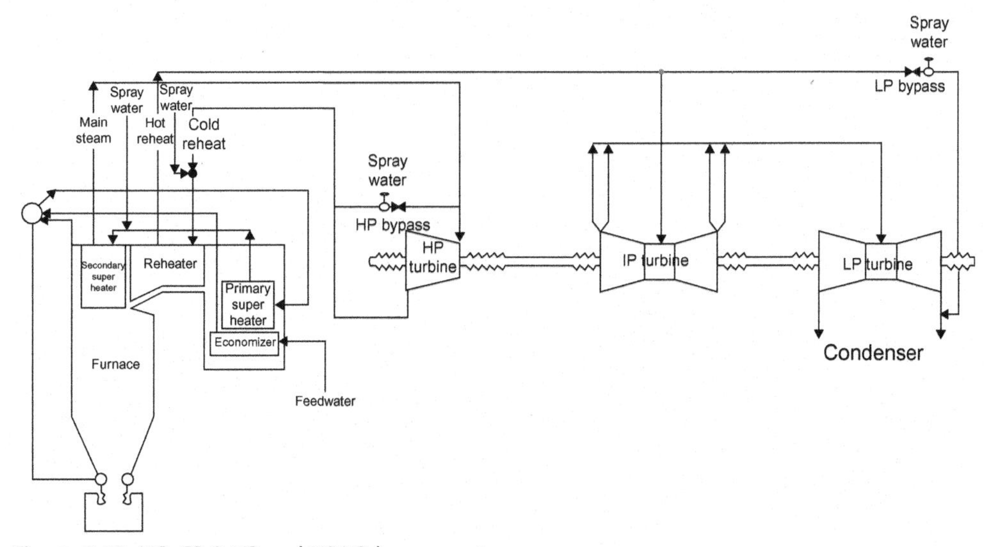

Figure 9.15 MS, CR & HR and HP/LP bypass systems.

to reheater, attemperators are provided to control the reheat steam temperature to a level equal to or higher than the main steam temperature. The reheated steam, known as *hot reheat steam*, is directed from the steam generator to the intermediate-pressure (IP) turbine. This steam expands in IP and low-pressure (LP) turbines to carry out additional mechanical work and then exhausts into the condenser, where exhaust steam is condensed and condensate is collected in the hotwell.

9.2.11 HP/LP bypass system (Figure 9.15)

A high-pressure/low-pressure (HP/LP) turbine bypass system is used to facilitate fast start-up and to allow quick reloading of the unit following a turbine trip. The HP bypass system takes steam from the main steam line, upstream of the HP turbine inlet emergency stop valves (ESVs), and discharges it to the cold reheat line through a pressure reducing valve and a de-superheater. Likewise, the LP bypass system takes steam from the hot reheat line ahead of the IP turbine inlet interceptor valves (IVs) and discharges it to the condenser through another pressure reducing valve and a de-superheater.

The HP/LP bypass system allows operation of the steam generator independent of the steam turbine. When a unit generates full load steam does not pass through this system. Under part load power generation a portion of the steam generated in the steam generator flows through the steam turbine and the remaining portion may pass through the HP/LP bypass system to maintain the rated temperature of the superheated steam, which is attained at a load of about 60% boiler maximum continuous rating (BMCR) (which is the stable load for boiler control) and above.

The HP/LP bypass system performs the following functions:

i. Enables boiler and turbine to operate independently of each other during start-up and shut-down
ii. Allows quick reloading of the unit following a turbine trip
iii. Controls the boiler outlet pressure for constancy by tracking this pressure with an adjustable higher differential pressure setting
iv. Opens quickly to relieve the system from high steam pressure in the event of full load or large load throw-off and also in the event of generator trip
v. Closes quickly whenever protections of the HP/LP bypass system operate

9.2.12 Steam drain system

The primary purpose of providing a steam drain system is to establish circulation of steam for gradual warm-up of main steam, cold reheat, and hot reheat steam pipelines as well as turbine stop and control valve bodies. It is imperative that the process of warming-up thus discussed is completed prior to turbine start-up. The steam drain system also prevents accumulation of water in the pipelines at start-up and under all operating conditions. Drains from this system are transferred to the condenser.

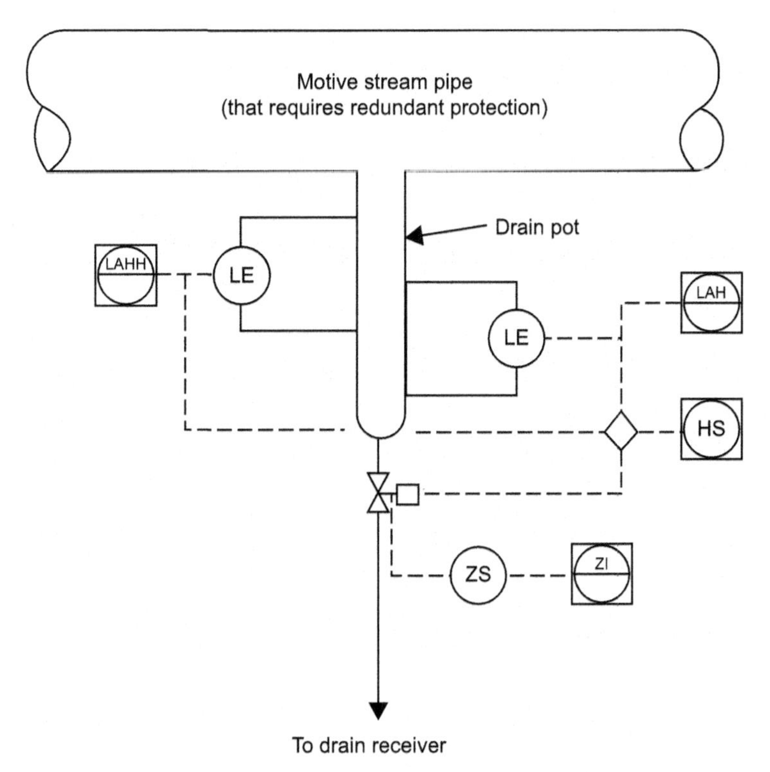

Figure 9.16 Typical drain system of motive steam pipe [ASME TDP-1-2006]. *Source: Fig. 5, P11. CH3 Design Recommendations. Article 3.3.3. ASME TDP-1-2006: Recommended Practices for the Prevention of Water Damage to Steam Turbines Used for Electric Power Generation. Fossil-Fueled Plants.*

Drain pots are provided for draining condensate from cold reheat pipelines and other steam pipelines prone to water accumulation from which there is concern of water/wet steam entering the hot turbine. Entry of water/wet steam would cause serious damage to steam turbines and should be prevented by all means. (For detailed treatment on the prevention of water damage to steam turbines, readers can refer to ASME TDP-1: Recommended Practices for the Prevention of Water Damage to Steam Turbines Used for Electric Power Generation—Fossil-Fueled Plants.)

A typical drain system for a motive steam pipe is shown in Figure 9.16.

9.2.13 Extraction steam system (Figure 9.17)

Prior to introducing feedwater to a steam generator both condensate and feedwater are pre-heated in feedwater heaters including deaerating heater by extracting steam from the turbine to achieve regenerative heating and deaeration of condensate and feedwater. Pre-heating of feedwater and condensate in stages allows maximum removal of heat stored in extraction steam, resulting in overall improvement of plant

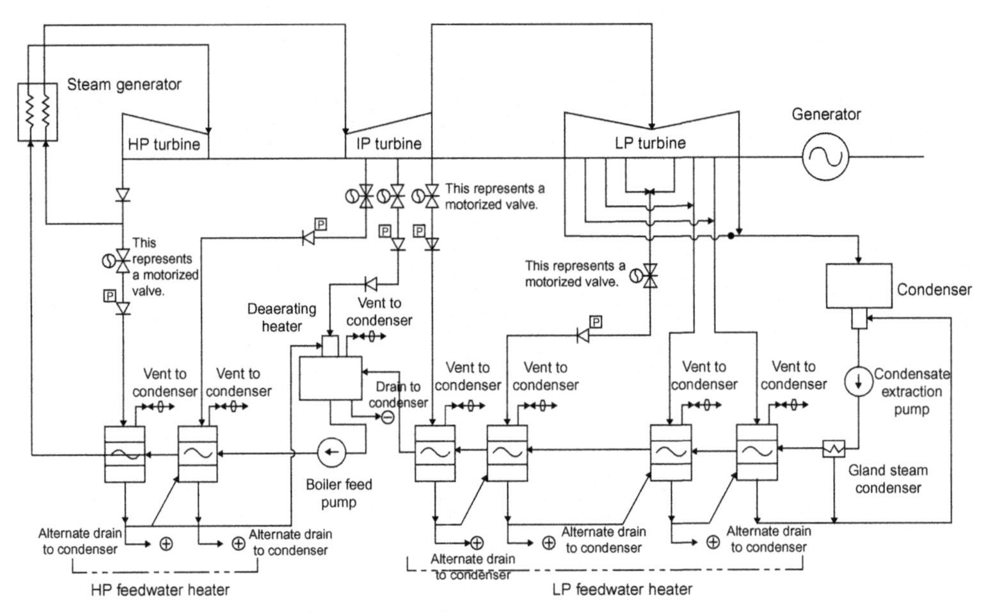

Figure 9.17 Extraction steam, heater drains & vents system.

efficiency. (For detailed treatment on Regenerative Rankine cycle refer to Chapter 1, *Steam Power Plant Cycles*.)

The extraction steam system receives steam from the cold reheat header and IP and LP turbines. In very large steam turbines, in general, with a unit capacity of 800 MW or higher, the extraction point is also provided in the HP turbine. The extraction steam system has two distinct sections: the high-pressure (HP) section and the low-pressure (LP) section. In the HP section steam is extracted from the cold reheat line and IP turbine to pre-heat the feedwater, while the LP section uses extraction steam from the IP turbine exhaust and LP turbine to pre-heat the condensate.

Each extraction steam line is provided with one power-assisted non-return valve (NRV) and a motorized isolation valve (Figure 9.18). The extraction line to the deaerating heater also contains one additional power assisted non-return valve (Figure 9.19). In the event of a high level in any of these heaters the non-return valve and isolation valve should close automatically. The non-return valve also provides automatic protection against reverse flow on rapid load reduction or in the event of turbine trip. Extraction steam lines to the lowest pressure LP heaters may not be provided with any isolation, since more often than not these heaters are located in the condenser neck.

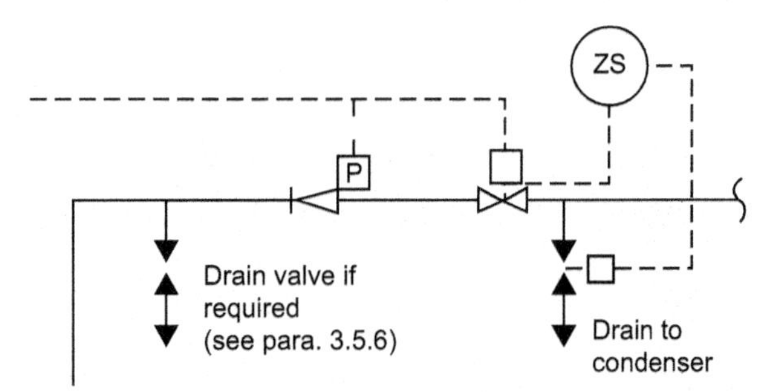

Figure 9.18 Non-return valve & isolation valve on extraction steam line [ASME TDP-1-206]. *Source: Fig. 6, P15. CH3 Design Recommendations. Article 3.5.1.2. ASME TDP-1-2006: Recommended Practices for the Prevention of Water Damage to Steam Turbines Used for Electric Power Generation. Fossil-Fueled Plants.*

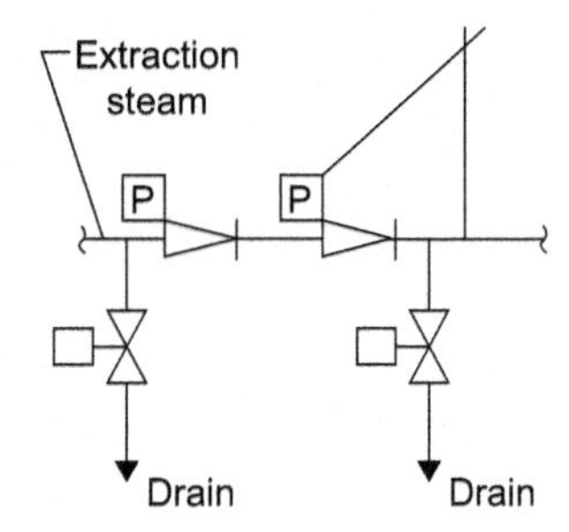

Figure 9.19 Two non-return valves on extraction steam line to deaerating heater. *Source: Fig. 10, P20. CH3 Design Recommendations. Article 3.6. ASME TDP-1-2006: Recommended Practices for the Prevention of Water Damage to Steam Turbines Used for Electric Power Generation. Fossil-Fueled Plants.*

9.2.14 Heater drains and vents system (Figure 9.17)

The condensed steam accumulated in each closed feedwater heater shell needs to be separated from its heat source, the extraction steam, and be cascaded to the next lower-pressure heater as the heater drain. In the event that next lower-pressure heater is not available or the level in the heater under consideration reaches a high level, the drain from the affected HP heater should be diverted to the deaerating heater and the drain from the affected LP heater diverted to the condenser. An alternate drain to the condenser from each heater is also provided. All heater drains are the gravity type.

Each heater is provided with a starting vent line and a normal vent line. The starting vent line is terminated with an isolation valve, which is opened to release trapped gases from the heater to atmosphere, prior to putting it in service. This valve remains closed during normal operation. The normal vent line is provided with an orifice for flow control and is connected with the condenser shell. This feedwater heater vent line

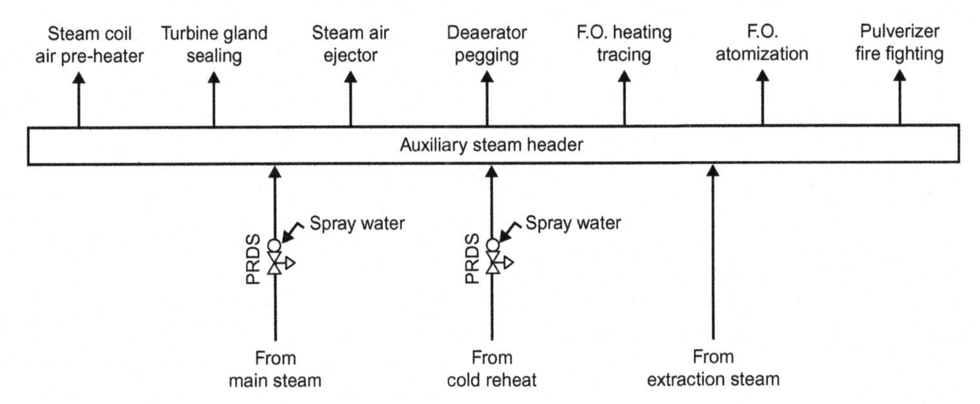

Figure 9.20 Auxiliary steam system.

always remains open when the heater is in service. The heater vent system removes non-condensable gases from each feedwater heater shell and deaerating heater.

9.2.15 Auxiliary steam system (Figure 9.20)

The auxiliary steam system provides steam at suitable pressure and temperature (usually 1.6 MPa and 493–503 K) to various plant auxiliaries during all modes of plant operation, including cold start-up and low load operation. The system uses steam from the extraction steam system during normal operation. However, during cold start-up of a unit when steam from either the extraction steam system or from the cold reheat line is not available or pressure is inadequate, the auxiliary steam header receives supply from main steam line. In the event that pressure in the cold reheat line becomes adequate to feed the auxiliary steam header the supply is switched over from the main steam line to the cold reheat line. A suitable pressure-reducing and de-superheating (PRDS) arrangement is provided while the auxiliary steam is sourced from the cold reheat line or main steam line.

Some typical uses of steam from the auxiliary steam header are as follows:

- Turbine gland sealing
- Steam air ejectors
- Deaerator pegging during start-up and low load operation
- Fuel oil (F.O.) storage tank and day tank heating, F.O. line tracing, F.O. heaters
- F.O. atomization and burner scavenging
- Steam coil air pre-heater (SCAPH or SCAH)
- Pulverizer firefighting system

9.2.16 Turbine gland sealing steam system (Figure 9.21)

Turbine gland seals are provided to eliminate the possibility of leakage of steam from the turbine through the turbine glands under pressure and ingress of atmospheric air

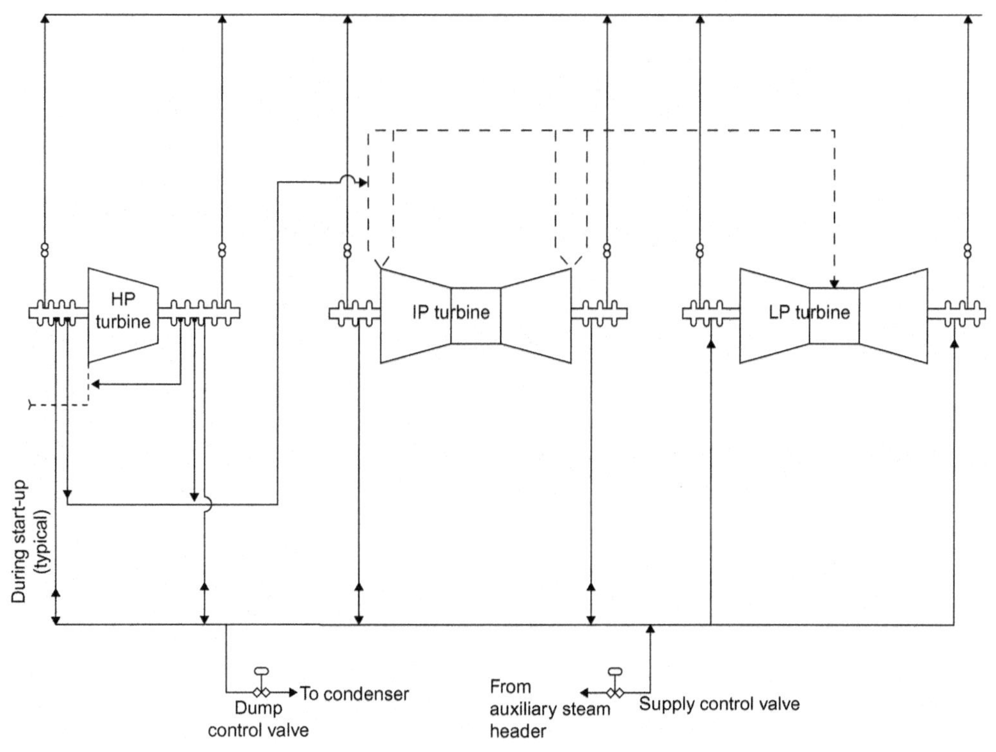

Figure 9.21 Turbine gland sealing steam system.

into the turbine under vacuum. While the leakage of steam causes a reduction in cycle efficiency, entry of cold air may cause thermal shock to hot turbine internals.

All turbine glands are continuously sealed by steam from the turbine gland steam header. During normal operation, leak-off steam tapped from different sections in the pressure-side stages of turbine glands is used to seal the glands of low-pressure stages. Airside leakage is cooled in the gland steam cooler/condenser, and the condensate is drained to the condenser hotwell. Pressure in the turbine gland sealing system is maintained at slightly above atmospheric pressure.

9.2.17 Condenser air evacuation system

The Carnot cycle proved that in the steam-water cycle the lower the heat sink temperature the higher the cycle efficiency. This means the condenser pressure should be as low as possible. The condenser pressure is lowered to sub-atmospheric condition by evacuating air from the condenser shell as well as from the internal area of the connected LP turbine. This evacuation may be realized either by using a vacuum pump

Table 9.2 Rapid evacuation equipment capacity

Sl. No.	Exhaust steam from LP turbine to be condensed, kg/s	Required capacity of the evacuating equipment, Nm^3/s
1.	63–126	0.153
2.	126–252	0.307
3.	252–378	0.460
4.	378–504	0.613
5.	504–630	0.767

or with the help of a steam jet-air ejector. Either of these vacuum-creating devices sucks air from the condenser shell and discharges it to the atmosphere.

Steam to the ejector is supplied from the auxiliary steam header during all modes of operation. Condensed steam from the ejector is recycled back to the condensate system.

Prior to starting a steam turbine it is best to evacuate the LP turbine and condenser rapidly to reduce the condenser pressure from atmospheric to a lower value. This is achieved by using either a non-condensing type single-stage starting air ejector or a vacuum pump. The Heat Exchange Institute (HEI) recommends that for the evacuation of air from atmospheric pressure to 33.86 kPa absolute pressure Hg in about 1800 s the capacity of the evacuating equipment should be as given in Table 9.2. (Note: As per the HEI, the standard condition corresponds to pressure 101.3 kPa (14.7 psia) and temperature 294 K (70°F)).

9.2.18 Condensate, condensate make-up, and dump system (Figure 9.22)

Steam after expansion in the LP turbine is condensed in the condenser. The main condensate is collected in the hotwell, which is then pumped by condensate extraction pumps (CEP) to the deaerator through steam jet-air ejectors (not required if the vacuum pump is used), gland steam condenser, and LP feedwater heaters, where condensate is successively heated by extraction steam. During this process, the steam condensate collected from different heaters/coolers in the cycle joins the main condensate flow, is heated in the low-pressure heaters, and is finally stored in the feedwater storage tank of the deaerating heater or deaerator (Figure 9.22). A common minimum flow recirculation line (in large plants each CEP may be provided with a dedicated recirculation line) is provided downstream of gland steam condenser to the hotwell to protect the running pump/s in the event that throughput to deaerating heater falls below a minimum flow limit. Each of the LP feedwater heaters is provided with individual isolation valves as well as a bypass valve.

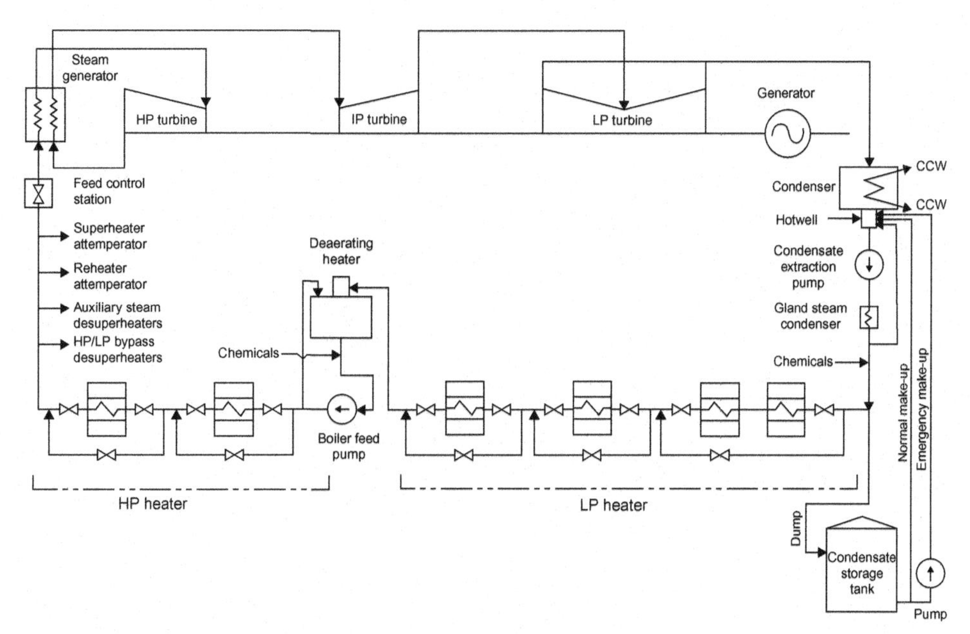

Figure 9.22 Condensate, condensate make-up, and dump system and feedwater system.

Chemicals (Section 9.2.20) are dozed in the condensate line to maintain pH and conductivity of condensate water. Dissolved oxygen (DO) in condensate downstream of the hotwell should not exceed 15 ppb (parts per billion) or 15 μg/kg.

In the case of fluctuation in the cycle, condensate will be transferred to and from a condensate water storage tank as required. The level in the hotwell is maintained by adding make-up water to the condenser or dumping condensate from the condensate cycle. Normally, in the event of a low level in the hotwell condensate water will flow from the storage tank to the hotwell by the static head. If this supply fails to restore the level an emergency supply through a pump will be cut-in. Once the level becomes high in the hotwell the condensate will be dumped to the storage tank.

Regarding the minimum hotwell storage capacity, the HEI recommends, "....volume sufficient to contain all of the condensate produced in the condenser in a period of 60 s under conditions of design steam load."

Nevertheless, common industry practice for sizing of the condenser hotwell in many countries, which is also followed by Central Electricity Authority (CEA) of India, is "180 s storage capacity (between normal level and low-low level of the hotwell) of total design flow with the turbine operating at valve wide open (VWO) condition, zero percent make-up and condenser design pressure."

The deaerator was designed to provide deaerated water to the steam generator. It is a direct contact type heater provided with a deaerating chamber and a feedwater storage tank of adequate capacity (discussed in the following). In the deaerating chamber condensate is sprayed from the top and steam flows in a counter-flow direction to the condensate and thus by scrubbing action non-condensable gases are released. Water in the storage tank is normally blanketed with steam to minimize absorption of oxygen. Dissolved oxygen in feedwater downstream of deaerating heater should not exceed 5 ml/l or 7 ppb (parts per billion) or 7 μg/kg.

The feedwater storage tank capacity as recommended by the HEI should be "...600 s to overflow as determined by the manufacturer's design criteria." The overflow level is also defined by the HEI "...shall not be closer to the top of the vessel than 85% of the tank diameter."

The usual industry practice in many countries including recommendation of the CEA, however, is: "the feedwater storage tank capacity (between normal level and low-low level) shall be 360 s of BMCR with a filling factor of 0.66. It shall be designed for maximum incoming steam flow under HP/LP bypass operation when LP heaters remain isolated."

9.2.19 Feedwater system (Figure 9.22)

The purpose of the feedwater system is to provide an adequate flow of properly heated and conditioned water to the boiler and maintain the boiler drum level in accordance with the boiler load. This system also sends water to the boiler superheater attemperators, reheater attemperators, auxiliary steam de-superheaters, HP/LP Bypass de-superheaters, etc.

Under normal operating conditions, feedwater flows from the deaerator feed storage tank, which is the beginning of the feedwater system, to the suction of boiler feed pumps (BFP). High-pressure discharge from BFP flows through the HP feedwater heaters before entering the economizer through the feed regulating/control station, which regulates feedwater flow to the boiler to maintain boiler drum level.

Each boiler feed pump is provided with a minimum flow recirculation line for protection of the pump, which ensures the minimum flow through pump when the throughput to the boiler is stopped or falls below the minimum flow limit. Each of the HP feedwater heaters is provided with individual isolation valves as well as a bypass valve. Chemicals (see Section 9.2.20) are dozed at the suction of boiler feed pump to maintain pH and dissolved oxygen of feedwater.

The HP feedwater heaters receive extraction steam from the cold reheat line and IP turbine (Figure 9.17). The steam and feedwater flow through the HP heaters in a counter-flow arrangement. Feedwater successively flows through sub-cooling (for final cooling of drain condensate), condensing (for condensing steam), and separate de-superheating (for de-superheating extraction steam) zones provided in the HP heaters and absorb heat from the extraction steam as it passes through

9.2.20 Chemical feed system

The chemical feed system provides a way to introduce chemicals into the plant steam/water cycle to reduce formation of scale and corrosion in the boiler, feedwater system, condensate system, and closed cycle cooling water system. The chemical feed system has two distinct units: a LP dozing system and a HP dozing system.

In LP dozing system ammonia/morpholine and hydrazine are used. Ammonia/morpholine is injected into the condensate supply header to maintain a pH of the condensate between 9.0 and 9.3 to inhibit corrosion. Hydrazine is injected at the boiler feed pump suction line for scavenging of oxygen and for secondary pH control. The presence of oxygen leads to corrosion of boiler internal surfaces. The consequent effect of dissolved oxygen in condensate and feedwater is pitting of internal surfaces. Hydrazine, in conjunction with mechanical facilities of feedwater deaeration, controls the chemistry of water and steam in the power cycle. Morpholine is also introduced in the make-up line to the hotwell to maintain the pH of water. Analyses of treated quality of condensate and feedwater are given in Table 9.3 and Table 9.4, respectively.

Calcium and magnesium salts in the feedwater if not controlled get deposited on the boiler heat transfer surfaces and form a hard, tightly adhering scale. The HP

Table 9.3 Analysis of condensate

Contaminants	Unit	Normal condition
Total Dissolved Solids	μg/l	<100
Silica (as SiO2)	μg/l	<25
Chloride (as Cl)	μg/l	<10
Sodium (as Na)	μg/l	<10
Total Ferric Iron	μg/l	<50
Total Copper	μg/l	<5
pH	–	9.0–9.3
Conductivity (after the cation-exchanger) (at 25°C)	μS/cm	<0.1

Table 9.4 Analysis of feedwater

Contaminants	Unit	Normal condition
Hardness	μmol/l	≤1
Dissolved Oxygen	μg/l	≤7
Iron	μg/l	≤20
Copper	μg/l	≤5
Hydrazine	μg/l	10–50
pH	–	8.8–9.3
Oil Content	mg/l	≤0.3

Table 9.5 Analysis of superheated steam

Contaminants	Unit	Normal condition
Sodium	μg/l	≤10
Silica (as SiO2)	μg/l	≤20
Iron (as Fe)	μg/l	≤20
Copper (as Cu)	μg/l	≤5
Conductivity (after the cation-exchanger)	μs/cm	≤0.3

dozing system uses trisodium phosphate (Na_3PO_4) and sometimes caustic soda ($NaOH$) that are injected into the boiler drum to remove these salts and to maintain the pH in the boiler water. In the presence of phosphate ions (PO_4) and at high temperature and pH, these calcium and magnesium salts become soft, loosely adhering scales, which are removed by boiler water blow-down. Analysis of treated quality of superheated steam is presented in Table 9.5.

9.2.21 Generator hydrogen cooling system

Each side of the generator enclosure is provided with stator end shields to keep the enclosure gas-tight. When the machine is running heat is generated within the generator enclosure. Some cooling medium needs to be circulated through the generator stator and rotor to dissipate the heat and prevent an increase in conductor temperature to within a permissible limit. In small-sized generators this cooling medium is air. This air is circulated by axial fans installed on both sides of the rotor shaft ends. The air flows through the rotor conductors, exits radially, and mixes with fresh cold air. The mixed air then cools the stator core (Figure 9.23), and the hot air in turn is cooled by circulating cooling water.

As the unit size increases an air-cooled generator becomes huge. To restrict the size of the generator hydrogen gas in lieu of air is used as the cooling agent principally because the density of hydrogen is 1/14th the density of air at a given pressure and temperature and the thermal conductivity of hydrogen is almost 7 times that of air. The combination of these two benefits permits a reduction of about 20% in the amount of active material of a generator for a given output and a given temperature rise of windings.

The pressure of hydrogen is normally maintained between 0.2−0.4 MPa, but could be raised to 0.6 MPa. The purity of hydrogen for circulation has to be 99.7% or better; otherwise it could be explosive.

In a typical design an axial-flow fan circulates hydrogen within the generator (Figure 9.24). The fan draws hot gas from one end, passes it through coolers, and then cooled gas is passed over the rotor winding, stator core, and stator end winding space. Hydrogen in turn is cooled by circulating the cooling water through the hydrogen coolers. Thus, the generator temperature is kept within the maximum permissible limit.

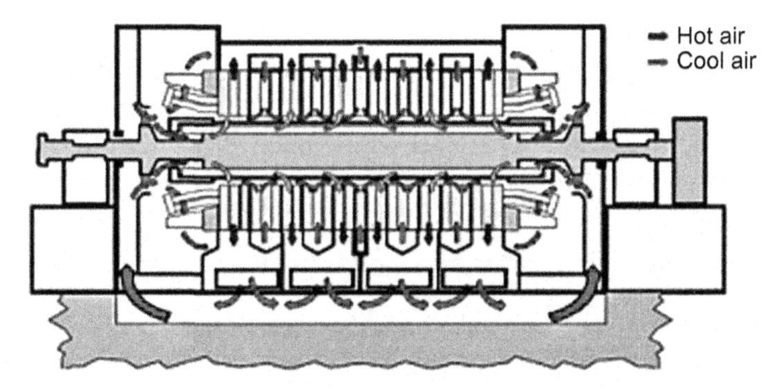

Figure 9.23 Air-cooled generator. *Source: http://www.e-cooling.de/en-elektrische-maschinen-erwaermung-kuehlung.htm.*

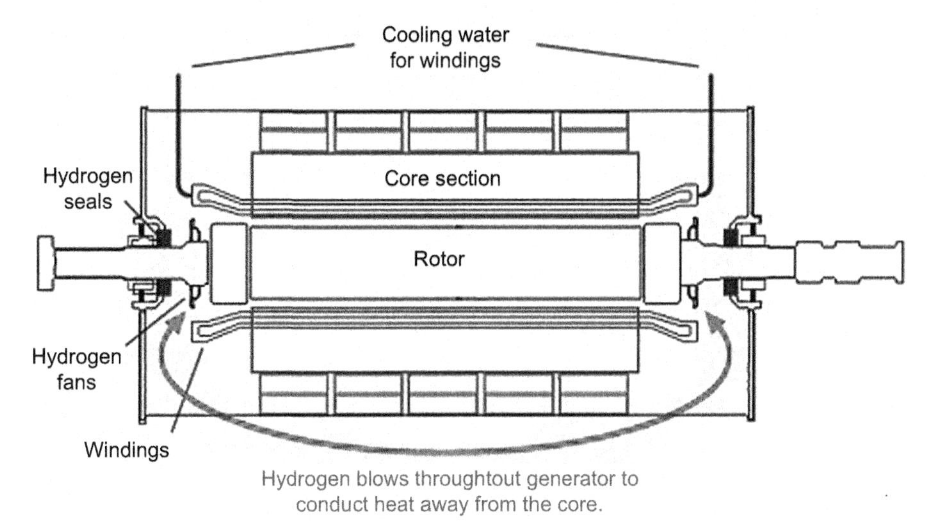

Figure 9.24 Generator hydrogen cooling system. *Source: http://electricalengineeringdesigns.blogspot.in/2012_06_01_archive.html.*

9.2.22 Generator stator cooling water system (Figure 9.25)

Heat generated in the stator winding is dissipated by circulating cooling water through stator conductors. DM water or distilled water may be used as the coolant. The dissolved oxygen content of the coolant must not be more than 80 ppb ($\mu g/l$) and the pH of the coolant should be maintained between 8.0 and 8.5 by dozing chemicals, otherwise copper corrosion could result.

DM water is supplied to the stator cooling water tank, from which a pump drawing water passes it through the cooler and filter and then supplies the water to stator conductors. The return from the conductors is drained to the stator cooling water tank.

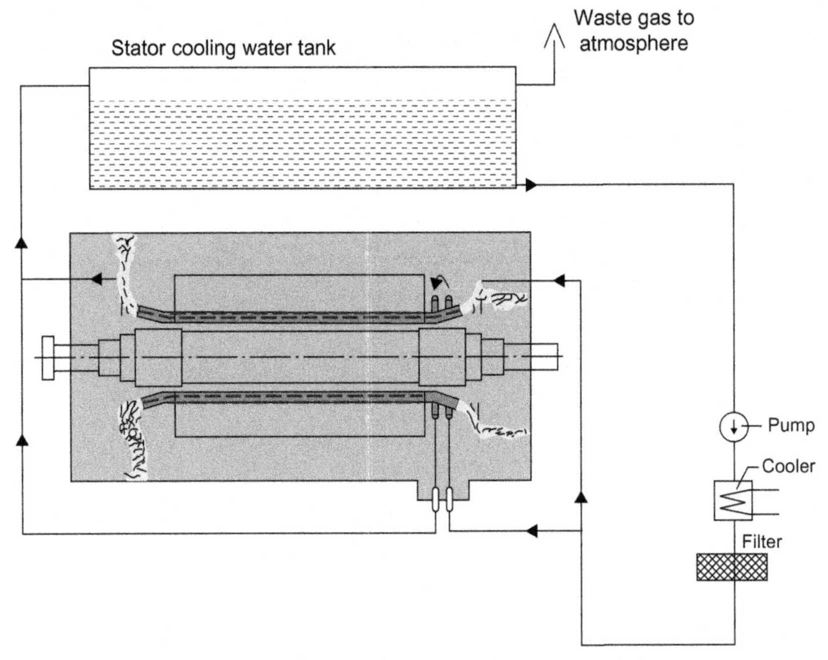

Figure 9.25 Generator stator cooling water system.

9.2.23 Generator seal oil system

As discussed in Section 9.2.21 the rotor winding of large generators is generally cooled by hydrogen, hence hydrogen may escape through shaft seals where the rotor comes out of the enclosure even though each side of the generator enclosure is gas-tight. To prevent such leakage of hydrogen a continuous film of oil is maintained between the rotor and stationary seal rings. Seal oil also prevents entry of air into hydrogen-cooled enclosure. The pressure of seal oil on both air side and hydrogen side is maintained slightly higher than the hydrogen pressure (Figure 9.26) to restrict entry of hydrogen and air through seal rings on either side.

Figure 9.27 shows a simplified seal oil flow diagram that comprises two separate seal oil circuits, i.e., air-side seal oil circuit and hydrogen-side seal oil circuit. One seal oil pump receives suction from the seal oil storage tank and supplies oil to the air-side shaft seals through the cooler and filter. Drainage from the air-side seal ring is returned to the seal oil storage tank.

Another seal oil pump draws seal oil from the seal oil tank and supplies the oil to the hydrogen-side shaft seals through a cooler and filter. Drainage from the hydrogen-side seal ring is returned to the seal oil tank. The seal oil tank is connected with the seal oil storage tank through an over-flow float valve. The seal oil on the hydrogen side, being at a higher pressure than hydrogen, prevents the continuous loss of hydrogen.

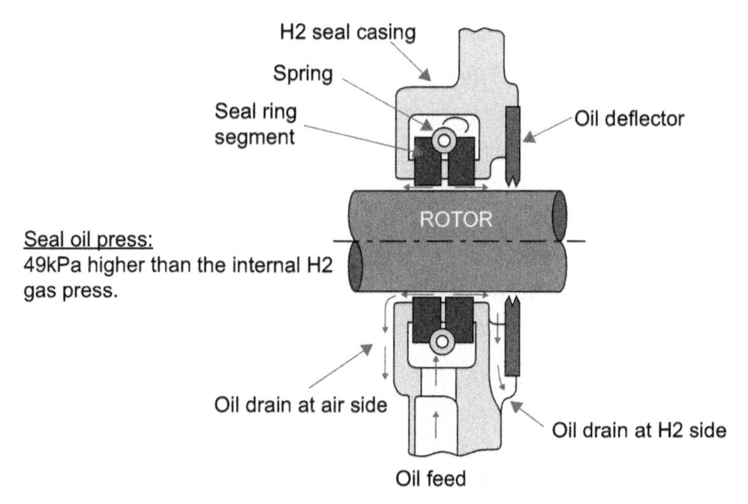

Figure 9.26 Generator seal ring. *Source: http://anto-hendarto.blogspot.in/2013/06/mass-transfer-from-water-vapor-to.html.*

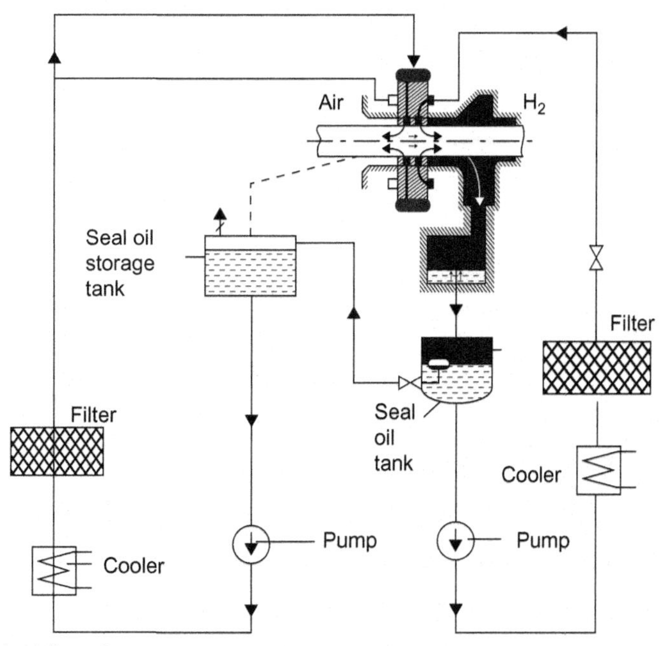

Figure 9.27 Seal oil flow diagram.

9.2.24 Fuel oil pressurizing and heating system (Figure 9.28)

Fuel oil is the secondary fuel of large pulverized coal-fired boilers. The main purpose of the fuel oil system is to facilitate the start-up of boiler, since pulverized coal on its own is unable to ignite and requires ignition energy from an external source.

Fuel oil is required for flame stabilization at low-load (usually between 30% and 40% BMCR) operation with pulverized coal. The system is also capable of ensuring low-load (up to about 30% BMCR) operation of boiler without coal firing. The fuel oil system can generally handle two different types of oil, i.e., light diesel oil (LDO) and heavy fuel oil (HFO). The functional requirement of the above two oils is as follows:

i. The primary function of LDO is to light-up the boiler during cold start-up.
ii. The HFO provides "hot start-up of boiler," "flame stabilization," and "low load operation."

The LDO system consists of high-pressure screw or gear pumps and a pressure control re-circulation valve, provided on common discharge header of pumps, for maintaining constant pressure. Pumps are also provided with strainers at their suction. These pumps take suction from the LDO tanks. In this system the return line from the burner end is not required, since viscosity of this oil is compatible with that required for proper firing, as recommended by manufacturers of burners, and thus do not require prior warm-up.

The HFO pumping and heating system comprises high-pressure screw pumps and dedicated fuel oil heaters. The pumps take suction from HFO tanks. Strainers are

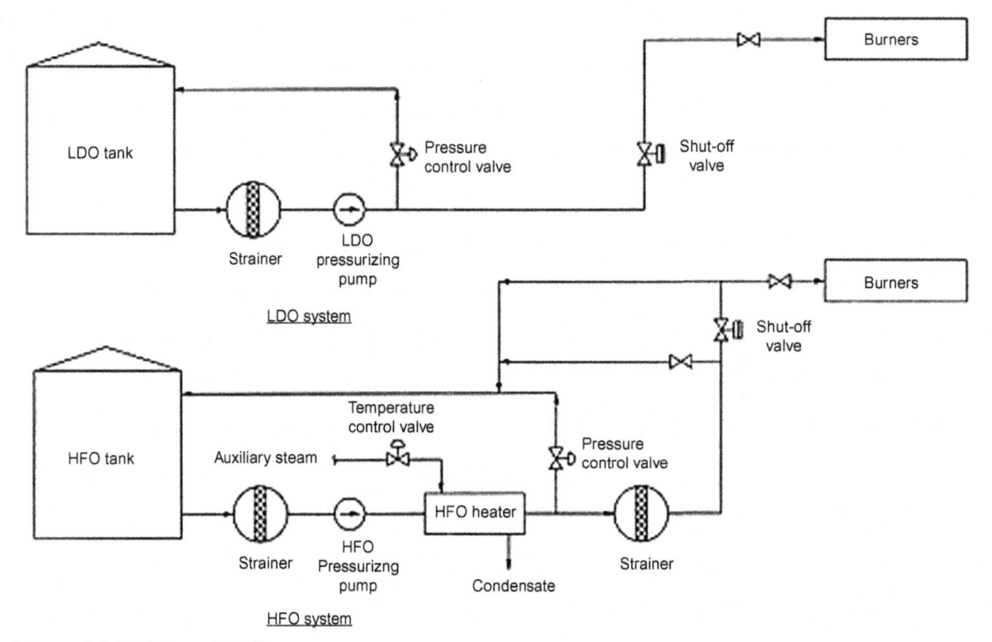

Figure 9.28 LDO and HFO systems.

provided at the suction of pumps as well as downstream of the HFO heaters. A pressure control recirculation valve is provided on common discharge header of pumps, which maintains constant pressure downstream of heaters.

Since it is quite viscous, HFO is required to be pre-heated to attain requisite viscosity at burners, as per manufacturers' recommendations, for proper atomization and combustion. Fuel oil heater uses steam from the auxiliary steam header and is controlled by a temperature control valve at the inlet to the heater to maintain constant fuel oil temperature at heater outlet.

Typical specification of HFO, LDO, and HSD (high-speed diesel) is presented in Table 9.6.

Table 9.6 Typical specification of fuel oils

Characteristics	Heavy fuel oil (HFO)	Light diesel oil (LDO)	High speed diesel oil (HSD)
Acidity, inorganic	NIL	NIL	NIL
Acidity, total mg of KOH/g, max.	–	–	0.50
Ash, % by weight, max.	0.1	0.02	0.01
Gross calorific value, MJ/kg	41.86	41.86	–
Specific gravity at 288 K	Not limited but to be reported	0.85	–
Flash point, K, min. (Pensky-Martens Close Cup)	339	339	311 Abel Close Cup
Pour Point, K	–	285 (Winter) 291 (Summer)	279 max.
Kinematic viscosity in centistokes	370 at 323 K	2.5 to 15.7 at 311 K	2.0 to 7.5 at 311 K
Sediment, % by weight, max.	0.25	0.10	0.05
Total sediments, mg per 100 ml, max.	–	–	1.00
Total sulphur, % by weight, max.	4.5	1.8	1.0
Water Content, % by volume, max.	1.0	0.25	0.5
Carbon residue (Ramsbottom), % by weight, max.	–	1.5	0.20

Kinematic Viscosity:

1 Stoke = 407 seconds Redwood Viscosity No. 1 at 311 K

= 462.7 seconds Saybolt Universal Viscosity at 323 K

= 13.16 Engler Degrees

$$1 \text{ Stoke} = 10^{-4}\frac{m^2}{s}$$

The HFO at the required pressure and temperature is supplied to the burners. A return line, from the fuel oil header supplying to the burners, is provided to route excess oil back to the oil tank and also to maintain hot oil circulation up to the boiler front for use in burners in the case of emergency. The HFO supply and return lines are insulated and either steam traced or electric traced to keep the lines warm at all times. Similarly, pumps and strainers are steam/electric jacketed.

9.2.25 Coal-handling system (CHS) (Figure 9.29)

Coal is the primary fuel of a pulverized fuel-fired boiler. Coal is received either at the track hopper area or at the wagon tippler complex by railway wagons. The track hopper is provided with in-motion weighbridges, while the rotary type wagon tipplers are provided with integral type weighbridges. For unloading and feeding the coal to the bunker, conveyors are arranged along with crushers, transfer points, etc. Conveyors are provided with rollers, idlers, and in-line-magnetic-separators to remove any tramp iron in the coal.

For the purpose of unloading coal from railway wagons coal is collected either at track hoppers or at wagon tippler hoppers. The coal unloaded in hoppers is sent to to coal crusher via conveyors. From the crusher coal is taken to the crushed-coal storage area through another conveyor system. From this storage area coal is reclaimed by either bulldozers and/or stacker-reclaimers and transferred to powerhouse terminal hopper, from which the coal is fed to each coal bunker of the powerhouse by a separate coal conveyor through mobile trippers.

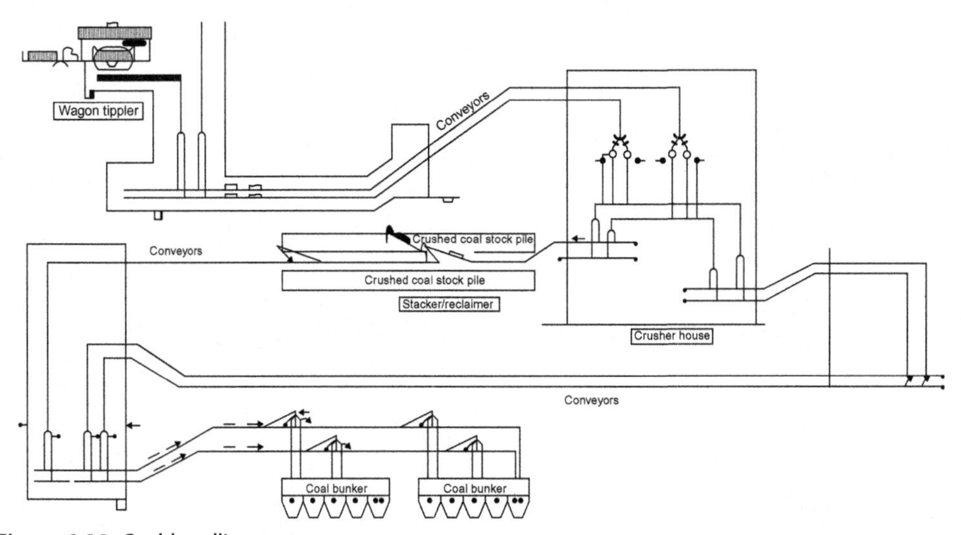

Figure 9.29 Coal-handling system.

From the coal bunkers coal is fed to the pulverizers through raw coal feeders. Pulverized coal from each pulverizer is carried by hot primary air and is fed to the coal burners of the boiler. Secondary air is then supplied near the burners for complete combustion of pulverized coal. (For more details on the pulverized coal-fired system see Chapter 4.) For improving operating conditions in the coal-handling area, a separate arrangement is provided for dust extraction, dust suppression, and ventilation.

9.2.26 Ash-handling system (AHS)

In large steam generators using low-grade coal, with 33–45% ash content, collection and disposal of ash poses a serious problem. In a typical 500 MW coal-fired unit production of ash per diem varies from about 3.0×10^6 kg to about 5.0×10^6 kg, depending on the quality of these low-grade coals.

Coal ash may be broadly classified into three parts. The heaviest part, clinker, is collected at the furnace bottom at the bottom ash hopper. The remainder is carried away by flue gas, of which coarser ash in grit form is collected in economizer hoppers and air pre-heater hoppers. Finer ash in powdered form is further carried away by flue gas to either electrostatic precipitator (ESP) and/or bag filter, where ash is separated from flue gas and collected in hoppers.

Ash collected at the bottom of the furnace is called *bottom ash* (BA) and constitutes about 15–20% of total ash generated, but at times could be as low as 10%. The balance, 80–85% of the total ash generated, is called *fly ash* (FA), of which around 5% is *coarse ash*.

The function of the ash-handling system is to remove ash collected at various hoppers, as described above, and to dispose it conveniently to an allocated disposal area away from the main powerhouse. While collection and disposal of bottom ash and coarse ash is normally continuous, that of fly ash is carried out periodically and sequentially.

In earlier days, the ash-handling system used to be the wet type. Today, plants use "zero leakage systems" to comply with environmental protection regulations, which usually means dry ash-handling systems. In a wet ash disposal system the bottom ash is collected in water-impounded hoppers located just below the boiler furnace and then is removed by jet pump/eductors/hydro-ejectors (Figure 9.30).

Coarse ash and fly ash in wet ash-handling are collected either by wetting water jets or by a hydro-pneumatic system. All the slurry ash is then taken to an ash slurry sump, from which this slurry is pumped into an ash pond for disposal.

In a dry ash-handling system bottom ash and coarse ash are continuously disposed of through a metallic conveyor located beneath the boiler and typically transported to an intermediate bottom ash silo. This metallic conveyor normally moves

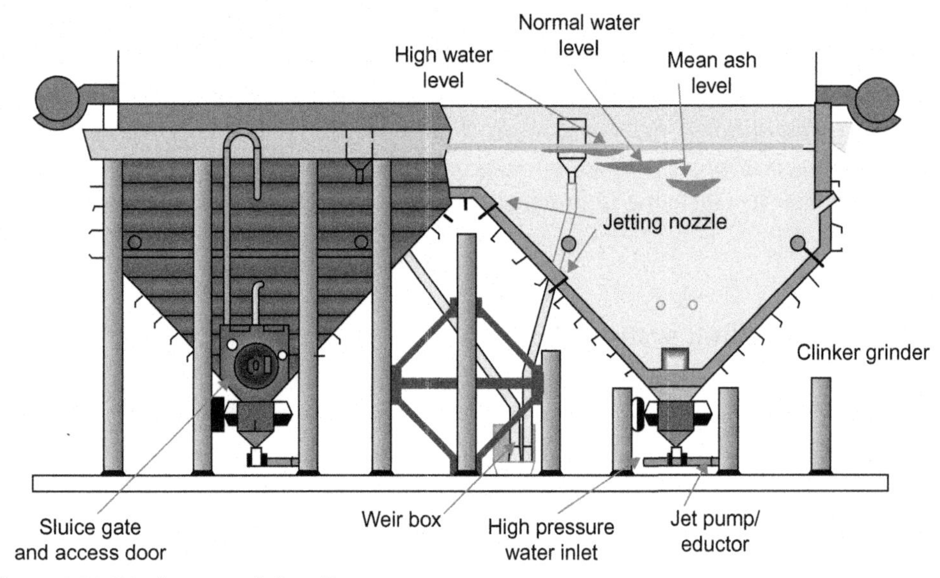

Figure 9.30 Wet bottom ash-handling system.

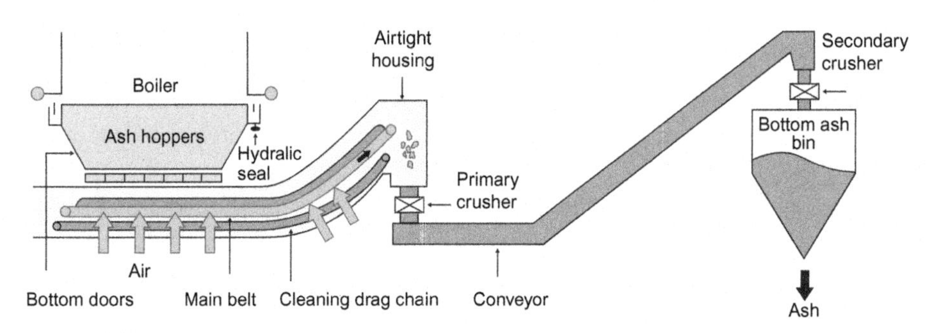

Figure 9.31 Dry bottom ash-handling system. *Source: DRY BOTTOM ASH EXTRACTOR.*

very slow; this is juxtaposed with flow of cold atmospheric air through the conveyor to the furnace due to furnace draft. This allows cooling of bottom ash prior to its disposal to the silo. Discharge from the bottom ash silo is sent pneumatically to the final bottom ash silo (Figure 9.31) for onward disposal through road tankers, trucks, etc. The bottom ash hopper is provided with gates. In the event that the metallic conveyor is under maintenance, evacuation of the bottom ash is isolated with the help of these gates.

Dry fly ash collected from the ESP/bag filter hoppers is transported pneumatically to a remote silo via an intermediate surge hopper. From the remote silo fly ash is disposed of through road tankers, trucks, etc.

Whenever disposal through trucks/tankers fails an emergency dense phase slurry disposal system, known as high concentration slurry disposal (HCSD) system, acts as a backup. In this system ash from the remote silos is discharged to a tank, mixed with water to form dense slurry. Slurry from this tank is disposed of continuously to an ash disposal area through HCSD pumps.

9.2.27 Steam water analysis system (SWAS)

As noted in the discussion on chemical feed system in order to reduce problems of scale and corrosion in the boiler, feedwater, and condensate systems, chemicals need to be dozed into the plant steam/water cycle. To quantify the amount of particular chemical to be dozed it is essential to monitor the chemistry of the steam/water.

Chemical analyzers are usually installed in a central place for better control of the steam/water chemistry and for ease of maintenance. Stainless-steel sample lines are drawn to a central sample table, where the samples are cooled in sample coolers, their pressure is reduced in pressure reducing valves, and the flow is maintained prior to bringing the samples into contact with their respective analyzers.

Various types of chemical analyzers are provided in a thermal power plant to check purity, assess carryover of silica and sodium, detect oxygen ingress, monitor pH, etc. Some common analyzers along with their location for monitoring are shown in Table 9.7.

9.3 ELECTRICAL SYSTEMS

Each unit has a unique electrical distribution system, as discussed in this section. A typical distribution system normally comprises 6.6 kV switchgears, 415 V switchgears/motor control centers (MCCs), 240 V AC and 220 V DC distribution panels, 240 V AC lighting panels, ± 24 V DC system, etc.

Table 9.7 List of analyzers and their monitoring location

Sl. No.	Analyzers	Fluid and its location
1.	Conductivity, Silica, Sodium	Superheated steam
2.	Conductivity, Silica, pH	Boiler drum water or blow-down water
3.	Conductivity, Silica, pH, Dissolved oxygen, Residual hydrazine	Feedwater at economizer inlet
4.	pH, Dissolved oxygen, Sodium	Condensate at CEP discharge
5.	Specific conductivity, Cation conductivity	Condensate in hotwell
6.	Conductivity, pH	Condenser make-up water
7.	Conductivity, Dissolved oxygen	Stator cooling water

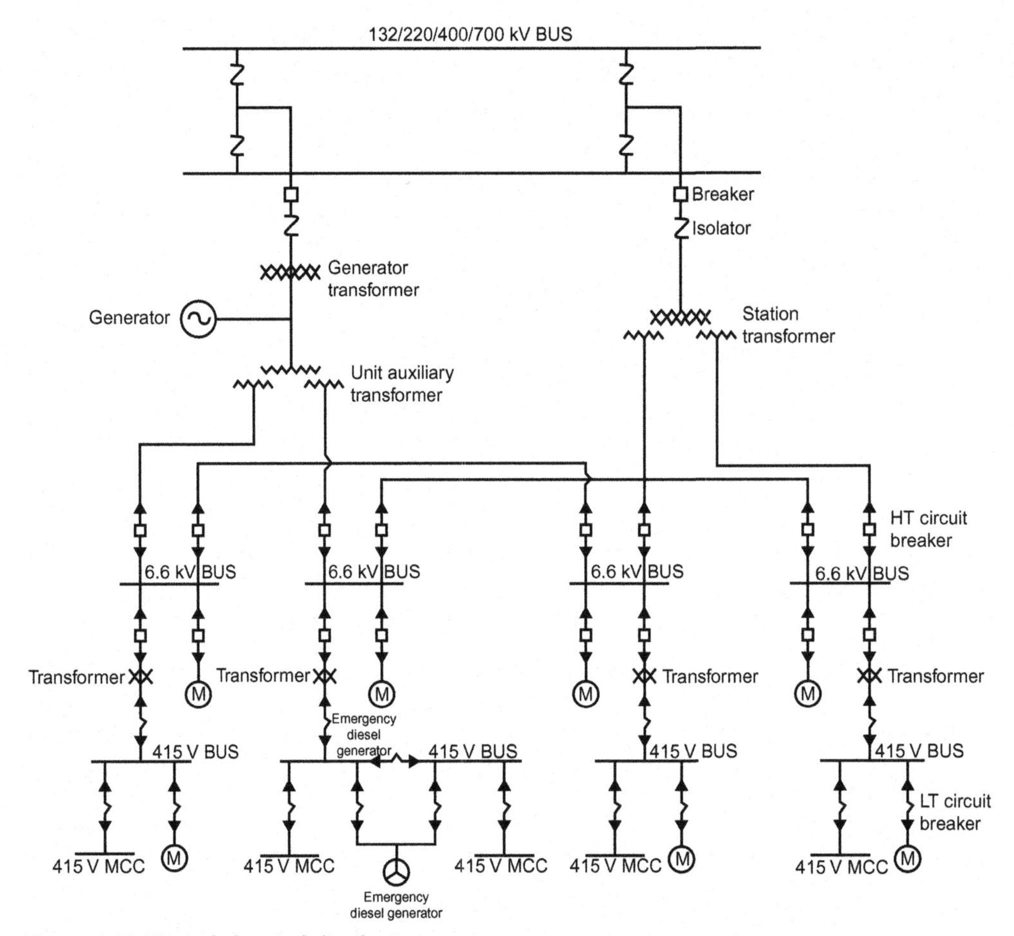

Figure 9.32 Typical electrical distribution system.

The unit auxiliary transformer supplies electric power to all loads, control circuits, and electrical instrumentation serving the unit. Common loads like coal-handling, ash-handling, heating-ventilation and air conditioning (HVAC), lighting, etc., are fed from station transformers. The circulating water system and intake pump house loads are usually not integrated with the unit auxiliary power distribution system and are supplied through independent transformers. Figure 9.32 shows a typical distribution system for a high voltage (132/220/400/700 kV), medium voltage (6.6 kV), and low voltage (415 V) supply.

9.3.1 Main generator system

All generators, also known as alternators, used in various power stations are connected to the grid (or infinite bus-bar) in parallel. These generators may be of different capacities, but their voltage and frequency essentially have to be identical and constant.

The main generator system transmits generator output power to a high-voltage (132/220/400/700 kV) grid through generator transformers (GT), isolators, and generator circuit breakers. This system also protects and monitors plant output and the station auxiliary power source with appropriate alarms. The system boundaries include:

- Unit auxiliary transformers (UATS)
- Station transformers (STS)
- Isolated phase bus
- Generator transformer (GT)
- Switchyard interconnections
- Control and protective relaying schemes

9.3.2 Medium-voltage system

Power received from either the generator or grid is supplied to 6.6 kV switchgears through unit auxiliary transformers (UATs) or station transformers (STs) and high-tension (HT) circuit breakers. 6.6 kV switchgears in turn distribute power to the plant's 415 V electrical system and large electrical drives.

9.3.3 Low-voltage system

415 V distribution systems receive supply from 6.6 kV switchgears through transformers and low-tension (LT) circuit breakers to 415 V MCCs. These MCCs supply power to all 415 V loads including small electrical drives.

9.3.4 DC power system

The DC power system provides reliable power to those loads, which need to remain in service even with a loss of AC power. 220 V DC power is supplied to DC motors and other large loads such as emergency lighting, plant inverters, etc. ± 24 V DC power caters to the control and instrumentation system. Some of the major DC loads include:

- Computer
- Turbine generator instrumentation and control
- TG emergency lube oil pump and emergency seal oil pumps
- Combustion and burner controls
- Miscellaneous controls
- Circuit breaker operations
- Fire protection
- DC emergency lighting
- Annunciations and indicating lights

9.3.5 Essential power system

The essential power system provides power from the "emergency diesel generator" to auxiliary loads required to permit safe shut-down of the unit in the event of plant blackout (i.e., loss of STs and UATs). Essential power is also provided to auxiliaries and services required for personnel safety and minimum plant maintenance during the blackout.

The following major loads are generally connected to the essential power system:

- Lubricating oil pumps of major auxiliaries
- Turbine auxiliary oil pump, turning gear and jacking oil pump
- Emergency lighting
- Selected building ventilation fans
- Selected sump pumps
- Fire pump
- Battery chargers

9.3.6 Uninterruptible power supply (UPS) system

The UPS system furnishes a reliable source of 240 V AC to equipment vital for plant operation and emergency shut-down. This system feeds the following loads:

- Burner management system (BMS)
- Coordinated control system
- TG electro-hydraulic control system
- Turbine supervisory instruments
- Computers
- Public address and plant communication system
- Fire protection system

CHAPTER 10

Automatic Control

10.1 INTRODUCTION

Control is the organization of activities that keep process variables constant over time. When plant operators perform this control it is called *manual process control* [1]. As plants become larger in size and more complicated, more errors occur because of plant operators not understanding the interactive actions among the controls of the plant's various equipment and systems. These interactive actions require so arduous degree of manipulation skill that it outperforms the capabilities of plant operators. In such cases, machines take over the rein and perform these actions effortlessly. Machines relieve operators from many monotonous and tedious actions and perform functions that are beyond the physical abilities of humans [2]. These comprehensive actions by machines are termed as *automatic control* that ensure [3]:

i. Control of processes more rapidly and consistently than human beings can
ii. Elimination of work not requiring human mental effort
iii. Work with a much higher skill
iv. Elimination of human error
v. Efficiency of operation

The function of automatic control systems is to control output variables. They are dynamic systems that change with time and are integrated systems of components. The fundamental tasks an automatic control system should perform are first to reduce the effect of load disturbances and second to guide the system variables to follow the desired set points. There are two concepts that provide the basis for automatic control strategies: feedback or closed-loop control and feed-forward or open-loop control [2]. Feedback control is the more commonly used technique, in which output signal/s in the flow stream are returned to the incoming control path to moderate any deviation of the measured variable from its desired status. This means that in a feedback control system the controlled variable tends to become equal to the desired value regardless of external disturbances. The effect of feedback is to make the controller aware of its control action.

Thermal Power Plant
DOI: http://dx.doi.org/10.1016/B978-0-12-801575-9.00010-X

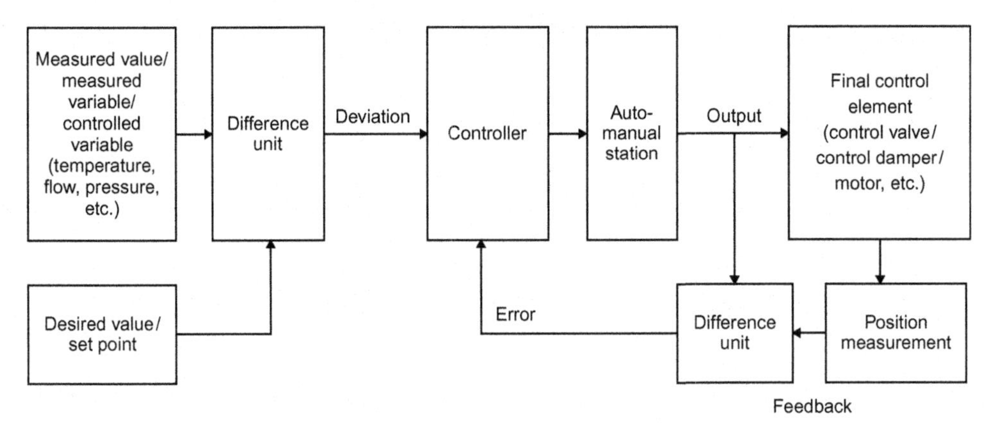

Figure 10.1 Conceptual automatic control diagram. *Source: FIG. 1, Page 16-33, Automatic Control, [4].*

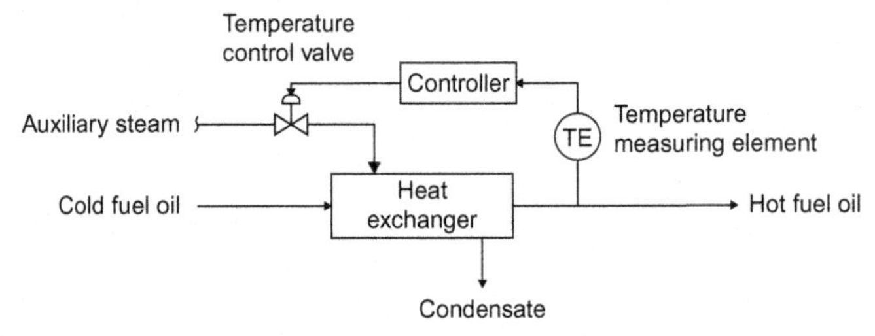

Figure 10.2 Flow diagram fuel oil heating system. *Source: FIG. 1-2, Page 6, Simple Fuel Oil Heating Process, [1].*

Thus, the feedback automatic control system is designed to maintain a desired value of any parameter by employing the following steps (Figure 10.1):

i. Measuring the existing value
ii. Comparing it to the desired or set value
iii. Employing the difference between the measured value and the set value to initiate the action needed to reduce the difference between the demand and output signals to zero

While the response of the feed-forward or open-loop signal is very fast, it's problem is that the output changes as the load changes on the output signal. In an open-loop system the controlled variable is not compared with the set value.

Figure 10.2 and Figure 10.3 show the flow and feedback control loop diagrams, respectively, of a simple fuel oil-heating process.

As the unit size increases the number of control loops becomes enormous. This chapter attempts to cover some of the major control loops pertaining to the steam generator or boiler, steam turbine, and regenerative systems to help both design and field engineers understand the intricacies of these control loops.

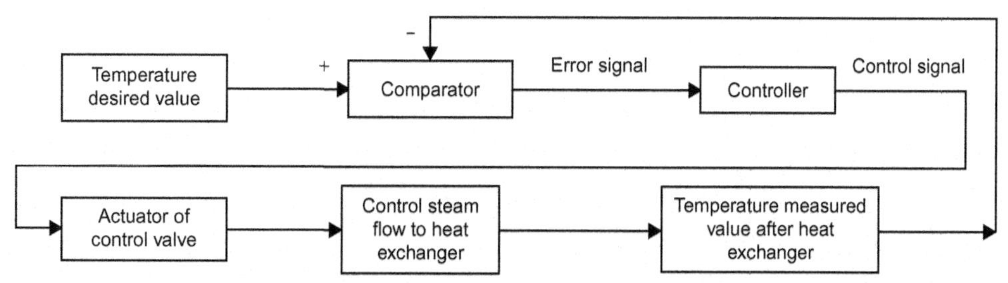

Figure 10.3 Control loop diagram fuel oil heating system. *Source: FIG. 1-3, Page 6, Feedback Control Loop Diagram, [1].*

10.2 AUTOMATIC BOILER CONTROL

The boiler is a utility supplier of the process, and as such must follow the needs of the plant in its demand for steam. The boiler and its control system must therefore be capable of satisfying rapid changes in load.

The independent variable in a boiler that generates saturated steam may be either temperature or pressure (but not both) at boiler outlet. While both temperature and pressure have to be controlled independently in a boiler generating superheated steam [3]. For operation and control of the boiler, monitoring of steam pressure is sufficient to detect any mismatch between fuel-energy input and energy content in the output steam. To obtain the desired outlet conditions of steam (pressure, temperature, flow, etc.), the amount of fuel, air, and water are adjusted. However, control of the boiler is not only limited to adjustment of fuel, air, and water. The overall boiler control system is divided into sub-control systems to meet the given requirements as follows [5]:

1. Provide heat release in accordance with steam demand
2. Proportion fuel to air for optimum combustion efficiency
3. Maintain constant furnace draft
4. Regulate feedwater flow
5. Control final steam temperature
6. Control final steam pressure

Although these sub-control systems appear to be independent, in practice they interact, since feedwater flow affects steam pressure, air flow affects steam temperature, and fuel flow affects drum level and furnace draft, etc. [6]. The integrated control system is shown in Figure 10.4.

10.2.1 Drum level/feedwater control

The function of this control is to regulate the flow of water to maintain the required level in the boiler drum. The control system will vary with the type and capacity

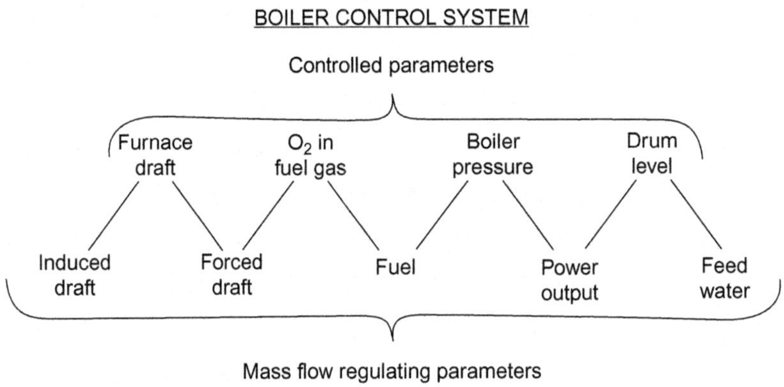

BOILER CONTROL SYSTEM

Controlled parameters

Furnace draft | O_2 in fuel gas | Boiler pressure | Drum level

Induced draft | Forced draft | Fuel | Power output | Feed water

Mass flow regulating parameters

Figure 10.4 Integrated boiler control system [6].

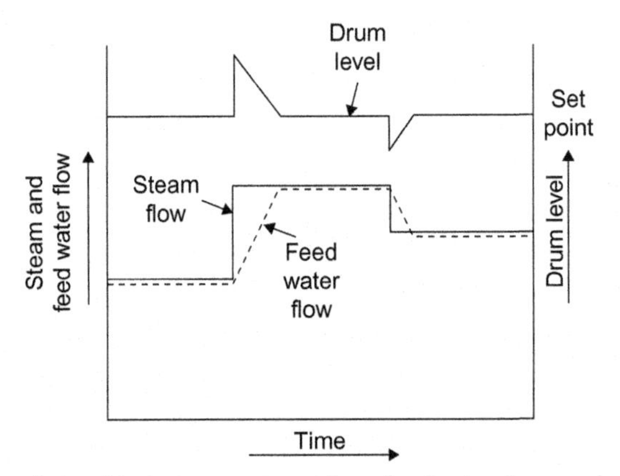

Figure 10.5 Response relationship between steam flow, feedwater flow, and drum level.

of the boiler as well as the characteristics of load. Figure 10.5 shows the response relationship between steam flow, feedwater flow, and drum level.

The drum level is controlled by modulating the feedwater control valves. During low-load operation (about 25–30% boiler maximum continuous rating (BMCR)), the low-load feed control valve is used to control the drum level. When the boiler load exceeds 25–30% BMCR, the control function is changed to the full-load feed control valve.

Single-element control In small boilers with relatively large drum storage volumes (about 120 s of storage capacity (between the normal level and low-low level) at BMCR) and slow-changing loads or during low-load operation of large boilers, feedwater flow is controlled by the drum level alone (Figure 10.6). The measured drum level signal is compared with its set point and the resulting error is fed to a controller to produce the feedwater flow control signal to regulate the low-load feed control valve. This control, however, is self-defeating. In the event of an increase in load, the drum pressure would fall with a consequent drop in density of fluid, and as a result

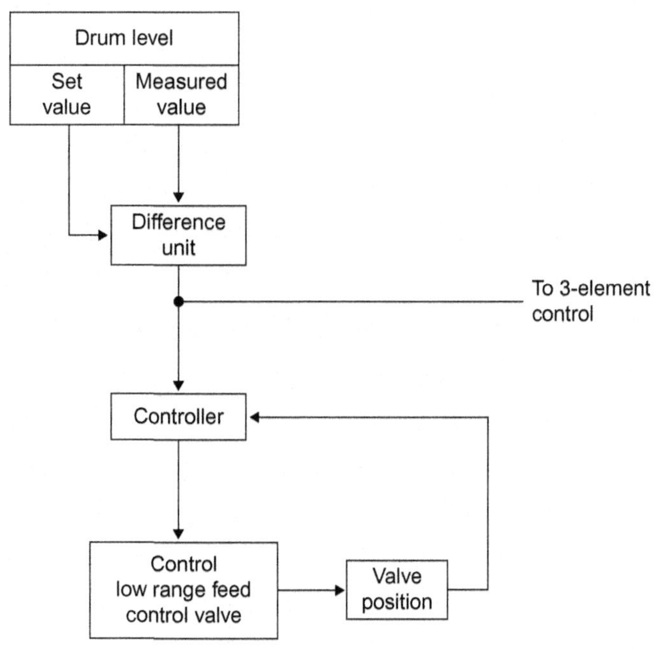

Figure 10.6 Single-element drum level control.

the water level would "swell" or increase, rather than fall, resulting in decrease of feedwater flow, instead of increasing. Conversely, a decrease in load would cause the drum pressure and eventually the density of the fluid to increase, leading to "shrink" or fall in the water level, rather than increase, resulting in an increase of the feedwater flow, instead of a decrease.

Three-element control For high-load operation of large boilers, with reduced drum sizes having reduced storage capacity (about 20–30s storage capacity or even less (between normal level and low-low level) at BMCR) or with severely fluctuating loads, a three-element control is used. The three elements are "steam flow," "feedwater flow," and "drum level." In a typical control system, the measured drum level signal is compared with its set point and the resulting error is summed with the pressure and temperature compensated steam flow signal to establish the demand signal for the feedwater flow. This demand signal is compared with the measured value of the temperature-compensated feedwater flow. The resulting error is fed to a controller to produce the feedwater flow control signal to regulate the full-load feed control valve. Further, since the drum storage capacity in this case is low, the effect of "swell" and "shrink" is severe. Hence, the necessary provision to take care of the "swell" and "shrink" under transient load change conditions is built into the control system as shown in the top-right column blocks of Figure 10.7 marked with an arrow pointing left.

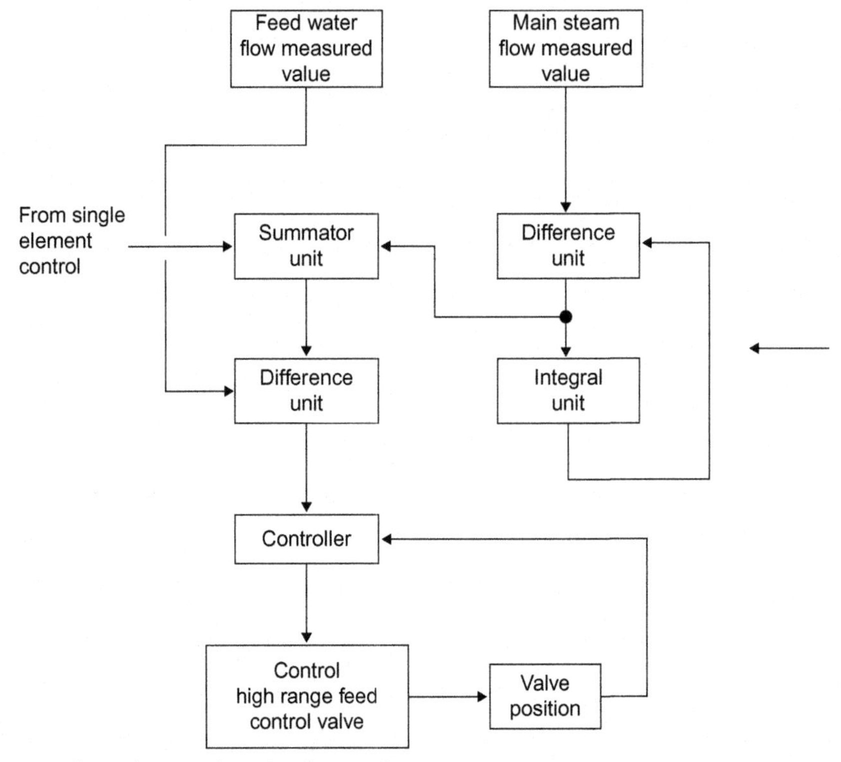

Figure 10.7 Three-element drum level control.

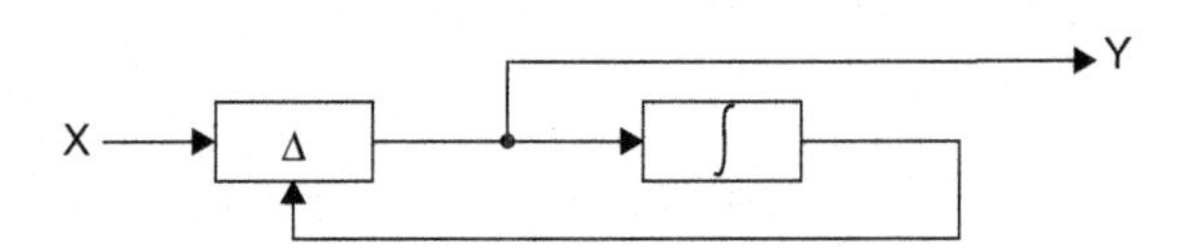

Where, 'X' represents 'main stem flow measured value'
and 'Y' represents input to summator unit of fig.10.7

Figure 10.8 Mathematical interpretation of swell and shrink of drum level.

The mathematical interpretation of the top-right column blocks of Figure 10.7 is as follows, taking into account the concept presented in Figure 10.8:

$$y = x - \int y dt \tag{10.1}$$

where

x: main steam flow measured value
y: input to "summator unit" of Figure 10.7

From Eq. 10.1, we get

$$y + \int y dt = x \tag{10.2}$$

The Laplace transformation of Eq. 10.2 results in

$$X = Y + \frac{1}{S}Y = \frac{1+S}{S}Y \tag{10.3}$$

or,

$$Y = \frac{S}{1+S}X = S * \left(1 - S + \frac{S^2}{2} - \ldots.\right) * X = \left(S - S^2 + \frac{S^3}{2} - \ldots.\right) * X \tag{10.4}$$

The inverse Laplace transformation of Eq. 10.4 yields

$$y = \frac{dx}{dt} - \frac{d^2x}{dt^2} + \frac{1}{2}\frac{d^3x}{dt^3} - \ldots \tag{10.5}$$

From Eq. 10.4 it may be concluded that any change in main steam flow would cause a delayed change in feed control valve position. (Note that the measured value of "fluid" flow in a plant is realized in "volume," while for all practical purposes "flow" has to be expressed as "mass." Hence, volume flow needs to be corrected with density. Density correction for a gas is accomplished with "pressure and temperature compensation," for liquid, however, only "temperature compensation" would suffice.)

10.2.2 Superheat/Reheat steam temperature control

For smooth, efficient plant operation, it is essential to control the steam generator outlet steam temperature within narrow bands. While a reduction in steam temperature from design results in loss of plant efficiency, an increase in steam temperature may result in overheating and failure of superheater and reheater tubes as well as turbine blades.

In practice, even at rated load it is difficult to maintain the rated steam temperature. As a result a ± 5 K deviation from the rated steam temperature is recommended in the ABMA (American Boiler Manufacturers Association) boiler specification. In a boiler the steam temperature may fluctuate because of a change in any of the variables: *load, excess air, burner operation, feedwater temperature, fuel quality, heating surface cleanliness, blow-down saturated steam bleed, etc.* Hence, in order to control steam temperature, it is necessary to handle load changes and operating variables by one or more of the following methods:

Excess air control Increasing excess air increases flue-gas velocity over tubes, which in turn raises heat transfer to the convective heat transfer surfaces. As a result, both the superheat and reheat steam temperature will increase. An increase in excess air or flame length reduces the radiant heat transfer but increases the

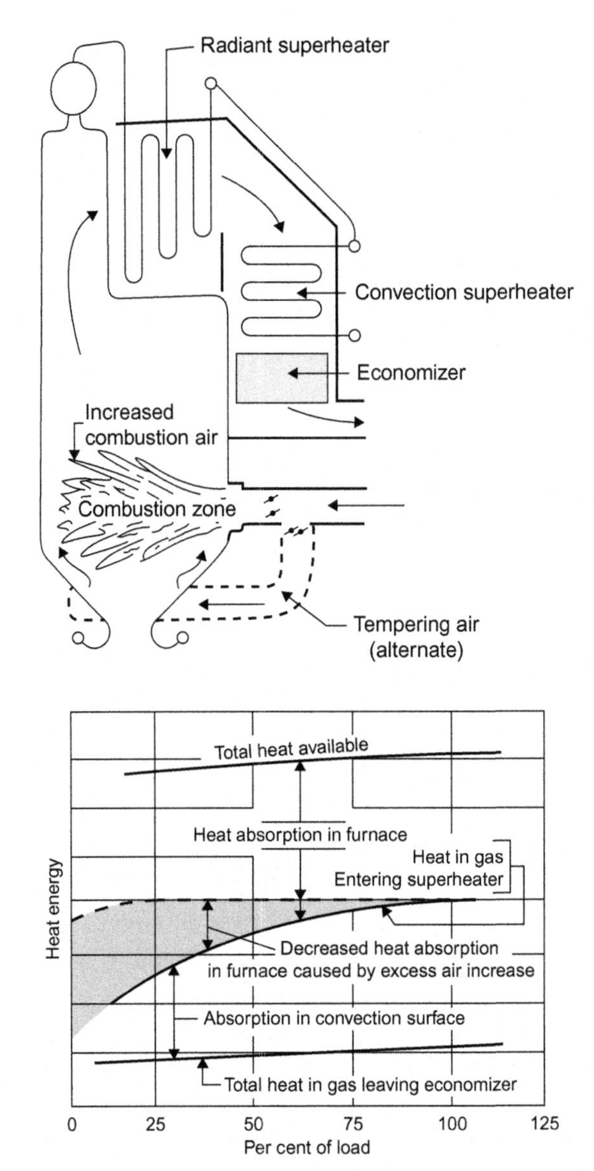

Figure 10.9 Steam temperature control by increased excess air. *Source: Fig. 16.13, P261. CH16: Superheaters, Reheaters and Attemperators [7].*

convective heat transfer. A reduction in excess air or flame length has the opposite effect. Increasing excess air, however, also increases exit gas temperature, which reduces the boiler efficiency. Hence, excess air is not used as a principal steam temperature control method. Figure 10.9 shows superheat steam temperature control by increased excess air.

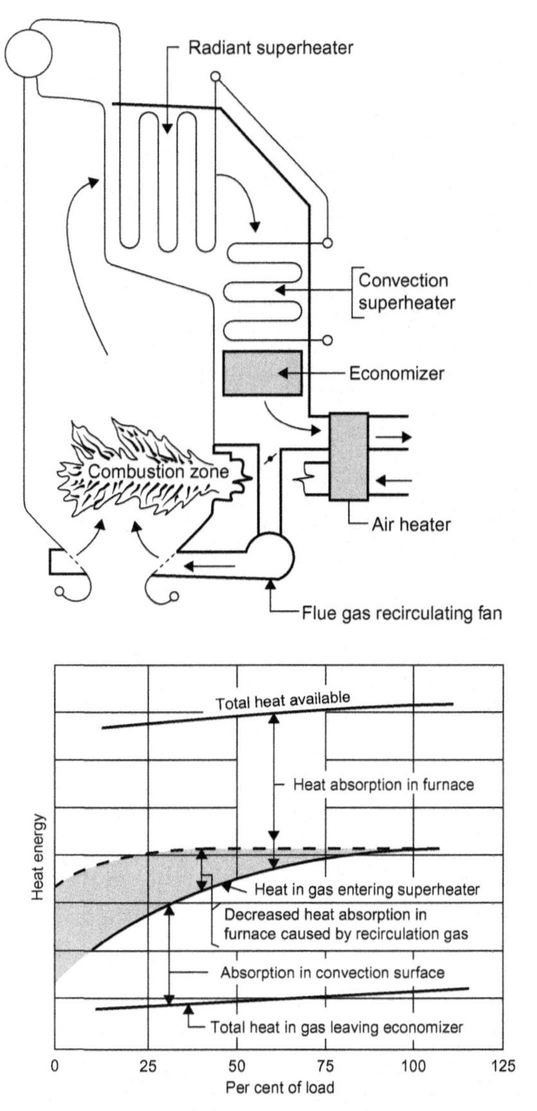

Figure 10.10 Steam temperature control by gas recirculation *Source: Fig. 16.14, P261. CH16: Superheaters, Reheaters and Attemperators [7].*

Gas recirculation control In this control some portion of flue gas is drawn from the low-temperature flue-gas zone and returned to the furnace to increase the convective heat transfer (Figure 10.10). Thus, steam temperature can be raised at lower loads. Recirculation of flue gas has the same effect on control steam temperature as does an increase in excess air.

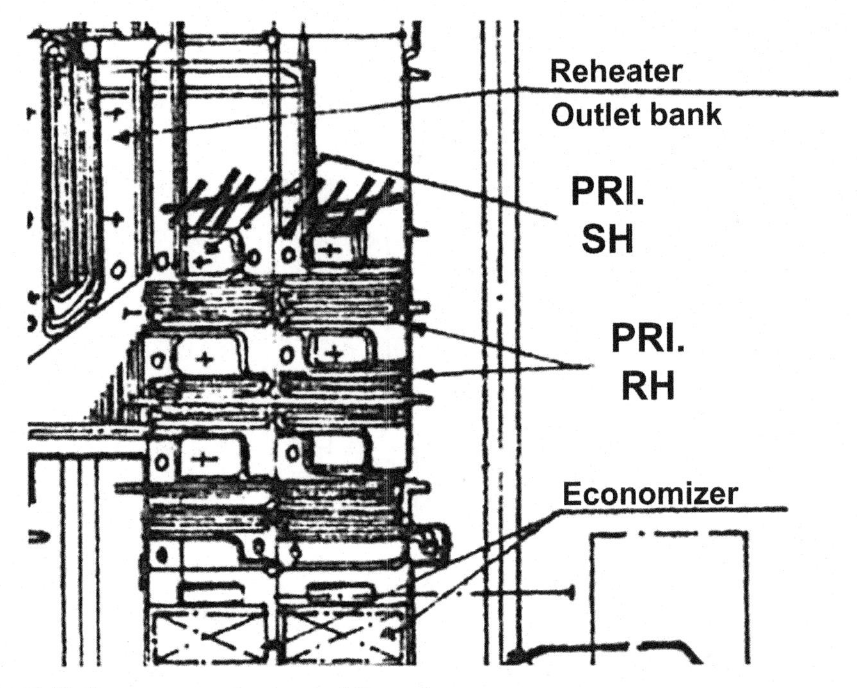

Figure 10.11 Steam temperature control by gas bypass.

Gas bypass control Steam temperature control by gas bypass is adopted in those boilers where boiler convection banks are arranged in the second pass in such a manner that the front parallel convection pass passage contains a reheat transfer surface, and the rear pass contains a superheat heat transfer surface (Figure 10.11). Each heat transfer surface is provided with gas bypass proportioning dampers.

When the reheat steam temperature becomes too low, the flue-gas flow through the reheat pass is increased by opening the reheat bypass damper further, consequently the flue-gas flow through the superheat pass reduces. Conversely, if the reheat steam temperature is too high, the reheat bypass proportioning dampers close, reducing the flue-gas flow through the reheat pass and more heat is then transferred to the superheater, where de-superheating spray water flow is increased to restrict the superheat steam temperature increase.

Adjustable burner control The flue-gas temperature at the furnace outlet could be maintained at varying steam generation by changing the relative furnace volume and water wall heat-absorbing surface by using either tilting burners or biased firing. Control of steam temperature with tilting burners is normally used in corner-fired boilers. With a change in burner tilt, the elevation of the fireball within the furnace changes, which in turn changes the heat absorption in the furnace with a consequent change in the quantity of heat delivered to the convective superheater and convective

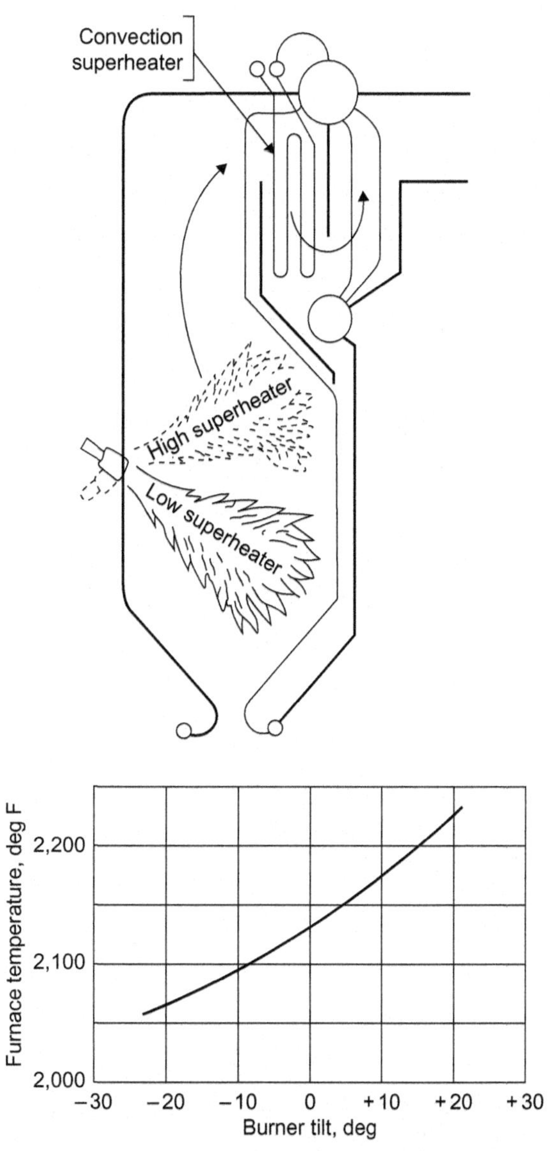

Figure 10.12 Steam temperature control by burner tilt. *Source: Fig. 16.17, P263. CH16: Superheaters, Reheaters and Attemperators [7].*

reheater. To raise the steam temperature, burners are tilted upward, and to reduce the steam temperature burners are tilted downward. Figure 10.12 shows the control of superheat/reheat steam temperature by burner tilt.

In a boiler where burners are located over the furnace wall, *biased firing* is adopted to control steam temperature (Figure 10.13). The method of steam temperature

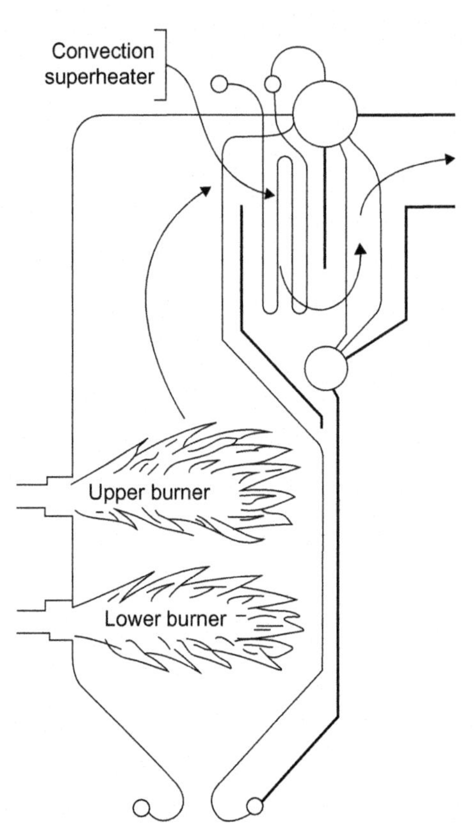

Figure 10.13 Steam temperature control by biased firing. *Source: Fig. 16.16, P262. CH16: Superheaters, Reheaters and Attemperators [7].*

control by biased firing is similar to that effected by tilting burners. To raise the steam temperature the upper-level burners are preferentially used, while to reduce the steam temperature the lower level burners are put into service.

Control of combination superheaters The control of a combination radiant-convective superheater is relatively simple because of its compensating characteristics (Figure 10.14). With an increase in steam demand, both fuel and air-flow increase, resulting in an increase in flue-gas flow. As a result, convective heat transfer coefficients increase. Thus, steam receives greater heat transfer per unit mass flow rate and the steam temperature increases with load.

The radiation heat transfer is proportional to the flame temperature, which is not strongly dependent on load. Thus, an increase in steam flow due to an increased load demand would result in a reduction on exit steam temperature, the opposite effect of a convection superheater.

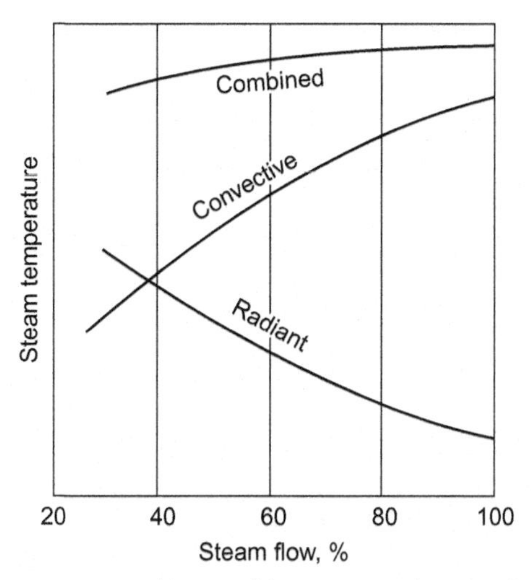

Figure 10.14 Steam temperature control by combination superheater. *Source: Fig. 44, Page 9-37, [4].*

Control of separately fired superheaters This control is not generally economical for a large utility boiler and is usually used in the chemical process industry. It is adopted in boilers having a separately fired superheater that permits an operating rate independent of the saturated steam boiler (Figure 10.15). The superheater is provided with its own burner, fans, combustion chamber, controls, etc. To achieve a reasonably flat steam temperature-load curve the firing rate is adjusted commensurate with a change in load.

Attemperation Attemperation is the process of reducing the steam temperature by removing heat from the superheated steam with the help of a desuperheating apparatus called an attemperator. There are two types of attemperators: surface attemperator (also known as a non-contact attemperator) and a spray or direct contact attemperator. The surface attemperator may further be classified as a header type, shell tube type, or a submerged or drum type.

Header type In this type the superheater inlet header is enlarged to accommodate the heat exchanger tubes carrying cool water (Figure 10.16). The header type attemperator has a limited control range because of the restricted diameter of the headers.

Shell tube type Steam temperature control by using a shell and tube type attemperator is sometimes adopted in relatively small-sized boilers from economic consideration. The control range of the shell tube type attemperator is limited. In this type, the shell and tube heat exchanger is fitted into a pressure vessel (Figure 10.17). A portion of the steam is taken from the primary superheater outlet by a control valve

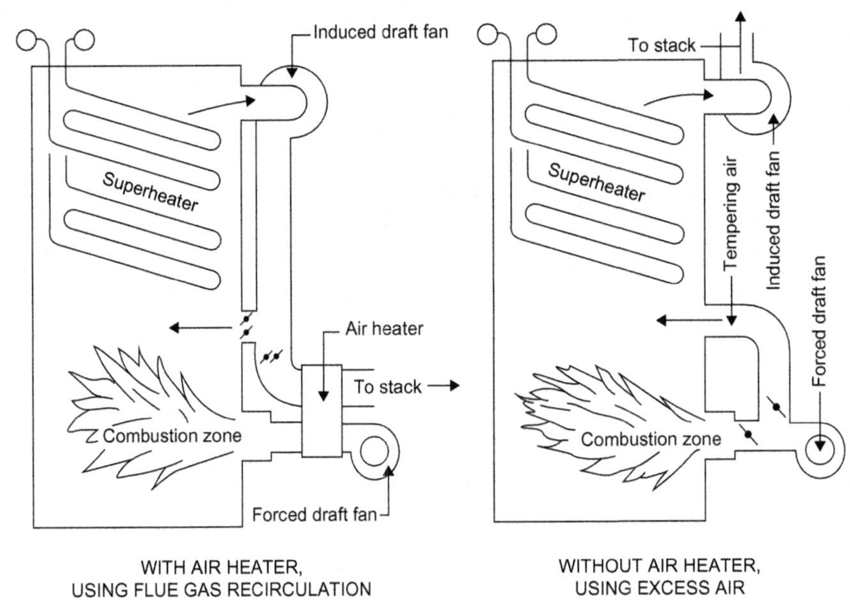

Figure 10.15 Steam temperature control by separately fired superheater *Source: Fig. 16.19, P264. CH16: Superheaters, Reheaters and Attemperators [7].*

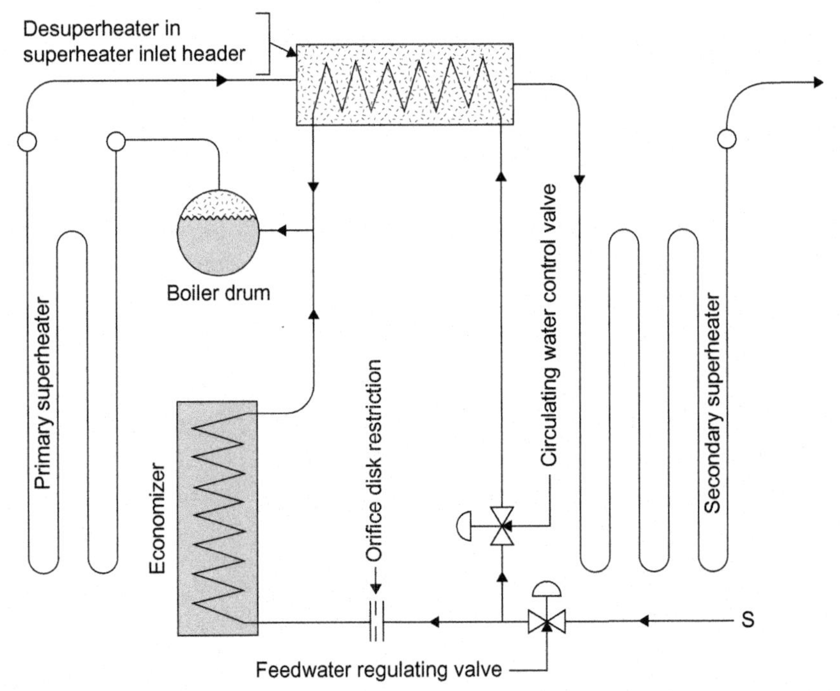

Figure 10.16 Header type attemperation. *Source: Fig. 16.23, P267. CH16: Superheaters, Reheaters and Attemperators [7].*

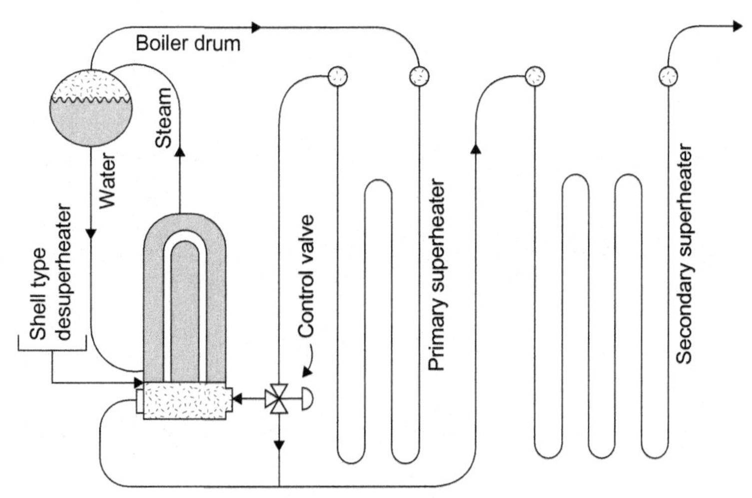

Figure 10.17 Shell and tube type attemperation. *Source: Fig. 16.25, P267. CH16: Superheaters, Reheaters and Attemperators [7].*

and it mixes with boiler water. Thus, a portion of the heat from the steam is transferred to the boiler water in a heat exchanger. The cooler steam coming from this heat exchanger remixes with the primary steam prior to entering the secondary superheater. Thus, the steam temperature is maintained by controlling the quantity of diverted steam.

Submerged type This type is similar to the shell tube type, except that the heat exchange between the diverted steam and boiler water takes place within the boiler drum (Figure 10.18).

Spray or direct contact attemperator In this type of attemperator the steam temperature is reduced by spraying low-temperature water from either high-pressure feedwater heater outlet or boiler-feed pump discharge header into the line between the low-temperature superheater outlet and the high-temperature superheater inlet (Figure 10.19). The spray nozzle injects water into the throat of a mixing venturi, where water mixes with high-velocity steam in the throat, vaporizes, and cools the steam (Figure 10.20). The spray type attemperator is by far the most satisfactory one, since it provides a rapid and sensitive means of steam temperature control. The effect of the spray attemperator is shown in Figure 10.21.

10.2.3 Combustion control

The function of a combustion control system is to control fuel and air input or firing rate to the furnace to maintain the superheat steam pressure at the turbine inlet at the desired value as dictated by load demand. However, it should be ensured that

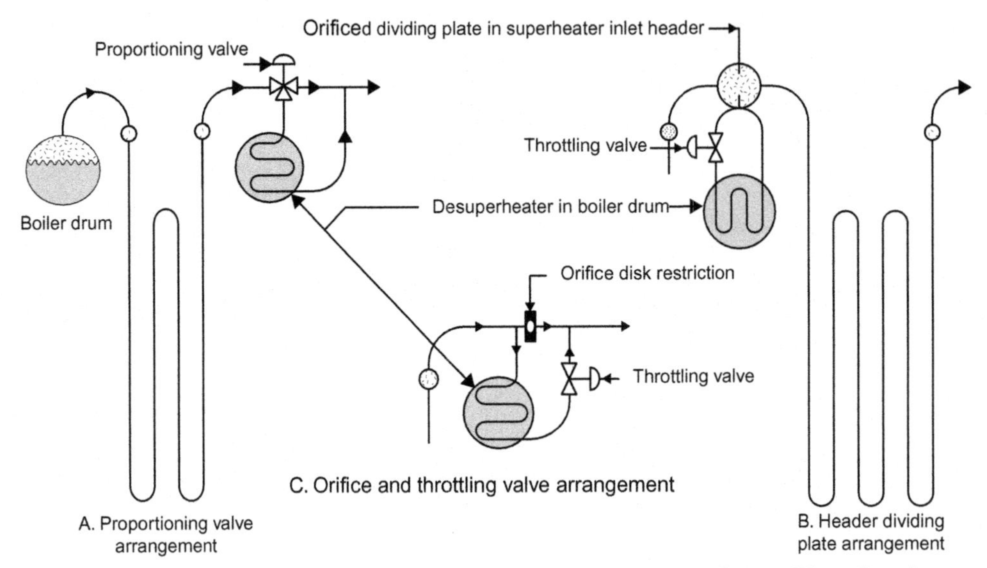

Figure 10.18 Submerged type attemperation. *Source: Fig. 16.27, P268. CH16: Superheaters, Reheaters and Attemperators [7].*

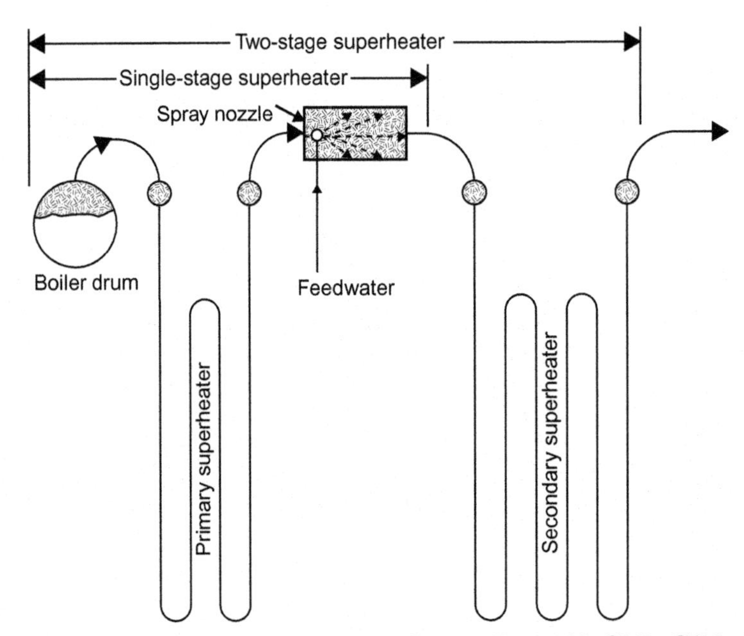

Figure 10.19 Direct contact type attemperation. *Source: Fig. 16.29, P269. CH16: Superheaters, Reheaters and Attemperators [7].*

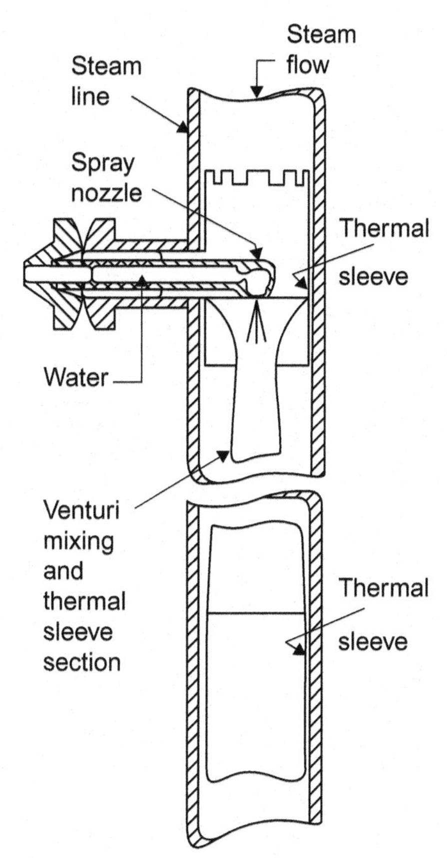

Figure 10.20 Spray nozzle. *Source: Fig. 46, Page 9-39, [4].*

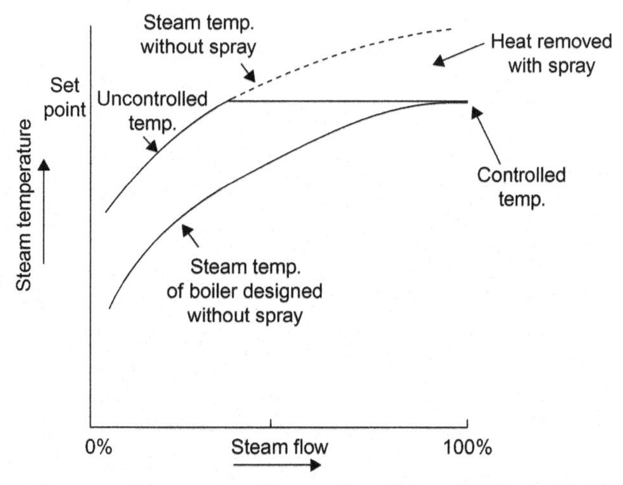

Figure 10.21 Effect of spray attemperator. *Source: Fig. 10.6qq, P1418. Article10.6: Control of Steam Boilers. CH X: Process Control Systems [8].*

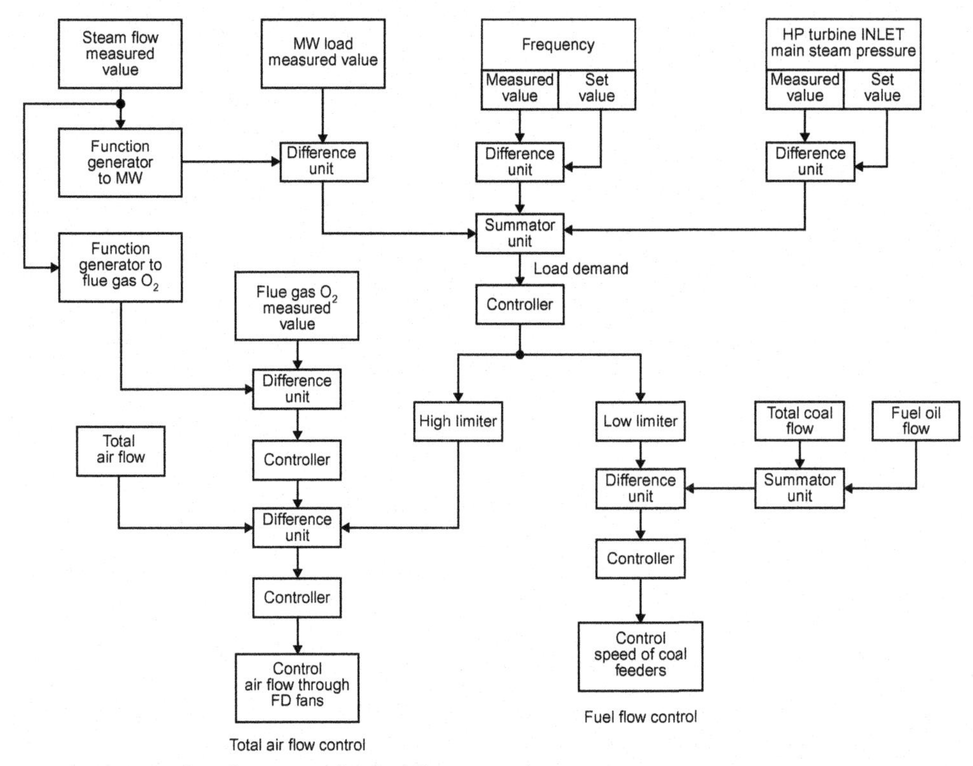

Figure 10.22 Total air flow control & fuel flow control.

from a flame stability point of view that the flow of combustion air to the furnace does not fall below a minimum limit. It should also be ensured that for safety the fuel addition be limited by the amount of available combustion air. The combustion control system comprises two subsystems: fuel flow control and total air-flow control (Figure 10.22).

The megawatt load signal is compared with a function of steam flow signal; the resultant error signal is then corrected by the system frequency deviation to generate the operating level of the system. The measured throttle (HP turbine inlet) pressure is compared with its set value. The throttle pressure error signal, corrected by the feed-forward MW error signal, is the load demand signal for subsequent fuel flow control and the total air-flow control. For safety an interlock is provided to ensure "air leading fuel on load increase" and "air lagging fuel on load decrease."

Fuel flow control The primary boiler fuels are coal, oil, and gas. Oil and gas involve the simplest controls, since they can be measured easily and flow is regulated by a control valve in the fuel line. On the other hand, since coal is an unmeasured fuel, the control system is open-loop, in which a fuel demand signal positions a coal-feeding device (speed variator of coal feeder) directly.

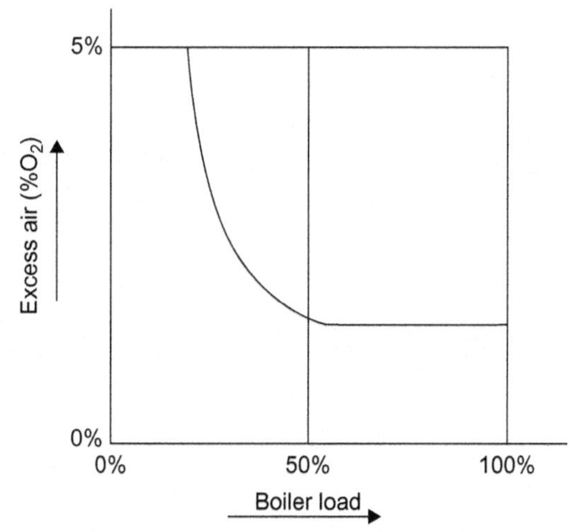

Figure 10.23 Excess air vs boiler load.

Total air-flow control To maintain proper fuel-air ratio, air flow should also be controlled per the load demand, either by varying the fan speed, by throttling fan inlet vane/damper/blade pitch, or by a combination of damper and speed control. The firing rate demand should be limited by the fuel flow to ensure an adequate air supply for safe combustion of fuel.

To ensure complete combustion of all the fuel, it is always necessary to supply more air than is required by proper fuel-air ratio, which may change due to variations in fuel calorific value, specific gravity, temperature, and other physical properties of either fuel or air. In most boilers, the desired excess air is not constant, but decreases as load is increased (Figure 10.23). Hence, it is essential to adjust the fuel-air ratio automatically. This is done by programming the %O_2 set point as a function of boiler load.

Maintaining the correct fuel-air ratio results in fuel savings and limits the following for safety purposes:

i. Fuel to available air flow
ii. Minimum air flow to match the minimum fuel flow or to other safe minimum limit (air flow should never be allowed to fall below 30%)
iii. Minimum fuel flow to maintain stable flame

10.2.4 Furnace pressure/draft control

To prevent hot gas leakage in balanced draft boilers, it is essential to maintain the furnace pressure (draft) to below the atmospheric pressure. This sub-atmospheric pressure should not be low enough to produce heat loss through air infiltration. Hence, in addition to air-flow control, the draft in the furnace is also controlled by varying either the fan speed, by throttling the inlet vane, or by a combination of these two (Figure 10.24).

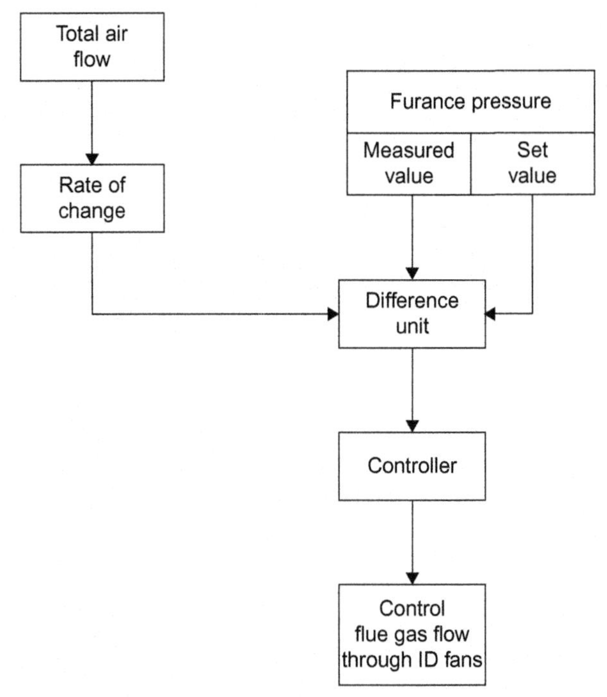

Figure 10.24 Furnace draft control.

10.2.5 Pulverizer outlet coal-air temperature and primary air-flow control

The main function of this control is to ensure efficient furnace performance, which is achieved by maintaining constant primary air flow through the pulverizer as well as by maintaining constant coal-air temperature at the pulverizer outlet. This control is ensured by modulating hot and cold primary air-flow dampers ahead of pulverizer (Figure 10.25). When the mass-flow rate of air is lower, both cold-air and hot-air dampers open to increase the mass-flow rate, keeping the outlet temperature at the set value. When the outlet temperature increases, the hot-air damper should close and the cold-air damper should open proportionately to bring down the temperature of the coal-air mixture at the pulverizer outlet, keeping the mass flow of air through the pulverizer the same (Figure 10.26).

10.3 AUTOMATIC CONTROL OF TURBINE

In Chapter 6, *Steam Turbines*, we learned that the speed and power output of a turbine are controlled by varying the steam input to the turbine. The steam input in turn is changed by modulating the opening of turbine inlet control valves, which are also

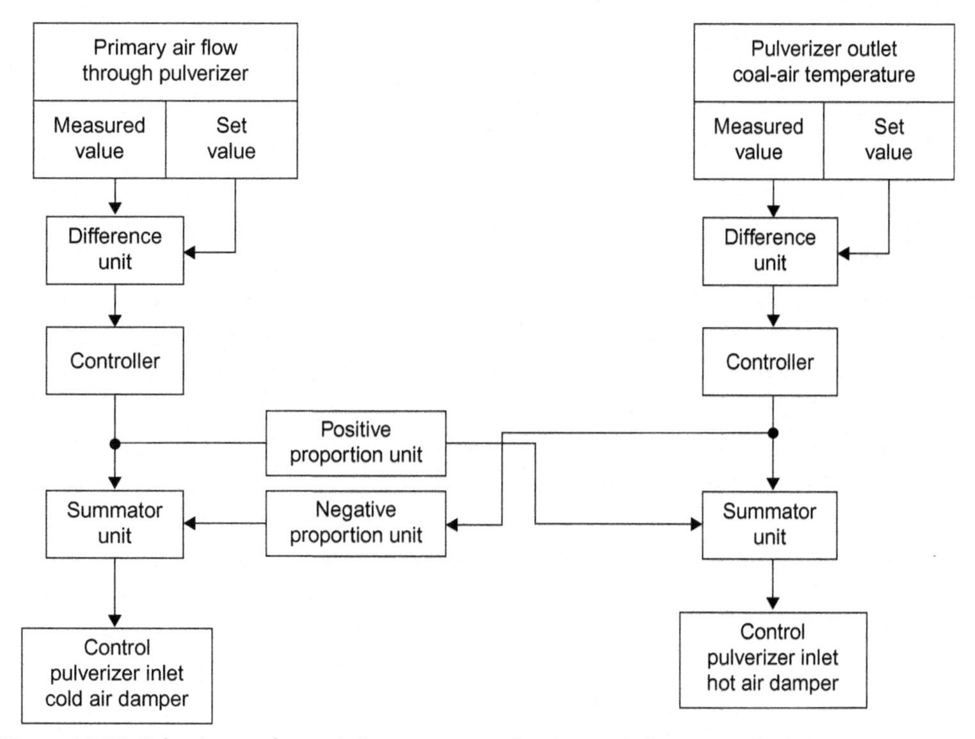

Figure 10.25 Pulverizer outlet coal-air temperature & primary air flow control.

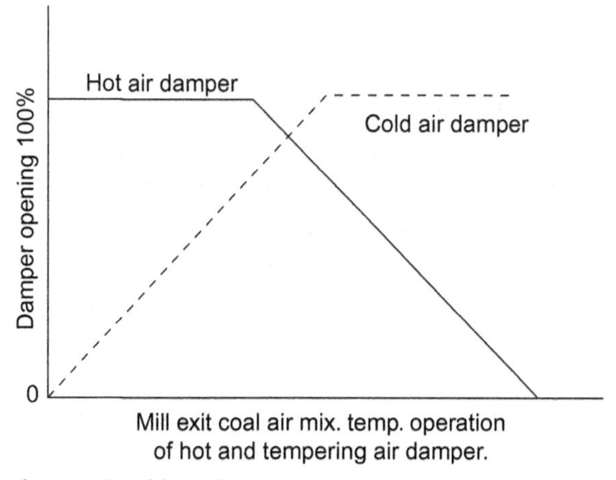

Figure 10.26 Hot air damper & cold air damper opening.

known as throttle valves. The opening of throttle valves is influenced by the turbine-governing system, per the steam flow demand. The turbine-governing system is comprised of three control loops: speed control loop, load control loop, and initial pressure control loop as shown in Figure 10.27. These loops are discussed in the following sections.

10.3.1 Speed controller

The speed controller compares the speed reference value with the actual speed of the turbine and accordingly provides an output for the throttle valve lift. The reference speed signal ensures that the highest permissible rate of speed is consistent with the safe operation of the turbine. The speed controller handles the following operations:

i. Start-up of the turbine
ii. Synchronization of the generator
iii. Minimum load operation
iv. Operation of the turbine in the complete power range when the load controller is out of service

10.3.2 Load controller

There are three elements, as follows, in this control loop:

i. Load limiter
ii. Frequency controller
iii. Load controller

The load controller receives the reference load signal from the load limiter and also from the frequency controller, depending on the frequency of grid. The reference load set value ensures that the highest permissible rate of load variation is consistent with the safe thermal stresses of the turbine, both during load increase as well as load decrease.

10.3.3 Initial steam pressure controller

In the event initial steam pressure at the turbine inlet falls below a preset value the initial steam pressure controller initiates the unloading of the turbine proportionally commensurate with the available boiler pressure. The loading of turbine begins only if the firing rate in boiler is increased.

10.4 COORDINATED CONTROL SYSTEM

The main function of the coordinated control system is to operate the steam generator and the steam turbine generator as an integrated (coordinated) unit to maintain all controllable parameters, as discussed in Section 10.2 and Section 10.3 above, within

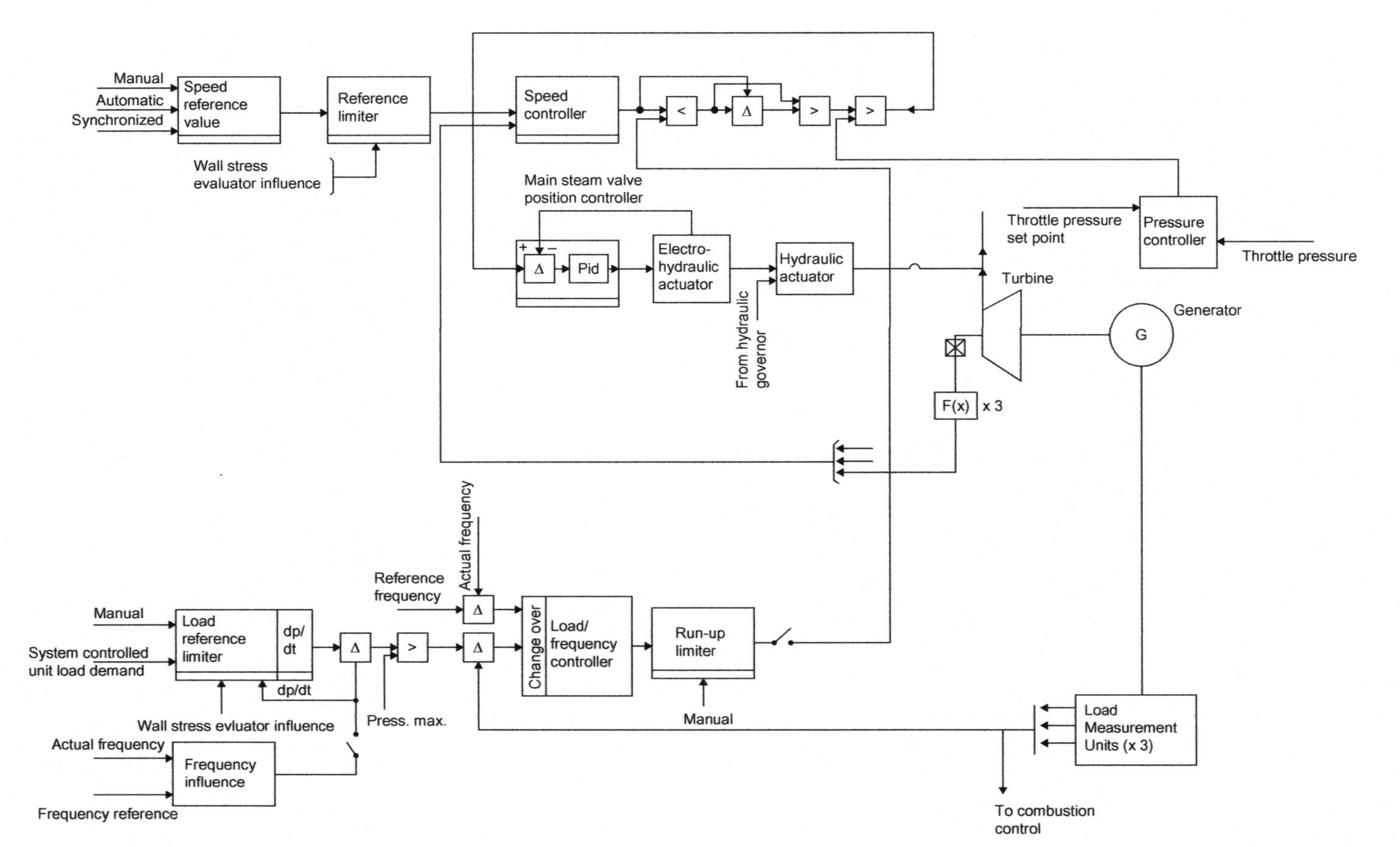

Figure 10.27 Turbine speed and load control loop.

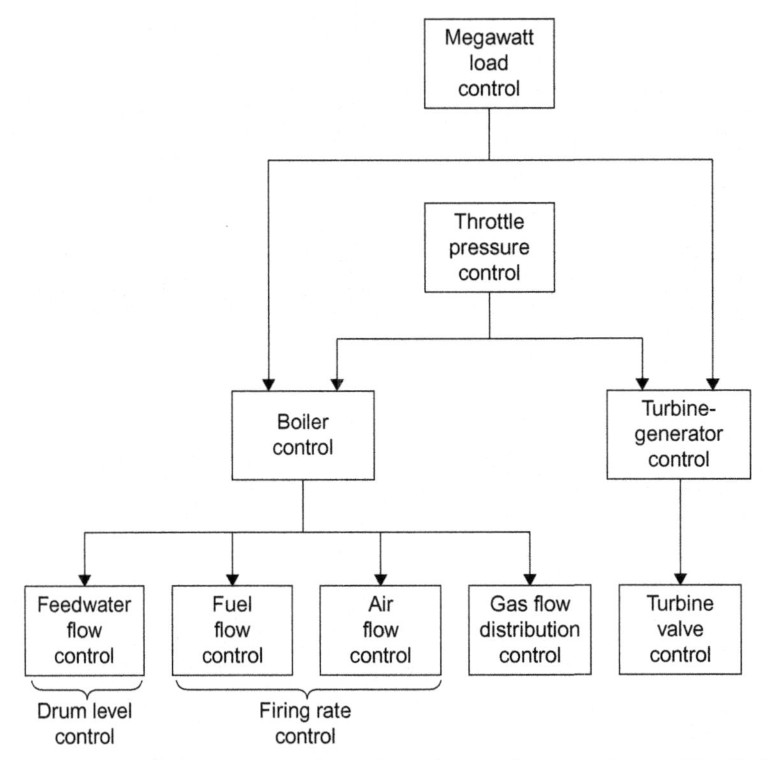

Figure 10.28 Basic control arrangement of coordinated control system. *Source: Fig. 14, P41-6. CH41. Controls for Fossil-Fuel Fired Steam Generating Plants [9].*

their preset limits. In this system, signals from the MW load demand and throttle steam pressure are transmitted to both the boiler master controller and the turbine-governing valve controller simultaneously (Figure 10.28). The boiler master controller transmits a single command to all boiler control loops discussed in Section 10.2 so that each controllable parameter adjusts itself within its limit at the same time.

Since the response time of the turbine-governing system is almost instantaneous, while the response time of the boiler combustion control is sluggish, the coordinated control system harnesses the advantage of the faster response to adjust any deviation in load demand at the quickest using the stored energy in the steam generator to maintain the throttle steam pressure. This mode of control takes advantage of the relative benefits of both "boiler following" mode and "turbine following" mode, as discussed in the following.

In the event that it is difficult to put a coordinated control system in service, the unit then operates in "boiler following" mode, which literally means the boiler is following the turbine. In this mode, any change in MW load demand is adjusted immediately by the turbine-governing valves. Any change in the position of the

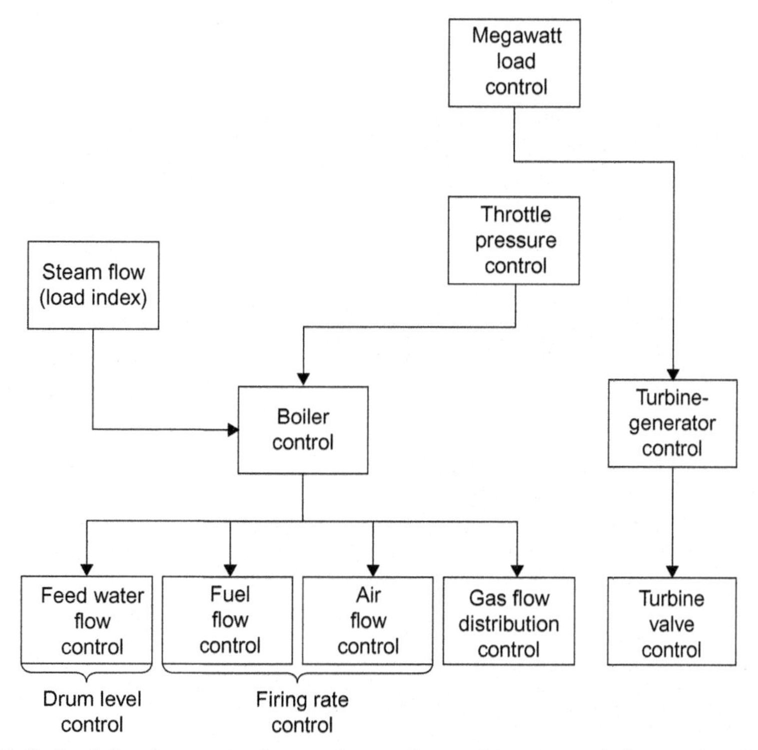

Figure 10.29 Boiler-following mode. *Source: Fig. 12, P41-5. CH41. Controls for Fossil-Fuel Fired Steam Generating Plants [9].*

governing valves causes a change in the throttle steam pressure, which in turn controls firing rate in the boiler by the combustion control system followed by other boiler control loops according to the new load (Figure 10.29). Thus, the new load is established and the throttle steam pressure is brought to its rated value.

If there is any failures in the operating turbine integral systems or any of its associated auxiliaries, "the change in MW load demand" signal is issued to the combustion control system first, instead of the turbine governor. Due to the change in firing rate, the throttle steam pressure will change, which is then taken care of by the turbine-governing system to restore the throttle steam pressure. This mode of operation is called the "turbine following" mode, i.e., the turbine is following the boiler (Figure 10.30). However, the response of this mode is slow, because turbine-governing system has to wait to take the corrective action until the firing control changes the energy output from the boiler. The advantage of this mode is that there is least variation in the throttle steam pressure and throttle steam temperature, among all the three modes discussed above.

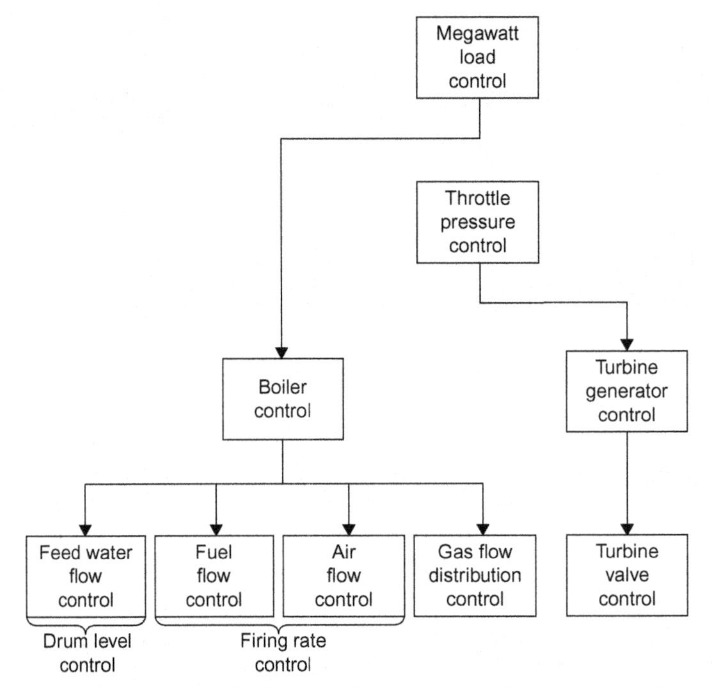

Figure 10.30 Turbine-following mode. *Source: Fig. 13, P41-6. CH41. Controls for Fossil-Fuel Fired Steam Generating Plants [9].*

10.5 AUTOMATIC CONTROL OF REGENERATIVE SYSTEM

The regenerative system in a utility power plant is the go-between that allows operation of the turbine and boiler as a single unit. The major control loops of this system are discussed in the following [5].

10.5.1 Feedwater heater level control (Figure 10.31)

The normal level in the feedwater heater is controlled by cascading the drain from the higher-pressure heater to the next lower-pressure heater with the help of the "normal level control valve." In the event that the level in the higher-pressure heater becomes high, due to a minor heater tube leak and/or malfunction of the normal level control valve, the heater drain is taken to the condenser through the high, also called alternate, level control valve.

If level in the heater still increases further and reaches the high-high level the following occurs:

i. Associated extraction line isolating valve and non-return valve are closed
ii. Drain valve on associated extraction line is opened
iii. Affected feedwater heater bypass valve is opened
iv. Affected feedwater heater water line inlet and outlet valves are closed

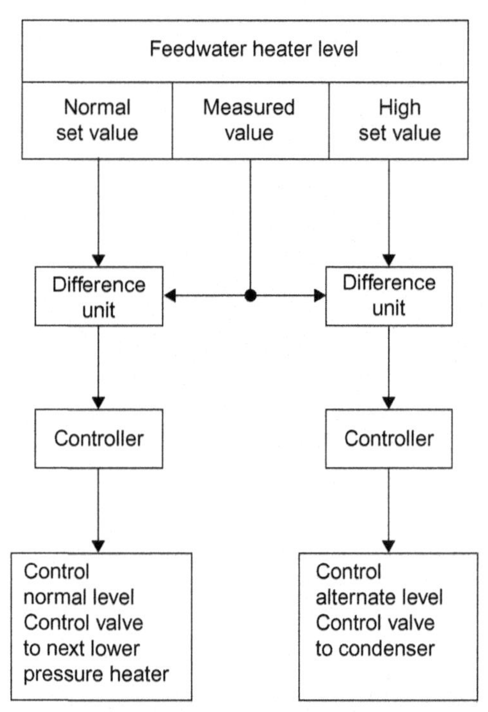

Figure 10.31 Feedwater heater level control loop.

10.5.2 Deaerator level control (Figure 10.32)

Three elements, i.e., "feedwater flow," "condensate flow," and "deaerator level," maintain the level in the deaerator. The feedwater flow signal anticipates the change in the deaerator level, while the condensate flow signal is a feedback signal that trims the control signal to maintain the desired deaerator level. The normal level in the deaerator is controlled by positioning the "deaerator level control valve," located on the condensate line leading to the deaerator. When the deaerator level increases to a high set position, the "deaerator overflow control valve" is opened to dump excess condensate.

In the event that the deaerator level becomes high-high, the signal is initiated to close the following valves:

i. Associated extraction line isolation valve and non-return valve/s
ii. Deaerator level control valve
iii. Steam supply valves from auxiliary steam header and cold reheat line

10.5.3 Condenser hotwell level control (Figure 10.33)

The level in the condenser hotwell is maintained through the "make-up control valve," also known as the "gravity make-up control valve," by supplying the de-mineralized water (DM) water to the hotwell through gravity. If this control valve fails to maintain

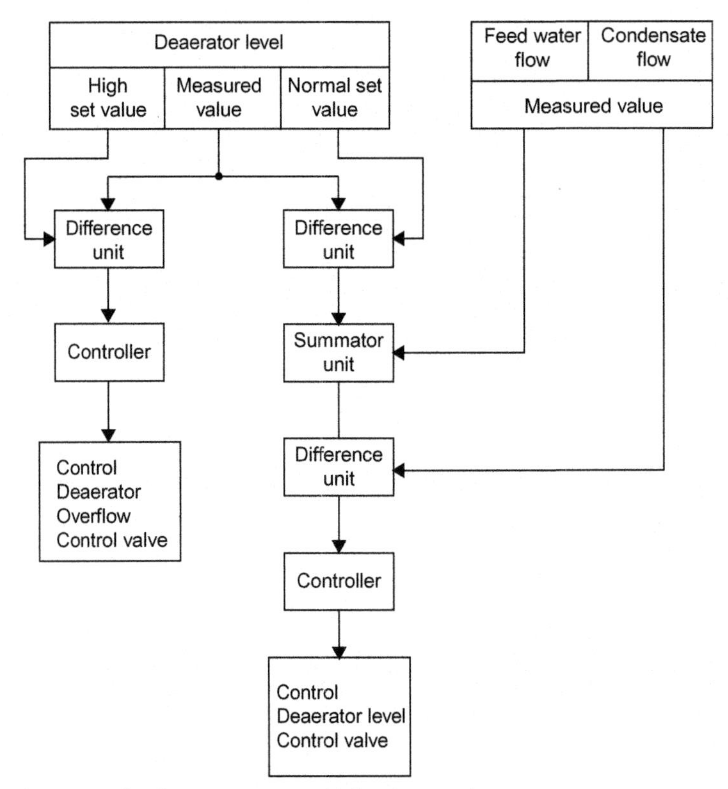

Figure 10.32 Deaerator feedwater storage tank level control.

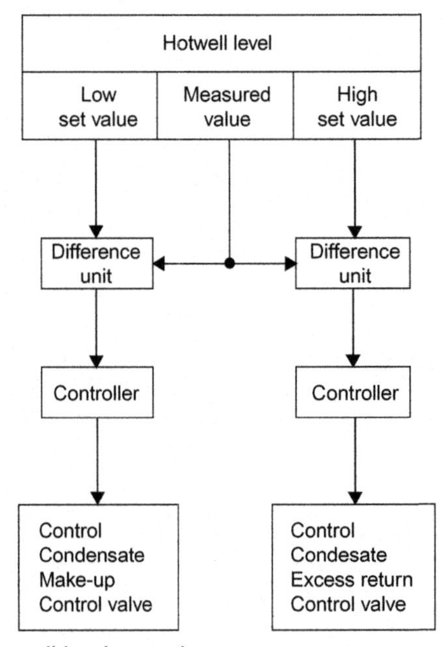

Figure 10.33 Condenser hotwell level control.

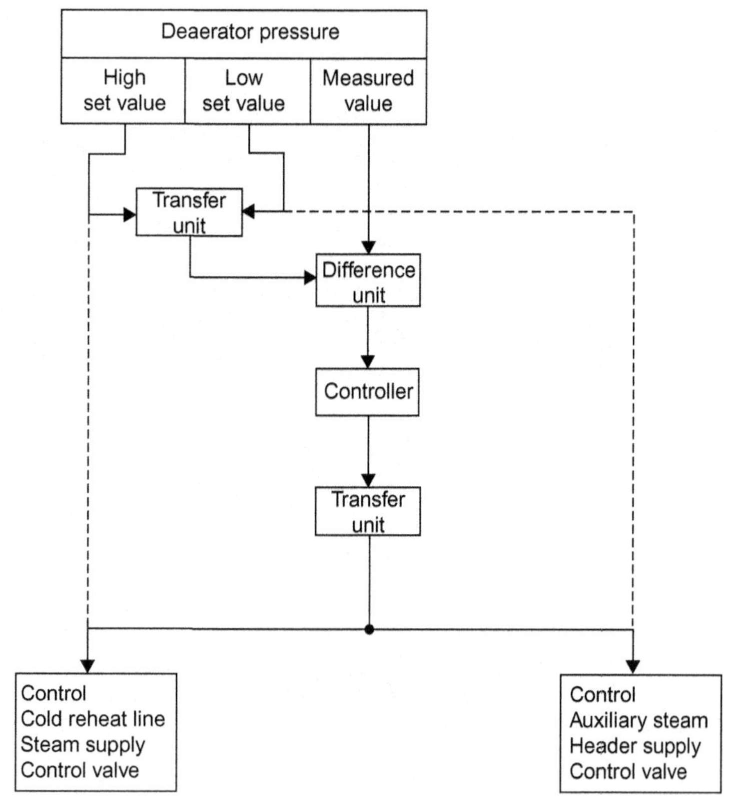

Figure 10.34 Deaerator pressure control.

the normal level, the pressurized DM water is pumped to the hotwell through an "emergency make-up control valve." If the hotwell level increases and reaches a high level, the "condensate excess return control valve" is modulated to dump the excess condensate from the condensate line. All running "condensate extraction pumps" should be tripped whenever the hotwell level becomes low-low.

10.5.4 Deaerator pressure control (Figure 10.34)

The pressure in the deaerator varies according to the turbine extraction steam pressure from a particular minimum load up to the full load. During start-up, steam to the deaerator is usually supplied from an auxiliary steam header and deaerator pressure is maintained at a pre-assigned minimum constant pressure, typically about 0.15 MPa. Once the machine is loaded, the pressure in the cold reheat line builds up. As this pressure becomes adequate, the steam supply to the deaerator is switched over from the auxiliary steam header to the cold reheat line to maintain another pre-assigned

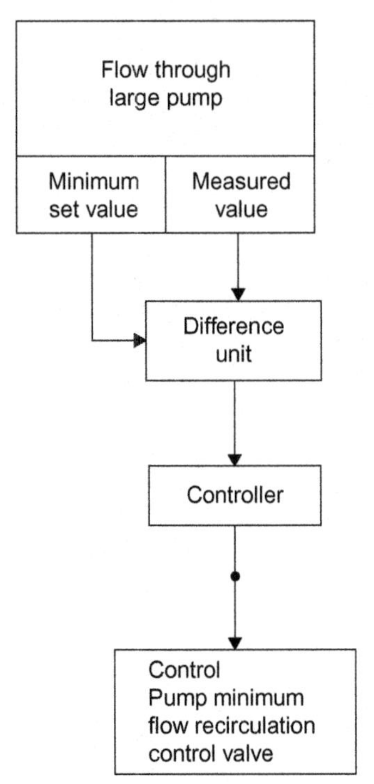

Figure 10.35 Pump minimum flow control.

constant pressure of about 0.35 MPa. As the turbine is loaded further, the steam supply to the deaerator is switched over from the cold reheat line to the turbine extraction steam line. Thereafter, the deaerator pressure varies with the turbine extraction steam pressure commensurate with turbine load.

10.5.5 Pump minimum flow recirculation control (Figure 10.35)

To ensure the safety of large pumps, e.g., boiler feed pump (BFP), condensate extraction pump (CEP), etc., it is essential to maintain a minimum flow through the pump at all modes of operation. This control valve is located on a separate line tapped from either the individual pump discharge or the common discharge header of pumps.

The minimum flow through a pump is maintained by using a pump minimum flow recirculation control valve, circulating fluid from the associated pump discharge line back to its source, i.e., the deaerator feedwater storage tank/condenser hotwell, etc. As long as the fluid flow to the process is at or below the safe minimum limit this control remains in service. When the fluid flow exceeds the safe limit, the pump minimum flow recirculation control valve is closed.

BIBLIOGRAPHY

[1] Johnson EF. Automatic process control. McGraw-Hill, Inc.; 1967.
[2] Raven FH. Automatic control engineering. McGraw-Hill Book Company, Inc. and Kogakusha Company Ltd.; 1961.
[3] Eckman DP. Automatic process control. Wiley Eastern Private Limited; 1972.
[4] Baumeister T, Marks LS, editors. Mechanical engineers' handbook. 6th ed. McGraw-Hill Book Company, Inc. and Kogakusha Company Ltd.; 1958.
[5] Central Electricity Authority of India. Standard design memoranda I C system volume I – control philosophy. Central Board of Irrigation and Power; 1984.
[6] Central Electricity Generating Board. Modern power station practice volume 6 – instrumentation, controls testing. London: Pergamon Press; 1971.
[7] Shields CD. BOILERS–types, characteristics and functions. New York: F. W. Dodge Corporation; 1961.
[8] Liptak Bela G, editor. Instrument Engineers' Handbook Vol. II: Process Control; 1970.
[9] Stultz SC, Kitto JB, editors. STEAM its generation and use. 41st ed. The Babcock and Wilcox Company; 2005.

CHAPTER 11

Interlock and Protection

11.1 INTRODUCTION

The *interlock and protection* system ensures the safety of equipment and personnel as well as the stable operation of a unit within permissible limits at all times. Many abnormalities can occur during normal operation of units, most of which can be corrected by operator intervention. However, operation of modern large power plants at times requires instant decision-making and corrective actions that may not be possible by operator action since human reaction time is inherently slow. A sluggish response on the part of operator may lead to catastrophic damage resulting from equipment failure, inadvertent error, and mishandling. To handle such situations the interlock and protection system initiates automatic corrective actions to stabilize the system quickly. This system also handles operations such as automatic cutting in and out of auxiliaries, systems, etc., in sync with changing load conditions. There are many examples of abnormalities that require immediate isolation of a unit for safety of equipment and/or minimizing damage to equipment. The protection scheme is usually designed to trip the equipment automatically with or without time delay, depending on the nature of emergency.

Protection to personnel and major equipment may be achieved by resorting to an intrinsically safe design and trained conduct of individual. However, individual conduct can not be predicted, since it is influenced by ignorance, fatigue, or negligence. Hence, it is essential that a design is intrinsically safe as far as possible so that no single failure in the system would prevent a mandatory shut-down. This aspect of design is called *fail-safe*. From this perspective, measurement, control, and protection systems are designed such that critical protection and control meet the "2 out of 3 logic" condition (which means if 2 inputs from a combination of 3 similar inputs of the same parameter are satisfied, then the logic will allow the output command to proceed). For less critical protection and control adopting "1 out of 2" (1 input from a combination of 2 similar inputs of same parameter) logic condition would suffice.

For guaranteeing safe operation various equipment, particularly larger equipment, of a power plant are provided with different type of interlock and protections. Most major equipment is provided with a start-interlock that prohibits it from starting until its all start permissive are satisfied. Similarly, trip-interlocks are provided for the protection of running equipment in the event of one or more off-normal operating

Thermal Power Plant
DOI: http://dx.doi.org/10.1016/B978-0-12-801575-9.00011-1

conditions and/or failure of associated equipment. Occurrence of any of these trip-interlocks and/or any off-normal condition, which may damage equipment and/or impair safety, should immediately cause the equipment to stop automatically.

In addition to incorporating protective devices, since not all conditions that cause damage are detected by any of the mandatory automatic trip devices, operating personnel must be knowledgeable of the limitations of automatic protection systems. Audio-visual alarms, supplementing interlocks, and other protections are provided to alert the operator when any equipment behaves abnormally or any of the operating parameters go beyond the preset permissible limits. Whenever any abnormal operating conditions occur and/or any equipment trip, corresponding "failure" alarm would alert the operator to take the applicable corrective actions.

The three major pieces of equipment in a steam power plant are the steam generator, steam turbine, and electric power generator or alternator. Each piece of equipment is provided with a dedicated hand-reset type lock-out relay through which tripping of associated equipment is initiated. Each of these three protection systems is independent of each other, but intertripping interlock among lockout relays is also provided to initiate isolation of the equipment, even if they are not directly affected, but isolation of which is necessary to ensure safety. A similar philosophy is adopted in gas turbine and diesel engine power plants.

The following sections cover the interlock and protection of a steam generator, steam turbine, gas turbine, diesel engine, and associated electric power generator or alternator. Interlock and protection of fuel firing equipment is also discussed, since they are integrally tied to start-up and shut-down of steam generators. No attempt, however, has been made to cover all types of protection, since that would be difficult to accommodate in a chapter. The scope of this book also does not cover interlock and protection of other auxiliary equipment associated with steam generator, steam turbine, gas turbine, diesel engine, and alternator.

11.2 STEAM GENERATOR

Steam generators today are large in size with a high steam-generating capacity and highly sophisticated firing system. These types of steam generators are more costly and vital to the power grid system than other types. With the increase in unit size and capacity, forced outages have greater significance not only for the larger loss of revenue but also for greater risk of injury and more damage to plants. Additionally, all steam generators, especially pulverized fuel-fired steam generators, are prone to explosion hazards due to unstable burning and fire-outs occurring in burners. All efforts should be made at both the design stage as well as during operation to ensure fail-safe operating conditions to prevent furnace explosion as well as furnace implosion.

An explosion in a container may occur when there is rapid pressure increase accompanied by combustion that propagates pressure waves. This pressure increase gets released by finding the least-resistant path in vent holes, explosion doors, etc. In the worst case scenario the pressure increase gets released by bursting the container. The primary reasons for such pressure build-up may be the result of one or a combination of the following:

i. Release of gases as a result of combustion
ii. Inadequate escape facilities for ventilation
iii. When the crest of pressure waves gets superimposed on the crest of fresh waves, the result is an accentuated effect

Explosion in a furnace may occur due to a variety of reasons. Some of the frequently occurring causes are [1]:

i. Improper purging of furnace, air, and gas paths
ii. Inaccurate ignition procedure
iii. Inadequate ignition procedure
iv. Fuel supply to the furnace is continued without ensuring proper combustion
v. Following a flame-out burners re-lit hurriedly
vi. Introduction of main fuel without ensuring adequate ignition energy

Statistics reveal that most furnace explosions are attributed to operator error, necessitating improvement in control and detection equipment. It is also known that operator actions are not fast and effective enough to execute parallel activities of multiple operating steps to ensure safe and proper supply and burning of fuel in the furnace. In large steam generators, where the fuel input rate is also high, major furnace explosions can result from the ignition of accumulated fuel even for 1–2 s. Statistics further reveal that the majority of explosions occur during start-up, shut-down, and low-load operations.

Implosion in a balanced draft fossil-fired furnace may occur if furnace negative pressure exceeds the structural stability of the furnace. From field experience it is found that implosions in some plants have occurred without main fuel firing, in some other plants prior to light-up of main fuel, and in some instances implosion occurred following a master fuel trip (MFT). From these experiences it may be concluded that during normal running a furnace is free from occurrence of any implosion.

Furnace implosion is usually induced by incorrect use of ID fan dampers, and/or rapid decay of furnace pressure either due to abrupt loss of supply of fuel to furnace or due to steep decay of furnace temperature following MFT [2].

(Note: Following an MFT the fuel supply to the furnace is stopped instantly, then the furnace gas temperature decreases rapidly to the air inlet temperature within approximately 2 s. However, due to reserve heat in the furnace, the gas temperature does not drop to that low a value but to an intermediate temperature of say, 813 K, rated temperature of steam.

During normal running of a steam generator pressure and temperature in the furnace are usually maintained at about (−) 0.1 kPa (g) {or 101.2 kPa (abs)} and 1673 K, respectively. Hence, after an MFT, applying ideal gas law, the pressure in the furnace reduces to

$$\frac{813}{1673} * 101.2 = 49.18 \text{ kPa(abs)} = -52.12 \text{ kPa(g)}$$

which is greatly higher than the recommendation of NFPA 85: Boiler and Combustion Systems Hazards Code discussed below.

The furnace pressure (draft) control system is designed to control furnace pressure at the desired set point in the combustion chamber by accurately positioning draft-regulating equipment (draft fan inlet vane/fan blade pitch/fan speed changing drive, etc.). Operating speed of furnace draft control equipment should not be less than that of air-flow control equipment to prevent negative pressure in the furnace from exceeding the danger limit.

Additional facilities, e.g., flame-scanning system, automatic burner control, and monitoring system including interpretation of adequacy of ignition energy, air and fuel metering, etc., are also provided to facilitate operators to secure safe and satisfactory performance.

In order to prevent permanent deformation of furnace structural members of all boilers, except fluidized bed boilers, due to yield or buckling arising as a result of either furnace explosion or furnace implosion, NFPA 85 recommends the following at the design stage:

i. The positive transient pressure should not be more than +8.7 kPa or the test block pressure of FD fan at ambient temperature, whichever is less

ii. The negative transient pressure should not be more than (−)8.7 kPa or the test block negative pressure of ID fan at ambient temperature, whichever is less

For fluidized bed boiler the NFPA 85 recommendation to prevent permanent deformation of furnace structural members is:

i. The positive transient pressure should be 1.67 times the predicted operating pressure or +8.7 kPa, whichever is greater, but the pressure should be restricted to the maximum head capability of the air supply fan at ambient temperature

ii. The negative transient pressure should not be more than (−)8.7 kPa or the test block negative pressure of ID fan at ambient temperature, whichever is less.

11.2.1 Burner management system (BMS)

In order to eliminate or to minimize boiler-furnace explosion/implosion hazards, the National Fire Protection Association (NFPA) of the United States has formulated some standards, a key feature of which is broadly identified as the Burner Management System (BMS).

The BMS should permit starting of equipment in sequence, when a set of preset conditions, necessary for the safety of equipment, are satisfied. When the equipment is started, the system will continuously monitor safe operating conditions determined in advance, warn operating personnel, and remove equipment from service in sequence again when the preset safe conditions are not met in practice.

BMS logic is designed so that no single failure in the system would prevent a mandatory boiler shut-down. The logic supervises overall furnace conditions, monitors all critical parameters of fuel firing system, and cuts out all fuel input to the furnace whenever dangerous conditions occur. Other requirements of the BMS are as follows:

1. An interlock system consisting of sequence interlocks, dictating the order in which the equipment must be operated, and also safety interlocks including mandatory trips
2. Dependable flame-monitoring and scanning devices both for oil and coal flame. In the event of complete/partial loss of ignition in the furnace, immediate action to be taken to stop fuel supply
3. A reliable start-up including *furnace purge*
4. Fuel oil start-stop control and oil firing supervision
5. Pulverizer and coal-feeder start-stop control and coal-firing supervision
6. Maintain minimum total air-flow supply to boiler at 30%, even when fuel load is less than 30%
7. Prevent accumulation of unburned fuel in furnace, which can ignite and lead to furnace explosion
8. Secondary air damper control and supervision
9. Intrinsic protection of boiler leading to safe shut-down

The flame scanner sighting head may be sensitive either to broadband ultraviolet radiations peculiar to hydrocarbon (coal, oil and gas) combustions or to the brightness level in the red band of the spectrum of oil and coal flame or combination of these two, but must not respond to radiations from hot refractory or slag. Visible light flame scanners are also popular.

A comprehensive BMS of a pulverized coal fired boiler includes control, protection, and intertripping interlocks of the following firing equipment and their associated equipment:

a. Induced draft (ID) fans, forced draft (FD) fans, and their associated dampers
b. Coal and oil burners with associated dampers
c. Pulverizers
d. Seal air fans
e. Primary air (PA) fans and associated dampers
f. Coal feeders

11.2.2 Sequence of operation

Earlier safety was ensured by sequential operation of various boiler auxiliaries. The sequence of starting and stopping used to be designed to ensure that unless the conveying system, which removes the products of combustion from the furnace, was in service, those feeding fuel and air to the furnace would not run. As such, the starting sequence of various boiler auxiliaries is configured as follows:

- **i.** Rotary/Regenerative air heaters (if provided)
- **ii.** Induced draft (ID) fans
- **iii.** Forced draft (FD) fans
- **iv.** Primary air (PA) fans
- **v.** Seal air fans
- **vi.** Pulverizers
- **vii.** Coal feeders

The stopping sequence of the above boiler auxiliaries should be in the reverse order of the starting sequence. The failure of any equipment in the above order should cause immediate stop of all succeeding equipment. Selective interlocking should ensure that with only one of the two ID fans running one FD fan could be operated. Similarly, failure of one of the two FD/PA fans should initiate preferential cut out of running pulverizers as recommended by the equipment manufacturer.

11.2.3 Furnace purge interlock

Prior to admitting fuel to an unfired boiler, it is essential to ensure the furnace is free from gaseous or suspended combustible matters. The sequence of activities of furnace purge cycle is as follows:

- **i.** Establish AC and DC power supply to BMS and associated equipment
- **ii.** Ensure master fuel relay (MFR) is in tripped condition
- **iii.** Verify that none of the MFR tripped commands (described below) are present
- **iv.** Open air and flue gas path isolating dampers, wind box dampers, and other control dampers to ensure open-flow path from the inlet of FD fans, through the furnace, to ID fans and the chimney
- **v.** Place all burner air registers in purge position or ensure all auxiliary air damper positions maintain adequate wind box to furnace differential pressure
- **vi.** Ensure all burner headers, igniter header, individual burner, and individual igniter shut-off valve are closed
- **vii.** Verify that all PA fans are not running
- **viii.** Check all flame scanners show "no flame" condition
- **ix.** Establish a minimum of 25% of design full load mass air flow through the furnace
- **x.** For pulverized coal-fired steam generators ensure mass air flow through the furnace is not greater than of 40% of design full load mass air flow

If all above conditions are satisfied "Purge Ready" indication comes ON. Initiate command for "Purge Start." Such action will maintain above air flow for a period of at least 5 minutes. On completion of above time period, "Purge Complete" indication switches ON. The furnace may now be treated to be free from any combustible matter and is ready for fuel firing.

11.2.4 Boiler light-up

After the purge is complete, the "MFR Reset" may be initiated provided all operating conditions of the boiler are normal and "MFR Trip" conditions, presented below, are absent. The boiler is now ready for light-up. (Detailed treatment of start-up and shut-down of steam generators is given in Chapter 12.)

11.2.5 Master fuel trip (MFT)/boiler trip

In accordance with NFPA 85, there are certain minimum required interlocks that should be provided to ensure basic furnace protection of a multiple burner boiler. Furthermore, in order to establish integrated operation of boiler-turbine-generator, these minimum required interlocks are supplemented by some additional interlocks for the protection of the furnace. Occurrence of any one of emergency conditions, as follows, will trip automatically *master fuel relay* (MFR, industrially known as 86MF relay):

i. Loss of all ID fans
ii. Loss of all FD fans
iii. Total air flow decreases below purge rate air flow by 5% of design full load air flow (Whenever total air flow falls below purge rate air flow, removal of combustibles and products of combustion from the furnace gets impaired.)
iv. Furnace pressure in excess of prescribed operating pressure (In the event of excessive increase in furnace pressure MFR has to be tripped so as to obviate furnace explosion.)
v. Furnace negative pressure in excess of prescribed operating negative pressure (In case of excessive fall in furnace pressure MFR should be tripped in order to avoid furnace implosion.)
vi. Boiler drum level (applicable to drum boiler) in excess of prescribed operating level (Excessive increase in boiler drum water level may interfere with the operation of the internal devices in the drum, which separate moisture from the steam, resulting in carryover of water into superheater or steam turbine causing mechanical damage.)
vii. Boiler drum level (applicable to drum boiler) falling below prescribed operating level (Excessive fall in boiler drum water level may uncover boiler riser wall tubes and expose them to furnace heat without adequate water cooling, resulting in riser tube burn-out unless action is taken either to restore the supply of feed-water or to kill the fire, i.e., trip the MFR.)
viii. Igniter fuel trip

ix. First pulverizer burners fail to ignite
x. Last pulverizer in service tripped
xi. All fuel inputs to furnace are shut off
xii. Loss of all flame
(Admission of fuel in the furnace at any time without ensuring proper combustion is a potential hazard as such fuel input to the furnace must be immediately cut out in case of loss of flame, i.e., loss of fire.)
xiii. Partial loss of flame that results in a hazardous accumulation of unburned fuel
xiv. Loss of energy supply for combustion control, burner control, or interlock systems
xv. All source of 6.6kV (HT power to major auxiliaries) supply tripped
xvi. In a reheat boiler if the following conditions occur simultaneously:
- **a.** One or both ESV and IV of steam turbine are closed
- **b.** HP bypass valve closed
- **c.** Any load carrying oil burner or coal mill in service.

xvii. Manual trip

Tripping of MFR will initiate the following:

i. Oil burner tripping sequence
ii. PA fan tripping sequence
iii. Seal air fan tripping sequence
iv. Pulverizer tripping sequence
v. Closing fuel oil shut-off valve
vi. Maintaining air register position
vii. Closing HP/LP bypass station
viii. Closing superheater attemperation spray water supply
ix. Closing reheat attemperation spray water supply
x. Tripping high voltage supply to electrostatic precipitator
xi. Rapid unloading of alternator if its load is above unloading limit. If the alternator load remains above the unloading limit even after expiry of a preset time, initiate tripping of alternator.

When the steam generator is tripped follow these steps:

1. Check and ensure that all fuel to furnace has been completely cut off
2. Maintain drum level (DO NOT take water to drum if level in drum is not visible. It may happen when water wall tube ruptures, that will damage the tubes.)
3. Purge the furnace and shut down draft fans (except in the case there is change of furnace getting pressurized, e.g., due to tube rupture, etc., keep one ID fan running with furnace maintained in suction)
4. Check and ensure SH, RH attemperator spray water supply is completely cut off and follow shut-down procedure of steam generator (see Chapter 12)

11.3 FUEL-FIRING EQUIPMENT

11.3.1 Fuel oil shut-off valve

This valve may be opened provided the following conditions are satisfied:

i. MFR is reset
ii. All burner oil valves are close

The shut-off valve will close if any one of the following conditions occurs

i. Fuel oil pressure after shut off valve is low
ii. Fuel oil temperature is low (applicable to heavy fuel oil only)
iii. MFR trips

11.3.2 Pulverizer

Pulverizer may be started provided all of the following conditions are fulfilled:

i. Unit DC power is available
ii. No pulverizer trip command is present
iii. Pulverizer ignition energy is available
iv. Pulverizer discharge gate/valve open
v. Pulverizer seal air valve open
vi. Pulverizer cold air damper open
vii. Tramp iron hopper valve open
viii. Coal feeder inlet and coal-bunker outlet gates are open
ix. Pulverizer outlet temperature less than a high limit recommended by the manufacturer
x. Pulverizer lubricating oil pressure, if provided, is healthy
xi. Primary air pressure adequate
xii. Differential pressure between pulverizer under bowl and seal air header is adequate
xiii. Air flow is adjusted between 30% and 40% of boiler maximum continuous rating (BMCR) flow
xiv. For tangential fired steam generators, either air and fuel nozzle tilts are horizontal or any coal feeder is ON

OR

For wall-fired steam generators, the corresponding pulverizer air registers are in oil firing position

Starting of pulverizer should permit associated coal feeder to start
Pulverizer should trip automatically in the event of any of the following conditions:

i. MFR is tripped
ii. Pulverizer discharge gate/valve is not open

iii. Loss of elevation/unit DC power
iv. Loss of ignition energy
v. Lubricating oil pressure, if applicable, falls below a permissible limit
vi. Loss of one of the two running PA fans should cause preferential pulverizer tripping recommended by the manufacturer
vii. Loss of both PA fans
viii. Upon detection of loss of flame

Stopping/tripping of pulverizer should initiate tripping of associated coal feeder.

11.3.3 Coal feeder

The coal feeder may be started provided all of the following conditions are satisfied:

i. Elevation DC power is available
ii. Ignition energy is available
iii. Associated pulverizer is running
iv. Pulverizer hot air damper/gate is open
v. Pulverizer inside temperature is adequate as recommended by the manufacturer
vi. Fuel air damper is closed (tangential fired steam generators)

OR

Burner air register is at coal firing position (wall-fired steam generators)

Starting of the coal feeder should initiate the following:

i. Release associated feeder speed and fuel air to auto control
ii. Check pulverized fuel flame after expiry of trial time
iii. If the coal flame is established, that ensures the pulverized fuel burner is established which in turn allows oil burner administrative shut-down
iv. If coal flame is not established within a preset time, initiate the pulverizer trip sequence.

The coal feeder should trip automatically in the event of any of the following conditions:

i. Failure of any of the start permissive
ii. Associated pulverizer is tripped
iii. Coal flow has not been established within a preset time of starting of the coal feeder
iv. Upon detection of loss of flame

Stopping/tripping of the coal feeder should initiate the following:

i. Cut out auto signal to associated feeder speed and fuel-air
ii. Emergency shut-down of pulverizer and coal firing system (see Chapter 12)

11.4 STEAM TURBINE

Prior to admitting steam to turbine, it is essential that following minimum conditions are met:

- **i.** AC and DC power supply to turbine protection, control and associated equipment are available
- **ii.** None of the turbine-tripped commands (described below) are present
- **iii.** Condensate extraction pumps are running
- **iv.** Boiler feed pumps are running
- **v.** Manufacturer's recommended steam conditions, i.e., pressure and temperature of steam at turbine inlet are available
- **vi.** Turbine lube oil pressure is adequate
- **vii.** Condenser vacuum meets manufacturer's recommended value
- **viii.** All associated equipment and systems are functional

Turbine is now ready for admitting steam.
(Detail treatment on the start-up and shut-down of steam turbines is given in Chapter 12.)

A running turbine may trip due to variety of reasons, e.g., malfunction or trouble in one or more of its auxiliaries and/or operating conditions, abnormal conditions in the boiler, or trouble in the unit electrical system. Tripping of a turbine is initiated through the *Turbine Lockout Relay* (industrially known as 86T relay).

Turbine protection system is independent of other protection systems of the plant. The governor and the protection system are constructed in such way that failure of any component will not prevent the turbine from being safely shut-down.

The turbine protection system may be actuated either electrically or hydraulically. In the event of electric trip, the turbine lockout relay should actuate and energize turbine trip solenoid. Both trip solenoid and hydraulic trip system act on the hydraulic control system, which causes fall in control/trip oil pressure leading to closure of HP turbine emergency stop valves (ESVs), HP and IP turbine control valves and IP turbine interceptor valves (IVs), thereby shutting off steam supply to turbine.

When steam supply to turbine is insufficient to drive the generator, motoring will follow, which is detrimental to the health of turbine blades because of loss of ventilation effect of steam flow with consequent overheating of blades. Under such condition it is essential to trip the *unit lockout relay* (industrially known as 86U relay).

In order to ensure safe running of the turbine following automatic trip conditions of turbine are generally provided:

1. Axial shift high

   (In case of excessive axial shift of the rotor due to thrust, blades and glands of turbine will get subjected to mechanical damage.)

2. Lube oil pressure very low

   (In the event of lube oil system failure, bearings would quickly lead to wiping of the journal and thrust bearings).
3. Trip oil pressure very low
4. Condenser vacuum very low

   (If the condenser vacuum falls below a lower set value, turbine exhaust hood temperature will increase, which may cause damage to the last stage blades due to stress from centrifugal force juxtaposed with thermal stress. Condenser tubes will also get damaged since under the influence of temperature they will expand and may get buckled.)
5. Turbine over-speed
6. Boiler drum level very high

   (Excessive increase in boiler drum water level results in carryover of water into steam turbine. Water droplets accompanying steam impinge high speed turbine blades and damage the blades severely impairing efficiency of the turbine. Hence, on occurrence of carryover of water, steam supply to turbine should be cut off.)
7. Main steam pressure very low
8. Main steam temperature very low

   (On occurrence of conditions 7/8, there is an apprehension of moisture carryover to turbine, leading to mechanical damage to turbine blades.)
9. Main steam/hot reheat steam temperature very high

   (To prevent creep failure, this protection is provided.)
10. Turbine/generator bearing temperature very high

    (This protection is provided to protect bearing babbit metal from wiping.)
11. Turbine/generator bearing drain oil temperature very high

    (Turbine lube oil, being a mixture of hydrocarbons, has a tendency to oxidize. If the oxidation rate is high due to higher temperature sludge will form, which gets deposited in various parts of lubrication system, creating blockage and leading to oil starvation and overheating. Furthermore, with increase in oil temperature, viscosity of the oil will become low resulting in discontinuities in the bearing oil film with consequent damage to bearings.)
12. Turbine/generator bearing vibration very high
13. Excessive differential expansion
14. Frequency low

    (Continuous operation of a turbine with a frequency lower than a recommended value may reduce the life of some of the blades, since an under-frequency condition may be the resonance condition for some of the stages resulting in higher stresses in these blades. Nevertheless, tripping a turbine under low frequency is not desirable since this will cause further stress to other running units, which may lead to total power failure.)

15. HP feedwater heater level very high

    (In the event the level in any of the HP feedwater heaters exceeds a preset very high limit, there may be a possibility of water entering the turbine through associated extraction steam line causing serious water damage necessitating complete shut-off of steam supply to turbine.)
16. Main oil tank level very low (optional)
17. Fire protection operated
18. Master fuel relay operated
19. Generator lock-out relay operated.

Tripping of Turbine should initiate the following:

- **i.** Close emergency stop valves and HP control valves
- **ii.** Close interceptor valves and IP control valves
- **iii.** Close turbine inlet isolating valve/s (if any)
- **iv.** Close extraction steam non-return valves (NRVs)
- **v.** Close heater inlet extraction steam valves
- **vi.** Open HP casing drain valve/s
- **vii.** Open extraction steam line drain valves
- **viii.** Open HP-LP bypass valves
- **ix.** Trip 6.6 kV (HT supply to major auxiliaries) unit bus incomer breakers
- **x.** Trip generator field breaker
- **xi.** Trip generator circuit breaker
- **xii.** Trip bus-tie circuit breaker

Immediately after the turbine trip, check and ensure that

1. Turbine speed is actually falling down
2. Lube oil supply to turbine/generator bearings is adequate
3. Auxiliary oil pump has taken start automatically. If this pump fails to start then start the same manually
4. All extraction line NRVs and isolating valves in the extraction lines near feedwater heaters have closed
5. Seal oil supply to generator seals is adequate

If the cause of the trip is such that the unit can be brought back shortly, it is not needed to break condenser vacuum. However, if it is necessary to bring the rotor to a quick stop, break the vacuum by opening vacuum breaker valve and admitting atmospheric air to the condenser. Initiate emergency shut-down of boiler (discussed in Chapter 12). When the turbine is coasting down before it comes to stand still put it on turning gear for uniform cooling of hot turbine rotor.

11.5 GAS TURBINE

The interlock and protection system of gas turbine performs following functions prior to startup, during startup, running, shut-down and cool-down processes:

i. Prior to starting, it checks all vital systems of turbine, generator, and associated auxiliaries are in a healthy condition
ii. During start-up operation, this system maintains all checks and permits to start the turbine
iii. The system monitors various protective circuits, and permits the unit to continue if the operating parameters are within limits
iv. At any stage of the process, if a vital system fails and indicates that the turbine is not within safe run up parameters, it will trip the machine
v. During normal operation, the system provides an overall protection to the turbine against any failure of protective circuits and trips the machine.

Prior to starting the gas turbine the following minimum conditions must be met:

i. AC and DC power supply to turbine protection, control and associated equipment are available
ii. None of the gas turbine tripped commands are present
iii. Fuel gas pressure at gas turbine fuel gas skid inlet is adequate
iv. Fuel oil pressure at gas turbine fuel oil skid inlet is adequate
v. Lube oil flow through sight glasses are within normal limit
vi. Lube oil pressure is equal to or above the manufacturer's recommended value
vii. Electrical control and protection systems are in service
viii. Annunciations, instruments, all indicating lamps are healthy
ix. Auxiliary power supply is at the rated voltage and frequency
x. All associated equipment and systems are in service.

Once above conditions are satisfied the gas turbine is purged with at least 8% of full-load mass air flow for a period of not less than 5 minutes. The gas turbine is now ready for starting.

(Further details on start-up and shut-down of gas turbine are given in Chapter 12.)

To prevent damage to gas turbine and auxiliaries from abnormal operating conditions arising out of malfunction of equipment or system, a number of protective devices are provided. The protective function supervises parameters of the gas turbine in all phases of operation. The sensors are monitored continuously for malfunctions.

Whenever such a malfunction does occur, an alarm is given. The protective system includes over-speed, over temperature, flame detection, high vibration, etc., as described below:

Over-speed protection The over speed protection actuates whenever speed of the rotor exceeds a preset limit causing closure of fuel gas/fuel oil trip shut-off valve.

Over-temperature protection The gas turbine unit is protected against any possible damage caused to the machine in the event turbine outlet gas temperature exceeds a predetermined limit. In such case the over-temperature protection system trips the turbine to arrest any further temperature increase.

Flame detection and protection Flame detection device avoids accidental flooding of combustors with unburned fuel due to loss of flame or improper combustion. This device is used to detect flame in the combustors and to trip the turbine in case of flame failure.

Vibration protection The vibration protection system of the gas turbine unit comprises vibration transducers fitted on various bearing housings and relative shaft vibration transducers at journal bearings. In the event the vibration becomes excessive, the unit will trip to avoid any damage.

Lube oil pressure and temperature protection In the case gas turbine lubricating oil pressure drops below a preset low value, the protective system trips the unit. This is to prevent the turbine from being run with inadequate oil supply to bearings and other accessories. Likewise, bearing temperature also trips the unit should the temperature of bearings exceed preset limit.

Hydraulic and pneumatic protection In the event of collapse in control oil/air pressure in piping systems, the hydraulic/pneumatic protection system brings vital system components to a safe operating condition and trips the turbine.

Electrical interlocks and protections The protections provided for generator, transformers, motors, etc., are against unhealthy system conditions or to isolate faulty equipment from the system to minimize the extent of damage through the fault. In the event any of the electrical protections, linked with the connected generator or connected electrical system, operates it trips the turbine.

Immediately after the gas turbine trip, check and ensure that

1. Gas turbine speed is actually falling down
2. Lube oil supply to gas turbine/generator bearings is adequate

11.6 DIESEL ENGINE

Like all prime movers diesel engines are also not free from emergencies, occurrence of which should cause starting-up and/or running of engines to stop.

Before the diesel engine is started the following checks should be made:

i. AC and DC power supply to diesel engine protection, control, and associated equipment are available
ii. None of the tripped commands are present
iii. Fuel oil pressure at diesel engine inlet is adequate
iv. Lube oil pressure is equal to or above the manufacturer's recommended value
v. Starting air cylinders air pressure is at or above the recommended value
vi. Water to the external cooling circuit is adequate
vii. Shaft barring arrangement is engaged
viii. Electrical control and protection systems are in service
ix. Annunciations, instruments, all indicating lamps are healthy
x. Auxiliary power supply is at the rated voltage and frequency
xi. All associated equipment and systems are in service

The diesel engine is now ready for starting.

(Detail treatment on the start-up and shut-down of diesel engines js given in Chapter 12.)

Emergency conditions that trip the diesel engine are discussed below:

Incomplete oil circulation (at starting time) When the engine start command is initiated, the lube oil and fuel oil pumps for pre-start priming will operate. If the lube oil pressure does not build up to the preset value during priming period, starting-up of engines is to stop.

Air starting fault After the oil pressure reaches the preset value air is admitted to diesel engine pipeline. If air pressure fails to reach the preset value start-up of engine is to be aborted.

Lube oil pressure drop While the engine is running, if the lube oil pressure drops below a preset value running engine has to be stopped.

Fuel oil pressure drop In the event fuel oil pressure drops below a preset value while the engine is running, the engine has to be stopped.

Inadequate cooling water pressure During normal engine operation, if the cooling water pressure drops below the low limiting value, the engine should stop.

Speed of the started engine fails to increase Once command for starting the engine has been given, and normal priming is achieved, the engine speed should increase automatically to warming up speed. If the speed fails to increase the supply of starting air as well as the lube oil and fuel oil priming pumps should be stopped.

Engine over-speed If the speed of the engine reaches or exceeds its over-speed limit starting-up and/or running of the engine needs to be stopped.

Tripping of generator electrical protections In the event of operation of any of the electrical protections, pertaining to the connected generator or connected electrical system, starting-up and/or running of the engine needs to be stopped.

Immediately after the tripping of diesel engine, check and ensure that

1. Engine speed is actually falling down
2. Lube oil supply to engine/generator bearings is adequate

11.7 GENERATOR (ALTERNATOR)

As discussed in Chapter 9, *Systems Of Large Power Station*, power generated by the generator is supplied to the grid/bus-bar through the generator transformer and generator circuit breaker. By closing the generator circuit breaker the generator gets connected to the grid – this is called *synchronization*. Prior to synchronizing minimum following conditions should be fulfilled:

i. AC and DC power supply to generator protection, control and associated equipment are available
ii. Generator seal oil system, if there be any, is in service
iii. Generator hydrogen cooling system, if there be any, is in service
iv. Generator stator cooling water system, if there be any, is in service
v. None of the generator-tripped commands are present
vi. Excitation system provides permissive
vii. Voltage developed by generator matches with that of the grid,
viii. Phase sequence of voltage of the generator must be same as that of the grid
ix. The speed (revolutions per minute or more precisely revolutions per second) of the prime mover (steam turbine/gas turbine/diesel engine) to be connected must match the frequency (cycles per second/Hertz) of the grid.

Tripping of the generator essentially means disconnecting it from the grid by opening the generator circuit breaker.

The generator is tripped for faults, which may cause severe damage to it and/or the connected prime mover (steam turbine/gas turbine/diesel engine) and needs immediate isolation of the unit from the grid. These faults generally include the fault inside and outside the protective zone of the generator, generator transformer, and unit auxiliary transformer, and in the case of a steam turbine, low lubricating oil pressure, high axial shift, etc. The tripping of the generator, initiated through *generator lockout relay* (industrially known as 86G relay), in turn ensures complete shutting down of the prime mover and the boiler (applicable to steam power plant only). The generator is tripped automatically on occurrence of any of the following typical conditions, but may not be restricted to these only:

i. Generator over voltage
ii. Generator under voltage
iii. Generator over current
iv. Generator over frequency
v. Generator under frequency
vi. Generator volts/Hertz high (over flux)
vii. Generator differential protection
viii. Generator loss of excitation
ix. Generator over-excitation
x. Excitation system failure
xi. Generator stator earth fault
xii. Generator rotor second earth fault
xiii. Generator split phase protection for inter turn fault
xiv. Generator pole-slip protection
xv. Generator back up impedance protection
xvi. Generator negative phase sequence protection
xvii. Generator rotor high temperature protection
xviii. 132/220/400/700 kV over current protection
xix. Fault in the station bus
xx. Bus differential protection
xxi. Overall differential protection
xxii. Generator transformer oil temperature high
xxiii. Generator transformer winding temperature high
xxiv. Generator transformer negative sequence current very high
xxv. Generator transformer over current protection
xxvi. Generator transformer standby earth fault
xxvii. Generator transformer restricted earth fault
xxviii. Generator transformer Buchholz protection
xxix. Unit auxiliary transformer on-line tap changer (OLTC) Buchholz protection
xxx. Unit auxiliary transformer Buchholz protection
xxxi. Unit auxiliary transformer differential protection
xxxii. Unit auxiliary transformer over current protection
xxxiii. Unit auxiliary transformer restricted earth fault protection
xxxiv. Generator reverse power relay operation

xxxv. Steam turbine trips causing generator low forward power relay to operate (applicable to steam power plant only)
xxxvi. MFR trips with generator load above unloading limit (applicable to steam power plant only)
xxxvii. Emergency manual trip

Tripping of generator should initiate the tripping of:

i. Prime mover (steam turbine/gas turbine/diesel engine) lockout relay
ii. Master fuel relay (applicable to steam power plant only)
iii. HT (6.6kV) switchgear unit incomer breakers
iv. Generator automatic voltage regulator (AVR)
v. Generator field breaker

Immediately after the generator trip, check and ensure that

1. Generator circuit breaker is open
2. Field breaker is switched off
3. Unit HT (6.6 kV) switchgear breaker is open
4. Seal oil supply, if applicable, to generator seals is adequate
5. Generator hydrogen cooling system and stator cooling water system are in service, if these systems are provided
6. Prime mover (steam turbine/gas turbine/diesel engine) speed is falling down
7. Lube oil supply to bearings of generator/prime mover (steam turbine/gas turbine/diesel engine) is adequate
8. Auxiliary oil pump has taken start automatically, if not, start the same manually
9. All the extraction line NRVs and isolating valves in the extraction lines near feed-water heaters have closed (applicable to steam power plant only)
10. Check and ensure that all fuel to furnace/gas turbine/diesel engine (as applicable) has been completely cut off
11. Maintain condenser vacuum (applicable to steam power plant only)
12. Keep condensate extraction pump, boiler feed pump, circulating water pump running (applicable to steam power plant only)
13. Maintain drum level (applicable to sub-critical steam power plant only)
14. Purge the furnace and shut-down draft fans (applicable to steam power plant only)
15. Check and ensure SH, RH attemperator spray water supply is completely cut-off (applicable to steam power plant only).

BIBLIOGRAPHY

[1] NFPA 85 Boiler and Combustion Systems Hazards Code; 2004.
[2] Durrant OW, Lansing EG. Furnace Implosions and Explosions. COMBUSTION; September 1976.

CHAPTER 12

Start-Up and Shut-Down

12.1 INTRODUCTION

In addition to the steam generator and steam turbine, a thermal power plant essentially consists of equipment such as pumps, fans, blowers, compressors, etc., and systems of pipelines including valves, hangers, and supports. The start-up and shut-down of a unit thus comprises starting, running, and stopping of this equipment and charging, filling, warming, and isolating of the piping systems. Since covering all these activities in a single chapter is unrealistic, here we will focus on the start-up and shut-down of major equipment such as the steam generator, steam turbine, gas turbine, and diesel engine.

12.2 START-UP OF STEAM GENERATOR

12.2.1 Line-up of steam generator before start-up

Before the first start-up or start-up after overhaul the following activities/actions should be carried out or verified by operating personnel:

1. All works in boiler furnace have been completed
2. All works in air and gas ducts have been completed
3. Unit has been chemically cleaned
4. Steam lines have been blown
5. Safety valves on drum, superheater and reheater headers have been set
6. Boiler furnace, air pre-heaters, economizer, electrostatic precipitators, flue gas desulfurization (FGD) equipment, reduction of nitrogen oxides (DeNOx) equipment, induced draft (ID), forced draft (FD) and primary air (PA) fans, gas and air ducts are all cleared of men and foreign material
7. Boiler access doors, observation doors, and ash hopper doors are closed
8. Boiler drum manhole covers are properly seated
9. Boiler drum vent valves are open
10. Bottom ash hopper and hopper seal trough are filled with water; open the continuous make-up water to these hoppers and ensure proper overflow (this condition is applicable to a wet bottom ash system)
11. The following dampers are open:
    a. Air heater air inlet/outlet
    b. Air heater gas inlet/outlet

Thermal Power Plant
DOI: http://dx.doi.org/10.1016/B978-0-12-801575-9.00012-3

12. The following dampers are closed:
    a. Over-fire air dampers
    b. Pulverizer hot air shut-off gates
    c. Pulverizer hot air control dampers
13. Secondary air dampers are open or modulating
14. Superheater start-up vent valves are in a partially open position
15. Superheater header manual vent valves and drain valves are open full
16. Reheater start-up vent valves are in partially open position
17. Reheater header manual vent valves and drain valves are open full
18. Superheater attemperator spray water supply block and control valves are closed
19. Reheater attemperator spray water supply block and control valves are closed
20. Boiler drum pressure and level transmitter are in operable condition
21. Isolation valves of boiler drum level gauge glasses and level transmitters are open and their blow-down valves are closed
22. Steam pressure gauge isolation valves and impulse line root valves are open
23. Feedwater, boiler drum water, saturated steam, and superheated steam sample line isolation valves are open
24. Chemical feed system inlet isolation valves to drum are open
25. Continuous blow-down line isolation valves are open and emergency blow-down line valves are closed
26. Continuous blow-down tank vent valve is open
27. Water wall drain header drain valves are closed
28. Drain valves at feedwater inlet to economizer inlet header are closed
29. Feedwater inlet valve to economizer inlet header is open
30. Coal feeder inlet and coal-bunker outlet gates are closed
31. Alarm annunciation system and data acquisition system (DAS) are available
32. Distributed control system (DCS) is available

12.2.2 Start-up

The sequence of start-up is as follows:

1. Start secondary regenerative air pre-heaters (if provided)
2. Start ID fans; verify that their inlet and outlet dampers open
3. Start FD fans; verify that their outlet dampers open
4. Adjust flow through ID fans and FD fans to permit a purge air flow of at least 30% of total air flow
5. Purge the furnace as discussed in Chapter 11

   The steam generator is now ready for light-up.

12.2.3 Oil burner lighting

Lighting of oil burners includes the following:

i. Place all burner air registers to pre-light position

ii. Verify that burner atomizing valve is open
iii. Start the minimum number of oil burners
iv. Initiate light-up sequence of the lowest elevation of igniters and warm-up oil burners; oil burner is inserted
v. Flame scanner logic supports oil flame indication

With insertion of oil burner, the following gas/electric igniter start sequence is initiated.

12.2.4 Gas igniter

i. At the start of the ignition sequence of a gas igniter, the igniter gas valve will open and a spark timer will energize. After a preset time period the igniter flame is checked.
ii. Should this flame be established, the main fuel oil shut-off valve and the burner fuel oil valve is signaled to open
iii. If burner oil flame is not detected at the end of the preset time period, the igniter gas valve and the burner fuel oil valve will close and an igniter fault will be signaled.

12.2.5 Electric igniter

i. At the initiation of the igniter start sequence, the igniter gets inserted and when the igniter is fully inserted a spark timer will energize
ii. When igniter energize signal is received, the main fuel oil shut-off valve and burner fuel oil valve will open; a trial (an 'in-built checking circuit') for ignition timer will operate, at the end of which the oil flame is checked
iii. If the burner oil flame is not established at the end of the preset time period, the oil burner shut-down sequence is initiated

If the oil flame is okay the following sequence is initiated:

i. Burner oil valve is kept in the open position
ii. Atomizing air/steam valve is kept in open position
iii. Oil burner is kept in inserted position
iv. Igniter is de-energized
v. Oil burner firing is signaled

12.2.6 Pressure raising

1. Regulate oil flow to increase the boiler water temperature and verify that the furnace exit gas temperature remains below 813 K in order to protect superheater, reheater, etc., through which coolant fluid flow is not yet established.
2. Gradually more oil guns are put into service according to the rate of pressure rise recommended by the boiler manufacturer
3. Check oil flame and stack; in a cold boiler the furnace tends to be smoky in the beginning, but it will clear up as the furnace warms up, however, the flame

should be clear and bright. After adequate warm-up the stack should also be clear, and in no case should white smoke be allowed from the stack

4. Close the drum and superheater vent valves when a copious amount of steam issues out of them, at about 0.2 MPa drum pressure; close all the super heater drain valves; the starting vent valve should be kept partially open
5. Monitor drum metal temperature
6. Monitor the furnace exit gas temperature; in no case should it be more than 813 K to protect re-heater tubes, which have no cooling steam flow until now
7. Maintain drum level at normal (applicable to drum boiler)
8. Watch bottom ash hopper for proper overflow for removal of the accumulated unburned oil
9. With rise of steam pressure and flow it is necessary to feed water to boiler continuously
10. When steady flow of feedwater is established through the economizer, put the water side of the high-pressure feedwater heaters into service
11. Start PA fans; verify that their outlet dampers are open
12. Open coal feeder inlet and coal-bunker outlet gates
13. Start the pulverizer serving a lower elevation of coal nozzles
14. Open the hot-air shut-off gate and bring the pulverizer up to the required operating temperature (typically 343 K) as recommended by the boiler manufacturer
15. Maintain the fuel feed at minimum, consistent with stable ignition
16. Regulate hot and cold air dampers to hold pulverizer outlet temperature (typically 343 K) as recommended by the manufacturer
17. With increased firing rate, it may be required to open attemperator spray water
18. Ensure that air flow to the furnace corresponds to the firing rate at all times
19. Place additional pulverizers in service according to unit load demand
20. Remove fuel oil guns and igniters from service when unit-firing conditions are stabilized
21. Adjust feedwater supply to the boiler as required to maintain normal drum water level (applicable to drum boiler)

12.3 NORMAL SHUT-DOWN OF STEAM GENERATOR

When shutting down the steam generator, the reduction of the firing rate is determined by the reduction in the steam turbine steam demand:

1. Gradually reduce load on the unit, reduce the firing rate according to the decreasing steam flow.
2. Put the "combustion control" and "superheat and reheat temperature control" on manual.
3. When coal feeder rating on all pulverizers is reduced to 40% of MCR, start the fuel oil guns associated with the upper-most pulverizer in service.

4. Gradually reduce coal feeder rating further. When a minimum feeder rating of about 25% load is reached, close the hot air shut-off gate.
5. Close the superheater and reheater attemperator spray water supply block valves.

12.3.1 Controlled shut-down of pulverizer and coal-firing system

Controlled shut-down of the pulverizer and coal-firing system is accomplished as follows:

1. Reduce associated coal feeder speed to a minimum.
2. Close the associated hot air damper/gate.
3. When the pulverizer outlet coal-air temperature falls to a preset low limit (typically 323 K) as recommended by the manufacturer shut down the associated coal feeder.
4. Allow the pulverizer to run to ensure that it is completely empty by noting the load of the mill motor and shut it down.
5. When the coal feeder rating on all remaining pulverizers reaches 40%, take the second pulverizer, supplying the next higher elevation, out of service.
6. In the same manner continue taking out pulverizers at consecutive lower elevations.
7. When down to two pulverizers start the associated oil burners before reducing load on either of the associated coal feeders lower than 50% MCR.
8. Operate all sootblowers.
9. Reduce the air flow according to the fuel reduction until 30% of maximum air flow is reached. Further lowering of air flow should only be done after fire-out of boiler.
10. After the last pulverizer has been shut down, remove fuel oil guns as discussed in the following.

12.3.2 Oil burner controlled shut-down

The shut-down sequence of the oil system may be initiated either "through burner stop command" or "by oil flame failure."

i. Close oil valve
ii. Open scavenging steam valve
iii. Start scavenge timer
iv. Verify burner scavenging is in progress
v. Close atomizing steam valve
vi. Retract oil burner when scavenge time is complete

The shut-down sequence of the steam generator proceeds as follows:

1. Check and ensure all fuel to the furnace is cut off
2. Purge the furnace
3. Shut down ID fans, FD fans, and air heaters
4. Close superheater outlet shut-off valves

5. Open re-heater vent and drain valves
6. Stop chemical dosing to the drum
7. Maintain drum water level (applicable to drum boiler) near normal by adding feedwater as required
8. Take boiler feed pumps out of service
9. Maintain deaerator feedwater storage tank level near normal by adding condensate water as required
10. Take condensate extraction pumps out of service
11. Open drum vent valves and superheater vent and drain valves when drum pressure falls to 0.2 MPa
12. Boiler may be emptied when the water temperature falls below 368 K or to a value recommended by manufacturer

12.4 EMERGENCY SHUT-DOWN OF STEAM GENERATOR

A variety of conditions can arise with steam generators, such as boiler steam/water tube rupture, that may require its shut-down.

12.4.1 Emergency shut-down of pulverizer and coal-firing system

Any pulverizer and coal-firing system should be shut-down on an emergency basis if:

i. Flame goes out in any two of four pulverized fuel burners in a group with the pulverizer running
ii. Pulverizer lubricating oil pressure is low
iii. MFR (master fuel relay) trips
iv. Both PA fans trip
v. Signal from preferential cut-out of mills as recommended by manufacturer

Emergency shut-down of pulverizer and coal-firing system will initiate:

i. Tripping of coal feeder
ii. Tripping of pulverizer
iii. Tripping of PA fans
iv. If oil burner is not in operation, bring the air register of pulverizer group to the pre-light-up position

12.4.2 Oil burner emergency shut-down

The oil burner shuts down automatically if:

i. MFR trips
ii. Oil header pressure is low
iii. Atomizing air/steam pressure is low
iv. Burner oil temperature is low (for heavy fuel oil-firing only)

The following sequence of activities will follow during the shut-down of the oil system:

i. Igniter is de-energized
ii. Oil valve is closed
iii. Atomizing air/steam valve is closed
iv. Oil burner is retracted

The shut-down sequence of the steam generator proceeds as follows:

1. Stop pulverizers one by one and then oil burners
2. MFT should be operated manually without hesitation in case of any doubt about flame stability
3. Purge the furnace normally, if possible; if not possible, purge the furnace at the earliest opportunity
4. Keep ID fan/s running to maintain furnace in suction in order to prevent any pressurization due to tube rupture, etc.
5. Maintain drum level (applicable to drum boiler) at normal (*Note*: Do not feed water into drum, if the steam generator is tripped due to loss of drum water. Such feeding will cause damage to the furnace tubes.)
6. Cool steam generator as quickly as conditions permitted by the manufacturer
7. Proceed to normal shut down of the turbine and auxiliaries

12.4.3 Steam turbine stops/trips

In the event that the steam turbine stops or trips due to any trouble lying with steam turbine then it is not essential to stop the steam generator. Proceed as follows:

1. Shut down the steam generator and keep it boxed and bottled up so that it can be used on short notice
2. Do not reduce drum pressure
3. Close all superheater drain and vent valves; reheater vent and drain valves are left open
4. Close hot air shut-off gates to pulverizers
5. Stop coal feeders
6. Stop pulverizers when they are empty
7. Verify that all fires are extinguished
8. Keep ID fans, FD fans, and air heaters running
9. As the unit cools down and the water shrinks, add make-up water intermittently to steam generator to prevent drum level from dropping below the visibility limit of the drum level gauge glass

12.5 START-UP OF STEAM TURBINE

Steam turbine start-up may be categorized as cold, warm, hot, and very hot start-up. This categorization follows industry practice based on the thermal conditions of steam turbines, usually the metal temperature of the HP turbine inner casing, at the time of start-up. It is also common in the industry to classify the start-ups on the basis of time since previous shut-down. The various categories of start-up are defined as follows (IEC 60045-1:1991: Steam Turbines Part-1: Specification and/or BS EN 60045-1:1993: Guide for Steam Turbine Procurement) [1,2].

12.5.1 Cold start-up

This category of start-up is usually followed after a 72-hour shut-down period of the steam turbine when the metal temperature of the HP turbine inner casing is expected to drop below approximately 40% of full-load metal temperature of the casing.

12.5.2 Warm start-up

Warm start-up is usually followed after a shut-down period of between 10 and 72 hours when the metal temperature of the HP turbine inner casing is between 40% and 80% of full-load metal temperature of the casing.

12.5.3 Hot start-up

Hot start-up is followed after a shut-down period of 10 hours. The metal temperature of the HP turbine inner casing is above approximately 80% of full-load metal temperature of the casing.

12.5.4 Very hot start-up

When a steam turbine is started within 1 hour after shut-down period, this is treated as very hot start-up. During this start-up, metal temperature of the casing remains at or near its full-load metal temperature.

12.5.5 Line-up of steam turbine before start-up

Before the first start-up or the start-up after a overhaul of steam turbines carry out or verify following activities/actions:

- **i.** Steam generator has been started up and is ready for supplying steam to steam turbine
- **ii.** Alternator, generator circuit breaker, generator transformer, excitation system, automatic voltage regulator, and associated electrical systems are ready for synchronizing with the grid
- **iii.** Condenser cooling water inlet valves are closed
- **iv.** Condenser cooling water outlet valves are open
- **v.** Condenser water box vent valves are open
- **vi.** Condenser support springs, if provided, are free
- **vii.** Circulating water (CW) pumps are running

viii. Establish condenser cooling water system by opening CW pump discharge valves
ix. Slowly open condenser inlet valves to vent condenser water boxes
x. Close water box vent valves after water issues profusely through them
xi. Open steam turbine oil cooler cooling water inlet and outlet valves; establish cooling water flow through the coolers
xii. Check steam turbine main oil tank level is normal as recommended by the manufacturer
xiii. Check main oil tank level alarm-annunciations are active
xiv. Open suction and discharge valves of *AC auxiliary oil pump* and *DC emergency oil pump*. Prime these pumps by opening respective air vent valves
xv. Start auxiliary oil pump and gradually fill up bearing oil system and governing oil system
xvi. When the systems are filled-up, check oil level in main oil tank
xvii. Replenish oil into main oil tank to maintain normal level, if required
xviii. Verify pressure in lube oil system after oil coolers and at levels of bearing axis is adequate as recommended by the manufacturer
xix. Verify sufficient quantity of oil drains out from all bearings
xx. Verify temperature of lube oil is within permissible limits (313–318 K)
xxi. Switch off auxiliary oil pump
xxii. Check automatic starting of auxiliary oil pump and emergency oil pump
xxiii. Start *jacking oil pump*; put steam turbine on *turning gear*
xxiv. Check automatic tripping of turning gear at low lube oil pressure and at low jacking oil pressure per manufacturer's guidance
xxv. Check the protection and governing system as follows:
- **a.** Open emergency stop valve/s (ESVs) of HP turbine, interceptor valve/s (IVs) of IP turbine, and control valves of both HP turbine and IP turbine; check that their opening is smooth
- **b.** Trip steam turbine manually through its shut-down switch and verify all above valves are fully closed
- **c.** Normalize the turbine shut-down switch; ensure that no valves open again
- **d.** Open emergency stop valve/s, interceptor valve/s, and control valves
- **e.** Verify that emergency stop valve/s, interceptor valve/s, and control valves close on actuation of the *turbine high axial shift*
- **f.** Verify that emergency stop valve/s, interceptor valve/s, and control valves close on actuation of *turbine low bearing oil header pressure*

xxvi. Put steam turbine on turning gear
xxvii. Place generator H_2 cooling system, if provided, in service

12.5.6 Line-up of condensate system

i. Supply DM water and fill up condenser hot well up to 2/3 in the gauge glass

ii. Open inlet and outlet valves of low-pressure (LP) feedwater heaters; close bypass valves of LP heaters
iii. Open inlet and outlet valves of steam-air ejector, if provided
iv. Open suction valves of *condensate extraction pumps* (CEP); prime these pumps by opening respective air vent valves
v. Start one condensate extraction pump (CEP) with its discharge valve shut
vi. Slowly open discharge valve as the pump speeds up
vii. Open CEP common minimum flow recirculation control valve
viii. Supply sealing water to glands of all valves connected with vacuum service

12.5.7 Line-up of miscellaneous valves

i. Isolation valves of all pressure gauges are open
ii. Root valves of impulse lines, installed on steam lines, condensate line, oil lines, and cooling water lines, are open
iii. Emergency stop valve/s (ESV/s) of HP turbine is/are closed
iv. Interceptor valve/s (IV/s) of IP turbine is/are closed
v. Control valves of both HP turbine and IP turbine are closed
vi. Non-return valves at HP turbine exhaust are closed
vii. Non-return valves on turbine extraction steam lines are closed
viii. Drain/blow-down valves on main steam piping are closed
ix. Vent valves before HP bypass valve are closed
x. Drain/blow-down valves on cold and hot reheat piping are closed
xi. Drain valves for draining HP turbine casing are closed
xii. Before-seat drain valves and after-seat drain valves of non-return valves on turbine extraction steam lines are closed
xiii. Warm-up drain valves for emergency stop valve/s and control valves are open

12.5.8 Increase condenser vacuum

It is known from Chapter 9, *Power Plant Systems*, that evacuating air from the condenser may be realized either with the help of a steam jet air ejector or by utilizing a vacuum pump. Increase the condenser vacuum as detailed below:

In the event a steam jet-air ejector is used:

i. Slowly charge the ejector steam header and warm it up
ii. Start the steam-air ejector by opening its steam supply valve
iii. Open the isolating valves on air evacuating line connecting condenser and steam-air ejector
iv. As soon as the vacuum starts building up in the system apply the gland sealing steam, at a pressure slightly above atmospheric, to the front and rear gland seals of each of HP, IP, and LP turbines

v. Verify that the vacuum stabilizes or tends to stabilize at a level recommended by the manufacturer
vi. Keep the steam-air ejector running to maintain vacuum
vii. Check low vacuum trip of turbine

When vacuum pump is used:

i. Start the vacuum pump
ii. Open the isolating valves on air evacuating line connecting condenser and suction line of vacuum pump
iii. As soon as the vacuum starts building up in the system apply the gland sealing steam, at a pressure slightly above atmospheric, to the front and rear gland seals of each of the HP, IP, and LP turbines
iv. Verify that vacuum stabilizes or tends to stabilize at a level recommended by the manufacturer
v. Keep vacuum pump running to maintain vacuum
vi. Check low vacuum trip of turbine

12.5.9 Sequence of activities for cold start-up (follow Cold Start-up Curve as furnished or recommended by the manufacturer)

1. Open drain valves on main steam line/s
2. Open vent valves before HP bypass valves
3. Gradually open superheater outlet isolation valve/s
4. At the start of heating of main steam line/s condensed water will get drained, but gradually steam will come out through drain lines (that can be sensed by audible sound of steam escape), which indicates adequate warm-up of main steam line/s; close drain valves
5. When steam comes out through vent lines before HP bypass valves close vent valves
6. Gradually open HP bypass valve/s fully, ensuring that temperature control loop is functioning normally
7. Also ensure that LP bypass system is functioning normally
8. Open drain valves on cold reheat and hot reheat lines located as close as possible to steam turbine
9. Steam flow thus established through the bypass system and drain lines should assist in heating cold reheat and hot reheat lines
10. At the start of heating cold reheat and hot reheat lines condensed water will be drained, but gradually steam will come out through the drain lines, indicating adequate warm-up of cold reheat and hot reheat lines
11. Close drain valves; this should assist in increasing steam parameters

12. Once the main steam, cold reheat, and hot reheat lines are adequately warmed up increase the steam temperature at the steam generator outlet to a value higher than the minimum temperature requirement of the steam turbine as recommended by the manufacturer
13. Verify the H_2 purity in generator, if provided, is above 94% or as recommended by the manufacturer
14. Put generator seal oil system, if provided, in service
15. Put generator stator cooling water system, if provided, in service
16. Verify that degree of superheat of main steam upstream of HP turbine is more than 50 K
17. Do not allow entry of steam into steam turbine until recommended matching of stem temperature is accomplished
18. Follow the manufacturer's guidelines for matching steam-metal differential temperature
19. Open turbine casing drain valves
20. Open emergency stop valve/s (ESV/s) of HP turbine or verify that these valves open automatically through turbine start-up system
21. Open interceptor valve/s (IV/s) of IP turbine or verify that these valves open automatically through turbine start-up system
22. Admit steam to turbine and increase speed to predetermined value as per guidelines of the manufacturer for warming up of turbine casings
23. During warm-up and start-up of the steam turbine, observe following parameters remain at or within predetermined limits:
    a. Steam-metal differential temperature
    b. Differential temperature between upper and lower halves of turbine casing
    c. Differential expansions of turbine
    d. Bearing temperature and vibration
    e. Steam parameters
    f. Operating parameters of condensing system
    g. Operating parameters of turbine oil system
24. Maintain warm-up speed for soaking purposes
25. Once casings are warmed-up increase turbine speed and start rolling
26. Stop turning gear or verify that it stops automatically once turbine speed exceeds preset limit
27. Gradually increase the speed at a rate recommended by the manufacturer by allowing more steam to enter the steam turbine provided following parameters are within permissible limits:
    a. Steam-metal differential temperature margin
    b. Differential temperature between upper and lower halves of turbine casing
    c. Differential expansion of casings
    d. Bearing temperature and vibration
    e. Turbine exhaust hood temperature

28. On reaching the rated speed leave the steam turbine for a recommended duration to ensure proper soaking of the turbine
29. Place automatic voltage regulator (AVR) on "auto"
30. Switch on field breaker
31. Synchronize generator with the grid (this is discussed in detail under Section 11.7, Chapter 11, *Interlock and Protection*) and take a block load
32. Close drain valves of HP turbine casing
33. Close warming up drain valves for emergency stop valve/s and control valves
34. Open before-seat drain valves and after-seat drain valves of non-return valves on turbine extraction steam lines
35. Increase load at a predetermined rate
36. Open extraction steam supply block valves; place LP feedwater heaters in service
37. Increase main steam temperature and pressure following the manufacturer's guidelines
38. Close HP-LP bypass valves
39. Place HP feedwater heaters in service
40. Increase load at a predetermined rate to a load according to the system demand once the main steam temperature and pressure reach the rated values
41. Observe the following parameters remain at or within predetermined limits:
    a. Differential expansions of turbine
    b. Bearing temperature and vibration
    c. Steam parameters
    d. Operating parameters of all associated systems, e.g., condensate, feedwater, extraction steam, and turbine lube oil, control oil, hydrogen, seal oil, stator cooling water, circulating water, etc.

12.5.10 Sequence of activities for warm start-up (follow Warm Start-Up Curve as recommended by the manufacturer)

1. Auxiliary equipment should be started in the same manner and order as for cold start-up
2. Check that all auxiliary equipment is working satisfactorily
3. Check the condition of all protections and interlocks
4. Ensure that the boiler has been lit and steam flow established through the HP-LP bypass system
5. Using the manufacturer recommended start-up curve increase the desired steam temperature and pressure for rolling the turbine
6. Admit steam to the turbine and start steam rolling
7. Observe the following parameters are within the permissible limits:
    a. Steam-metal differential temperature
    b. Differential temperature between upper and lower halves of turbine casing
    c. Differential expansions of turbine
    d. Bearing temperature and vibration

   e. Steam parameters
   f. Operating parameters of condensing system
   g. Operating parameters of turbine oil system.
8. Maintain warm-up speed for soaking purposes
9. Once casings are warmed up increase turbine speed and start rolling
10. Stop turning gear or verify that it stops automatically once turbine speed exceeds preset limit
11. Close all drain valves
12. Gradually increase the speed to the rated value at a rate recommended by the manufacturer by allowing more steam to enter the turbine, provided the following parameters are within the permissible limits:
    a. Steam-metal differential temperature margin
    b. Differential temperature between upper and lower halves of turbine casing
    c. Differential expansion of casings
    d. Bearing temperature and, vibration
    e. Turbine exhaust hood temperature.
13. Place automatic voltage regulator (AVR) on "auto"
14. Switch on field breaker
15. Synchronize generator with the grid and take a block load
16. Open extraction steam supply block valves; place LP feedwater heaters in service
17. Increase main steam temperature and pressure following manufacturer's guidelines
18. Close HP-LP bypass valves
19. Place HP feedwater heaters in service
20. Increase load at a predetermined rate to a load according to the system demand once the main steam temperature and pressure reach the rated values
21. Observe following parameters remain at or within predetermined limits:
    a. Differential expansions of turbine
    b. Bearing temperature and vibration
    c. Steam parameters
    d. Operating parameters of all associated systems, e.g., condensate, feedwater, extraction steam, and turbine lube oil, control oil, hydrogen, seal oil, stator cooling water, circulating water, etc.

12.5.11 Sequence of activities for hot start-up (follow Hot Start-Up Curve as recommended by the manufacturer)

The procedure for hot start-up is similar to the procedure for warm start-up with the following differences:

1. Before rolling, the HP turbine inlet pressure and temperature will be much higher than those for warm start-up; from hot start-up curve determine the desired steam temperature and pressure for rolling the turbine

2. Admit steam to the turbine and start steam rolling
3. Observe the following parameters are within the permissible limits:
   a. Steam-metal differential temperature
   b. Differential temperature between upper and lower halves of turbine casing
   c. Differential expansions of turbine
   d. Bearing temperature and vibration
   e. Steam parameters
   f. Operating parameters of condensing system
   g. Operating parameters of turbine oil system.
4. Stop turning gear or verify that it stops automatically once turbine speed exceeds preset limit
5. Close all drain valves
6. Gradually increase the speed to the rated value at a rate recommended by the manufacturer by allowing more steam to enter the turbine provided following parameters are within permissible limits:
   a. Steam-metal differential temperature margin
   b. Differential temperature between upper and lower halves of turbine casing
   c. Differential expansion of casings
   d. Bearing temperature and vibration
   e. Turbine exhaust hood temperature.
7. Place automatic voltage regulator (AVR) on "auto"
8. Switch on field breaker
9. Synchronize generator with the grid and take a block load
10. Open extraction steam supply block valves; place LP feedwater heaters in service
11. Increase main steam temperature and pressure following manufacturer's guidelines
12. Close HP-LP bypass valves
13. Place HP feedwater heaters in service
14. Increase load at a predetermined rate to a load according to the system demand once main steam temperature and pressure reach rated value
15. Observe the following parameters remain at or within the predetermined limits:
    a. Differential expansions of turbine
    b. Bearing temperature and vibration
    c. Steam parameters.
    d. Operating parameters of all associated systems, e.g., condensate, feedwater, extraction steam, and turbine lube oil, control oil, hydrogen, seal oil, stator cooling water, circulating water, etc.

12.5.12 Sequence of activities for very hot start-up (follow Very Hot Start-Up Curve as recommended by the manufacturer)

The procedure for very hot start-up of steam turbine, as given below, is quite different from the previous three start-ups, since the metal temperature of the casing remains at or near its full-load metal temperature.

1. Before rolling, the HP turbine inlet pressure and temperature will be near to or at their rated conditions as per very hot start-up curve
2. Admit steam to the turbine and gradually increase the speed to the rated value at a rate recommended by the manufacturer by allowing more steam to enter the turbine, provided the following parameters are within the permissible limits:
   a. Differential expansions of turbine
   b. Bearing temperature and vibration
   c. Steam parameters
   d. Operating parameters of all associated systems, e.g., condensate, feedwater, extraction steam, and turbine lube oil, control oil, hydrogen, seal oil, stator cooling water, circulating water, etc.
3. Stop turning gear or verify that it stops automatically once turbine speed exceeds preset limit
4. Close all drain valves
5. Place automatic voltage regulator (AVR) on "auto"
6. Switch on field breaker
7. Synchronize generator with the grid and increase load at a predetermined rate to a load according to the system demand
8. Open extraction steam supply block valves; place LP feedwater heaters in service
9. Close HP-LP bypass valves
10. Place HP feedwater heaters in service.

12.6 NORMAL SHUT-DOWN OF STEAM TURBINE

1. Inform the boiler house that the turbine is being shut down
2. Ensure auxiliary oil pump and emergency oil pump are available and are ready for operation
3. Ensure the HP-LP bypass controllers are ON
4. Check for non-seizure of ESVs and IVs
5. Unload the turbine by gradually closing the control valves; the unloading should be carried out at a rate determined by the manufacturer's guidelines
6. During unloading always keep watch on the following parameters to ensure they remain within the permissible limits:
   i. Differential contractions of turbine, especially of HP rotor

   ii. Bearing vibration
   iii. Steam-metal differential temperature margin
   iv. Differential temperature between upper and lower halves of turbine casing
7. For minimizing the contraction of the HP rotor during shut-down admit main steam to HP front gland, if recommended by the manufacturer
8. Start the auxiliary oil pump and ensure that oil pressure in lube oil system is normal
9. Maintain lube oil temperature
10. After unloading the turbine to no load, trip the turbine manually
11. Ensure the ESV, IV, and control valves get closed
12. Ensure the extraction steam line block valves closed
13. Ensure the non-return valves at the HP turbine exhaust are closed
14. Ensure the generator is isolated through reverse power/low forward power relay
15. Verify that the field breaker is switched off
16. With the tripping of generator circuit breaker the HP-LP bypass system becomes operative; this may be used to stabilize boiler conditions and maintain steam flow through reheater until the boiler is shut down
17. Start jacking oil pump
18. When rotor speed falls to a predetermined low value, put the rotor on turning gear manually or turning gear may cut-in automatically. The rotor should remain on turning gear till temperature of HP casing falls below a recommended limit
19. Stop jacking oil pump
20. Follow the manufacturer's guidelines to open drain valves as follows:
   i. HP turbine casing
   ii. Before-seat and after-seat drain valves of non-return valves on turbine extraction steam lines
   iii. Warm up drain valves for emergency stop valve/s and control valves
21. Close the HP-LP bypass valves
22. Stop the boiler firing per the recommendation of the boiler manufacturer and bottle up the boiler
23. Break the condenser vacuum once the fire in the boiler is killed
24. When the vacuum falls to zero, stop the gland steam supply
25. Stop the condensate extraction pump
26. The CW pump may be stopped when the temperature of the exhaust part of the LP turbine falls to 328 K or to a value recommended by the manufacturer
27. Open the drain valves on main steam piping
28. Open drain valves on cold and hot reheat piping

29. As the unit cools down and water shrinks, add make-up water intermittently to the boiler to prevent the drum level from dropping below the visibility limit of drum level gauge glass.

12.7 EMERGENCY SHUT-DOWN OF STEAM TURBINE

An emergency shut-down of a steam turbine may become imminent due to variety of off-normal conditions either inside the steam turbine, such as unusual noise, etc., or in its associated equipment/systems, e.g., quality of stator cooling water deviating from its recommended limit.

1. Manually operate emergency trip push buttons
2. Ensure the ESVs, IVs, and control valves of both the HP and IP turbines get closed
3. Ensure the extraction steam line block valves are closed
4. Ensure the non-return valves at HP turbine exhaust are closed
5. Isolate the generator and switch off the field breaker
6. Make sure the HP-LP bypass valves have opened
7. Start the auxiliary oil pump and ensure the oil pressure in lube oil system is normal
8. Maintain the lube oil temperature
9. Start the jacking oil pump
10. When the rotor speed falls to a predetermined low value, put the rotor on turning gear manually or turning gear may cut-in automatically
11. Break the condenser vacuum
12. Shut down the boiler and keep it boxed and bottled up
13. Close HP-LP bypass valves
14. Following the manufacturer's guideline open the drain valves as follows:
    - **i.** HP turbine casing
    - **ii.** Before-seat and after-seat drain valves of non-return valves on turbine extraction steam lines
    - **iii.** Warm up drain valves for emergency stop valve/s and control valves
    - **iv.** Drain valves on main steam piping
    - **v.** Drain valves on cold and hot reheat piping.
15. When the vacuum falls to zero, stop the turbine gland seal steam supply
16. Stop the condensate extraction pump
17. The CW pump may be stopped when the temperature of the exhaust part of the LP turbine falls to 328 K or to a value recommended by the manufacturer
18. The rotor should remain on the turning gear until the temperature of the HP casing falls below a recommended limit
19. Stop the jacking oil pump

20. As the unit cools down and water shrinks, add make-up water intermittently to the boiler to prevent the drum level from dropping below the visibility limit of the drum level gauge glass
21. Do not restart the boiler until identification and rectification of defects of turbine.

12.8 APS

From the previous discussion it is clear that start-up and (normal) shut-down of a large pulverized coal-fired unit involves a huge number of parallel activities that are realized through multiple actions by operators. In the process if any errors occur, due to erroneous operation and/or lack of proper understanding of sequence of operation, catastrophic damage to equipment, safety of the unit, loss of time, loss of generation, and as an ultimate loss of revenue could occur. To relieve operators from having to perform such arduous tasks, there is APS, a computer-aided start-up and (normal) shut-down system.

APS, which stands for automatic plant start-up and shut-down system, ensures the automatic start-up and shut-down of a unit according to start-up/shut-down program by a 'single command' by an operator. It reduces degree of operator error, start-up/shut-down time of each unit, and simultaneously enhances safety. APS is based on integrated unit-level-based control philosophy that can be achieved with minimum operator intervention.

If for some reason the APS sequence gets stalled the operator may either take the necessary corrective action or consciously bypass that step to ensure that the continuity of the start-up/shut-down process does not get impaired. However, operator intervention may be inevitable in the event of any or all of the following conditions:

i. Areas or equipment that cannot be automated
ii. On occurrence of spurious signal
iii. There is lack of required instrumentation
iv. Inputs fail to satisfy desired requirement
v. Realizing specific requirement of APS

The APS sequence is conveniently hooked up to major equipment logic sequences, such as the burner management system, automatic turbine start-up system, sequential control system, balance of plant operating system, etc., such that APS appears as a stand-alone program for both the steam generator and the steam turbine and generator.

Before discussing the details of APS, it should be noted that it is not the first start-up of any plant. During the first start-up of a plant there are many areas that may operate in parallel and demand the special attention of the operating personnel. In these cases, more vigilance is needed to observe the individual or interactive behavior of the various systems and equipment before the sequence of activities may proceed.

In contrast, APS is executed in a step-by-step manner. It is like any "automatic control loop" that can be put into service once all associated systems had been started, trialed, tested, and stabilized. Thus, before initiating the APS program there are certain "prerequisite" conditions/areas that need to be handled by the operator.

Readers are advised to become familiar with the previous sections as well as the previous chapters to better understand the intricacies of APS.

12.8.1 Automatic start-up

The first step in the start-up program is to ensure that all prerequisites are established. It must be ensured that all associated systems, e.g., compressed air system, raw water system, DM water system, CW, ACW, and CCW systems, DCS, etc., are in service and stabilized. It must also be ensured that all associated electric power supply systems are in service and stabilized. Furthermore, since emergency power acts as back-up to power control centers, uninterrupted power supply systems (UPS), and DC distribution boards, the status of emergency power supply is also prerequisites to ensure that in the event of unit power failure the progress of the APS sequence will not get interrupted.

These conditions, nevertheless, do not initiate the start-up program. There are many more prerequisite conditions that need to be satisfied before issuing the start-up command. These prerequisites, as follows, are neither prioritized nor put in sequential order, but should all be fulfilled for automatic start-up to progress:

1. Firefighting system is ready to be put into service
2. Power receiving substation is energized
3. LT (415V) start-up switchgears are energized
4. Common power centers are energized
5. Water treatment switchgear is energized
6. Unit power centers are energized
7. Common distribution boards are energized
8. 125V DC distribution board is energized
9. UPS is energized
10. DCS is in service
11. Emergency DG and associated emergency switchgear are commissioned
12. Lighting and electric service power system is energized
13. Communication system is in service
14. Air-conditioning plant is in service
15. Generator step-up transformer is ready to be put into service
16. Unit auxiliary transformer is ready to be put into service
17. Generator circuit breaker is ready to be put into service
18. Switchyard substation is ready to be put into service
19. Excitation panel is energized

20. AVR is on auto
21. Alarm annunciation system and data acquisition system (DAS) are in service
22. Adequate storage of raw water is available
23. Adequate storage of DM water is available
24. Auxiliary steam is available at adequate pressure
25. DMW system is lined-up
26. ACW system is lined-up
27. CCW system is lined-up
28. Instrument air system is lined-up
29. Service air system is lined-up
30. CW system is lined-up
31. Condensate system is lined-up
32. Feedwater system is lined-up
33. HFO system is lined-up and is ready to be put into service
34. Atomizing steam system is lined-up
35. Atomizing burner valves open and are ready to be put into service
36. Boiler superheater outlet stop valves are closed
37. Boiler superheater header drain valves are open full
38. Boiler reheater header drain valves are open full
39. Boiler waterwall header drain valves are open full
40. Boiler superheater outlet (SHO) vent valve is open full
41. Boiler reheater outlet (RHO) vent valve is open full
42. Feedwater inlet valve to economizer inlet header is open
43. Drain valves at feedwater inlet to economizer inlet header are closed
44. Drain valves on Main Steam piping are closed
45. Drain valves on Hot Reheat piping are closed
46. Drain valves on Cold Reheat piping are closed
47. ESP system is lined-up
48. FGD system is lined-up
49. DeNOx system is lined-up
50. Debris filter on CW supply pipe line to condenser is in service
51. Condenser on-load tube cleaning system is in service
52. Drum level (applicable to drum boiler) is maintained normal through auto-control
53. Deaerator feedwater storage tank level is maintained normal through auto-control
54. Hotwell level is maintained normal through auto-control
55. DMW transfer pump recirculation control valve is on auto
56. CCCW head tank DMW inlet control valve is on auto
57. CCCW header pressure control valve is on auto
58. Condensate transfer pump recirculation control valve is on auto

59. Condensate storage tank DMW inlet control valve is on auto
60. Turbine seal steam header supply control valve from auxiliary steam system is on auto
61. Turbine seal steam header drain control valve is on auto
62. HFO forwarding pump recirculation control valve is on auto
63. Atomizing steam pressure control valve near burner is on auto
64. Deaerator pressure control valve is on auto
65. CEP minimum flow recirculation control valve is on auto
66. BFP minimum flow recirculation control valve is on auto
67. Turbine lube oil (LO) pressure control valve is on auto
68. Turbine LO temperature control valve is on auto
69. Turbine control oil pressure control valve is on auto
70. Turbine control oil temperature control valve is on auto
71. Generator seal oil pressure control valve is on auto
72. Generator H_2 pressure control valve is on auto
73. Feedwater low load control valve is on auto
74. Feedwater full load control valve is on auto
75. Auxiliary steam header pressure control valve is on auto
76. Auxiliary steam header temperature control valve is on auto
77. HP/LP bypass pressure control valve is on auto
78. HP/LP bypass temperature control valve is on auto
79. Turbine lube oil (LO) system is ready to be put into service
80. Turbine jacking oil (JO) system is ready to be put into service
81. Turbine control fluid system is ready to be put into service
82. Generator seal oil system is ready to be put into service
83. Generator H_2 system is ready to be put into service
84. Generator stator cooling water system is ready to be put into service
85. HP turbine emergency stop valves are closed
86. HP turbine control valves are closed
87. IP turbine interceptor valve/s (IVs) are closed
88. IP turbine control valves are closed
89. Coal bunkers have enough storage of coal
90. Coal feeders are ready to be put into service
91. Pulverizers are ready to be put into service.

[Note that the term "lined-up" appears in many places in this list. It essentially means that before initiating the sequence to progress the associated system/s must ensure free flow through it/them. Some examples of "lined-up" conditions are:

i. a valve that is to remain open (or closed) before initiating the sequence, then the status of this valve should be ensured by "open" (or closed) limit switch

ii. a damper that is to remain open (or closed) before initiating the sequence, then the status of this damper should be ensured by "open" (or closed) limit switch
iii. a storage tank from which fluid has to be taken to system, then it is to be ensured that level in this tank is adequate
iv. a sink, where the fluid will flow, must be in a position to receive the fluid
v. an electric power supply source must be in energized condition and be available whenever required, etc.]

Along with the prerequisites, the program determines the status of start-up condition (i.e., cold, warm, hot, or very hot start-up) in line with the guidelines described in Section 12.5 (and shown in Figure 12.1).

Depending on the type of start-up condition, some permissive conditions are either bypassed or may need to be repeated. Once all prerequisites are satisfied and the status of permissive condition is determined, the operator initiates the single "start-up sequence" command to the "APS start-up program" for it to advance automatically and thereby the main sequence is activated (Figure 12.2). As the program advances, it issues commands to various equipment to start and then checks back to determine whether the command was realized. If satisfied, the program advances further sequentially from the source of the DM water up to the loading of turbine.

The APS flow diagram of a typical pulverized coal-fired power plant is very complex and large in size, which is difficult to accommodate in a single chapter. Therefore, a general step-by-step automatic start-up sequence is given in Table 12.1 for simplicity. While preparing Table 12.1 the input criteria at many steps essentially resembles the check-back criteria of the previous step.

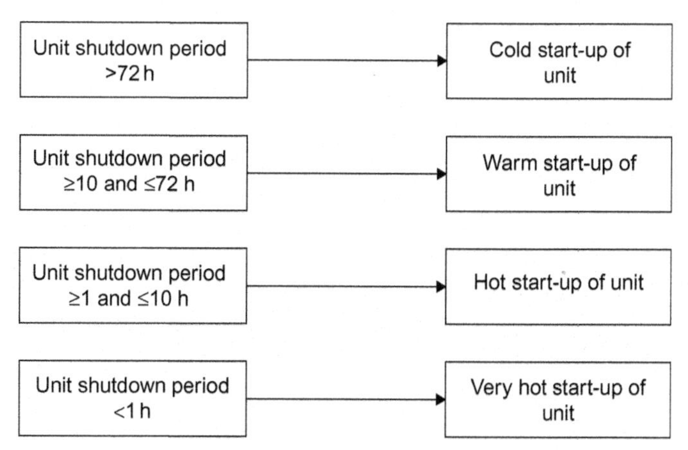

Figure 12.1 Start-up conditions.

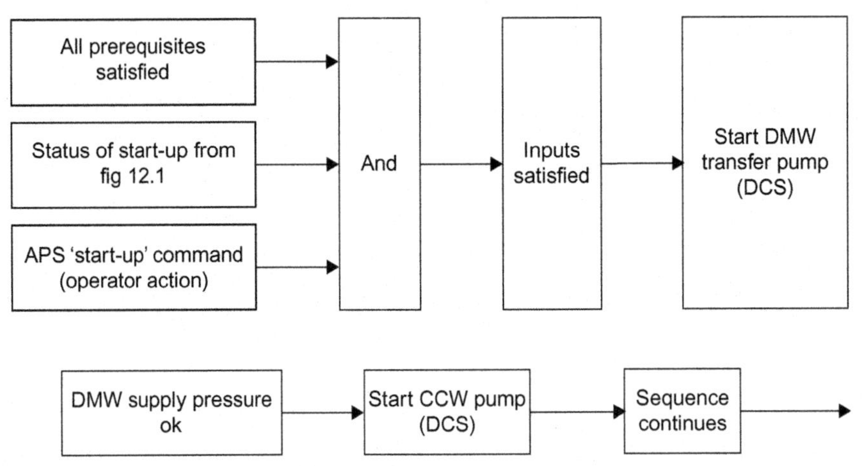

Figure 12.2 Start command initiated by operator.

12.8.2 Automatic (Normal) shut-down program

For normal shut-down of a unit the turbine is unloaded first, and simultaneously the HP/LP bypass is operated. With the bypass in service once the turbine load is reduced to 60% (which is the stable load for boiler control) unloading of boiler takes place according to the load on the turbine and the opening of the HP/LP bypass valves. Once the turbine is fully unloaded, the boiler starts unloading to zero load.

As in case of the automatic start-up program certain "prerequisite" conditions, as follows, need to be fulfilled by operator action before initiating the shut-down command. This equipment may not remain in service during normal operation, but before initiating automatic (normal) shut-down of the unit, it must be available.

1. Turbine auxiliary oil pump is available
2. Turbine emergency oil pump is available
3. Turbine JO pump is available
4. Turbine control oil pump is ON
5. Turning gear is available

(*Note*: During turbine shut-down, "below a certain turbine speed" the turning gear comes into service and remains in service till HP turbine casing temperature is above a low temperature limit as recommended by the manufacturer. The turning gear is stopped only when "HP turbine casing temperature falls below the recommended low limit.")

The step-by-step sequence of the automatic (normal) shut-down of a typical pulverized coal-fired power plant is given in Table 12.2.

(*Note*: If the following equipment are installed 'unit wise' they may also be stopped at Step 25:

Instrument and service air compressors, CW pumps, and DMW pumps.)

Table 12.1 Step-by-step sequence of automatic start-up of a unit

Step no.	Input critera	Command issued	Check-back criteria
1	i. Prerequisites satisfied ii. Status of coldstart-up condition identified iii. Once above two conditions are realized "auto start-up sequence of cold start-up" command is initiated by operator	Start DMW transfer pump	DMW supply header pressure OK
2	DMW supply header pressure OK	Start closed cycle cooling water (CCCW) pump	CCCW supply header pressure OK
3	CCCW supply header pressure OK	Start instrument air (IA) compressor	IA supply header pressure OK
4	IA supply header pressure OK	Start HFO pump	HFO supply pressure near burner OK
5	HFO supply pressure near burner OK	Start water box vacuum pump	Water box vacuum pump ON
6	Water box vacuum pump ON	Open condenser CW outlet valve	Condenser CW outlet valve open
7	Condenser CW outlet valve open	Open condenser CW inlet valve	Condenser CW inlet valve open
8	Condenser CW inlet valve open	Open turbine gland seal supply and drain control valves	Turbine gland seal steam header pressure OK
9	Turbine gland seal steam header pressure OK OR i. Prerequisites satisfied ii. Status of warm start-up condition identified (bypass step 1 to step 8)	Start condenser vacuum pump/s	Condenser vacuum OK

(*Continued*)

Table 12.1 (Continued)

Step no.	Input critera	Command issued	Check-back criteria
	iii. Once above two conditions are realized "auto start-up sequence of warm start-up" command is initiated by operator		
10	i. Condenser vacuum OK ii. Master Fuel Trip Relay (MFT) reset	Initiate boiler draft group start sequence	Regenerative Air pre-heaters, ID and FD fans running
11	Regenerative Air heaters, ID and FD fans running Inlet and outlet dampers of ID fans open Outlet dampers of FD fans open	Adjust flow through ID and FD fans to permit at least 30% total air flow	Total air flow $\geq$ 30%
12	Total air flow $\geq$ 30%	Initiate boiler purge	Purge complete
13	Purge complete	i. Put furnace draft control on "auto" ii. Place all burner air registers to pre-light position iii. Initiate light-up sequence of the lowest elevation of igniters and warm-up oil burners	Initiate igniter start sequence
14	Igniter energized	i. Open the main fuel oil shut-off valve ii. Open the burner fuel oil valve	Burner flame established
15	Burner flame established OR i. Prerequisites satisfied ii. Status of hot start-up condition identified (bypass step 1 to step 14) iii. Once above two conditions are realized "auto start-up sequence of hot start-up" command is initiated by operator	i. Regulate HFO flow to increase boiler water temperature ii. Open MS header drain valve iii. Open HRH header drain valve iv. Open CRH header drain valve	i. Furnace exit gas temperature $\leq$ 813 K ii. MS, HRH and CRH drain valves open

16	MS, HRH, and CRH drain valves open	Increase steam pressure at a rate as recommended by boiler manufacture	Super heater outlet steam pressure $\geq$ 0.2 MPa
17	i. Drum steam pressure $\geq$ 0.2 MPa ii. Drum metal temperature OK iii. Drum level OK	i. Close boiler superheater header drain valves ii. Close boiler reheater header drain valves iii. Close boiler waterwall header drain valves iv. Close boiler SHO vent valve v. Close boiler RHO vent valve.	i. Boiler superheater header drain valves close ii. Boiler reheater header drain valves close iii. Boiler waterwall header drain valves close iv. Boiler SHO vent valve close v. Boiler RHO vent valve closed.
18	Check-back criteria of Step 17 OK	i. Open SHO stop valve ii. Start turbine JO pump iii. Start turbine LO pump	i. SHO stop valve open ii. Turbine JO pressure OK iii. Turbine LO pressure OK
19	Check-back criteria of Step 18 OK	Start turning gear	Turning gear ON
20	Turning gear ON OR i. Prerequisites satisfied ii. Status of very hot start-up condition identified (bypass step 1 to step 19) iii. Once above two conditions are realized "auto start-up sequence of very hot start-up" command is initiated by operator	Start turbine control oil pump	Control oil pressure OK
21	i. Control oil pressure OK ii. Cold start-up status realized	HPLP bypass to maintain steam pressure and temperature at the inlet to both HP turbine and IP turbine as per the Cold Start-up Curve	Initiate turbine cold start-up
22	i. Control oil pressure OK ii. Warm start-up status realized	HPLP bypass to maintain steam pressure and temperature at the inlet to both HP turbine and IP turbine as per Warm Start-up Curve	Initiate turbine warm start-up

(*Continued*)

Table 12.1 (Continued)

Step no.	Input critera	Command issued	Check-back criteria
23	i. Control oil pressure OK ii. Hot start-up status realized	HPLP bypass to maintain steam pressure and temperature at the inlet to both HP turbine and IP turbine as per Hot Start-up Curve	Initiate turbine hot start-up
24	i. Control oil pressure OK ii. Very hot start-up status realized	HPLP bypass to maintain steam pressure and temperature at the inlet to both HP turbine and IP turbine as per Very hot Start-up Curve	Initiate turbine very hot start-up
25	i. turbine cold start-up initiated (check-back criteria of Step 21) or turbine warm start-up initiated (check-back criteria of Step 22) or turbine hot start-up initiated (check-back criteria of Step 23) or turbine very hot start-up initiated (check-back criteria of Step 24) ii. IP turbine inlet pressure and temperature OK as per either the cold start-up curve or the warm start-up curve or the hot start-up curve or the very hot start-up curve iii. Differential temperature between inlet steam and IP turbine metal within recommended limit	Open interceptor valves (IVs) of IP turbine	IVs open

26	IVs open	i. Close MS header drain valve ii. Close HRH header drain valve iii. Auto operation of CRH header drain valve	i. MS header drain valve close ii. HRH header drain valve closed iii. CRH header drain valve operating on "auto"
27	Check-back criteria of Step 26 OK	Modulate IP control valve	IP control valve not closed
28	i. IP control valve not closed ii. HP turbine inlet steam pressure and temperature OK as per the start-up curve iii. HPT casing top-bottom differential temperature OK iv. Differential expansion of turbine OK v. Turbine bearing temperature OK vi. Turbine bearing vibration OK vii. Turbine exhaust hood temperature OK	Increase turbine speed as per turbine start-up curve	Turbine speed > minimum speed (turning gear speed)
29	Check-back criteria of Step 28 OK	Stop turning gear	Turning gear stopped
30	Turning gear stopped	i. Stop jacking oil pump ii. Increase turbine speed as per turbine start-up curve	Turbine speed 100%
31	i. Turbine speed 100% ii. Speed set point at 100% iii. Load limit set point > 100% iv. Generator seal oil pressure OK v. Generator H_2 pressure OK vi. Generator stator cooling water pressure OK vii. Generator H_2 purity OK viii. Generator voltage OK ix. Excitation voltage OK	Synchronize turbine	Turbine synchronized

(Continued)

Table 12.1 (Continued)

Step no.	Input critera	Command issued	Check-back criteria
32	Turbine synchronized	Energize LT switchgear	LT switchgear energized
33	i. LT switchgear energized ii. Rotor stress within limit	Modulate HP control valve	HP control valve not closed
34	HP control valve not closed	Switchover turbine load control to "auto" and load turbine	Turbine load 12–15%
35	Turbine load 12–15%	Start pre-heating of extraction steam lines (operator action)	Extraction steam lines pre-heating OK
36	Extraction steam lines pre-heating OK	Initiate extraction steam start-up sequence (operator action)	Extraction steam start-up sequence OK
37	i. Extraction steam start-up sequence OK ii. Turbine load 30%	i. Check steam line drain valves closed ii. Check extraction steam line drain valves closed iii. Start PA fans iv. Open bunker outlet and coal feeder inlet gates v. Start pulverizer serving a lower elevation of coal nozzles	i. Steam line drain valves ii. Extraction steam line drain valves closed iii. PA fans ON iv. Pulverizer ON
38	Check-back criteria of Step 37 OK	i. Open hot air shut-off gate and bring pulverizer up to the required operating temperature ii. Maintain fuel feed at minimum consistent with stable ignition iii. Switchover drum level control to "auto"	i. Pulverizer outlet temperature OK ii. Fuel feed minimum iii. Drum level control on "auto"
39	Check-back criteria of Step 38 OK	i. Regulate hot and cold air dampers to maintain pulverizer outlet temperature ii. Increase load	i. Pulverizer outlet temperature OK ii. Turbine load > 40%

40	Turbine load > 40%	i. Place additional pulverizers in service ii. Remove fuel oil guns and igniters from service iii. Increase load	i. Fuel oil guns and igniters removed ii. Turbine load > 60%
41	Turbine load > 60%	i. Switchover SH steam temperature control to "auto" ii. Switchover HR steam temperature control to "auto" iii. Switchover combustion control to "auto"	i. SH steam temperature control on "auto" ii. RH steam temperature control on "auto" iii. Combustion control on "auto" iv. Throttle steam pressure and temperature reached rated value
42	i. Check-back criteria of Step 41 OK ii. Turbine load equal to or greater than load demand	Switchover coordinated control to "auto"	Coordinated control on "auto"
43	Coordinated control on "auto"		AUTOMATIC UNIT START-UP SEQUENCE COMPLETE

Table 12.2 Step-by-step sequence of automatic shut-down of a unit

Step no.	Input critera	Command issued	Check-back criteria
1	i. Prerequisites satisfied ii. Once above condition is realized "auto shut-down sequence start" command is initiated by operator	i. Switchover coordinated control to manual ii. HP/LP bypass becomes operative iii. Unloading of turbine starts as per recommended rate by gradual closure of turbine control valves	i. Coordinated control on "manual" ii. HP/LP bypass in service iii. Turbine unloading in progress
2	i. Turbine unloading in progress ii. HP turbine steam-metal differential temperature OK iii. HPT casing top-bottom differential temperature OK iv. Differential contraction of turbine OK v. Turbine bearing temperature OK vi. Turbine bearing vibration OK vii. Turbine LO pressure OK viii. Turbine LO temperature is within safe limit	Unloading of turbine continued	Turbine load $< 60\%$
3	Turbine load $< 60\%$	i. Gradually reduce boiler load as recommended rate ii. Unloading of turbine continued	i. Boiler unloading in progress ii. Turbine unloaded fully
4	Turbine unloaded full	i. Switchover turbine speed control to manual ii. Start turbine auxiliary pump iii. Trip turbine	i. Turbine speed control on manual ii. Turbine auxiliary oil pump ON iii. Turbine tripped

5	i. Turbine tripped ii. Turbine LO pressure OK iii. ESVs of HP turbine closed iv. HP control valves closed v. IVs of IP turbine closed vi. IP control valves closed vii. HP turbine exhaust non-return valves closed viii. Extraction steam line block valves closed	Operate reverse/low forward power relay	Reverse/low forward power relay operated
6	Reverse/low forward power relay operated	Open generator transformer circuit breaker	Generator transformer circuit breaker open
7	i. Generator transformer circuit breaker open ii. Turbine speed falling below a preset limit	Switch-off field breaker	Field breaker off
8	i. Field breaker off ii. Excitation voltage zero	Start jacking oil pump	Jacking oil pressure OK
9	i. Jacking oil pressure OK ii. Turbine speed falling below minimum preset limit	Start turning gear	Turning gear ON
10	i. Turning gear ON ii. HP turbine casing temperature reaches preset low limit	i. Open turbine drain valves ii. Stop turning gear iii. Stop JO pump iv. Stop auxiliary oil pump	i. Turbine drain valves open ii. Turning gear stopped iii. Turbine JO pump stopped iv. Turbine auxiliary oil pump stopped

(*Continued*)

Table 12.2 (Continued)

Step no.	Input critera	Command issued	Check-back criteria
11	i. Boiler unloading in progress ii. Boiler load < 60%	i. Switchover combustion control to manual ii. Switchover superheat steam temperature control to manual iii. Switchover reheat steam temperature control to manual	i. Combustion control on manual ii. Superheat steam temperature control on manual iii. Reheat steam temperature control on manual
12	Pulverizer outlet coal-air temperature falls to preset low limit	i. Reduce air flow through pulverizer to its preset minimum limit ii. Close hot air damper of pulverizer	i. Air flow through pulverizer is minimum ii. Hot air damper of pulverizer closed
13	Boiler load < 40%	i. Start oil burners associated with the uppermost pulverizer in service ii. Gradually reduce coal feeder load to their minimum loading of 25% iii. Close hot air shut-off gates	i. Oil flame established ii. Coal feeder load 25% iii. Hot air shut-off gates closed iv. Pulverizer motor current reduced to "no-load" current
14	i. Oil flame established ii. Pulverizer motor current reduced to "no-load" current	Switch-off pulverizer as per preferential cut-out of pulverizers recommended by manufacturer	Pulverizers OFF
15	Pulverizers OFF	i. Stop PA fans ii. Stop coal feeders	i. PA fans OFF ii. Coal feeders OFF
16	Boiler load < 25%	i. Reduce total air flow-down to 30% ii. Close SH attemperator block valve iii. Close RH attemperator block valve	i. Total air flow 30% ii. SH attemperator block valve closed

			iii. RH attemperator block valve closed
17	Boiler load < 15%	Trip Master Fuel Trip Relay (MFR)	i. HFO trip valve closed ii. LFO trip valve closed iii. HFO burner valves closed iv. LFO burner valves closed v. Furnace "no flame"
18	Check-back criteria of step 17 OK	i. Close HP/LP bypass valves ii. Close SHO stop valves iii. Open RHO vent and drain valves iv. Close burner atomizing valve v. Open burner scavenge valve	i. HP/LP bypass valves closed ii. SHO stop valves closed iii. RHO vent and drain valves open iv. Burner atomizing valve closed v. Burner scavenge valve open
19	Burner scavenge time expired	Close burner scavenge valve	Burner scavenge valve closed
20	i. Burner scavenge valve closed ii. Burner atomizing valve closed iii. Drum level OK	i. Purge furnace ii. Switchover HFO temperature control to manual iii. Switchover HFO pressure control to manual iv. Switchover LFO pressure control to manual	i. Purge time expired ii. Stop HFO forwarding pump iii. Stop LFO pump
21	Purge complete	i. Switchover all boiler controls to manual ii. Stop FD fans iii. Stop ID fans iv. Stop regenerative air pre-heaters	i. All boiler controls on manual ii. FD fans OFF iii. ID fans OFF iv. Regenerative air pre-heaters OFF
22	SHO pressure < 0.2 MPa	Open SHO vent and drain valves	SHO vent and drain valves open

(Continued)

Table 12.2 (Continued)

Step no.	Input critera	Command issued	Check-back criteria
23	SHO vent and drain valves open	i. Open MS line drain valves ii. Open HRH line drain valves iii. Open CRH line drain valves	i. MS line drain valves open ii. HRH line drain valves open iii. CRH line drain valves open
24	i. Check-back criteria of step 23 OK ii. Drum level normal	Stop boiler feed pumps (BFPs)	Boiler feed pumps OFF
25	i. BFPs OFF ii. LP turbine exhaust temperature $<$ 328 K	i. Close condenser inlet CW valves ii. Close condenser outlet CW valves	i. Condenser inlet CW valves closed ii. Condenser outlet CW valves closed
26	i. Check-back criteria of step 25 OK ii. Deaerator feedwater storage tank level normal	Stop condensate extraction pumps (CEPs)	CEPs OFF
27	i. CEPs OFF ii. Hotwell level normal	Stop vacuum pumps	Vacuum pumps OFF
28	i. Check-back criteria of step 10 ii. Vacuum pumps OFF		AUTOMATIC UNIT (NORMAL) SHUT-DOWN SEQUENCE COMPLETE

12.9 START-UP OF GAS TURBINE

The operation of a modern gas turbine is a fully automated process, with various auxiliaries of the turbine automatically switched in and out of operation in a sequential manner. There is no operator intervention required except for the selection of fuel and the setting of load. The operations take place step-by-step in a pre-set order.

12.9.1 Pre-start checks

Before starting the gas turbine, the following must be inspected and checked:

1. Gas turbine and its components are in a state of tidiness and no rags or tools are lying in or around them
2. Maintenance work (if any) on different parts of the unit has been completed and all permit to work cards have been returned by authorized maintenance personnel duly signed
3. Inspect all parts of the plant subjected to maintenance work and make sure that all maintenance personnel have been withdrawn and the plant is ready for start-up
4. "Danger" tags from valves and switchgears, if there be any, are removed
5. All relevant valves in fuel gas conditioning skid as well as fuel oil treatment system are open
6. Drain all drain points of fuel gas conditioning skid
7. Gas pressure at gas turbine fuel gas skid inlet is adequate
8. Oil pressure at gas turbine fuel oil skid inlet is adequate
9. Check that air intake filter house is free of accumulated water either by opening drain plugs or by visual inspection
10. Starting device is switched on
11. Turbine lube oil tank level is at normal or above
12. Lube oil pressure drop across strainers, filters, and also oil flow through sight glasses are within normal limit
13. Lube oil pressure is equal to or above the recommended value
14. All systems, i.e., fuel gas, fuel oil, lube oil, etc., are tight and no visible leakage exists
15. Earthing is in order
16. All control and power fuses are in place
17. Battery voltage and charging equipment are in order
18. All control and selector switches are in proper sequence for start-up program
19. Electrical control and protection systems are not isolated and are in service
20. Annunciations, instruments, all indicating lamps are healthy
21. Operation from control room is in order

22. Check that "generator transformer", "unit auxiliary transformer", various electrical boards and power supply systems, etc., are in "ready to operate" position
23. Auxiliary power supply is at the rated voltage and frequency
24. Fire-detection system and automatic CO_2 fire-extinguishing system are in order
25. Portable fire extinguishers are in place around machinery space, fuel oil skid, fuel gas skid, etc.

Once these conditions are satisfied the gas turbine is purged with at least 8% of full-load mass air flow for a period of no less than 5 minutes. The gas turbine is now ready to start.

12.9.2 Start-up, Synchronizing, and Loading

Start-up of a gas turbine essentially means regulation of the fuel flow to the gas turbine at the initiation of a start signal. During this period the fuel/air ratio is maintained so the outlet gas temperature does not exceed permissible limits at maximum acceleration.

The typical gas turbine start-up and synchronization procedure is as follows:

1. Check that field breaker is OFF and AVR is operative
2. Ensure starting power from an external source
3. Gas turbine auxiliaries are thus placed in service
4. Gas turbine set is then accelerated with the help of starting device until the turbine is ignited and self-sustaining speed is reached
5. Governing system then takes over and accelerates the set to synchronous speed
6. After the machine has attained the synchronous speed, the synchronizing device adapts the turbine speed to match the grid frequency; the machine voltage is then brought to the rated value
7. While connecting to the grid, the parameters of frequency, phase angle, and voltage are adjusted to suit the grid; the generator is then synchronized with the grid by closing the generator circuit breaker.

12.9.3 Normal operation

During loading and subsequent operation of the machine, the primary responsibility of the operator is to observe and record the operational parameters of the main unit as well as its auxiliaries and ancillaries and to take corrective measures when required to control the machine. The operator must also keep a close watch on the performance of the generator exciter, AVR, and load regulator.

During start-up and normal operation, the following parameters should remain at or within the predetermined limits:

1. Bearing temperature and vibration
2. Operating parameters of turbine oil system

3. Compressor discharge pressure
4. Turbine exhaust temperature
5. Wheel space temperature
6. Fuel pressure
7. Temperatures of stator winding and rotor winding

12.10 NORMAL SHUT-DOWN OF GAS TURBINE

Before stopping the gas turbine unload the generator first and then initiate stop command. Normal shut-down of gas turbine is initiated by a stop signal resulting in total cut-off of fuel flow to the gas turbine.

12.11 EMERGENCY SHUT-DOWN OF GAS TURBINE

To protect the gas turbine from serious damage in the event of failure of open- and/or closed-loop control systems emergency shut-down of gas turbine is followed to shut-off the fuel supply to gas turbine without waiting for unloading the unit.

12.11.1 Unit cool-down and shut-down checks (normal or in the case of an emergency)

Immediately following a shut-down, the gas turbine should be put on barring gear for a period of predetermined time to assist in uniform cooling of the turbine rotor. Uniform cooling of the turbine rotor prevents rotor bowing and resultant rubbing and unbalance, which might otherwise occur on subsequent starting attempts.

The barring device is interlocked with a lubricating oil and jacking oil system and will not operate until the lube oil and jacking oil supply to the bearings is established. The minimum cool-down time depends on the ambient temperature and air draft. However, if the machine is not going to remain idle for a prolonged period, it is advisable that the barring operation be continued until a subsequent starting attempt is made.

The following checks have to be made after the machine is shut down:

i. Lube oil pump is maintaining good flow of oil at each bearing
ii. Battery charger voltage and current are normal
iii. Investigate any faults found during running
iv. Perform any repair/maintenance, as required
v. Open exhaust duct drain cock
vi. Fuel gas and fuel oil source is isolated and secured
vii. All compartment doors are closed

12.12 START-UP OF DIESEL ENGINE

12.12.1 Preparing the diesel engine for start-up during normal operation

Before the diesel engine is started the following checks/activities should be done:

1. Ensure there are no foreign objects on the diesel-generator and its accessories
2. Governor speed scale indicator pointer is in STOP position
3. Shaft barring arrangement is engaged with the shaft and locked in that position
4. All safety devices, lower crankcase, main drive, and drives for major accessories travel freely
5. Condition of joints and fastening of diesel engine auxiliary system pipelines is good; make sure that all cocks, valves, and flaps are in the operating position
6. Admit water to the external cooling circuit and make sure the water system joints are not leaking
7. Oil and fuel filters cut in for the operation of both sections
8. Check for the absence of water in the circulating oil tank and in the oil cooler (ensured by taking samples from the lower part of tank and cooler)
9. Bleed air out of the upper points of diesel engine cooling system, fill the system with internal circuit water, and pour oil and fuel into the respective systems up to the rated levels
10. Scavenge condensate out of starting air cylinders and charge them with air up to a pressure recommended by the manufacturer
11. Level of oil in the speed governor is up to the center of oil level indicator
12. Set the filter and cooler priming pipeline valve in the position for shutting off flow into the tank and ensuring priming of filter and cooler
13. Start automatic heating of oil and water
14. When warming up and priming the diesel engine with water, check the internal cooling circuit system for absence of leakage and also check the sealing of pipe unions, starting valves, injectors, etc.
15. Cocks on the receiver and exhaust manifold oil drain pipes are open and are not choked; clean pipes, if necessary
16. Engage in independent priming and scavenging oil electric pumps
17. Air pipeline valves should be closed and fuel feed limiter shut-off valve is open
18. Leave the electric pumps engaged, open indicator valves, prime the diesel engine with oil, and at the same time turn the crankshafts by means of shaft barring arrangement; check supply oil pressure to the diesel engine
19. While priming oil and cranking the diesel engine, check supply of oil to all the bearings, to the main drive and the drives for major and minor accessories; check

flow of oil out of piston cooling pipes, and check also proper sealing of joints of oil pipelines in the engine crankcase, starting valves, and starting system

20. Having accomplished the check, close air pipeline valves, and open shut-off valves on the main starting valves
21. Prime the high-pressure fuel injection pumps and engine system
22. Supply air to thermo regulators; make sure that valves of thermo-regulators move smoothly and without jamming
23. Prime diesel engine with oil
24. Crank diesel engine with air

The diesel engine is now ready for *starting*.

When the diesel engine is running, monitor the following parameters to ensure that they remain within their recommended limits:

i. Diesel engine speed
ii. Temperature of exhaust gases in manifolds
iii. Engine oil inlet temperature
iv. Engine oil outlet temperature
v. Engine internal circuit water inlet temperature
vi. Engine internal circuit water outlet temperature
vii. Engine oil inlet pressure
viii. Engine internal circuit water inlet pressure
ix. Fuel oil supply pressure
x. Pressure of external circuit water downstream of pump
xi. Pressure of air fed to thermo-regulators
xii. Level of oil in the circulating tank
xiii. Temperature of exhaust gases in the cylinders
xiv. Temperatures of bearings, stator winding, and rotor winding

12.13 NORMAL SHUT-DOWN OF DIESEL ENGINE

Before stopping the diesel engine after its continuous rated-power operation, unload the generator first then initiate stop command. Once the stop command is initiated the engine speed is reduced to a value as recommended by the manufacturer. The engine continues to run at this speed until it is cooled down sufficiently. After the oil temperature drops to about 328 K the fuel supply to the engine is cut-off to bring down the speed to zero.

12.14 EMERGENCY SHUT-DOWN OF DIESEL ENGINE

Stopping a diesel engine in an emergency can be achieved, without cooling the engine, by initiating the emergency stop command and cutting the engine load. Right after the engine is stopped (normally or in the case of emergency), do the following:

a. Open cocks on the drain pipes of the receiver and exhaust manifolds
b. Make sure that water outlet temperature is within 303 and 323 K and the oil outlet temperature is within 313 and 323 K. If the temperatures are higher, prime water and oil systems using respective pumps until the required range of temperature is reached
c. Pump oil from the diesel engine sump and system scavenging pipeline into the circulating tank using a oil scavenging pump
d. Close the engine air supply pipeline valve and also the valve that admits air to the thermo-regulators

BIBLIOGRAPHY

[1] IEC 60045-1:1991: Steam Turbines Part-1: Specification.
[2] BS EN 60045-1:1993: Guide for Steam Turbine Procurement.

CHAPTER 13

Abnormal Operating Conditions

13.1 INTRODUCTION

Abnormal operating condition occur in running equipment, in a system or in a unit when its various operating parameters, i.e., pressure, temperature, flow rate, voltage, current, etc., exceed the rated design limits for normal operating conditions. When an abnormal condition occurs, it may not pose an immediate threat to the life of personnel or safety of plant and equipment, since it is usually a non-emergency condition [1], but it can lead to catastrophe if not taken seriously and addressed appropriately.

When any abnormal condition occurs operating personnel should investigate the reason/s for such occurrence following the steps described below (not necessarily in this order) and take whatever corrective actions are needed:

- **i.** Whether any safety device has operated
- **ii.** Whether there is unintended closure or opening of any valve in the system
- **iii.** Whether any component has malfunctioned
- **iv.** Whether there is any failure in control/monitoring signal communication
- **v.** Whether there is any mal-operation by operating personnel, etc.

Once conditions are returned to normal, equipment and/or unit may be operated in a regular mode.

During normal running of the unit, operating personnel often experience various types of abnormal operating conditions. Some of these abnormal conditions are simple in nature and may be corrected by operator action easily, but some conditions may be complex and require simultaneous manipulation of many parallel activities to correct. Due to its limited scope, this chapter only addresses some of the complex abnormal operating conditions pertaining to steam generators, steam turbines, gas turbines, and diesel engines. This chapter also discusses general electrical faults with a specific focus on generators, transformers, and induction motors. The following sections deal with probable causes of abnormal conditions and addresses suitable corrective actions to overcome some of these conditions.

13.2 ABNORMAL OPERATING CONDITIONS OF STEAM GENERATORS

Maintenance of a stable furnace without imposing undue restriction on plant operation and without endangering the safety of plant and equipment at all times is of

Thermal Power Plant
DOI: http://dx.doi.org/10.1016/B978-0-12-801575-9.00013-5

paramount importance in the operation of steam generators or boilers. For a natural circulation boiler, the drum level is one of the most important parameters that must be continuously monitored and maintained throughout the period of operation [2]. The automatic control system of a steam generator will normally be able to maintain specified limits of various parameters, i.e., furnace draft, drum level, steam temperature, steam pressure, etc., but when it fails to do so, audio-visual alarms must alert the operator. (More details on automatic control are found in Chapter 10).

13.2.1 High drum level

An abnormal increase in drum water level may be caused by (a) sudden load demand, (b) malfunction of feed-regulating system, and/or (c) safety valve blowing. In the event of a "drum level high" alarm the operator should immediately check the feed-regulating system and change the control from auto to manual, if required. Boiler drum drain valves may be opened to restore the normal water level in the drum. An excessive increase in drum water level may result in carryover of water into the superheater or steam turbine, resulting in erosion/corrosion of the steam path. Necessary steps have to be taken by the operator to prevent such occurrence. Interlocks should be provided for tripping the *master fuel relay* (MFR) [3] and the *turbine lock-out relay* if the level becomes excessively high.

13.2.2 Low drum level

A high-pressure boiler has insignificant water capacity compared to its steaming rate. Any interruption of feed to the boiler for any reason, for a couple of minutes, sometimes even for less than a minute, will cause the disappearance of the water level from the gauge glass/level indicator mounted on the drum. This could expose the wall tubes to furnace heat without water cooling, which may lead to disaster, unless timely action is taken either to restore the feed or to kill the fire. In the event the water level in the drum gauge glass/level indicator vanishes unnoticed, the operator must kill the fire without taking any chance and maintain feedwater flow until the boiler is sufficiently cooled.

13.2.3 Loss of boiler feed pump

The loss of water in the drum may be caused by a loss of boiler feed pump. The standby boiler feed pump should start immediately in the case of tripping of any of the running feed pumps. Prompt operator action may prevent outage of the unit and other consequences of loss of feedwater to the boiler. In the case of tripping of running feed pumps and failure to change over to standby, the load should be immediately reduced to restore the drum level. On the other hand, if the drum water level cannot be maintained, the fire will be immediately killed and the turbine tripped manually or through an interlock before the water level disappears from the drum gauge glass/level indicator.

13.2.4 Break-down of feed system/feed regulating station

Since the feed line is welded, its failure can be almost ruled out, although leakage through valve glands or cover joints cannot be eliminated. Immediate or planned shut-down of the unit will depend on the extent of leakage, but the unit has to be taken out of service for necessary repairs, in any case. In the case of failure of the operating feed regulating valve, the 100% emergency feed bypass valve must be cut in without delay and the failed one isolated, pending shut-down and rectification of the defect. In the event of failure of both valves, load on the unit should be immediately reduced and the water level in boiler-drum maintained by cutting in the low-load feed regulating valve until a planned shut-down can be arranged.

13.2.5 Heavy tube rupture

Failure of one or more tubes in a boiler can be detected by sound and either by an increase in make-up water requirement (indicating failure of water carrying tubes) or by an increased draft in the superheater or reheater sections (due to failure of superheater or reheater tubes).

Failure of tubes may occur due to various reasons, such as overheating, erosion by soot-blowing steam or ash particles, or by corrosion on the water and/or steam side. A pin-hole leakage on one or more tubes can usually be tolerated, since with a feed-regulating system with adequate margin, the unit may have a planned shut-down. However, continued operation of boiler under such conditions may cause considerable erosion damage to adjacent tubes as well, making a relatively simple and short repair job to become a major one. Hence, an operator has to apply his judgment to assess the extent of failure and to decide corrective action/s to be taken.

In the event of heavy tube rupture, maintaining boiler drum water level may prove difficult. Heavy leakage of water into the furnace may also cause disturbance in combustion, even loss of fire. In this case, the unit must be immediately shut down. Fuel feed and load will have to be stopped immediately by actuating the MFR emergency trip push button. The induced draft (ID) fan damper should be opened to maintain the draft in the furnace to cope with heavy evolution of steam in the furnace due to leakage of water from boiler tube/s.

Damage to tubes due to any cause has to be repaired. Repair work of a damaged tube consists of cutting off the tube section sufficiently above and below the damage and then inserting and re-welding a new tube section.

13.2.6 High superheat temperature

The superheat temperature may increase abnormally due to malfunction of the temperature regulation system and/or sudden increase in heat input to superheater elements. The attemperators provided for automatic control of the superheater outlet steam temperature normally are able to maintain the steam temperature within the specified limits.

Heat input to the superheater may increase gradually over a period of time due to changes in the furnace heat absorption pattern resulting from slagging, etc., or an increase in heat input may be sharp due to late/secondary combustion, high excess air, sudden increase in fuel burning rate, etc. In the case of slagging, it is necessary to operate the soot blowers. In other cases, it is necessary to adjust fuel and air suitably for proper combustion. Should the automatic control fail to regulate manual control should be immediately restored for the safety of equipment. An audio-visual alarm annunciation should appear in the event of "high superheat temperature."

13.2.7 Low superheat temperature

Low superheater outlet steam temperature results in increased moisture content in the steam going to the turbine later stages with consequent risk of increased erosion rate and hence should be avoided. A sudden drop in superheat temperature also causes quenching of hot metal surfaces resulting in the development of abnormal stresses in metal surfaces of superheaters, headers, main steam pipes, turbine stop valves, steam chests, high-pressure turbine inlet, and first stage region. The contraction of the turbine rotor due to sudden chilling may lead to severe rub and vibration.

The superheater outlet steam temperature may drop sharply due to water carryover from boiler and/or due to malfunction of the attemperator system. In either case, immediate action must be taken to reduce carry over and bring back steam temperature within specified limits. A gradual decrease in attemperator water flow and/or superheated steam temperature over a period of time is an indication of fouled superheater elements and should be taken care of by periodic soot blowing. It is obvious that exposing a turbine to an abnormally high rate of steam temperature fall or to run it at a substantial load with an abnormally low superheat steam temperature for an extended period is undesirable. In either case, steps should immediately be taken to restore safe conditions. Interlocks have to be incorporated to trip the turbine at a preset low temperature. An audio-visual alarm annunciation should be provided indicating "low superheat temperature."

13.2.8 High and low reheat temperature

The reheat steam temperature should be kept within specified preset limits. While high reheat temperature will cause damage to reheater metal, low reheat temperature will result in temperature strain in reheater metal and steam turbines. Audio-visual alarm annunciation are provided for "high"/"low" reheat temperature. In the event of a high or low reheat temperature alarm, the operator should check the regulator, and if it fails to operate automatically, manual control should be restored.

Reheat steam temperature may exceed a predetermined high limit due to either malfunction of reheat temperature regulator, or high gas temperature in the reheater zone. A high reheat temperature is automatically regulated in two stages. First, the control is accomplished by regulating the burner tilt or by gas bypass or by any other means specific to a typical boiler design. If even after such corrective action, reheat steam temperature still increases emergency spray water attemperators should be used. If attemperators fail to operate automatically the spray water attemperator should be operated manually.

The reheater outlet steam temperature may drop sharply due to water carryover from the attemperator system and/or due to malfunction of attemperator system. Temperature may also drop due to fouled reheater tubes. In the case of slagging in reheater tubes, the operator should activate soot blowers to remove deposits in reheater tube elements. For low reheat temperature, the operator should stop feeding spray water, if already in service, and regulate the burner tilt/gas bypass to bring up the temperature.

13.2.9 Loss of ID and FD fans

Loss of both ID or both forced draft (FD) fans or all four fans at a time is a rare phenomenon and is usually caused by an electrical problem involving complete loss of the unit. In either case, the MFR will be actuated through interlock [3] and the turbine will also be tripped preceding investigation of the malfunction. However, if it is possible to restart the fans immediately, it may be possible to relight the fire after purging. Automatic tripping of the FD fan on tripping of associated ID fan or vice-versa, i.e., tripping of ID fan with FD fan (when all four fans are running), should be provided in order to stabilize the draft quickly. In either case, the load on the turbine should be reduced to about 60% of the maximum rating, which approximately corresponds to the maximum rating of one set of fans.

The operator's goal should be either to ensure rapid stabilization of the furnace by manipulation of ID/FD fan controls, and cutting in oil support if necessary, and/or to prevent rapid loss of pressure and temperature of the main steam by shedding load in excess of 60% of the maximum rating and by adjusting steam temperature controls.

13.2.10 Flame instability

The stability of a flame is primarily dependent on the ability of each burner to burn the fuel successfully. Each gas/fuel oil/pulverized coal burner should be provided with at least one flame detector/scanner and continuously monitored individually to initiate the basic signal for operation of flame failure protection.

Flame in a furnace is in a state of dynamic equilibrium, so that during steady state operation, the furnace draft remains constant. In contrast, frequent and/or excessive variation of the furnace draft indicates lack of equilibrium resulting from an unstable flame condition in the furnace. From a practical point of view a balanced draft furnace is considered as stable if the furnace draft is maintained within (-) 0.05 kPa to (-) 0.10 kPa under normal operating conditions.

Furnace instability in a boiler may generally result from the following conditions:

1. Sudden change in firing rate due to sharp change in load demand
2. Malfunction of ID and/or FD fan/s regulating system
3. Loss of any draught fan group
4. Air-rich or fuel-rich mixture causing improper "fuel-air" ratio leading to either incomplete or secondary combustion
5. Sudden change in quality of fuel
6. Interruption in fuel-oil firing due to improper atomization and/or failure in supply of this fuel
7. Non-uniform pulverized coal flow resulting from defect in pulverizer, choked fuel pipe, arching in bunker, etc.

In the event of flame instability, it is best to promptly transfer both fuel and air controls to "manual" and adjust them manually until a stable fire is established. While this process of manual control is in progress it may be necessary to restrict the maximum/minimum load and fluctuation of load on the unit temporarily. It may also be necessary to cut in supplementary fuel to service to stabilize flame. Once the flame becomes stable, the cause of instability can be investigated thoroughly and necessary corrective measures taken.

13.2.11 Flame failure trip

The signal from the flame failure trip system is used to trip the boiler or affected fuel burner or group of burners. Upon detection of loss of a burner flame in a gas/fuel oil fired boiler, that individual burner safety shut-off valve on the gas/fuel oil supply line will be closed automatically. Loss of flame in a gas/fuel oil fired boiler furnace or partial loss of flame, which create hazardous conditions, will initiate an automatic master fuel trip [3].

In a pulverized coal-fired boiler stability is also dependent on the number of pulverizers in service at any time as coal flame from one burner supports the flame in an adjacent burner to some extent. Flame failure protection logic is designed to protect the furnace from danger of explosion, depending on the number of burners in a pulverizer where the flame has been lost and the number of pulverizers in service. Upon detection of loss of flame on a predetermined number or arrangement of burners served by a pulverizer group, that pulverizer and coal feeder will be automatically tripped. Upon loss of flame in a zone or in the entire furnace, the MFR trip will be initiated automatically [3].

13.2.12 Loss of fire in boiler

Loss of fire in a boiler may result from a variety of reasons as follows:

- **i.** Disturbance in fuel-burning equipment, leading to marked disturbance and/or interruption of fuel flow to furnace
- **ii.** Disturbance in the draft system, leading to sudden change in fuel-air ratio and/or flow pattern in the combustion zone of one or more burners
- **iii.** Extraneous reasons, such as abrupt change in fuel-burning rate due to abnormal load fluctuation, sudden change in quality of coal, and/or flame instability from heavy boiler tube leakage, etc.

Since, admission of fuel in a furnace at any time without proper combustion is a potential hazard the fuel supply to the furnace must be immediately cut out in case of loss of fire. As a result immediate actions, as follows, should be taken to prevent damage to boiler:

- **a.** Fuel input to the furnace must be cut out
- **b.** Manually trip MFR through its emergency trip push button if it does not occur automatically
- **c.** Following the MFR trip, ID and FD fan controls should be transferred to "manual" and furnace draft adjusted to normal value; under no circumstances should the air flow be increased
- **d.** To stop drop in steam temperatures switch superheater and reheater attemperation to "manual" and shut attemperation spray water supplies, if they have not been shut-off automatically following MFR trip, so as to arrest drop in steam temperatures
- **e.** Keep the boiler drum level within specified limits manually by overriding automatic control.

13.2.13 Loss of fuel oil pressure

Loss of fuel oil pressure in the inlet header to the oil burners can occur due to (i) reduced oil supply, and/or (ii) increased fuel take-off.

The oil supply may fall for the following reasons:

- **i.** Lack of adequate oil in oil tank
- **ii.** Inadequate pressure of oil at the suction of fuel oil pump/s due to choked suction filter and/or low temperature (high viscosity) of oil resulting in flow difficulties
- **iii.** Improper operation of fuel oil pump or its relief and/or pressure-regulating valve
- **iv.** Undue high restriction in any section of the piping due to malfunction/mal-operation of valves, like the oil-pressure control valve, etc.

Abnormally high take-off may result from any rupture in fuel oil pipelines and/or abnormally high oil flow through one or more of the burners due to lack of orifice/sprayer plate of proper size, excessive recirculation and/or draining of oil, etc.

In the case of loss of fuel oil pressure, the fuel oil trip shut-off valve/s in the oil inlet header and oil burners should be interlocked to shut off automatically [3]. During low load operation, it may be best to shut down the unit because the flame may become unstable for lack of oil support at low load. The defect should be investigated and rectified and the boiler may be lighted after purging. Operation of oil burners with oil pressure below a specified lower limit will result in inadequate atomization and should be avoided.

13.2.14 Oil burner problems

Failure or improper operation of the oil burner may arise due to the following:

1. Building up of carbon deposit on burner tips
2. Mechanical jamming of burner
3. Poor jointing in the oil system
4. Mechanical jamming of combustion air dampers
5. Inadequate oil pressure or temperature
6. Inadequate atomizing media pressure
7. Flame failure devices that fail to work due to slag and dust

To obviate these problems, the operator may resort to the following:

1. Before putting an oil burner into service, it should be cleaned and properly assembled and tightened
2. When a burner is being cut out of service and oil is shut off, the burner should be scavenged with steam or compressed air to remove all oil remaining in the gun and the nozzle
3. An idle gun should not be left inserted
4. In the event of carbon deposit, gun and tip assemblies should be removed, cleaned thoroughly, and then reassembled
5. Adequate oil pressure/temperature, atomizing media pressure, and instrument air pressure should always be ensured
6. Lubrication of all moving parts should be checked periodically
7. Check burner cylinder seals, compression fittings, etc., and replace if necessary
8. Oil samples should be tested periodically at the laboratory to check viscosity-temperature relation
9. Periodically check external piping and valves for evidence of leaks, etc.
10. Check that oil/steam hoses are hanging freely without any sharp bend
11. Constant attention is required for flame-failure devices to ensure that the sensitivity of these is not impaired by carbon deposits or dust on the viewing windows.

13.2.15 Air heater fire

Air heater fire is caused mainly by the accumulation of combustible matter, chiefly oil droplets and unburned pulverized coal, which impair heat transfer across heating elements resulting in an increase of gas side temperature and consequent ignition of the combustible matter. Air heater fires are indicated by a sharp increase in temperature of both air and flue gas leaving the air heater. The chance of air heater fire occurring in a particular boiler increases with the increase in the number of start-stop operations.

Whenever there is an indication of an air heater fire, steps should be taken immediately to investigate. If there is fire, the affected air heater should be isolated by closing all dampers, doors, etc., to starve the fire of oxygen. If the fire is in initial stages, i.e., only the combustible deposits are burning, it may go out. However, if the metal has started burning, the temperature will be in the order of 2500–3100 K [4]. In such cases, the location of the fire should be flooded with huge quantities of water. After the fire has been extinguished completely and the unit cooled, the air heater is thoroughly cleaned and the necessary repairs done.

In addition to the above, pulverized coal-fired boilers also suffer from the following typical abnormal conditions.

13.2.16 Loss of Primary Air (PA) fan

Sudden tripping of a PA fan will in general result in marked disturbance in the furnace and may affect flame stability, apart from drop in boiler outlet steam pressure and temperature. In the event of loss of one of the running PA fans, the first action has to be automatic preferential tripping of running pulverizers as recommended by manufacturers. Thereafter, urgent action should be taken to stabilize the flame by putting oil support followed by load runback commensurate with the number of pulverizers in service. This will prevent an abnormal drop in boiler outlet steam pressure and temperature. At the same time all out effort should be made to cut in the standby pulverizer/s as earliest as possible to restore the unit output.

13.2.17 Loss of coal feeder

Loss of a coal feeder causes raw coal feed to associated puverizer to shut off. Since in a direct firing system, fuel is fired according to the requirement of the steam demand and reserve fuel in a pulverizer is very limited, a loss of feed will cause unbalance in operating pulverizer/s, thereby impairing the furnace stability. Prompt operator action is essential to prevent this and restore normalcy without imposing undue restrictions on the operating load requirement.

If the cause of feeder trip is the loss of one group of draft fans, bring down the load to about 60% with careful control of furnace draft, drum level, and steam temperature. If the feeder is tripped due to jamming, a faulty condition in electrical supply, or overload, shut down the associated pulverizer after determining that it is empty (indicated by dying flame). Light oil burner/s or put in-service standby pulverizer/s to restore the load. The defective coal feeder can then be inspected and defect rectified.

13.2.18 Loss of pulverizer

Whenever there is a loss of a pulverizer, the operator should light oil burner/s to help the operating group of pulverizers to stabilize the flame. At the same time, the operator should bring down the load matching to the capability of the running puverizer/s. Effort should be made to cut in standby pulverizer/s depending on draft fan group capability. Faults in electric supply, if there are any, can then be inspected and rectified. In the case of jamming in the pulverizer internals, the affected pulverizer should be cooled and cleaned and prepared for the next operation.

After tripping of a pulverizer it is recommended to adopt the following corrective actions [3]:

> *"a pulverizer that is tripped under load will be inerted as established by equipment manufacturer, and maintained under an inert atmosphere until confirmation that no burning or smouldering fuel exists in the pulverizer or the fuel is removed. Inerting media may be any one of CO_2, Steam or N_2. For pulverizers that are tripped and inerted while containing a charge of fuel, following procedure will be used to clear fuel from the pulverizer:*
>
> 1. *Start one of the pulverizers*
> 2. *Isolate from the furnace all shut-down or tripped pulverizers*
> 3. *Continue to operate the pulverizer until empty*
> 4. *When the operating pulverizer is empty, proceed to another tripped and inerted pulverizer and repeat the procedure until all are cleared of fuel"*

13.2.19 Fire in idle pulverizer

Whenever there is an emergency shut-down of an operating pulverizer, it will contain a substantial amount of hot, partially pulverized coal. If allowed to stand for a considerable period of time, the coal remaining inside may ignite. Therefore, either this pulverizer should be put back into service as soon as possible, or it should be emptied manually.

In the event of a fire inside an empty pulverizer, prevent smoking to develop, keep fire extinguishers handy, and take the following corrective steps:

i. Close the cold air damper. Check that the hot air shut-off damper, mill outlet coal-air dampers, and fuel pipe dampers are all closed

ii. Open windbox doors partially and drench the interior with a discharge from a fire extinguisher or with steam. Repeat this step every 10 minutes or so until the fire is extinguished

iii. Open the access door carefully and allow any gases that may have been generated during the period of fire to escape

iv. When the fire is extinguished, the residual fuel is cooled and after determining that poisonous gases, if any, have escaped, the pulverizer should be cleaned out manually, making certain that no trace of fire or smoldering fuel is left

(Note: Air should never be used for cleaning pulverized fuel equipment suspected of containing fire. The use of the vacuum-cleaning equipment could lead to trouble in other parts of the equipment. Injection of CO_2 may also cause chilling in grinding elements and cracks in keyways.) [3]

v. Inspect and clean out the pulverizer windbox, air ducts, fuel pipes, primary air fan, and feeder of any residual or half-burnt material

vi. Check the lubricants, and replace if there is any evidence of carbonizing

vii. Inspect the pulverizer and if satisfied that it is completely clear of traces of fire, it may be put into operation according to the normal procedure.

13.2.20 Fire in operating pulverizer

Fire in an operating pulverizer is a rare phenomenon. Nevertheless, if fire outbreaks it must be extinguished promptly and adequately without allowing it to cause any damage or even interruption in operation. Immediate steps to fight this type of fire include smothering it by reducing the supply of oxygen to a point where combustion is no longer supported.

The outbreak of fire in an operating pulverizer may occur for the following reasons:

1. Too high an inlet air temperature
2. Burning fuel from a storage dump or a raw coal bunker being fed to the pulverizer
3. Pulverizer when idle might have been subjected to welding or burning and subsequently boxed up with smoldering debris inside and released for operation

Fire in an operating pulverizer can be detected by a sudden increase in outlet coal-air temperature. If this occurs the operator should take the following steps:

a. If hot air and cold air dampers have not automatically changed over already to give maximum cold air and minimum hot air, carry out this operation manually

b. If the rise in outlet temperature could not be arrested even after switching over hot air and cold air dampers following above step and returns to normal within 15 minutes, proceed as follows:

1. Shut down the pulverizer and associated coal feeder.
2. Isolate the mill and close the hot air and cold air dampers.
3. Follow the same steps as given for the "fire in idle pulverizer."

c. Manually switch the feeder control to the highest speed and increase the amount of coal in the pulverizer to the point where there is a possibility of pulverizer choking. If the pulverizer is already operating at its maximum capacity, ensure that the fuel feed is not interrupted.

d. In the event that the raw coal feed needs to be interrupted, immediately shut down the pulverizer and the coal feeder. Then close all associated dampers and valves.

e. Reduce the air supply to maintain a rich fuel air mixture in the pulverizer until it is certain the fire has been extinguished. Thereafter, restore the normal air-fuel ratio.
f. If the increase in pulverizer outlet temperature can be arrested and brought to normal within a few minutes continue operation until the normal outlet temperature is restored. Then reduce the pulverizer output to meet the operating demand.

13.2.21 Plugged ash hopper

In the event of a plugged ash hopper condition an alarm should alert the operator. A "plugged hopper" may occur when cleaning ash from a hopper and ash arches over the hopper outlet preventing flow. The operator has to then open the fly ash hopper valve of the plugged hopper branch and the fly ash extraction valve of the plugged hopper. Then removing the poke-hole cover of the extraction valve and inserting a poking rod inside the arching that formed to break it. Once cleaned the operator has to start cleaning from the next hopper. After poking and breaking the arches from all hoppers, the sequential operation can be restarted.

13.3 ABNORMAL OPERATING CONDITIONS OF STEAM TURBINES

A running steam turbine may encounter various abnormalities both during steady-state operation as well as under transient operating conditions. Each of these problems has serious effects on the turbine heat rate and its reliability. Problems generally encountered during steady-state operation include:

a. Off-frequency operation
b. Off-nominal voltage operation
c. Overloading
d. High exhaust pressure
e. Acid corrosion
f. Steam turbine blade failure
g. Boiler carry-over deposition on steam path parts

Problems that arise in the transition condition are more critical in nature and may cause damage to the turbine. Some of these critical problems include:

1. Water damage to blades due to back flow from heaters
2. Heater flooding
3. Steam hammer in turbine piping
4. Problems of HP turbine stage when load drops to no-load
5. Loss of feed tank water level
6. Loss of vacuum
7. High exhaust hood temperature

8. Loss of oil pressure
9. High differential expansion
10. High bearing vibration
11. High axial shift
12. Temporary increase in turbine speed/turbine over-speed
13. Performance deterioration due to erosion, corrosion

13.3.1 Off-frequency operation

Sustained operation at more than 1% below the rated frequency causes the long and slender LP stage buckets to vibrate in several different modes and frequencies. When bucket natural frequencies are stimulated by operating condition, they display resonant conditions, where bucket failures are likely to occur.

13.3.2 Off-nominal voltage operation

Longer periods of under-voltage operation can lead to excessive heating of generator components, whereas over-voltage may damage the generator's magnetic components.

13.3.3 Overloading

Overloading increases the steam flow. High flow and high pressure overstresses the parts of the unit and increases the potential for damage.

13.3.4 High exhaust pressure trouble

Operations at excessive exhaust pressure may lead to high exhaust loss, unstable operation, and ultimately last stage bucket failure.

13.3.5 Acid corrosion

Acids are formed in the boiler by thermal decomposition of natural or synthetic substances entering the feedwater circuit with make-up water. They can also be produced in the event of cooling water leakage in the condenser. The result is corrosion in the turbine steam flow path.

13.3.6 Steam turbine blade failure

This failure is generally fatigue failure resulting from excessive vibratory-stress levels. This occurs in the initial wet zone of condensing steam turbines.

13.3.7 Boiler carry-over deposition on steam path parts

Boiler carry-over may result in deposits on steam-path parts, which can have a significant effect on unit efficiency, capacity, and reliability.

13.3.8 Water damage to blades due to backflow from heaters

On sudden drop of load, backflow of water and steam from the heaters to the turbine may take place, which may cause significant impingement of fluid droplets on turbine blade. Blade erosion is one of the effects of such impingement. Backflow of steam from extraction steam lines to the turbine may result in turbine over-speed as well [5].

13.3.9 Heater flooding

Excessively high levels of condensate in the feedwater heater shell and consequent flooding of heater tube coils may result chiefly from either inadequate draining of condensate due to malfunction of the heater shell level regulation system, or heavy tube rupture. Backflow of water from heaters to the turbine may cause blade erosion and also quenching of turbine blades leading to thermal stresses in blades. In the case of a high level in any of the feedwater heaters, the affected heater should be completely isolated both on steam and water sides pending inspection and rectification of defects [5].

13.3.10 Steam hammer in turbine piping

Pressure waves, produced by stopping steam flow in a turbine with a rapidly closing valve, is called steam hammer, and these waves propagate through the turbine. The steam hammer in turn reduces the efficiency and reliability of the turbine.

13.3.11 Problems in HP turbine stage when load drops to no-load

When a turbine-generator loses full load and remains in service carrying auxiliaries, or just spins with no-load, the turbine is subjected to a sudden drop in temperature, which causes high thermal stresses in the initial stages of the HP turbine and consumes some of the cyclic life of the turbine's high-temperature components. At the same time, the last stages of the turbine get overheated due to churning and whirling of the steam, which again leads to high thermal stresses.

13.3.12 Loss of feed tank water level

In the case of the feedwater tank level falling below the preset emergency level the boiler-feed pump may get vapor bound with consequent risk of damage, which should be avoided at all times. In spite of corrective actions, if the feed tank level remains uncontrolled, it is advisable to shut down the unit along with boiler feed pumps as quick as possible.

13.3.13 Loss of vacuum

Loss of vacuum in the condenser is generally caused by any or a combination of the following conditions:

a. Partial or a complete loss of cooling water
b. High condenser hotwell level
c. Leakage in vacuum system
d. Loss of gland seal steam
e. Malfunction of vacuum creating equipment, i.e., air ejector/vacuum pump
f. Fouled condenser tubes

The rapidity of a decrease in vacuum due to causes (a) to (e) depends on the severity of the cause, whereas in case of (f), the decrease in vacuum is generally gradual over time. The remedy, of course, is isolating the cause of the decrease in vacuum and fixing it.

13.3.13.1 Partial or complete loss of cooling water

In the event of loss of one of the running circulating water (CW) pumps it is necessary to bring down the load on the turbine so that it is within the capacity of running CW pumps. The cause of failure of the pump should be investigated and rectified, after which it should be put back to service and the unit loaded in conformity with load demand following manufacturer's instructions. In the event of loss of all CW pumps, and inability to restore any of them within a short period of time, the unit has to be shut down. However, such a contingency is, in all probability, rare but may happen due to a loss of auxiliary power.

13.3.13.2 High condenser level

Too high a level in condenser floods the condenser tubes and restricts passage of exhaust steam inside the condenser interfering with heat transfer causing high back-pressure or in extreme cases may even flood the air suction pipes of vacuum creating equipment. In the case of malfunction of either the condensate pump or the level controller or excessive inflow of make-up water beyond the requirement, the standby pump should be put on load or the respective out-of-order control valve should be bypassed or put on manual control as needed to restore the condensate level to normal value.

13.3.13.3 Leakage in vacuum system

The condenser vacuum may be affected by the entry of air into the condenser due to leakage in pipes and fittings. Pending investigation of the leakage, the standby vacuum-creating equipment may be put in service if necessary.

13.3.13.4 Loss of gland seal steam

Loss of steam to the turbine gland seals may occur in the case of failure of the gland seal steam pressure control valve. The operator should bypass the control valve or place it in manual control and establish seal steam supply at turbine gland seals at recommended pressure.

13.3.13.5 Malfunction of air ejectors

The ejector performance is affected by either high/low supply steam pressure to the ejector or by high condensate level in the ejector. In such cases, put the starting air ejector in service immediately along with the running main air ejector to prevent a decrease in vacuum. Care should be taken to ensure the steam valve opens first and then the air intake valve. Put the standby air ejector in service. After vacuum is brought back to normal, take out the defective air ejector by closing the air valve first followed by the steam valve, taking care to watch the vacuum at all times. Then take out the starting air ejector, first closing the air valve and then the steam valve.

13.3.13.6 Malfunction of vacuum pump

The vacuum pump performance is affected by failure of either the "electric power supply to the pump" or the "seal water supply to the pump." In either case start the standby pump, restore the vacuum, and then stop the defective pump.

13.3.13.7 Fouled condenser tubes

Fouling of condenser tubes should be taken care of by periodic isolation and cleaning of each the condenser, the frequency of which is best determined by actual conditions.

13.3.14 High exhaust hood temperature

The likely causes of high exhaust hood temperature are either low condenser vacuum or continuous operation at a very low steam flow. This problem is likely to happen at the time of start-up, especially if the unit is not loaded for a long time after full speed has been attained. Normally a unit should be synchronized quickly and load applied. The operator should investigate the reason and take the necessary action to restore it to normal. However, when the temperature exceeds the preset limit, the unit will be tripped immediately pending identification and rectification of defects, if any.

13.3.15 Loss of lube/control oil pressure

Since the requirements of lubrication and/or control oil of the turbine are met by a shaft-driven centrifugal oil pump, a complete loss of oil pressure is a rare phenomenon, but may occur due to failure of oil pump.

Loss of control oil pressure may also result from a failure or rupture of piping, joints, etc., in a high-pressure control oil system. With a turbine running normally and oil supplied by a shaft-driven centrifugal oil pump, the turbine will trip through an interlock in the case of a major fault in the control oil system.

In the case of complete loss of lubricating oil supply, the rotor must be brought to complete rest in the shortest time possible. The low lube oil pressure protection must trip the turbine followed by killing of vacuum by opening the vacuum breaker. The rotor should not be put on the turning gear until oil supply to turning gear, bearings, and seals has been restored.

If significant time elapses between the rotor coming to a standstill and restoration of oil supply, and if there is a sign of serious rub or shaft warp or bearing run out, the rotor should not be turned. In that case it is advisable to completely cool down the rotor and casing to ambient temperature. The oil supply to bearings should, however, be kept on in order to cool them.

13.3.16 High differential expansion

In the case of excessive differential expansion, metal-to-metal contact may occur between stationary and rotating parts of the turbine because of close running clearances. Hence, it is necessary to maintain the high-pressure turbine (HPT), intermediate-pressure turbine (IPT), and low-pressure turbine (LPT) differential expansions within the predetermined limits.

An increase in differential expansion may be caused by too fast a change in inlet steam temperature or due to too rapid change in steam flow to the turbine. Hence, in normal running conditions, maintaining the steam temperature as close as possible to the metal temperature and keeping the rate of change of temperature to a minimum will help maintain the turbine differential expansion within the limits.

During a cold start-up when the steam temperature is much higher than the metal temperature the expansion tends to go toward positive, i.e., the rotor expands faster than the cylinder, even though provisions are made for warming up the turbine before rolling. Hence, the recommended start-up procedures regarding inlet parameters, rate of raise of parameters, and speed and soaking of the turbine should be strictly adhered to. Similarly, during a hot start when the rotor tends to contract, following the recommended start-up procedures will help keep the differential expansion within the limits. During shut-down the rotor tends to contract, i.e. the cylinder expands faster than the rotor, and it can be controlled if it is a planned shut-down by gradually unloading and reducing the parameters. However, during a trip out, to assist in maintaining differential expansion additional provisions have to be made, such as rotor heating by supplying steam to the HPT and IPT front seal glands.

13.3.17 High bearing vibration

Vibration in the turbine bearing during rolling and loading should be taken seriously and when it exceeds the permissible limit, the turbine should be tripped and the vacuum broken. While rolling, critical speeds should be crossed as quickly as possible without pause.

13.3.18 High axial shift

Undue increase in axial shift will lead to overloading of the thrust pads of thrust bearing and eventual failure. It also contributes to bearing vibration and abnormal differential expansion. Protection will be incorporated to trip the turbine and break condenser vacuum at a limiting value of axial shift.

All the time during start-up, loading and normal running avoid sharp changes in load and steam parameters.

13.3.19 Temporary increase in turbine speed/turbine over-speed

Turbine speed may increase, temporarily exceeding its limit in the event of large load throw off. Temporary increase in turbine speed may cause serious damage to the turbine from shaft or blade distortion. Bearing overheating and gland-seal-damages are other results of temporary over-speed.

Over-speed of the turbine occurs when the HP turbine inlet emergency stop valve/s (ESV), IP turbine inlet interceptor valve/s (IV), and/or HPT/IPT inlet control valves fail to close after turbine tripping or steam finding its way into the turbine through extraction lines due to non-closure of "non return valves" or through any other lines connected to the turbine.

Over-speed is likely to cause destruction of the turbine, harm to the personnel and damage to the machinery in the vicinity. Hence, a schedule of testing of proper closing of ESV, IV, and extraction line "non-return valves" should be created and followed strictly. Actual over-speed tests should also be carried out as and when possible.

13.3.20 Performance deterioration due to erosion and corrosion

During the normal running of a turbine for prolonged period its performance deteriorates due to erosion, corrosion, salt deposition on turbine blades, and also due to change in turbine clearances. This deterioration usually starts after about 3 months following the first synchronization of the turbine. On average, the turbine heat rate increases by approximately 2% per 10 years of operation. The rate of increase of such deterioration in turbine performance, however, may be abated by overhauling the turbine every 5 or 6 years. Figure 13.1 shows the performance of a turbine over its economic life of 30 years when adopting a routine overhaul program. Item (1) of this figure starts with the initial level of turbine performance. Item (2) depicts the rate of deterioration in turbine performance between

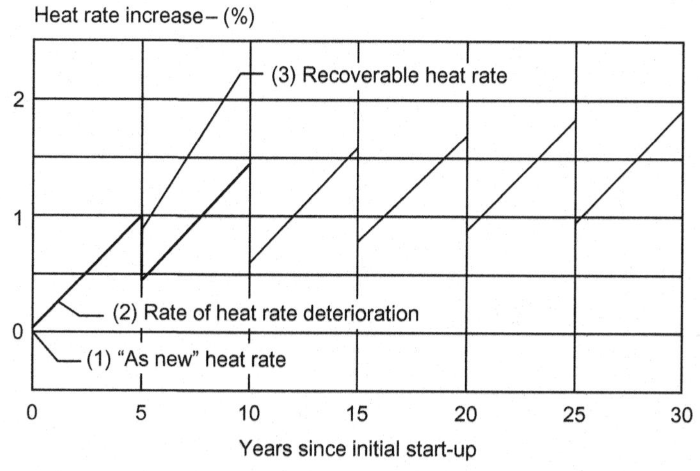

Figure 13.1 Recovery in turbine heat rate after overhauling. *Source: Fig. 1. P1. Ref. No.GER 3750A. TAB 23: Steam Turbine Sustained Efficiency by P. Schofield [6].*

Table 13.1 Base factor

N, Months of Normal Operation	0	12	24	36	48
BF, Base Factor %	0	1	1.5	1.9	2.2

MW = Rated output of the turbine.
f = 1.0 for fossil units.
P = Initial pressure of the turbine, MPa.

consecutive overhauls. The recovery of a certain portion of such deterioration following a scheduled overhaul is shown in Item (3) of the figure [6].

To determine the "estimated deterioration in the heat rate of turbine" during the initial stage of its operation different approaches, as follows, are usually followed:

a. One of the approaches lays down the following [7]:

$$\%\ Deterioration = \frac{BF}{\log MW} f \left(\frac{P}{16.55} \right)^{0.5}$$

where

BF = Base Factor, %, which follows Table 13.1.

b. Table 13.2 presents another approach [8].

Table 13.2 Deterioration in heat rate

Turbine rating, MW	Months of normal operation after synchronization		
	2 to 12 months	12 to 24 months	
≤150	0.1	0.06	
>150	$0.1\left(\frac{150}{MW}\right)^{0.5}$	$0.06\left(\frac{150}{MW}\right)^{0.5}$	% Per Month

13.4 ABNORMAL OPERATING CONDITIONS OF GAS TURBINES

In the event of any abnormal operating condition arising in a gas turbine that may lead to consequent drop in gas turbine load. In-depth investigation is needed to locate the actual problem area to take the necessary corrective action/s. Some of the areas to be investigated include:

1. Whether all relevant valves in gas conditioning skid (GCS) are open
2. Whether condensate accumulated in the GCS is draining properly
3. Whether gas pressure after pressure-reducing valve is adequate
4. Whether air intake filter house is free from accumulated water
5. Whether air intake filter-cleaning blowers are ready to come into service
6. Whether exhaust duct is free from condensate
7. Whether there is adequate oil in lube oil tank
8. Whether lubrication, temperature, and vibration of all bearings of running equipment are in order
9. Whether cooling water pressure and flow to the lube oil cooler and generator air cooler are adequate
10. Whether there is visible leakage from various systems, i.e., fuel gas, fuel oil, lube oil, cooling water, etc.
11. Whether auxiliary power supply is at the rated voltage and frequency
12. Whether DC power supply to the control system and DC motors is in order
13. Whether protections are isolated
14. Whether availability of healthy monitoring and annunciation systems is ensured
15. Whether fire detection system and automatic CO_2 fire extinguishing system are in order
16. Finally, inspect turbine and its operating parameters

In addition to the above areas the following sections discuss some of the abnormal symptoms of gas turbine operation and their probable causes [9].

When any abnormal condition occurs operating personnel should investigate the reason/s for such occurrence and take applicable corrective actions.

13.4.1 Loss of fuel pressure

Probable causes are:

i. Malfunctioning of fuel control system
ii. Choked fuel strainers and filters
iii. Faulty operation of governing valves

13.4.2 Drop in lube oil pressure

Probable causes are:

i. Lube oil reservoir level low
ii. Substantial leakage in the system
iii. Lube oil filter choked
iv. Lube oil cooler choked/rupture of tube
v. Main lube oil pump malfunction
vi. High lube oil temperature

13.4.3 Drop in compressor discharge pressure

Probable causes are:

i. Dirty compressor blades due to foreign matter and ingress of deposits
ii. Partially chocked intake air separation filter
iii. Erosion of the compressor blades

13.4.4 Ignition failure

Probable causes are:

i. Malfunctioning of electrical control system
ii. Defective ignition torches
iii. Dirty or worn out burner nozzles resulting in poor injection
iv. Damaged or distorted combustor liner
v. Faulty flame monitors
vi. Loss of ignition gas

13.4.5 High wheel space temperature

Probable causes are:

i. Blockage in the cooling air lines
ii. Faulty thermocouples
iii. Worn or deformed wheel space seals due to rubbing or bowed shaft

13.4.6 High turbine exhaust temperature

Probable causes are:

- **i.** Defective thermocouples
- **ii.** Air compressor capacity deterioration
- **iii.** Damaged or deformed turbine blades, excessive wear of blades due to corrosion, erosion, fouling, and increased blade tip clearance
- **iv.** Exhaust temperature controller out of calibration
- **v.** Bad combustion

13.4.7 High differential in turbine exhaust temperature

Probable causes are:

- **i.** Faulty exhaust thermocouple
- **ii.** Uneven fuel flow to individual burner due to partially chocked or worn out burner nozzles
- **iii.** Damaged or destroyed combustor liners

13.4.8 Abnormal vibration

Probable causes are:

- **i.** Turbine shaft misalignment
- **ii.** Damaged or worn out bearing (excessive clearance)
- **iii.** Shaft coupling loose or misaligned
- **iv.** Turbine or compressor blade damaged
- **v.** Bowed turbine shaft
- **vi.** Malfunction of inlet guide vane and bleed valves
- **vii.** Exhaust temperature controller out of calibration
- **viii.** Instrument malfunction

13.5 ABNORMAL OPERATING CONDITIONS OF DIESEL ENGINES

During the normal service life of a diesel engine, emergencies may occur due to failure or partial failure of any component or system causing premature or even sudden engine failure. Operation of the engine thereafter becomes hazardous unless due care is taken. A few typical examples of such problems are discussed in the following [10, 11].

13.5.1 Contamination of engine oil with glycol and water

Engine motor oil may get contaminated with glycol and/or water as a result of defective seals, blown head gaskets, cracked cylinder heads, corrosion damage, and cavitation. Such contamination results in formation of sludge, deposits, oil flow restrictions, and filter blockage; impairs lubrication and oil cooling, and in extreme cases may lead to cold seizure of engines. To mitigate these problems maintenance and routine analysis of engine oil must be done. In the event that any contamination is detected it is best to change the engine oil or switch to a better quality lubricant.

13.5.2 Engine knocking

In Diesel engines, where fuel is injected into highly compressed air toward the end of the compression stroke, engine knocking is inevitable. In this type of engine there is always a short time lag from the beginning of fuel being injected to the initiation to an appreciable increase in compressive pressure that would cause combustion to start. Fuel accumulated during this short time interval will ignite first in areas of greater oxygen density. The heat and additional turbulence developed by this initial ignition in the combustion chamber cause very rapid combustion of the fresh charge of fuel. This uncontrolled combustion takes place at approximately constant volume, but with a sudden increase in pressure and temperature, leads to *diesel knock*. To mitigate the occurrence of knocking the following actions are usually taken at the design stage:

1. The compression ratio is raised substantially to produce a temperature within the combustion chamber considerably higher than the spontaneous ignition temperature of the fuel. Nevertheless, the upper limit of compression ratio is determined by the strength of the cylinder, the bearings and also by the stress developed in other parts
2. Degree of turbulence is reduced, but it lowers torque and thermal efficiency
3. Delay period is reduced

Fuel's ignition delay is measured by a scale called the *cetane* number. In a particular Diesel engine, higher cetane fuels will have shorter ignition delay periods than lower cetane fuels. When the engine is running the delay period could be reduced by applying additives, such as ethyl-nitrate, to the fuel, which helps in accelerating combustion process.

(Detail discussion on cetane number is available in CHAPTER 3: Fuels and Combustion.)

13.5.3 Diesel engine receiver safety valves operate

As soon as the diesel engine receiver safety valves operate, do the following:

a. Stop the engine immediately
b. Prime the engine with oil and water
c. Open cocks on the drain pipes of the receiver and exhaust manifolds
d. Pump oil from engine sump and system scavenging pipelines into the circulating tank by means of oil scavenging pump
e. Close engine air supply pipeline valve and also the valve that admits air to thermo-regulators

- **f.** After the engine has cooled down sufficiently, uncover all the inspection holes of the receiver and exhaust manifolds; remove oil accumulating in the receiver and manifolds, clean and scavenge drain pipes
- **g.** Examine the exhaust silencer, check and clean the silencer oil drain line, or remove fuel, if any, from exhaust path
- **h.** Inspect air coolers, compressor impeller stator part, and measure the axial play of turbo compressor rotor shaft
- **i.** After all the checks are accomplished, and faults eliminated, prepare the engine for starting

13.5.4 Safety valves on Diesel engine crank case coverings operate

In the case safety valves on the engine crankcase coverings operate, do the following:

- **a.** Stop the engine immediately
- **b.** Prime the engine with oil and water
- **c.** Open cocks on the drain pipes of the receiver and exhaust manifolds
- **d.** Pump oil from engine sump and system scavenging pipelines into the circulating tank by means of oil scavenging pump
- **e.** Close the engine air supply pipeline valve and also the valve that admits air to thermo-regulators
- **f.** After the engine has cooled down sufficiently, uncover all the inspection holes of the receiver and exhaust manifolds; remove oil accumulating in the receiver and manifolds, clean and scavenge drain pipes
- **g.** Do not open engine hand holes immediately
- **h.** Find the place of vapor flashing in the crankcase
- **i.** Detect and eliminate cause of oil vapor flashing per the manufacturer's recommendation

13.5.5 Diesel engine emergency condition warning system operates

As soon as warning signals are received, determine the problem with the help of a flashing signal.

- **a.** In the case oil pressure drops, stop the diesel engine immediately, and cut in the priming pump and prime the engine with water
  1. Make sure the oil pipeline is not damaged
  2. Check the level of oil in the circulating tank
  3. Disassemble and examine the engine oil pump valve

  Do not start the engine until the cause of oil pressure drop is detected and eliminated; cut in priming and scavenging pumps and prime the diesel engine with oil

b. In the case cooling water pressure in the internal cooling circuit drops, check the following:
   1. Check level of water in the expansion tank
   2. Check presence of air in the cooling system.
   If it is impossible to detect and eliminate the cause of the fault (and water temperature starts increasing), unload and stop the engine.

c. Should the fuel-pressure drop, do the following:
   1. Check level of fuel in service tanks
   2. Use stand-by fuel pump to pressure-test the system with fuel, and check proper sealing of pipelines of engine-driven fuel feed pump and shut-off valves.

13.6 GENERAL ELECTRICAL FAULTS

In a power station electrical faults are not a day-to-day phenomenon. Experience reveals that operating personnel in general never encountered some of these electrical faults even in their life time. Nevertheless, looking at the safety of an electrical equipment and consequent catastrophe that may occur, especially in critical equipment such as the generator, large transformers, large induction motors, etc., protective relays are provided to monitor and safeguard all electrical equipment. They must remain ready all the time in anticipation of a fault and act as an insurance against consequent damage resulting from a fault, even though many plant experts say that a relay operates far more number of times during testing and maintenance than during actual fault.

Current and voltage transformers are provided to facilitate protective relays to detect faults in electrical equipment. When faults occur these relays selectively remove only the faulty part from the rest of the system by tripping the associated circuit breakers. They must take necessary corrective action with utmost sensitivity, selectivity, and speed. These relays must be selective because there are certain operating conditions inherent to the operation of electrical equipment that are fundamentally abnormal, but that are not treated as such, e.g., the starting current of an induction motor, magnetizing inrush current of a transformer, and conditions arising from power swings. Under these circumstances a protective relay must ignore the high magnitude of currents and refrain from extending suitable protections against them consciously. Hence, the protective system has to be designed so that protective relays are capable of discriminating between normal operating conditions, abnormal operating conditions, and faults.

13.6.1 Generator

Abnormal operating conditions in a generator may occur under any or a combination of the following. If any of these conditions occur within the generator it must be isolated from the grid by opening the generator circuit breaker [12].

i. Over-voltage
ii. Under-voltage

iii. Over-current
iv. Over-frequency
v. Under frequency
vi. Volts/Hertz high (over-flux)
vii. Differential protection
viii. Loss of excitation
ix. Over-excitation
x. Stator earth fault
xi. Rotor second earth fault
xii. Split phase protection for inter turn fault
xiii. Pole-slip protection
xiv. Back up impedance protection
xv. Negative phase sequence protection
xvi. Rotor high temperature protection
xvii. Reverse power relay operation

Some of the above protections are discussed in following.

The reasons for over-voltage are normally due to over-speed of the machine and sometimes due to abnormal operation of automatic voltage regulator. Under these conditions the generator must be isolated from the grid.

Additionally, over-voltage is also generated by surge due to lightning, which is handled by providing surge arresters and surge capacitors. Transient over-voltage may be generated by switching surges across the contacts of circuit breakers or when the arcs are grounded. While switching surges can be reduced by using technologically advanced circuit breakers, arcing ground surges are mitigated by using resistance earthing.

Sustained over-current may result in overheating of the stator, leading to an increase in winding temperature. Before the increase in temperature can damage the winding, the generator should be isolated.

Loss of excitation may result in overheating of the rotor. Therefore, it is advised to isolate the generator on loss of excitation.

The negative phase sequence current may cause overheating of the stator, hence the generator needs to be isolated.

In the event of tripping of the prime mover, the machine draws power from the grid and the generator starts behaving like a motor. If the prime mover is a steam turbine the coolant (steam) flow through the turbine gets stopped, causing overheating of turbine blades, which may lead to distortion of low-pressure stage turbine blades. In the event that the prime mover is an engine, both external and internal combustion, load coming on the machine could be as high as 25% of rated load, which is a loss to the system. The natural solution is isolation of the generator.

(For further treatment on generator protection refer to, Chapter 11, *Interlock and Protection*.)

13.6.2 Transformers

Transformers play an important role in power supply and delivery systems and in maintaining the integrity of the whole electrical network in a power station. They come in various sizes and configurations. Since transformers are critical and expensive electrical equipment of a power station, any damage to the equipment may lead to unacceptable interruption in power transmission and distribution. Hence, transformers must be protected against adverse operating conditions by providing suitable protection such as over-excitation protection and temperature-based protection.

Depending on its size the failures generally encountered in a transformer are as follows [13]:

i. Winding failures resulting from short circuits
ii. Core faults
iii. On-line tap changer (OLTC) failures
iv. Overloading
v. Line surge
vi. Loose connections
vii. Oil contamination
viii. Faulty operation, such as improperly set controls, loss of coolant, accumulation of dirt and oil, corrosion, etc.
ix. Fire explosions, lightning

To handle these problems, transformers are provided with following protective arrangements:

1. Protective covers are provided to restrict access of rain water to the transformer oil-filled compartment
2. Dry air is continuously pumped into the gas space if humidity becomes severe
3. Above atmospheric gas pressure is maintained when oil-filled transformers are installed outside
4. When the transformer is put into service various temperatures and pressures should always be monitored; any parameter exceeding the recommended limit must isolate the transformer from service
5. Surge arresters are installed and connected to the transformer bushing/terminals

In addition to the above, protective relays are also provided to circumvent following abnormal operating conditions of transformers as applicable:

i. Oil temperature high
ii. Winding temperature high
iii. Negative sequence current very high
iv. Over-current

v. Standby earth fault
vi. Restricted earth fault
vii. Differential protection
viii. Buchholz protection

13.6.3 Induction motors

In a power station a huge number of induction motors is spread across the station. Abnormal operating conditions of an induction motor include the following:

i. Short-circuit in the stator winding: this condition may occur due to an inter-turn fault (between turns of the same phase), inter-phase fault (between different phases), or ground fault (between a phase winding and ground). In the event such a fault occurs the over-current protection must act instantaneously to trip the motor and protect the stator.
ii. Cracked rotor bars: cracks in rotor bars develop in motors that are frequently started due to thermal and mechanical stresses. Hence, operators must follow manufacturer's guidelines and restrict the number of starts per hour as recommended.
iii. Bearing failures: this fault occurs due to rotor unbalance, eccentricity, or misalignment between the motor shaft and associated load shaft. During the erection stage adequate precaution and surveillance must be exercised to restrict any misalignment between shafts within the recommended limit.
iv. Off-normal voltage quality: in the event the supplied voltage is abnormally high or low, the motor will suffer high losses. Too low a voltage may cause a motor to stall. Hence, operators must ensure that variation in supplied voltage to running motor remains within ±10% of rated voltage, and frequency remains within ±5% of the rated frequency.
v. Phase loss: this fault may occur whenever fuses in the supply line blow. Consequently, thermal overload relays operate and trip the motor circuit breaker.
vi. Mechanical overload: this fault may cause overheating or even stalling the motor. In this case thermal overload relays also operate to protect the motor.

BIBLIOGRAPHY

[1] United States Department of the Interior Bureau of Reclamation. Facilities Instructions, Standards and Techniques Volume 1–12. Abnormal Operations Generic Technical Guidelines for Power Stations. March 2003.
[2] British Electricity Institute. Modern power station practice volume G station operation and maintenance. London: Pergamon Press; 1991.
[3] NFPA 85. Boiler and combustion systems hazards code; 2004.

[4] Bozzuto C. (Ed.), Clean Combustion Technologies (fifth ed.). Alstom.
[5] ASME TDP-1-2006: Recommended Practices for the Prevention Water Damage to Steam Turbines Used for Electric Power Generation. Fossil-Fueled Plants.
[6] 37th GE Turbine State-Of-The-Art Technology Seminar, 1993.
[7] ANSI/ASME PTC 6 Report 1985 (R1997) – Guidance for evaluation of measurement uncertainty in performance tests of steam turbines.
[8] IEC 953-2{IS 14198 (Part 2)}. Rules for steam turbine thermal acceptance tests: part 2 method B – wide range of accuracy for various types and sizes of turbines.
[9] Boyce MP. Gas turbine engineering handbook. Gulf Publishing Company Book Division; 1982.
[10] Lilly L. C. R. Diesel Engine Reference Book. Butterworth and Co. (Publishers) Ltd.; 1984.
[11] Maleev VL. Internal combustion engine theory and design. McGraw-Hill Book Company, Inc.; 1945.
[12] IEEE Std. C37.102-2006, IEEE Guide for AC Generator Protection.
[13] IEEE Std. C37.91-2000 IEEE Guide for protective relay applications to power transformers.

CHAPTER 14

Air Pollution Control

14.1 INTRODUCTION

The problem of air pollution may be the worst curse of modern civilization. With the advent of industrialization human beings started to release more effluent into the atmosphere than nature could handle. Air pollution occurs when chemicals, fumes, particulate matter, odor, or biological materials introduced into air exceed safe limits and cause harm to the health of humans, animals, and other living organisms or damage to plants and inanimate objects. These materials can be in the form of solid particles, liquid droplets, or gases and are called pollutants. Pollutants, which are directly emitted from a plant or motor vehicles to atmosphere, e.g., ash, particulate matter, sulfur dioxide, or carbon monoxide, are called primary pollutants, while secondary pollutants result from interaction or chemical reaction of primary pollutants, e.g., formation of smog or depletion of stratospheric ozone layer.

Air pollutants are dispersed throughout the world's atmosphere in concentrations high enough to gradually cause serious health problems. Sulfur dioxide, nitrogen dioxide, and droplets containing solutions of sulfuric acid, sulphate, and nitrate salts descend to the earth's surface in wet form as rain or snow and in dry form as fog, dew, or solid particles. This phenomenon is generally called acid rain.

Air pollutants released from fossil fuel power plants include:

Particulate matter Particulate matter (PM) or more precisely suspended particulate matter (SPM) are fine solid particles suspended in a gas. Particulate matter is either naturally available in the atmosphere or can be manmade. Combustion of fossil fuels in power plants generates significant amounts of particulate matter.

When the presence of these fine particles in air exceeds a safe limit (Table 14.1 and Table 14.2) these particles cause health hazards. Long-term exposure to particulate matter results in heart disease, risk of decreased lung function, exacerbation of asthma, lung cancer, etc.

Carbon dioxide (CO_2) Carbon dioxide (CO_2) is one of the major pollutants in the atmosphere, but it is vital to living organisms. The major source of CO_2 into the atmosphere is combustion of fossil fuels.

Carbon dioxide is one of the "greenhouse" gases (other greenhouse gases in the earth's atmosphere are water vapor, methane, nitrous oxide, ozone, sulfur hexafluoride,

Thermal Power Plant
DOI: http://dx.doi.org/10.1016/B978-0-12-801575-9.00014-7

Table 14.1 Clean air act of the United States,1990

Pollutant	Primary standards		Secondary standards	
	Level	Averaging time	Level	Averaging time
Carbon Monoxide	9 ppm (10 mg/m^3)	8-hour(1)	None	
	35 ppm (40 mg/m^3)	1-hour(1)		
Lead	0.15 $\mu g/m^3$ (2)	Rolling 3-Month Average	Same as Primary	
	1.5 $\mu g/m^3$	Quarterly Average	Same as Primary	
Nitrogen Dioxide	0.053 ppm (100 $\mu g/m^3$)	Annual (Arithmetic Mean)	Same as Primary	
Particulate Matter (PM_{10})	150 $\mu g/m^3$	24-hour(3)	Same as Primary	
Particulate Matter ($PM_{2.5}$)	15.0 $\mu g/m^3$	Annual(4) (Arithmetic Mean)	Same as Primary	
	35 $\mu g/m^3$	24-hour(5)	Same as Primary	
Ozone	0.075 ppm (2008 std)	8-hour(6)	Same as Primary	
	0.08 ppm (1997 std)	8-hour(7)	Same as Primary	
	0.12 ppm	1-hour(8)	Same as Primary	
Sulfur Dioxide	0.03 ppm	Annual (Arithmetic Mean)	0.5 ppm (1300 $\mu g/m^3$)	3-hour(1)
	0.14 ppm	24-hour(1)		

(1)Not to be exceeded more than once per year.
(2)Final rule signed October 15, 2008.
(3)Not to be exceeded more than once per year on average over 3 years.
(4)To attain this standard, the 3-year average of the weighted annual mean PM2.5 concentrations from single or multiple community-oriented monitors must not exceed 15.0 $\mu g/m^3$.
(5)To attain this standard, the 3-year average of the 98th percentile of 24-hour concentrations at each population-oriented monitor within an area must not exceed 35 $\mu g/m^3$ (effective December 17, 2006).
(6)To attain this standard, the 3-year average of the fourth-highest daily maximum 8-hour average ozone concentrations measured at each monitor within an area over each year must not exceed 0.075 ppm. (effective May 27, 2008)
(7)(a) To attain this standard, the 3-year average of the fourth-highest daily maximum 8-hour average ozone concentrations measured at each monitor within an area over each year must not exceed 0.08 ppm.
(b) The 1997 standard—and the implementation rules for that standard – will remain in place for implementation purposes as EPA undertakes rulemaking to address the transition from the 1997 ozone standard to the 2008 ozone standard.
(8)(a) The standard is attained when the expected number of days per calendar year with maximum hourly average concentrations above 0.12 ppm is <1.
(b) As of June 15, 2005 EPA has revoked the 1-hour ozone standard in all areas except the fourteen 8-hour ozone non-attainment Early Action Compact (EAC) Areas. For one of the 14 EAC areas (Denver, CO), the 1-hour standard was revoked on November 20, 2008. For the other 13 EAC areas, the 1-hour standard was revoked on April 15, 2009.

Table 14.2 Central pollution control board notification 2009, INDIA

Pollutants	Time weighted	Concentration In ambient air, $\mu g/m^3$	
		Industrial, residential, rural, and other areas	Ecologically sensitive areas (notified by Central Government)
Sulfur Dioxide SO_2	Annual average*	50	20
	24 hours**	80	80
Nitrogen Dioxide NO_2	Annual average*	40	30
	24 hours**	80	80
Particulate Matter (size less than 10 μm) or PM_{10}	Annual average*	60	60
	24 hours**	100	100
Particulate Matter (size less than 2.5 μm) or $PM_{2.5}$	Annual average*	40	40
	24 hours**	60	60
Ozone, O_3	8 hours**	100	100
	1 hour	180	180
Lead	Annual average*	0.5	0.5
	24 hours**	1.0	1.0
Carbon Monoxide CO, mg/m^3	8 hours**	2.0	2.0
	1 hour	4.0	4.0
Ammonia NH_3	Annual average*	100	100
	24 hours**	400	400

*Annual Arithmetic Mean of minimum 104 measurements in a year taken twice a week 24 hourly at uniform interval.
**24 hourly or 8 hourly or 1 hourly monitored values, as applicable, shall be complied with 98% of the time in a year. 2% of the time, they may exceed the limits but not on two consecutive days of monitoring.

hydrofluorocarbons, perfluorocarbons, nitrogen trifluoride, etc.) that cause global warming. Global warming is the term that refers to the increase in the average surface air temperature of the planet as a result of build-up of these gases in the atmosphere. Greenhouse gases absorb and emit radiation from the atmosphere within the thermal infrared range.

Carbon dioxide is a good transmitter of sunlight, but partially restricts infrared radiation going back from the earth into space. This produces the so-called greenhouse effect that prevents cooling of the earth during the night and keeps the earth warm enough to support human, plant, and animal life. However, incessant release of manmade CO_2 to the atmosphere is so alarming that the concentration of CO_2 since the beginning of the Industrial Revolution (taken as the year 1750) has increased by about 43%, from 280 ppm in 1750 to 400 ppm in 2015 [1], in spite of continuous reduction of atmospheric concentration of CO_2 by photosynthesis. This in turn has reinforced the warming of the earth's surface further beyond the tolerable limit of earthly objects, both animate and inanimate. The Intergovernmental Panel on

Climate Change (IPCC) has predicted an average global increase in temperature of 1.4 K to 5.8 K between 1990 and 2100.

Carbon monoxide (CO) Carbon monoxide is generated when there is incomplete combustion of fuel. It is colorless and odorless, but is a very poisonous gas. However, due to the high efficiency of present-day steam generators, emission of carbon monoxide from power plants is not of much of a concern.

When inhaled, carbon monoxide enters the bloodstream and disrupts the supply of oxygen to the body's tissues, the results of which may become fatal if the CO concentration becomes 1000 ppm or more.

Sulfur oxides (SO_x) All fossil fuels, i.e., coal, oil, and natural gas, contain sulfur, and/or sulfur compounds. Sulfur in coal exists in three forms: organic, pyritic and sulfate. Sulfur that is an inherent constituent of coal is organic sulfur, which is considered as a non-removable impurity. Pyritic sulfur, also known as inorganic sulfur, occurs primarily as pyrite or mercasite. Sulfate sulfur usually exists in the form of calcium sulfate or iron sulfate. The sulfates represent a small fraction of the total sulfur in coal and have no significant role in the combustion process itself or in contributing to emissions. During combustion, both organic sulfur and pyrite are oxidized to sulfur dioxide (SO_2). Depending on combustion conditions, a small amount of sulfur trioxide (SO_3) may also be formed. Sulfur oxides in combination with moisture content of flue gas form acids (sulfuric acid: H_2SO_4) that may get deposited when combustion gas is cooled below its dew-point temperature, resulting in corrosion of air heaters, economizers, downstream flue-gas path, and stacks. Sulfur also contributes to clinkering and slagging and to spontaneous combustion of stored coal.

In the absence of a suitable emission control process SO_x, mainly SO_2, escapes to the atmosphere. Sulfur oxides emitted from the stack cause acid rain. Sufficiently high concentrations (10 ppm) of sulfur dioxide in the atmosphere irritates the upper respiratory tract of human beings and causes wheezing, breathing problems, exacerbation of asthma, etc. Hence, the use of sulfur-bearing fuels as a power source is a concern.

Nitrogen oxides (NO_X) Oxides of nitrogen are major air pollutants that have a significant detrimental effect on the atmosphere. During combustion of fossil fuels at high temperature, nitrogen present in coal as well as combustion air is converted to nitric oxide (NO) and nitrogen dioxide (NO_2) (together commonly referred to as NO_X). In the atmosphere, NO in the presence of sunlight oxidizes to NO_2. Nitrogen dioxide is toxic and is one of the major air pollutants. NO_X emissions are of concern because they are associated with acid rain, causing damage to aquatic ecosystems, corrosion of building materials, etc.

Exposure to high levels of NO_2 may lead to lung damage, asthma, respiratory problems, etc. NO_2 can enter into bloodstream, causing damage to tissues of the body due to disruption of oxygen supply much like carbon monoxide.

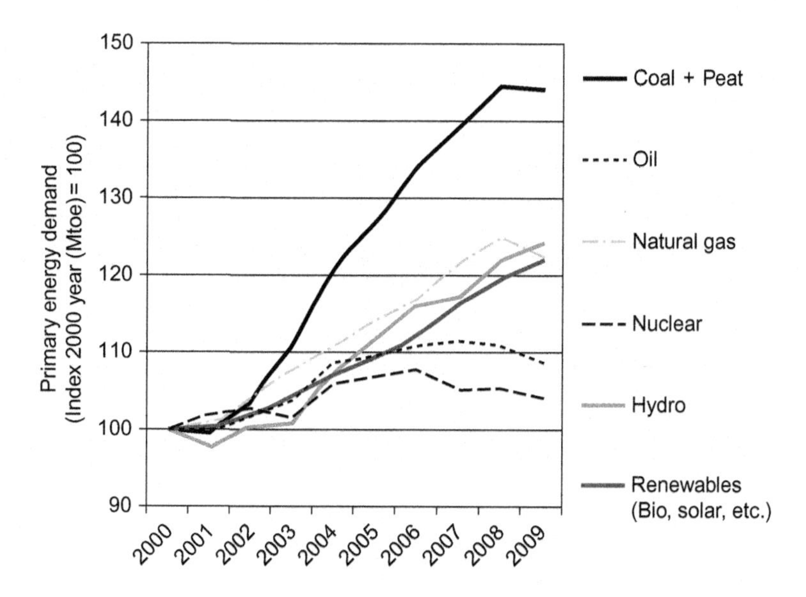

Figure 14.1 Primary energy demand [2].

The effects of these air pollutants is increasing continuously since the number and unit capacity of thermal power plants are increasing every day. Figure 14.1 shows how the primary energy demand is growing year after year. Therefore, with the growth of the industry, it is essential to take measures to help prevent air pollution.

Air pollution is so critical today that the feasibility of installing any new industry depends on the findings of an *Environmental Impact Assessment* (EIA) study along with any remedial measures proposed. Even at the bidding stage of a new project emission levels get the same priority if not more as price, efficiency, reliability, etc.

Thus, to prevent critical polluting environment, clean air legislation is now in force in many countries and often imposes severe restrictions on allowable emissions from stack. The permitted emission varies from country to country. Table 14.3, Table 14.4, and Table 14.5 present the emission guidelines of fossil fuel-fired steam power plants boiler as determined by the IFC—WB (International Finance Corporation-World Bank) and other countries.

In addition to stack emission limits discussed above, ambient air emission limits in various countries are more stringent to safeguard health of human beings, animals, plants, etc., as is evident from Table 14.1 that reflects emission limits in the United States and Table 14.2 which presents National Ambient Air Quality Standard of emission limits in India.

Table 14.3 Emission guidelines for particulate matter (mg/Nm^3)

FUEL	IFC-WB[a,b]/Plant capacity, MWth		United States	EU[a]/Plant capacity, MWth	CHINA[a]	INDIA/plant capacity, MW	
	NDA	DA					
Natural Gas	NA	NA					
Other Gaseous Fuels	50	30					
Liquid Fuels	50/ >50 to <600	30/>50 to <600					
Liquid Fuels	50/ ≥600	30/ ≥600					
Solid Fuels	50/ >50 to <600	30/ >50 to <600	6.4 gm/GJ	50/50 to 100	50	100–150[a]/200 to <500	
Solid Fuels	50/ ≥600	30/ ≥600		30/ >100		45[c] (maximum) with design coal and all fields in service/ sub-critical ≥500	80[c] (maximum) with maximum ash coal and one field of each pass of ESP out of service)/sub-critical ≥ 500
Solid Fuels						50[d] (maximum or as per MOEF"s requirement, whichever is more stringent with design coal and one field of each pass of ESP out of service)/ supercritical 660/ 800 MW	80[d] (maximum or as per MOEF"s requirement, whichever is more stringent with maximum ash coal and one field of each pass of ESP out of service)/ supercritical 660/ 800 MW

Notes: NDA = Non-Degraded Airshed, DA = Degraded Airshed (poor air quality), NA = Not Applicable.
MOEF = Ministry of Environment and Forests, India.
[a]IFC-WB (International Finance Corporation-World Bank) Environmental, Health and Safety (EHS) Guidelines THERMAL POWER PLANTS (December 19, 2008) [3].
[b]Dry Gas Excess O_2 Content for gaseous and liquid fuels is 3% and for solid fuels is 6%.
[c]CEA (Central Electricity Authority, India) Standard Technical Specification Main Plant Package of Sub-critical Thermal Power Project ≥500 MW (September 2008).
[d]CEA Standard Technical Features of BTG System for Supercritical 660/800 MW Thermal Units (July 2013).

Table 14.4 Emission guidelines for SO_2 (mg/Nm^3)

FUEL	IFC-WB[a,b]/Plant capacity, MWth		United States[a]	EU[a]/Plant capacity, MWth	CHINA[a]	INDIA/Plant Capacity, MW
	NDA	DA				
Natural Gas	NA	NA				
Other Gaseous Fuels	400	400				
Liquid Fuels	900–1500/ >50 to <600	400/ >50 to <600				
Liquid Fuels	200–850/ ≥600	200/ ≥600				
Solid Fuels	900–1500/ >50 to <600	400/ >50 to <600	640 gm/MWH (or 200 at 6% O_2 assuming 38% HHV Efficiency	850/50 to 100	400 (general) 800 for coal GCV <12,550 kJ/kg 1200 with mine-mouth plant burning	Stack Height, m[c] $H = 14 (Q)^{0.3}$ Q = Emission rate of SO_2, kg/h/ <200
Solid Fuels	200–850/ ≥600	200/ ≥600		200/ >100	S <0.5% coal	Stack Height 220 m[c]/200 to <500
Solid Fuels						Stack Height 275 m[c]/ ≥500

Notes: NDA = Non-Degraded Airshed, DA = Degraded Airshed (poor air quality), NA = Not Applicable.
MOEF = Ministry of Environment and Forests, India.
[a]IFC-WB (International Finance Corporation-World Bank) Environmental, Health and Safety (EHS) Guidelines THERMAL POWER PLANTS (December 19, 2008) [3].
[b]Dry Gas Excess O_2 Content for gaseous and liquid fuels is 3% and for solid fuels is 6%.
[c]Government of India's Extra Ordinary Gazette No. G.S.R. 801(E) dated December 31, 1993. General Standards for discharge of environmental pollutants Part-A: Effluents.

Table 14.5 Emission guidelines for NO_x (mg/Nm^3)

FUEL	IFC-WB[a,b]/Plant capacity, MWth		United States	EU[a]/Plant capacity, MWth	CHINA[a]	INDIA/ Plant capacity, MW
	NDA	DA				
Natural Gas	240	240				
Other Gaseous Fuels	240	240				
Liquid Fuels	400/ >50 to <600	200/ >50 to <600				
Liquid Fuels	400/ ≥600	200/ ≥600				
Solid Fuels	510 or up to 1100	200/ >50	450 gm/MWH	150/50 to 300		
Solid Fuels	If volatile matter of fuel <10%/ >50			200/ >300		

Notes: NDA = Non-Degraded Airshed, DA = Degraded Airshed (poor air quality), NA = Not Applicable.
MOEF = Ministry of Environment and Forests, India.
[a]IFC-WB (International Finance Corporation-World Bank) Environmental, Health and Safety (EHS) Guidelines THERMAL POWER PLANTS (December 19, 2008) [3].
[b]Dry Gas Excess O_2 Content for gaseous and liquid fuels is 3% and for solid fuels is 6%.

14.2 CONTROL OF SUSPENDED PARTICULATE EMISSION

Suspended PM in a gas stream may be separated by either a mechanical separating device or an electrical separating device. The mechanical separating device includes gravity settling, cyclone separation, scrubbers, fabric filters, etc. The electric separating device refers to electrostatic precipitation. An electrostatic precipitator (ESP) is one of the most widely used air pollution control devices since mechanical separating devices, barring bag filter, are unable to collect fine particles.

Both ESP and bag filters are capable of collecting very fine particles from 10 μ down to 0.05 μ size, with the highest collection efficiency over a wide range of contaminants as is evident from Figure 14.2. They can be used in applications where there is a large volume of contaminants. The performance of a bag filter is as reliable as an ESP and provides almost identical collection efficiency.

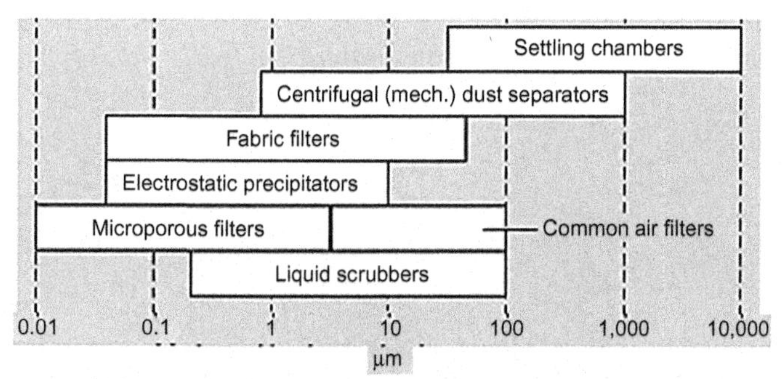

Figure 14.2 Contaminants and particle sizes that can be removed by ESP and mechanical separating devices. *Source: Figure 5-5, P5-5, Control of Power Plant Stack Emissions, [4].*

An ESP has relatively constant pressure drop and is a constant efficiency (as measured by the percentage removal of the inlet dust loading) device, depending on inlet loading and dust properties. Figure 14.3 shows a general view with part of a sectional arrangement of a typical ESP.

Bag filters are constant emission (as measured by the mass of the particulate emitted per unit of fuel fired) device and exhibit a varying pressure drop, depending on dust properties and the degree of cake thickness at any point of reference. A general view with a sectional arrangement of a typical bag filter is shown in Figure 14.4.

14.2.1 Control of suspended particulate matter (SPM) using ESP

Electrostatic precipitators depend on static electricity to do their work. The main components of an ESP are two sets of electrodes (Figure 14.5). The first is comprised of rows of electrically grounded vertical parallel plates, called collecting electrodes, between which the gas to be cleaned flows. The second set of electrodes is wires, called discharge electrodes, are centrally located between each pair of parallel plates. These electrodes may be of different shape and design. Figure 14.6 shows different types of collecting electrodes, while Figure 14.7 presents various types of discharge electrodes.

The wires carry a unidirectional, negatively charged, high-voltage DC current from an external source that generates unidirectional, non-uniform electric field. In a non-uniform electric field, the field intensity near the conductor is considerably higher than that on the periphery or on the plate. Therefore, as the voltage increases, the ionization takes place about the conductor. No ionization takes place near the periphery or on the plate. When the voltage is high enough, a blue luminous glow, called a *corona*, is produced around them. The corona is an indication of generation of negatively charged gas ions that travel from wires to the grounded collecting electrodes as a result of strong electric field between them. The field intensity that causes the corona to start may be calculated using the following formula [6]:

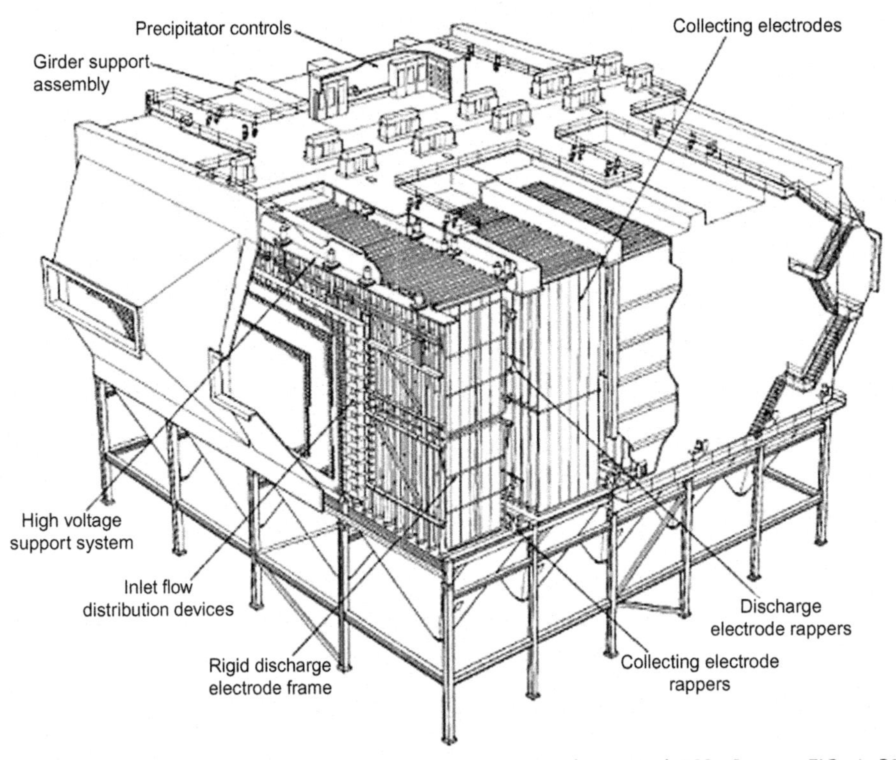

Figure 14.3 General view with part sectional arrangement of a typical ESP. *Source: FIG. 4, P33-4, CH33: PARTICULATE CONTROL [5].*

$$E_C = 3.04\left(\beta + 0.0311\sqrt{\frac{\beta}{r}}\right) * 10^6 \tag{14.1}$$

where

E_C = Field intensity, V/m
β = Ratio of density of gas under the working conditions to that under standard conditions (pressure: 101.3 kPa and temperature 298 K)

$$= \frac{(P_B - P_G) * 298}{101.3 * T_G} \tag{14.2}$$

where

P_B = Barometric pressure, kPa
P_G = Draft of gas in ESP, kPa
T_G = Temperature of gas in ESP, K
r = radius of discharge electrode, m

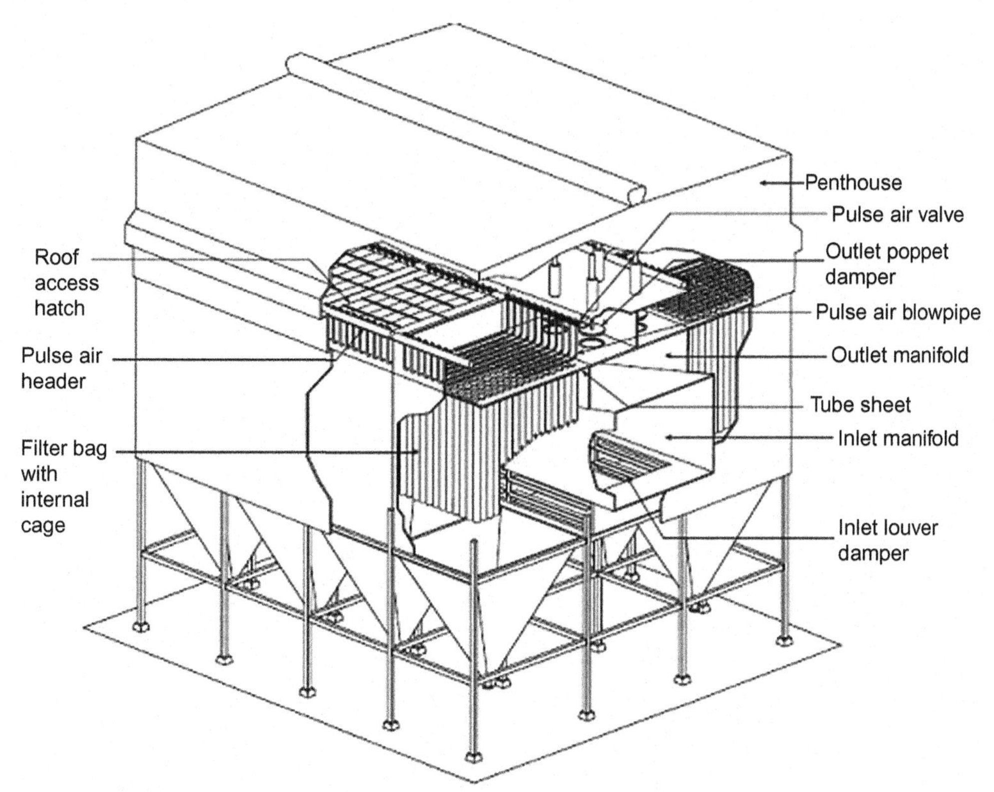

Figure 14.4 General view with sectional arrangement of a typical bag filter. *Source: FIG. 8, P33-9, CH33: PARTICULATE CONTROL [5].*

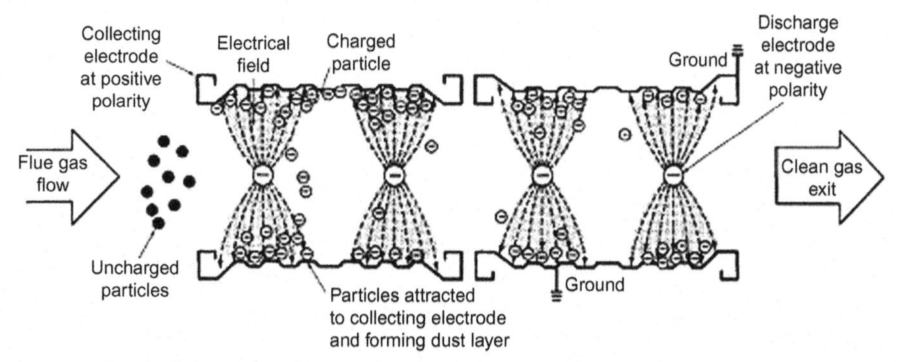

Figure 14.5 Charged particles between electrodes. *Source: FIG. 3, P33-3, CH33: PARTICULATE CONTROL [5].*

14.2.2 Fundamentals

It is known from basic physics that all bodies fall into two categories, i.e., conductors and dielectrics. A body in which electric charges may move freely throughout its volume is called a conductor. While a body that does not possess any such property and acts as an

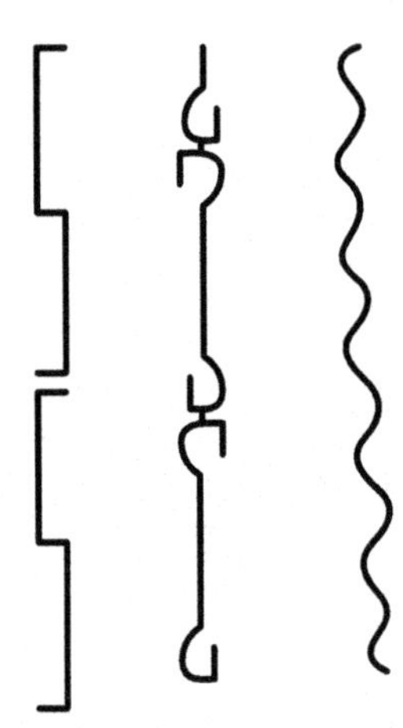

Figure 14.6 Types of collecting electrodes.

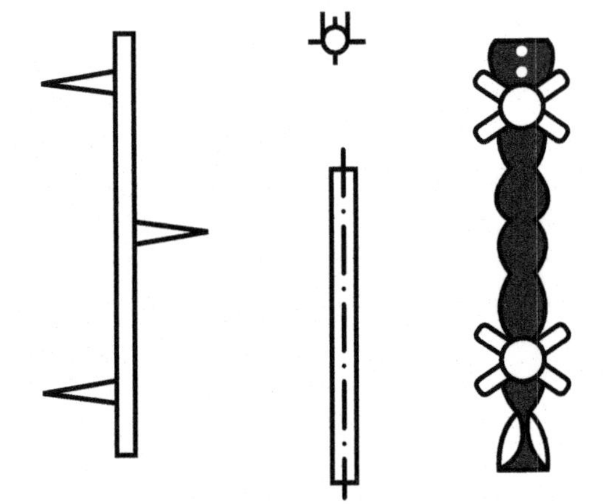

Figure 14.7 Types of discharge electrodes.

insulator is called dielectric. Gases under normal condition are insulators (dielectric), since they comprise electrically neutral atoms and molecules. However, when subjected to the influence of various artificial methods, i.e., X-rays, radioactive emissions, cosmic rays, strong heating, bombardment of gas molecules by rapidly moving electrons or ions, etc., these gases get ionized, i.e., split into electrons (negative ions) and positive ions, and become conductors. A combination of some of the free electrons of gases with their neutral gas molecules also results in the formation of negative ions.

Out of all artificial methods of ionization just described, the high-voltage direct-current corona is by far the most effective to ionize gas particles. In this method electrodes are connected to a voltage source and produce a non-uniform electric field.

When gases are passed between these electrodes, the free charges within gases move along lines of forces of the non-uniform electric field. Thus under the influence of strong electric field suspended particles in gas stream migrate toward collecting electrodes. On reaching the electrodes, particles release part of their charge and get collected. The collected dust is then discharged periodically by rapping and gets collected in storage hoppers. The whole process includes the following steps:

a. Charging of suspended particles
b. Collection of charged particles in an electric field
c. Removal of precipitated matters from collecting electrodes

The separation force acting on a particle is given by Coulomb's law, which states that force is proportional to the product of particle charge and the intensity of the collecting field. Hence,

$$F_C \infty E * q \tag{14.3}$$

where

F_C = Force, N
E = Field intensity, V/m
q = Particle charge, C (=A · s)

Although the Coulomb force steers particles to accelerate toward the collecting electrodes, inertial and viscous forces counteract the Coulomb force and oppose the motion of particles. For suspended particles, inertial forces are negligibly small and may be ignored. The viscous force of the gas is governed by Stokes' law and is expressed by Eq. 14.4. The resultant of the Coulomb force and Stokes (viscous) force determines velocities to be attained by a particle in the precipitation field. The velocity at which the particles move in the gas streams toward the collecting electrodes is called the *migration velocity* of the particle.

Thus, the viscous force,

$$F_S = 6\pi\mu r w \tag{14.4}$$

where

F_S = Frictional force
μ = Absolute viscosity of gas
r = Particle radius
w = Migration or drift velocity of a charged particle in an electric field and is given by Eq. 14.5 and Eq. 14.6

$$w = \frac{E_o * E_p * r}{\mu} \quad \text{(for particle size} > 1 \text{ micron)} \tag{14.5}$$

$$w = \frac{E_p}{\mu} \text{ (for sub-}\mu\text{ particles)} \tag{14.6}$$

where

w = Migration velocity
E_o = Charging field intensity
E_p = Collecting field intensity
r = Particle radius
μ = Absolute viscosity of gas

For $E_o = E_p = E$, and if w is expressed in cm/s, E in V/m, r in m, and μ in N.s/m^2, Eq. 14.5 and Eq. 14.6 change as follows [6]:

$$w = \frac{10^{-9} * E^2 * r}{\mu} \text{ (for particle size} > 1 \text{ micron)} \tag{14.7}$$

$$w = \frac{0.17 * 10^{-9} * E}{\mu} \text{ (for sub-}\mu\text{ particle)} \tag{14.8}$$

Equation 14.7 shows that the migration velocity of particles greater than 1 μ in diameter is determined by the particle size as well as by the product of field intensities and inversely by the gas viscosity. While from Eq. 14.8 it follows that sub-micron particles travel at velocities directly proportional to the field intensity and inversely proportional to the gas viscosity, but independent of their size. Since the migration velocity is dictated by gas viscosity, which in turn is directly proportional to gas temperature, it may be concluded that the migration velocity is sensitive to changes in gas temperature.

14.2.3 Factors to consider for sizing an ESP

Sizing of an ESP is determined by various factors or parameters, discussed in the following sections. The more accurate the values of these parameters can be predicted during the design stage the better performance of an ESP is expected in the field. Hence, the more the field feedback data available on the performance of a particular design of ESP, the more accurate these parameters can be predicted at the design stage that will ensure better performance of similar design of ESPs in future. Eventually the design of an ESP is more of an art than a science.

A single parameter, which is to be judiciously assumed prior to sizing an ESP, is effective migration velocity (EMV), which influences the collection efficiency of an

ESP. Deutsch–Anderson in 1922 developed the following formula for predicting the collection efficiency of an ESP,

$$\eta = \left\{1 - e^{-\frac{wf}{100}}\right\} * 100 \tag{14.9}$$

where

η = Collection efficiency of ESP, %
e = Base of natural logarithm
w = Effective migration or drift velocity (EMV), cm/s
f = Specific collection area (SCA), $m^2/m^3/s$
(=collecting electrode surface area, m^2, divided by actual volumetric wet gas flow, m^3/s)

From Eq. 14.9 it is evident that the factors that affect the performance of an ESP are EMV and SCA.

In the 1970s, a modified version of the Deutsch–Anderson equation, known as the Matts–Ohnfeldt equation, was developed to predicting better performance of an ESP. The modified equation is

$$\eta = \left\{1 - e^{-\left(\frac{w_k f}{100}\right)^k}\right\} * 100, \tag{14.10}$$

where

η = Collection efficiency of ESP, %
e = Base of natural logarithm
w_k = Apparent migration velocity, cm/s
f = Specific collection area (SCA), $m^2/m^3/s$, (as in Eq. 14.9)
k = Constant (the value of power "k" varies from 0.4 to 0.6, but is normally 0.5)

14.2.4 Effective migration velocity (EMV), *w*

From field experience it has been established that it is easier to collect larger particles than smaller particles, and the collection efficiency of an ESP decreases with a drop in particle size. Therefore, based on experimental studies the collection efficiency of an ESP at various operating conditions as well as for particle sizes ranging larger than 1 μ to sub-micron an overall effective migration velocity is estimated. The value of effective migration velocity is a measure of ease or difficulty of collection.

As explained earlier, the migration velocity is determined by the magnitude of particle size, the strength of electric field and the viscous force or drag of the particle. This theoretical value may become excessively high compared to actual value due to uneven gas flow, particle diffusion, electric wind, particle charging time, and re-entrainment.

Hence, during design stage, some characteristics/values of these factors are assumed to arrive at an acceptable effective migration velocity, which may get reduced significantly during actual operation. In practice actual migration velocity may deviate considerably from that estimated at design stage, which is usually based on empirical observations. In view of this, it is safe to select a low value of effective migration velocity, and even more so, for handling high resistivity fly ash (Table 14.6), since the migration velocity declines as resistivity increases (Figure 14.8).

Table 14.6 Effective migration velocity

Sl. no.	**Resistivity of fly ash, ohm-cm**	**Effective migration velocity, cm/s**	
		Electrode spacing (gap between two consecutive parallel collecting electrodes), mm	
		300	**400**
1.	10^{13} (Generally encountered while firing Indian coal).	3	4
2.	10^{10}	12	16

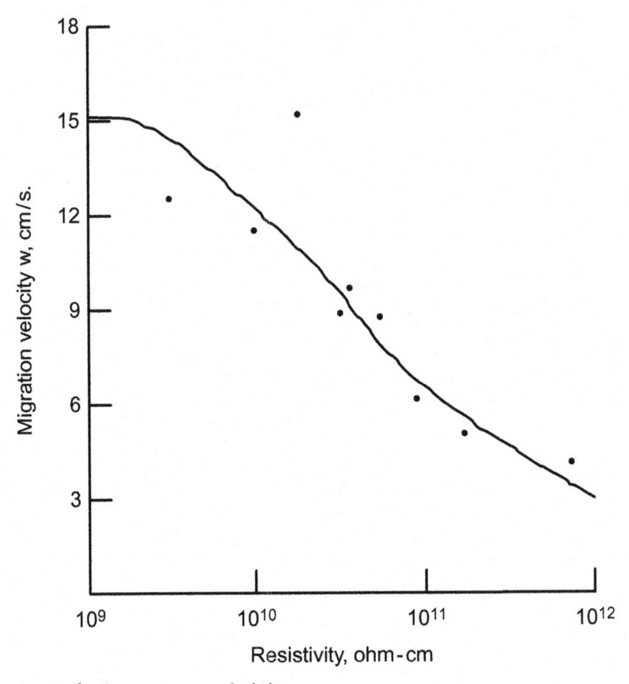

Figure 14.8 Migration velocity versus resistivity.

Table 14.7 Specific collection area

Sl. No.	Resistivity of fly ash, ohm-cm	Specific collection area, $m^2/m^3/s$	
		Electrode spacing (gap between two consecutive parallel collecting electrodes), mm	
		300	400
1.	10^{13} (Generally encountered while firing Indian coal).	200	150
2.	10^{10}	50	38

14.2.5 Specific collection area (SCA), f

The SCA is determined by dividing the total effective collecting electrode surface area by the actual volume flow rate of wet gas to be treated. The value of the SCA depends on physical, chemical, and electrical characteristics of the dust. Table 14.7 summarizes some typical values of SCA for different resistivity of fly ash.

In addition to the EMV and SCA other factors that influence ESP sizing are:

1. Aspect ratio
2. Number of fields in series
3. Length of a field
4. Treatment time
5. Electrode spacing
6. Automatic voltage control
7. Transformer rectifier (T/R) set
8. Hopper heating arrangement and storage volume
9. Rapping mechanism

14.2.6 Aspect ratio

This is the ratio of total collecting length to the collecting height of an electrode. If the length of collecting zone is too short compared with its height, some of the falling dust will be carried out of the precipitator before it reaches the hopper, resulting in re-entrainment of dust particles. As a result dust carryover to stack will increase. For good design, a designer should be comfortable with an aspect ratio of ESP of 2.5 or more.

14.2.7 Number of fields in series

To enhance the reliability of ESP performance, the longitudinal electrical section of gas flow path is divided into a number of independent electric fields. In practice a number of electric fields in series should be no less than 7, in case ash resistivity is too high number of fields in series can be even 9. With this arrangement, the ESP performance would not be impaired much, even with the loss of one electric field, particularly in the case of handling high resistivity ash.

The more the number of fields within a given longitudinal section, the smaller the size of each section, which in turn will provide the following advantages:

a. Electrode alignment and spacing are more accurate
b. Smaller rectifier sets needed that are inherently more stable under sparking conditions
c. Outage of one or two sections has smaller effect on collecting efficiency

14.2.8 Length of a field

The time required by each particle to migrate to collecting electrode from discharge electrode determines the length of a field of precipitator passage in the direction of gas flow. This migration time has to be less than the passage time of each particle while passing through a field of precipitator with the same velocity as gas for ensuring effective removal of PM in a field vis-à-vis and ESP. The length of a field of precipitator thus is ensured by [7]:

$$\frac{s}{w} \leq \frac{L}{V_g} \tag{14.10}$$

or

$$L \geq s\frac{V_g}{w} \tag{14.11}$$

where

s = Distance between discharge and collecting electrodes, m
w = Effective migration velocity, m/s
L = Length of a field, m
V_g = Gas velocity, m/s

14.2.9 Treatment time

The time taken by a particle to pass through the total length of the precipitator is known as the treatment time. For a good design the value of treatment time varies between 25 and 35 s for low-sulfur, low-sodium coals and between 7 and 20 s for high-sulfur, high-sodium coals [8].

14.2.10 Electrode spacing

Electrode spacing is double the gap between discharge electrodes and collecting electrodes, which means the gap between two parallel collecting electrodes. The electrode spacing considerably influences the migration velocity. For constant current density and field strength, an increase in electrode spacing facilitates choosing a higher migration velocity as is clear from Table 14.6. As a result, the total collection area is reduced substantially to achieve the desired collection efficiency, which is obvious from Eq. 14.9; thus the total capital expenditure gets reduced. A decrease in electrode spacing would result in lower migration velocity, resulting in higher capital expenditure.

To meet today's stringent pollution control regulations, the efficiency of an ESP has to be very high, about 99.98% or more, for handling high resistivity ash. Thus, from an economical perspective industry practice is to provide electrode spacing of the order of 300/350/400/450/500 mm [9].

14.2.11 Automatic voltage control

The collection efficiency of an ESP is determined by the precipitation voltage potential between the discharge electrode and collecting plate. The maximum precipitator efficiency may be achieved by maintaining the operating voltage in the individual field of the precipitator below, but close to the flashover limit (the point of sparking between discharge electrodes and collecting plates that switches off the plant). Further, as the flashover voltage varies with operating conditions of the precipitator, such as gas volume, dust content, gas composition, gas temperature, humidity, etc., continuous adjustment of precipitation voltage according to prevailing operating conditions is needed. Therefore, in order to avoid the frequent readjustment and possible trip-outs on manually operated sets, operating personnel often reduce the voltage and current to much lower than the flashover limit.

Automatic voltage control works in two steps: first, the voltage is gradually raised at a pre-set rate until flashover takes place and switches off the system. Then after a pre-set time delay, the system switches on automatically and the voltage is raised to slightly reduced level than the flashover limit. Thereafter, this mode of cycle gets repeated continuously.

14.2.12 Transformer rectifier (T/R) set

The high-voltage power needed to discharge the electrode system is supplied by a transformer rectifier set along with a controller. Two transformer rectifier sets are normally needed to power each precipitator field. In practice a medium voltage AC supply is regulated and transformed into a nominal voltage of 50–120 kV, rectified to a negative DC output then supplied to the discharge electrode system.

The current density usually is maintained over a range of 0.25–0.45 mA/m^2, from which the capacity of transformer rectifier set can be determined, since the collecting area of each field is already known.

14.2.13 Hopper heating and storage volume

In practice, dust collected near hopper walls will agglomerate and plug the dust-removing system. Hopper heating arrangement prevents formation of dust cakes and gets rid of the hopper plugging problem. Additionally, the bottom part of the hopper is sometimes provided with a stainless steel liner that ensures satisfactory flow of dust from it. Fluidizing pads are also fitted inside hoppers to ensure smooth flow of dust.

The dust hoppers must have adequate storage capacity to hold all the collected dust at a boiler maximum generating condition of the unit while firing the worst coal. An overfill hopper may result in loss of an electrical section. Therefore, the hopper storage volume should be adequate enough such that during the interval between two consecutive steps of removal of dust from the hopper usually (6–8 hrs) as per the industry practice, dust build-up will not bridge the high voltage frame.

14.2.14 Rapping mechanism

When dust builds up on plates, it deposits in a layer of increasing thickness with possible re-entry into the gas flow path, hence build-up ash needs to be removed periodically by hitting/rapping collecting electrodes with hammers. Based on the changing characteristics and collection rates of dust the intensity and frequency of rapping of individual collection plates varies from inlet to outlet of precipitator. Rapping should be carried out sequentially so that only a fraction of accumulated dust is disturbed at any one time.

There are two types of hammers: *drop hammer* and *tumbling hammer.* Drop hammers are either pneumatically driven or electrically driven. Each drop hammer is connected with a number of electrodes and is used to rap from the top of electrodes. Tumbling hammers are basically side hammers where a set of hammers is mounted on a horizontal shaft in a staggered fashion. As the shaft rotates one hammer at a time tumbles and hit a shock bar.

Excessive rapping of collecting electrodes is detrimental to the performance of an ESP as it causes re-entrainment of collected dust and reduces the life of internal components. In contrast, inadequate rapping results in build-up of a dielectric dust layer on the collecting plates, which is penetrated by corona current. As a result the dust layer reaches a level of resistivity that generates ionization or *back-corona* and causes dust particles to become positively charged and move toward negative high voltage discharge electrodes. Thus, localized excessive sparking occurs, reducing power input to the ESP, which in turn lowers the collection efficiency.

14.2.15 Factors that affect performance of ESP

The factors that determine the type, size, and efficiency of the electrostatic precipitator are discussed in following sections. During normal running if any of these factors varies substantially from the design, the performance of the ESP will be jeopardized, which may lead to violating a local SPM emission regulation.

i. Volume of gas flow
ii. Velocity of gas flow
iii. Temperature of flue gas
iv. Gas distribution
v. Dust resistivity
vi. Inlet dust concentration
vii. Weight of dust per unit volume of gas
viii. Basic properties of the gas and dust particles
ix. Disposal of collected gas

14.2.16 Volume of gas flow

The volume of gas to be treated is usually determined at the design stage of any project. Subsequently, if it is found that the actual volume of gas to be handled is much higher than the specified volume of gas, the following detrimental effects on ESP performance could be encountered:

- Achieved efficiency would reduce (Figure 14.9)
- Re-entrainment loss would increase
- Ash burden to hopper would increase
- Emission rate of SPM would increase.

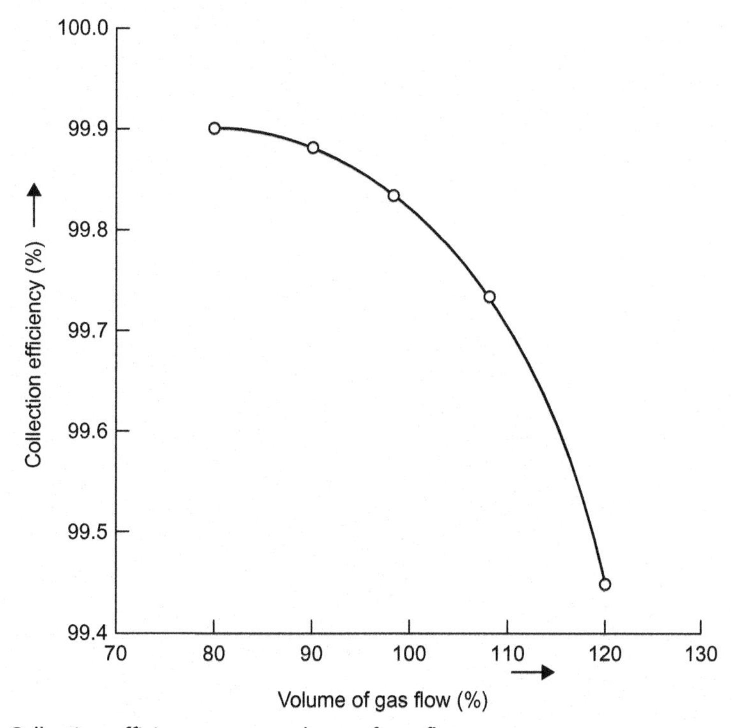

Figure 14.9 Collection efficiency versus volume of gas flow.

14.2.17 Velocity of gas flow

This is a critical design parameter and is calculated by dividing the actual volume flow rate of gas by the cross-sectional area of the precipitator. For low-sulfur, low-sodium coals, the gas velocity may be in the range of 0.5–1 m/s, and for high-sulfur, high-sodium coals, the gas velocity could be as high as 1.5 m/s [8]. At velocities greater than that, problems of re-entrainment of suspended particles are aggravated.

14.2.18 Temperature of flue gas

Temperature of flue gases passing through precipitator will affect the collectability of ash and removal of collected ash from collecting plates. When this temperature reaches the dew-point temperature moisture will form and cause dust particles to solidify into cakes. Further build up of these cakes may result in short-circuits or wire failures. In contrast, if the flue-gas temperature exceeds a high limit, then the gas volume will increase, leading to a reduction in collection efficiency. At higher temperature, the viscosity of the gas also increases, resulting in high viscous drag with an additional reduction in collection efficiency.

14.2.19 Gas distribution

While flowing through the precipitator the distribution of gas must be uniform all along its path, from the inlet up to its outlet. In sections where velocity is high, ash may be swept off collector plates and can show up in the stack as a visible plume, while at low velocity sections, although gas is more efficiently cleaned, the effect will be a reduction in overall collection efficiency. Hence, proper devices, e.g., perforated distribution plates, turning vanes, and baffles in the precipitator inlet and duct work, should be installed to prevent uneven gas distribution.

14.2.20 Dust resistivity

The electrostatic precipitator is very sensitive to the resistivity of ash or dust deposited on collecting plates. Experience has shown that coal ashes with resistivity between 10^8 and 2×10^{10} ohm-cm is normal for satisfactory and predictable precipitator design. Resistivity of ash that exceeds this limit becomes highly resistive to dust collection. The presence of constituents, such as SiO_2, Al_2O_3, CaO, etc., in fly ash causes high resistivity of ash that does not readily lose its charge when collected on plates, and the agglomerated ash is very difficult to get removed. When deep-enough deposits collect on the plate, back-corona (local sparking on dust layer) develops, causing the collection rate to reduce (Figure 14.10). A corona discharge occurs only in an inhomogeneous electric field with the electrodes definitely shaped and located [6].

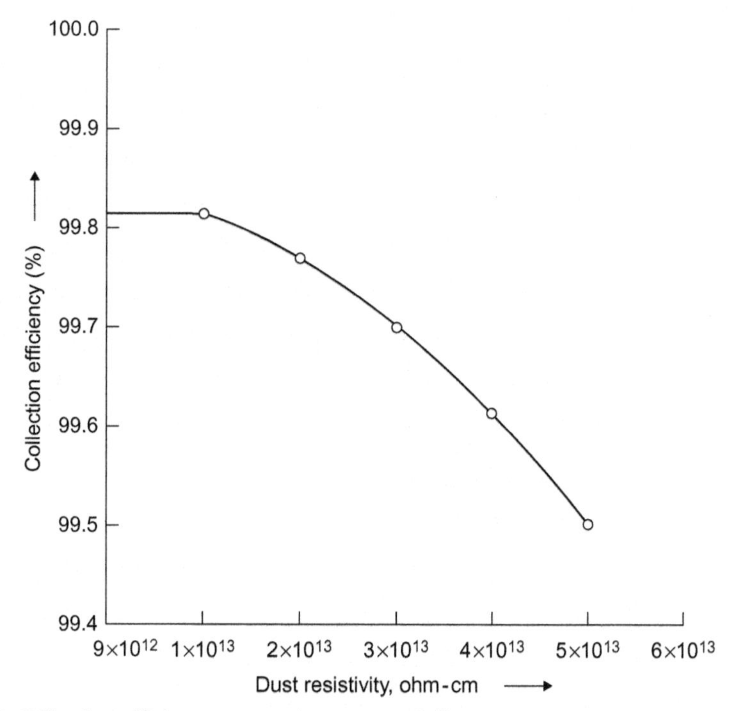

Figure 14.10 Collection efficiency versus dust responsibility.

The resistivity of ash is lowered by cooling the inlet gas to the ESP while simultaneously increasing the water vapor content or by introducing SO_3, NH_3, etc., into the flue gas. These additives are adsorbed onto fly ash particles and produce an electrically conductive surface condition, improving precipitation. The resistivity may be lowered by "hot" precipitation as well. Experimental results reveal that resistivity of ash is maximum at around 423 K (Figure 14.11).

14.2.21 Inlet dust concentration

At the time of design certain dust concentration at inlet to the ESP is determined. If this concentration changes during operation, there will be consequent change in dust collection efficiency (Figure 14.12).

14.2.22 Weight of dust per unit volume of gas

During the design stage the expected weight of fly ash per unit time may be calculated, considering analysis of a worst coal. If at a later stage it is found that the as-received coal has much higher ash content than a precipitator is designed to handle,

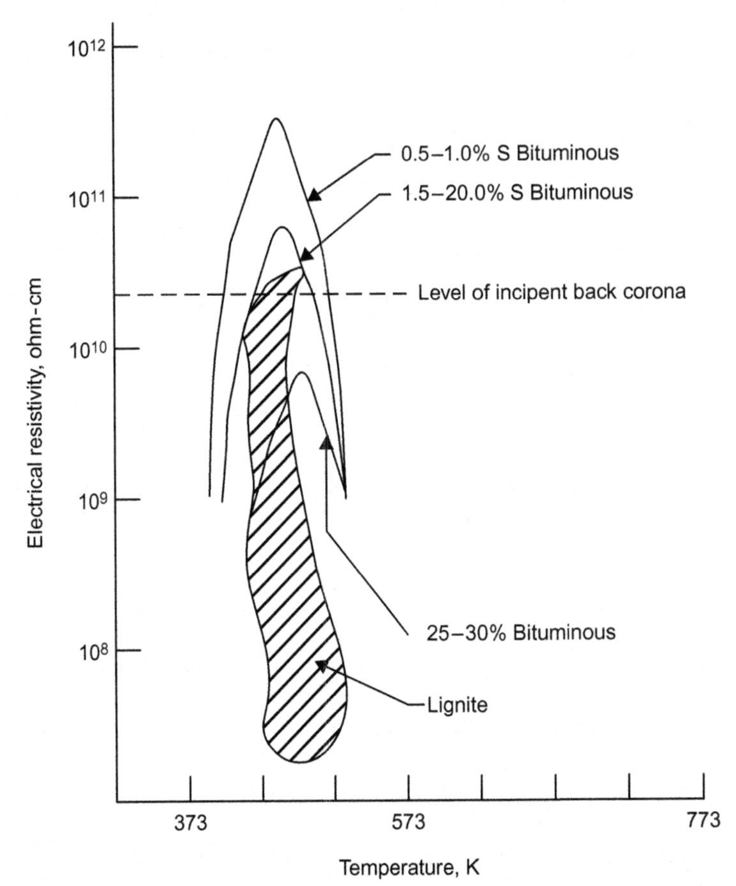

Figure 14.11 Electrical resistivity versus temperature. *Source: FIG. 10. P15.13. CH15: Control of Power Plant Stack Emission [10].*

then the ESP inlet dust concentration would increase, causing a fall in collection efficiency as shown in Figure 14.12.

14.2.23 Basic properties of the gas and dust particles

A precipitator is designed specifically for an ash produced by coal containing a certain amount of sulfur; if coal with different sulfur content is burned, the efficiency will vary. A coal with lower sulfur content may give satisfactory separation when the flue-gas temperature is increased. For corrosive gases and particles, the shell interior must be made of corrosion-resistant materials, such as lead for sulfuric acid mist and tile block for paper mill salt cake.

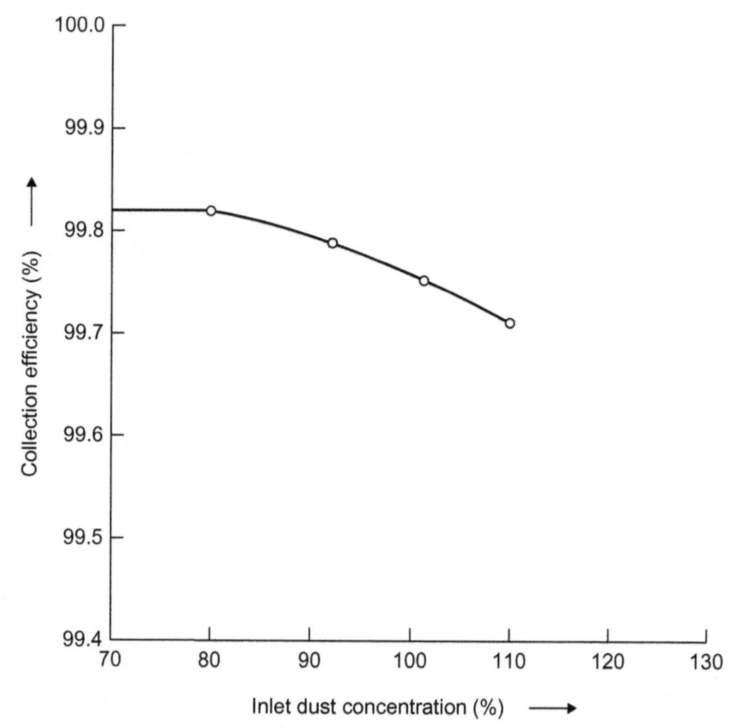

Figure 14.12 Collections efficiency versus inlet dust concentration.

The particle size also affects the settling rate. Particles larger than about 1 μ will settle too rapidly. Sub-micron particles (within the range of 0.1–0.5 μ) will settle very slowly and exhibit Brownian motions. The settling rate again becomes rapid for particle size less than 0.1 μ [8].

The optical properties of particles are of great importance in connection with gas cleaning and air pollution, because the degree of pollution is commonly judged by visual appearance of stack discharge.

14.2.24 Disposal of collected gas

An ESP is enclosed and insulated. If the hoppers are not insulated, ash near hopper walls will agglomerate as it cools. There is a tendency for the agglomerated mass to form cakes, which may plug the ash conveying system. If ash is not removed regularly from hoppers, it can build up to a point where it will bridge high voltage frame, causing a short-circuit and loss of an electrical section. Prolonged accumulation of ash may endanger ESP structure and cause catastrophic failures.

Example 14.1

For a sample of coal with the ultimate analysis and other combustion parameters as presented in EXAMPLE 3.4 and its Solution in Chapter 3, calculate the efficiency of the ESP if outlet dust concentration is to be maintained at a 150 mg/m^3 at 300 K dry basis. (Assume 80% generation of fly ash.)

Solution: From EXAMPLE 3.4 the following results are noted:

Excess air = 32%
Quantity of coal burned = 94 kg/s
Ash content of coal = 35.0%
Volume of dry flue gas at 300 K, $V_{FGD} = 597$ m^3/s
From given data, quantity of fly ash generated
$Q_{FA} = 94 \times 0.35 \times 0.8 = 26.32$ kg/s
Dust concentration at ESP inlet at 300 K dry basis,
$DCI = 26.32/597 = 44.10 \times 10^{-3}$ kg/m^3 = 44.10 gm/m^3

Therefore, the efficiency of the ESP is $\eta = \{(44.10 \times 10^3 - 150)/44.10 \times 10^3\} \times 100$ **= 99.66%**

Example 14.2

In the above example if the flue-gas temperature at the ESP inlet is maintained at 423 K and the effective migration velocity is 2.5 cm/s, calculate the total collecting area of the ESP using Deutsch–Anderson equation.

Solution: From EXAMPLE 3.4 it is noted that
Volume of wet flue gas at 423 K, $V_{FGW} = 924$ m^3/s
From Deutsch–Anderson equation, $\eta = \{1 - e^{-(wf)/100}\} \times 100$
or, $99.66 = \{1 - e^{-(2.5f)/100}\} \times 100$
or, $e^{(2.5f)/100} = 1/(1 - 0.9966)$
or, $0.025f = 5.684$
or, SCA, $f = 227.36$ m^2/m^3/s

Therefore, the **Total Collecting Area** = 227.36 × 924 = **210080 m^2**.

14.2.25 Control of SPM using a bag filter

A bag filter, also known as a fabric filter or baghouse, comprises multiple compartments (similar to ESP fields), each housing several vertical (2–10 m high) small diameter (120–400 mm dia) cloth or fabric bags (in place of electrodes in ESP) [11]. Depending on the temperature of the entering flue-gas the type of fabric material may be any one of Polypropylene (PP), Polyester (PES), Draylon T (PAC), Ryton (PPS), Nomex (APA), P84 (PI), Teflon (PTFE), Fiber glass (GLS), etc [4]. Of these materials, from an economical perspective, the most suitable one for handling fossil fuel-fired flue gas is Ryton, which can withstand a flue-gas temperature up to 453 K.

14.2.25.1 Principle of operation of a bag filter

On entering the baghouse, dust-burdened flue gas strikes a baffle plate, where larger dust particles fall by gravity (Figure 14.4). This partially cleaned dirty gas then enters each bag and leaves as clean gas from the top of the bag. PM adheres to the surface of cloth as cake and gets accumulated. These accumulated cakes are cleaned periodically by applying pulse-jet or reverse-jet air flow and get collected in hoppers. Dust particles collected in hoppers are removed periodically by a fly ash-handling system. The cakes may also be cleaned by shaking or rapping of bags.

14.2.26 Factors to consider for sizing a bag filter

Three factors that affect the sizing of bag filters are:

i. Air/Cloth Ratio ($m^3/s/m^2$): This is the ratio of flue-gas flow rate, m^3/s, to the total exposed surface area of clothes to flue gas, m^2. For pulse-jet bag filters the air/cloth ratio typically varies from 0.015 to 0.020 $m^3/s/m^2$, while for reverse-jet bag filters the value ranges between 0.0075 and 0.01167 $m^3/s/m^2$ [4].
ii. Pressure Drop: The pressure drop is the sum of pressure drop across clean filter bags, pressure drop that occurs across collected cakes, and that which takes place across the baghouse internals.
iii. Emission Rate: This is the emission limit that is to be complied with in line with the recommendation of local environmental control authority.

14.2.27 Factors that affect the performance of the bag filter

In addition to the air/cloth ratio other factors that may affect the efficiency of a bag filter are as follows:

i. Material of Fabric: Material selected should be capable of withstanding the maximum flue-gas temperature during operation, of resisting abrasion and of withstanding any chemical attack
ii. Temperature: Flue-gas temperature if exceeds the maximum withstand limit of the material chosen will cause detrimental effect on cloth to get damaged
iii. Re-entrainment: If the selected air/cloth ratio is too high or if the fabric open up for any reason the finer particles will get carried away along with the flue gas through the stack.

14.2.28 Comparison between ESP and bag filter

Table 14.8 summarizes comparative analysis of ESP and bag filter.

Table 14.8 Differences between ESP and bag filter [11]

Sl. no.	Parameter	ESP	Bag filter
1.	Pressure Drop	About 25 mmwc	100–200 mmwc
2.	Removal Efficiency	Constant	Variable
3.	Removal Emission	Variable	Constant
4.	High Temperature	Tolerant	Highly Sensitive
5.	Chemical Attack	Tolerant	Highly Sensitive
6.	Erosion	Tolerant	Highly Sensitive
7.	Abrasion	Tolerant	Highly Sensitive
8.	Ash Resistivity	Highly Sensitive	Tolerant
9.	Energy Consumption	Low	Extremely High
10.	Operation	Simple	Complex
11.	Maintenance	Easy	Difficult
12.	Service Life of Elements	Service life of electrodes is very long	Service life of fabric (Ryton) is about 20,000 hrs
13.	Cost of Replacing Elements	Very little	About a million USD every 2.5 years
14.	Capital Cost	Comparable	
15.	O and M Cost	Low	Very High

14.3 REDUCTION OF GREENHOUSE GAS EMISSION

Coal is an abundant fossil-fuel resource in many countries of the world and currently coal-fired stations account for over 40% of power produced globally. Coal will remain as a dominant fuel for power generation in those countries, where availability of coal is abundant. Even in countries where natural gas is abundantly available, coal-fired stations are attractive, since the cost of natural gas is exorbitantly high compared to the cost of coal. This is in spite of the fact that emission of CO_2 (which is a major greenhouse gas as explained earlier) per kWh from a coal-fired station is almost twice the amount emitted from a natural-gas-based combined cycle power plant. Therefore, in order to reduce CO_2 emissions from coal-fired power plants it is important that these plants produce power more efficiently. A study reveals that in a 660 MW power station, firing typical Indian coal, a gain in plant efficiency by 3% would result in reduction in CO_2 emissions by about 6% per kWh [12].

Hence, to achieve lower CO_2 emissions, coal-fired utility power plants should choose larger unit sizes and higher steam parameters. Eventually power plants operating with supercritical steam conditions play a vital role over sub-critical plants. In addition to efficient operation and less generation of CO_2 the added advantage of using supercritical technology is to emit less SO_X and NO_X.

Another option to mitigate CO_2 emissions is to adopt an "integrated gasification combined cycle" (IGCC) process for power generation. Since the IGCC process does not fall under the discussion of conventional thermal power plant, this chapter refrains from elaborating IGCC further.

(Detailed treatment of supercritical power plants was presented in Section 2.10, Chapter 2, *Steam Generators*).

14.4 METHODS OF SO_X EMISSION CONTROL

According to the guidelines laid out in the U.S. "Clean Air Act Amendments (CAAA)" of 1990, SO_2 emission from a coal-fired power plant stack to atmosphere is controlled by adopting any one or combination of following two methods [13]:

i. Switching over to low sulfur fuel
ii. Installing flue-gas desulfurization (FGD) system

During the initial stage, coal switching became more attractive to the utilities because of the cost of complying with the CAAA mandate. Subsequently it was realized by utilities that switching to low sulfur coal in existing boilers lead to various problems, as discussed in the following, since by and large low sulfur coals are low-ranking coals with higher ash content and lower heating value:

1. Combustion of low-ranking coal seriously affected heat transfer across various zones resulting in unacceptable generation of steam from boiler
2. Performance of pulverizer got worse
3. Lower heating value coal essentially required more total air to ensure complete combustion, thus capacity of existing ID and FD fans became inadequate
4. Higher flue-gas flow caused higher gas side erosion
5. The capacity of existing ash-handling system was inadequate for complete evacuation of ash generated

Utilities, however, observed that fuel switchover was only amenable for adoption in retrofit projects. Eventually FGD became attractive and continues to control SO_2 emission through stacks of pulverized coal-fired boilers to comply with local pollution control regulations in each nation.

SO_2 gas is acidic in nature and hence chemicals that may be used to remove SO_2 from flue gases must be alkaline, i.e., limestone ($CaCO_3$), hydrated lime $\{Ca(OH)_2\}$, magnesium hydroxide $\{Mg(OH)_2\}$, etc. A natural alkaline that may be used to absorb SO_2 is seawater.

Based on experience an FGD system can be classified into seven categories as discussed in the following sections.

14.4.1 Wet scrubber

A wet scrubber may be used for boiler applications with high- to low-sulfur coals. This is the most widely used FGD technology for SO_2 control. In wet scrubbers flue gas enters at the bottom of an absorber tower and exits as clean gas from the top. Both absorption and oxidation take place within the absorber tower. Water is sprayed at the top of the tower through a series of spray nozzles and air is injected through the tower from the bottom at a point lower than the entry point of flue gas. A slurry mixture of calcium-, sodium-, and ammonium-based sorbents, prepared in an external container, is injected into the tower to react with the SO_2 in the flue gas. From a techno-economic consideration sorbent that is widely used in operating wet scrubbers is limestone ($CaCO_3$). Next preferred sorbent is lime (CaO). Figure 14.13 shows the schematic arrangement of a typical wet scrubber system.

Limestone units generally operate in a forced-oxidation mode producing gypsum as a by-product. In this mode limestone reacts with SO_2 in flue gas producing calcium sulphite ($CaSO_3$). The injected air along with air contained in flue gas first oxidizes calcium sulphite producing calcium sulphate ($CaSO_4$), which in the next step reacts with incoming sprayed water to produce gypsum ($CaSO_4 \cdot 2H_2O$). The corresponding chemical reactions are:

$$CaCO_3 + SO_2 = CaSO_3 + CO_2$$
$$CaSO_3 + \tfrac{1}{2}O_2 + 2H_2O = CaSO_4 \cdot 2H_2O$$

In a magnesium-enhanced wet lime process, about 4–7% (by weight) of magnesium oxide (MgO) is added to quick lime (CaO). This mixture on reacting with incoming spray water results in a slurry of calcium hydroxide ($Ca(OH)_2$) and magnesium hydroxide ($Mg(OH)_2$). The magnesium-enhanced lime-scrubbing reagent reacts with the SO_2 in the flue gas producing hydrated calcium sulfite and hydrated magnesium sulfite. Thus this process would not yield gypsum as the end product as is evident from the following chemical reactions:

$$Ca(OH)_2 + SO_2 + H_2O = CaSO_3 \cdot 2H_2O$$
$$Mg(OH)_2 + SO_2 + H_2O = MgSO_3 \cdot 2H_2O$$

All limestone units are capable of removing 90–95% of SO_2 from flue gas. Magnesium-enhanced lime systems are considered as the most efficient SO_2 removal systems, where up to 98% removal can be ensured.

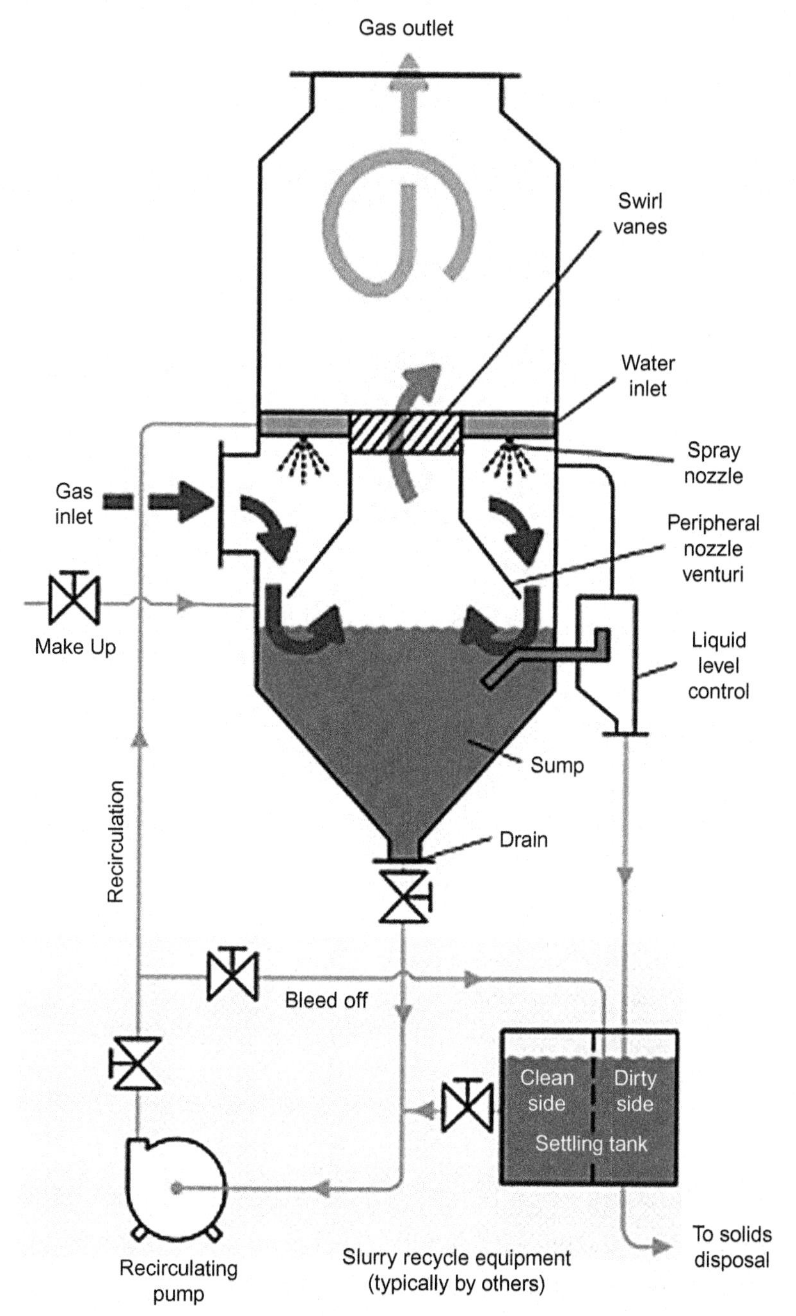

Figure 14.13 Wet scrubber. *Source: www.babcock.com/products/Pages/Wet-Particulate-Scrubbers.aspx.*

The advantages of wet scrubbers are:

1. Size is compact
2. SO_2 removal efficiency is high
3. Anticipated to run scale-free
4. Reliability and availability are high
5. May remove certain portion of flue-gas residual particulates downstream of ESP

The disadvantages of this method are:

1. Possibility of nozzle plugging
2. Droplet carryover may cause erosion, corrosion of downstream ducting
3. Flue gas may have to be reheated to improve ID fan performance
4. Due to additional pressure drop across the scrubber ID fans may require more power
5. Wastewater treatment is essential prior to discharge
6. Capital cost and operating cost are relatively high

14.4.2 Spray dry scrubber

Lime (CaO) is normally used as a sorbent in a spray dry scrubber. Prior to spraying into the absorber CaO is hydrated to $Ca(OH)_2$. Hydrated lime is directly injected as spray into the absorber tower, where moisture is evaporated by the heat of flue gas. SO_2 and SO_3 in flue gas react with hydrated lime within the absorber tower to form a dry mixture of calcium sulphate/sulphite. SO_2 removal efficiency of a spray dry scrubber falls within 90 and 95%. It is the second most widely used process to arrest SO_2.

The advantages of this process are:

1. Compared with the wet system, the dry system technique is simpler
2. Disposal of dry, powdery waste material is less complicated
3. Absence of wastewater treatment plant
4. Less expensive than wet scrubber
5. Recycle of spray-dryer solids can partly improve sorbent utilization

The disadvantages are:

1. Operational costs are higher
2. Efficiency of dry scrubber is less than that of wet scrubber
3. Suitable for units firing medium to low sulfur coal
4. Flue-gas volume handling capacity of dry scrubber tower is limited; hence, from techno-economic consideration this process is not viable to be adopted in large coal-fired power plants that would require several modules of such tower

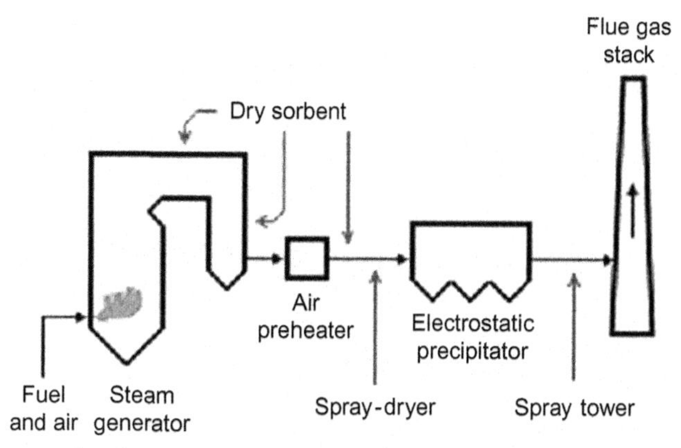

Figure 14.14 Location of sorbent injection and various FGDs. *Source: http://en.citizendium.org/wiki/Flue_gas_desulfurization.*

14.4.3 Sorbent injection system

A sorbent injection system is the simplest method of SO_2 removal from flue gas. Sorbent generally used in this system are limestone ($CaCO_3$), dolomite ($CaCO_3 \cdot MgCO_3$), lime (CaO), or hydrated lime ($Ca(OH)_2$). Based on location of injection of sorbent, this system is classified into furnace sorbent injection (FSI), economizer sorbent injection (ESI), duct sorbent injection (DSI) (Figure 14.14), and hybrid sorbent injection (HSI) process. In all these types the effectiveness of SO_2 removal depends on how evenly and quickly the sorbent is sprayed across the cross-section of flue-gas stream. The solid waste obtained downstream of each process is removed in the dust-collecting equipment.

In a *furnace/post furnace sorbent injection* system (FSI) powdered dry sorbent (usually limestone or hydrated lime) is sprayed into the entire cross-section in the upper part of the furnace, where the temperature is maintained above 1023 K but not more than 1523 K. When flue gas passes through this zone the SO_2 and O_2 content of the flue gas reacts with the sorbent to form $CaSO_4$.

The SO_2 removal efficiency that may be attained in FSI is 50% or higher based on even distribution of sorbent, fastness of application, ensuring optimum residence time (1 to 2 seconds), maintaining proper sorbent particle size (not more than 5 μ, but preferably less than 3 μ) and adopting Ca:S ratio between 2 and 3.

Sorbent used in an *economizer sorbent injection* process (ESI) is hydrated lime. When flue gas flows through the economizer zone its temperature remains between 573 K and 923 K. At this temperature range when sorbent is sprayed into the flue-gas stream $Ca(OH)_2$ reacts with SO_2 to produce $CaSO_3$ instead of $CaSO_4$.

In a *duct sorbent injection* (DSI) process sorbent is sprayed into the flue-gas stream at a temperature around 423 K, located downstream of air pre-heater. Along with the sorbent some water may also be sprayed to humidify the flue gas with the precaution that humidity always remains above the dew point. SO_2 removal efficiency of this process could be as high as 80%.

Adopting FSI in combination with DSI is called a *hybrid sorbent injection* (HSI) process. This combination results in higher sorbent utilization and greater SO_2 removal. With humidification SO_2 removal efficiency of HSI can be boosted up to 90%.

14.4.4 Dry scrubber

A circulating fluidized bed (CFB) involves a scrubbing technology known as a dry scrubber process. In this process, hydrated lime, limestone, or dolomite and water are intimately mixed with flue gas in a dedicated reaction chamber, which is the fuel-firing bed. Reacted solids get dried completely by the sensible heat of the flue gas. The process is easy to maintain and operate since it does not include high-maintenance mechanical equipment. The process can achieve about 95% SO_2 removal efficiency. Because of high SO_2 removal efficiency the Ca:S ratio (1.2−1.5) of the dry scrubber process is less than that of the sorbent injection system. (For more information, refer to Section 5.4, in Chapter 5, *Fluidized Bed Combustion Boilers*.)

14.4.5 Regenerable process

In a regenerable process SO_2 removal efficiency is very high (>95%), but the process suffers from high cost and high power consumption. In this process, once SO_2 is removed, the sorbent is regenerated for further use. As a result, this process provides an advantage of low sorbent make-up requirement. Another advantage of this process is it produces virtually no residual solid or liquid wastes. The sorbent may be regenerated either thermally by heating it to a temperature of about 923 K or chemically by bringing the sorbent in intimate contact with steam and a reducing gas, i.e., H_2S, H_2, and natural gas.

The process involves stripping of SO_2 from flue gas by scrubbing action of sodium sulfite solution or organic-amine liquid absorbent. Use of other chemical solutions is also possible, but currently this approach is at the research stage.

14.4.6 Combined SO_2/NO_x removal process

This process involves injection of ammonia in the flue gas upstream of a selective catalytic reduction (SCR) unit for NO_X control and injection of sorbent for SO_2 control. As a result this process is relatively complex and capital intensive and also requires lot of maintenance. This process still is in the development stage. Nevertheless, some field results show more than 70% SO_2 removal and about 90% NO_X removal by the combined process.

14.4.7 Seawater FGD

Seawater FGD is attractive to plants that are located near coastlines. Seawater is a source of natural alkalinity that can be used economically to arrest SO_2 from flue gas without the use of any other supplemental reagents. Seawater contains significant concentrations of alkaline ions including sodium, magnesium, potassium, calcium, carbonates, and bicarbonates. It also contains significant concentrations of chloride and sulphite ions. Desulfurization is accomplished by seawater scrubbing, and 90% SO_2 removal efficiency could be achieved by adopting this process.

Before entering into the seawater scrubber flue gas is cooled to its adiabatic saturation temperature of typically 365 K. The scrubber is packed with certain material proprietary in nature through which flue gas flows from the bottom and seawater is sprayed counter-current to the flow of flue gas in order to achieve effective SO_2 mass transfer and chemical absorption into the liquid phase. Thus, SO_2 in the flue gas is absorbed and converted to sulphite first. Sulphite is further oxidized to produce sulphate for safe disposal to the sea without jeopardizing the marine environment. For disposal of liquid effluent to sea, it should be ensured that liquid effluent is alkaline with a pH no less than 8.

14.5 METHODS OF NO_X EMISSION CONTROL

Oxides of nitrogen are generally referred to as NO_x, comprising both nitric oxide (NO) and nitrous oxide (NO_2). During the combustion of fossil fuel in a conventional boiler NO_x is formed of which NO accounts for 90–95% and the remainder is NO_2. NO_x is formed in the flame-zone vicinity by oxidation of nitrogen content of air (thermal NO_x) and nitrogen content of fuel (fuel NO_x) as well as by oxidation of intermediate hydrocarbons present in flames (prompt NO_x). The average temperature in the vicinity of the flame zone is about 1477 K with a peak temperature as high as about 1922 K.

The direct high-temperature reaction of atmospheric nitrogen with oxygen results in thermal NO_x. If the combustion temperature can be reduced by supplying less combustion air production rate of thermal NO_x will drop. Fuel NO_x formation, however, is not significant near the flame zone, and most fuel NO_x is formed in the secondary combustion zone along with the release of volatile matter. For burning natural gas fuel NO_x generation is insignificant. In a coal-fired power plant contribution of prompt NO_x is negligibly small. However, it is important in gas-turbine power plants where ultra-high level emission is considered.

NO_x is harmful in several ways – as a contributor to acid rain, as a destroyer of the ozone layer, and as a heat-trapping compound suspected of increasing atmospheric temperatures. At low elevations, NO_x also reacts with sunlight to create smog.

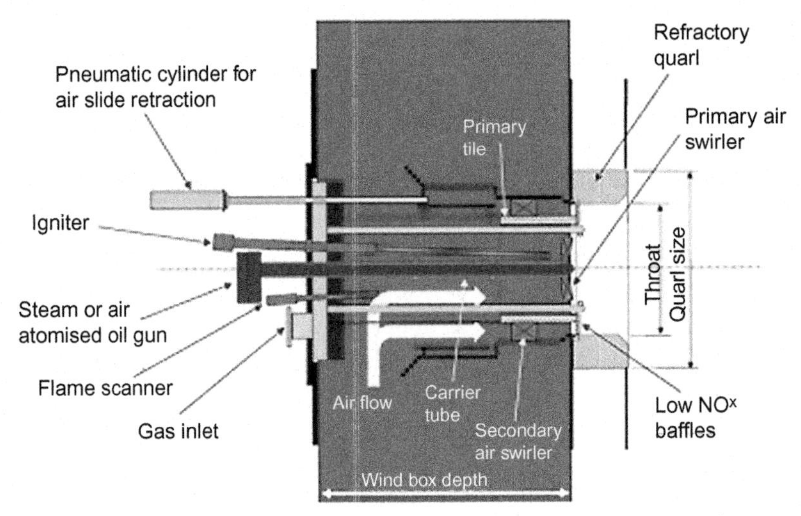

Figure 14.15 Low nox burner. *Source: http://www.hk-bs.com/eng/default.asp?mnuidx=40.*

To control NO_x emissions from fossil-fuel-fired combustors, in accordance with the US Clean Air Act Amendments of 1990 (CAAA), alternative techniques that are made available to utilities are use of low-NO_x burners (LNB), or adopting fuel biasing, or limiting the excess-air (LEA) in the combustion process, or application of over-fire air (OFA), or injecting steam or water into the flame, or recirculating flue gases. All these techniques seek redistribution of air and fuel, causing slow mixing of air and fuel, reduction of O_2 level in critical NO_x formation zones, and lowering of the amount of fuel burned at peak flame temperature. Other techniques that are available are back-end controls like selective catalytic reduction (SCR) and selective non-catalytic reduction (SNCR) systems [14].

The following sections discuss the techniques that are usually adopted.

14.5.1 Burner replacement

This method is normally adopted for retrofitting purposes. Old burners, which were not designed to meet currents standards for maximum NO_X emission, are replaced with low NO_x burners (LNBs). Figure 14.15 and Figure 14.16 show the cross-sectional details of typical LNBs from different manufacturers. These burners facilitate mixing of fuel and air in a controlled way to provide larger and more branched flames. As a result, the peak flame temperature is reduced, resulting in less formation of NO_x.

Within the flame four different processes take place as follows (Figure 14.17):

i. Initial combustion takes place in a fuel rich zone, where NO_x is formed
ii. Second process produces a reducing atmosphere where hydrocarbons are formed, which thereafter react with already formed NO_x

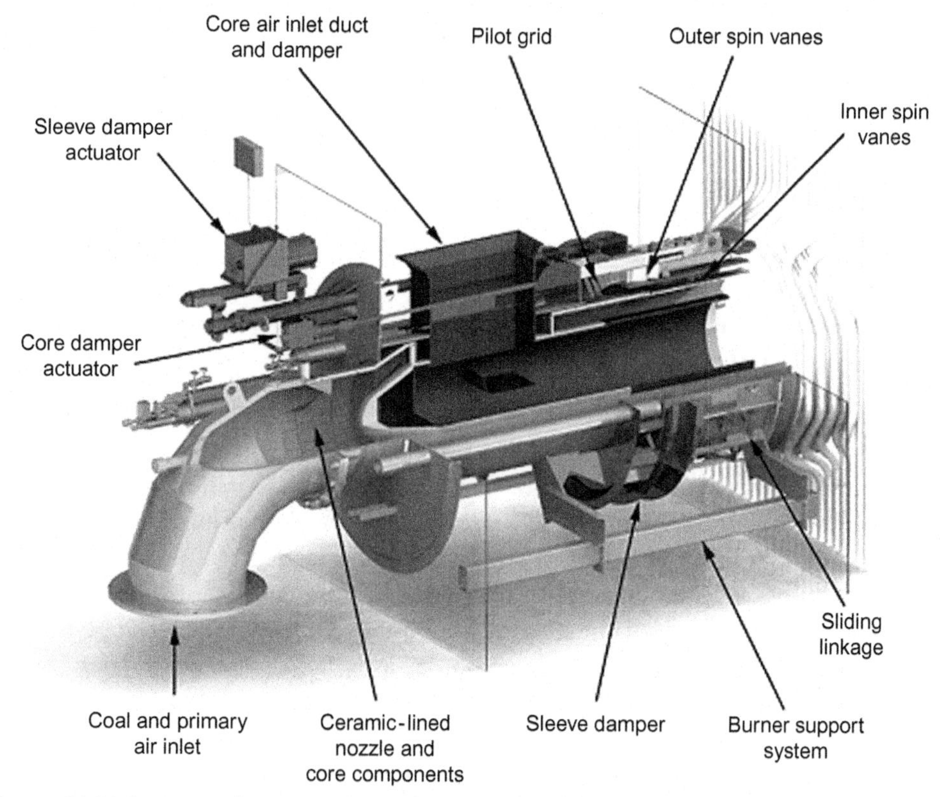

Figure 14.16 Low nox burner. *Source: http://www.babcock.com/products/Pages/AireJet-Low-NOx-Coal-Burner.aspx.*

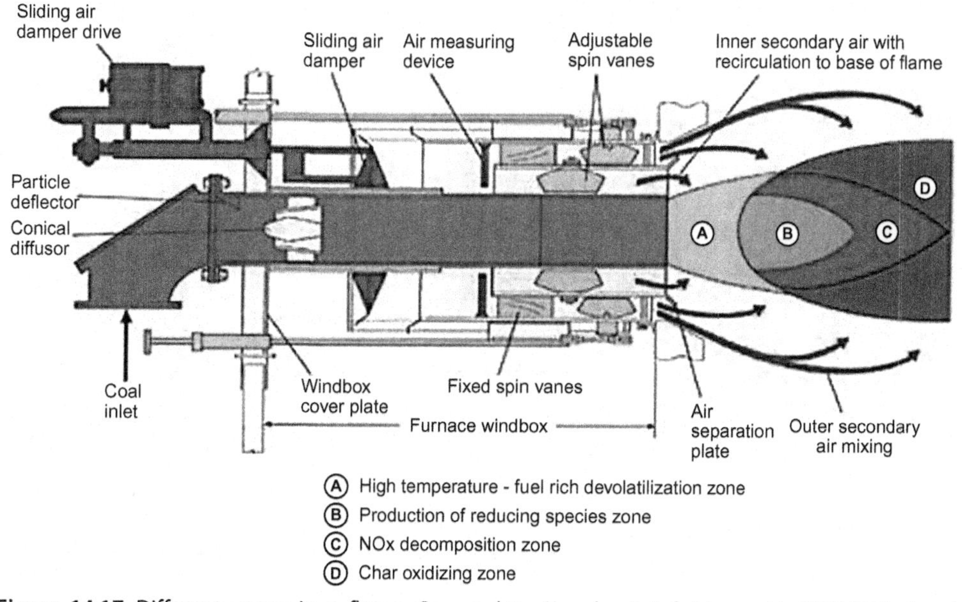

Figure 14.17 Different zones in a flame. *Source: http://mechanic-info.blogspot.in/2013/07/babcock-burner.html.*

iii. Final combustion occurs in an air-lean zone; as a result NO_x formation is minimized
iv. Char oxidizing zone

Use of LNBs, however, suffers from generating higher carbon loss in ash. The burner replacement method in isolation can not meet present-day emission level norm. To make this method effective LNBs are combined with combustion modification as discussed in the following.

14.5.2 Combustion modification

In order to ensure complete combustion with reduced NO_x emission, control of fuel and air supply is very important. This is accomplished at the burner as well as inside the furnace. Supply of secondary air and over-fire air facilitates vertical staged combustion. While supply of air at different points along the flame or introducing tertiary air at the burner ensures horizontal staging of combustion.

In addition to the above replacement of burners, adjustable angle over-fire air (OFA) dampers are also installed to quench flue-gas temperature, minimizing formation of NO_x. Furnace over-fire air helps in separating combustion air into primary and secondary air to achieve complete combustion. The primary air combines with fuel that produces a relatively low-temperature fuel-rich zone. Then the secondary air is introduced above the burners to complete the combustion also at a relatively low temperature. Thus, N_2 is produced in lieu of NO_x. The effectiveness of controlling NO_x formation by providing OFA dampers has already been "proven."

Flue-gas recirculation, like over-fire air, quenches the flame and at the same time reduces oxygen level, thereby NO_x formation is minimized.

Use of low excess air in the combustion region and then re-burning fuel-rich mixture above this region provides fuel staging. Fuel staging can also be achieved by taking the burner out of service (BOOS). This technique, however, is seldom practiced in pulverized coal-fired boilers. Fuel biasing is another method of fuel staging. In this method, the fuel supply is shifted from the upper-level burners to the lower levels or from central burners to burners located in the periphery. These types of arrangements create a fuel-rich zone in the lower level and peripheral burners and fuel-lean zone in the upper and central burners. As a result, the flame temperature is reduced, minimizing the formation of NO_x, and at the same time, improving the balance of oxygen concentration in the furnace.

NO_x formation is inherently minimized in furnaces with large cross-sections. Furnaces that are designed for high heat release rates are susceptible to high NO_x emissions, since an air-in leakage can quickly counter weigh low NO_x gain. The fundamental concept of low-temperature fluidized bed firing yields low NO_x emissions. (For additional information refer to Chapter 5, *Fluidized Bed Combustion Boilers*.)

The disadvantages of combustion modification method are:

i. It requires constant operator attention or a high degree of automatic control
ii. Any increase in excess air will lead to drop in unit efficiency
iii. In reducing combustion zones, instead of SO_2, H_2S may form, which is highly corrosive at furnace temperatures

14.5.3 Steam or water injection

In oil/gas-fired combustors flame quenching is performed through the use of water or steam injection, resulting in lower peak flame temperatures. This technique is mainly adopted in gas turbines. The added benefit of adopting this technique is a boost in power output as a result of greater mass flow. This technique, however, demands ultra-pure water, since even a minute presence of alkali in the combustor may destruct the gas turbine operating at an inlet temperature of 1423 K or above.

14.5.4 Selective catalytic reduction (SCR)

In the SCR process ammonia vapor is used as a reducing agent. This vapor is injected into the flue-gas stream in the presence of a catalyst, thereby converting NO_x into H_2O and N. In the absence of a catalyst these reactions take place at temperatures between 1143 and 1478 K, but in the presence of a suitable catalyst these reactions may take place at much lower temperatures of 573–673 K, normally prevalent at the economizer outlet. Common catalytic materials are aluminium oxide (Al_2O_3), vanadium pentoxide (V_2O_5), titanium dioxide (TiO_2), tungsten trioxide (WO_3), silicon dioxide (SiO_2), zeolites (e.g., alumina silicates), iron oxides, activated carbon, etc. In coal-fired power plants vanadium pentoxide in combination with titanium dioxide is mostly used as a catalyst. SCR system can accomplish over an 80% reduction of NO_X from flue gases [14].

In coal-fired power plants, the SCR system is placed in three typical locations: between ESP and FGD in low dust configuration (Figure 14.18) or between economizer and air heater in a high dust system, especially with dry bottom boilers (Figure 14.19), or after the FGD (Figure 14.20), primarily with wet bottom boilers with ash recirculation to avoid catalyst degradation.

14.5.5 Selective non-catalytic reduction (SNCR)

In the SNCR process a reagent, i.e., urea, ammonium hydroxide, anhydrous ammonia, or aqueous ammonia, is injected into flue gases in the furnace within the appropriate temperature zone, typically in the range of 1173–1373 K [14]. The NO_x and the reagent (urea, etc.) react to form N_2 and H_2O and do not require a catalyst. The process is relatively simple, but highly sensitive to ammonia slip. In the event of any ammonia slip it will react with SO_3 in flue gas to form ammonium bisulfate, which

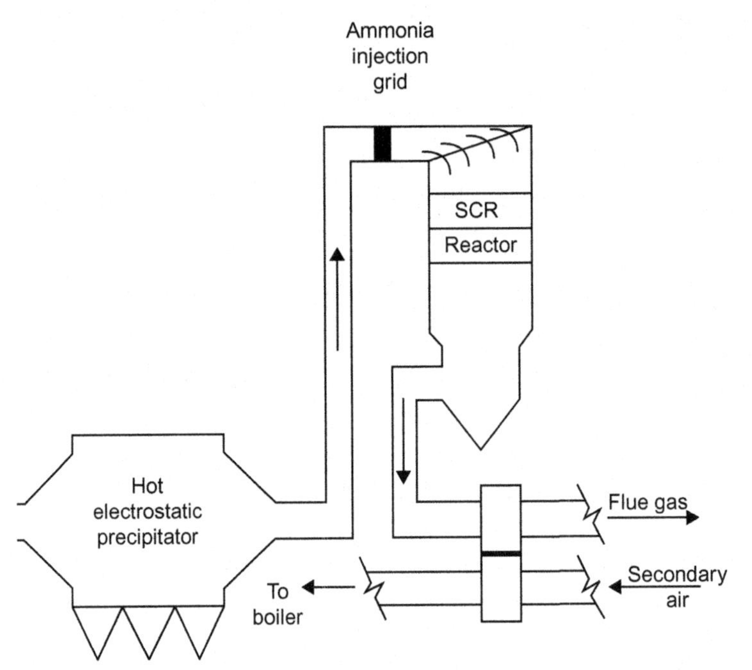

Figure 14.18 Location of SCR between ESP and FGD. *Source: http://www.epa.gov/ttncatc1/dir1/cs4-2ch2.pdf.*

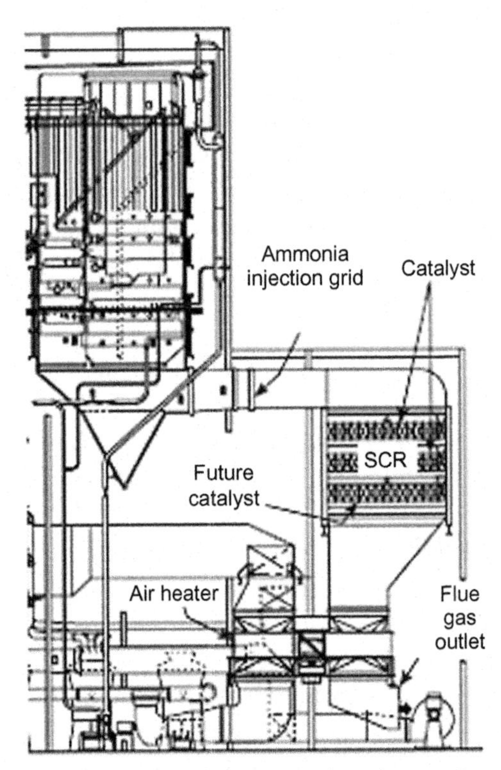

Figure 14.19 Location of SCR between economizer and air heater. *Source: FIG. 5, P 34-4, CH 34: Nitrogen Oxides Control [5].*

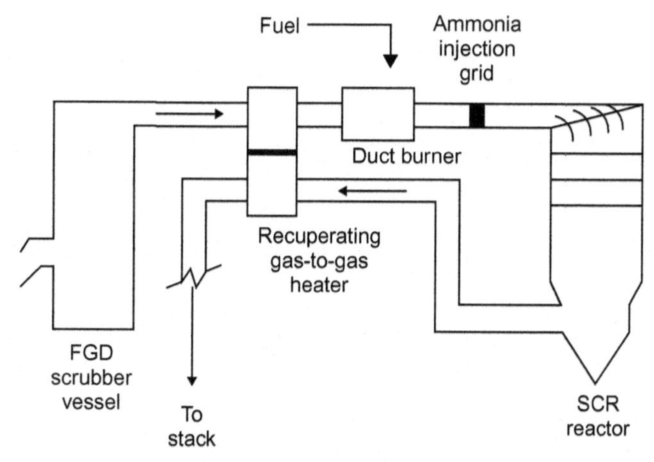

Figure 14.20 Location of SCR after FGD. *Source: http://www.epa.gov/ttncatc1/dir1/cs4-2ch2.pdf.*

would tend to precipitate at a temperature generally prevailing at air heater and lead to fouling and plugging of air heater. This process can continuously achieve 40–70% NO_x reduction. Generally, it is difficult to use SNCR in larger fossil-fired furnaces and in gas turbines, but it is widely applied to fluidized-bed-boilers in cases where further reduction of NO_x from flue gases is needed.

SNCR is an ideal retrofit technology and is compatible with other techniques (i.e., low- NO_x burners, OFA, FGR, gas reburning, etc.). Along with combustion modification, SNCR facilitates even higher reduction of NO_x.

The major disadvantage of the SNCR process is that it has a tendency to produce nitrous oxide (N_2O), which propagates the greenhouse effect.

14.6 REDUCTION OF MERCURY EMISSION

Mercury emission from coal-fired power plants to the environment is a pollutant of concern, since it is toxic and accumulates in the food chain, even though mercury is contained in coal as a trace element. Mercury is emitted from coal-fired power plants in three different forms: mercury oxides, elemental mercury, and mercury in particulate form. Mercury oxides are soluble in water, and thus could be captured in wet scrubbers, and the particulate mercury gets separated from flue gases through the collecting process in ESPs or bag filters. It is the elemental mercury that may escape to the atmosphere through the stack, but could be mitigated by injecting sorbents such as "powdered activated carbon" (PAC) [15].

From this discussion it is evident that conventional techniques that control other air pollutants, discussed above, in combination (Figure 14.21) can reduce emission of mercury from coal-fired power plants to safe airborne limit.

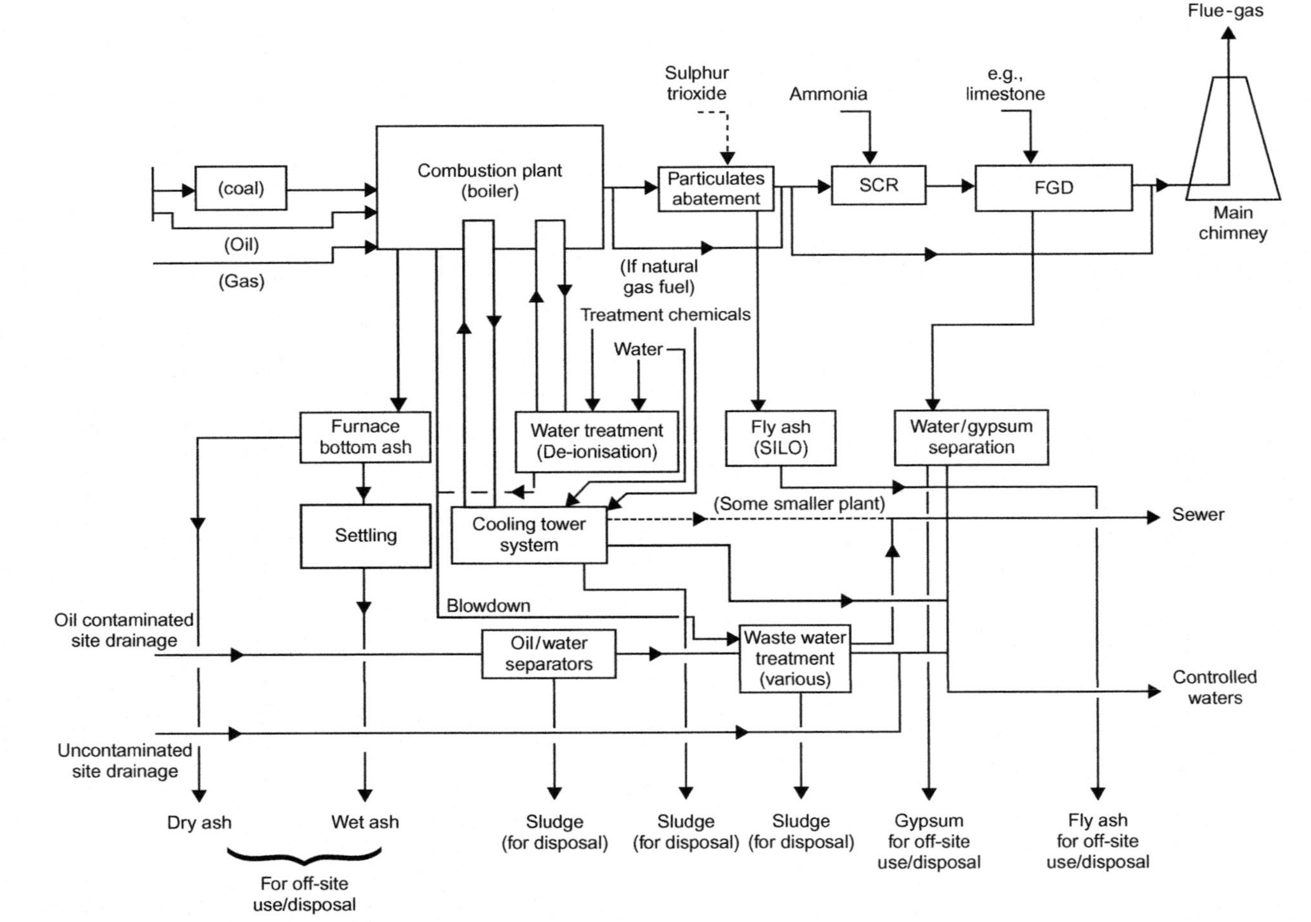

Figure 14.21 General flow diagram of a boiler with typical emission control devices. *Source: P30. Annex. A: General Description of Industry Activities [3].*

The United States was the first country where federal rule was put in place to reduce emission of "inherent mercury of coal" from new and existing coal-fired power plants.

14.7 PROBLEMS

14.1 In a coal-fired plant ultimate analysis of various elements of a sample of coal burned by weight was C = 44.2%, H = 3.4%, O = 7.5%, N = 1.5%, S = 0.4%, M = 8.0%, and A = 35%. Considering 8000 hrs of operation per annum determine the reduction in CO_2 emission from a 600 MW plant if the plant-operating configuration changes from sub-critical to supercritical. Given:

i. Efficiency of sub-critical unit = 38.62%
ii. Efficiency of supercritical unit = 41.61%
iii. Economic life of the plant = 30 years
iv. HHV of coal = 18.63 MJ/kg

(Note: Table 3.4 shows 1 kg of C produces 3.664 kg of CO_2)

(**Ans.:** 8.36×10^6 ton)

14.2 Products of combustion from a pulverized coal-fired boiler are traveling through an ESP at a temperature of 413 K. If the potential difference between the discharge electrodes and the collecting electrodes is maintained at 70 kV and the spacing between these electrodes is 150 mm, calculate the migration velocity of dust particles flowing through the ESP having a mean diameter of particles as (i) 0.48 μm and (ii) 4.8 μm. Given: viscosity of gas at 413 K is 3.01×10^{-5} N.s/m^2.
(**Ans.:** (i) 2.64 cm/s (ii) 17.39 cm/s)

14.3 Ultimate analysis of a sample of coal was found to contain various elements by weight as: C = 42.5%, H = 2.7%, O = 8.0%, N = 0.8%, S = 0.4%, M = 7.9%, and A = 37.7%. If the maximum dust concentration downstream of ESP is 100 mg/Nm^3, find the required SCA of this ESP using the Deutsch–Anderson equation. Given:

Quantity of coal burned = 110.76 kg/s
Quantity of fly ash = 80% of total ash generated
Moisture content of air at 60% RH and 300 K = 0.013 kg//kg dry air
Excess air = 32%
Effective migration velocity = 3.5 cm/s

(**Ans.:** SCA = 182.78 m^2/m^3/s)

14.4 In a coal-fired steam generator 60.67 kg of coal is burned per second with excess air of 32%. Flue gas, containing 90% of total ash generated, passes through an ESP at a temperature of 399 K. Emission downstream of the ESP is maintained within 50 mg/Nm^3. If the volume of dry flue gas and wet flue at the ESP inlet are 413.33 m^3/s and 444.60 m^3/s, respectively, determine the inlet dust burden, the Deutsch–Anderson efficiency, SCA, and total collecting area (TCA) of this ESP considering 3.15 cm/s effective migration velocity. Ultimate analysis of coal by weight is given as:

$$C = 33.68\%, \quad H = 2.42\%, \quad O = 4.68\%, \quad N = 0.73\%, \quad S = 0.49\%, \quad M = 10.00\% \text{ and } A = 48.00\%.$$

(**Ans.:** $DC_I = 92.68\ \text{gm/Nm}^3$, $\eta = 99.95\%$, $SCA = 241.3\ \text{m}^2/\text{m}^3/\text{s}$, $TCA = 107282\ \text{m}^2$)

14.5 The TCA of an ESP is noted to be $33880\ \text{m}^2$. Find the height and length of collecting electrode per field of this ESP if the aspect ratio is 2.5, number of fields in series is 7 and number of parallel collecting electrodes is 31. (**Note**: Barring two extreme electrodes, dust particles get collected on both sides of all intermediate electrodes.)

(**Ans.:** Height = 15.03 m Length per field: 5.37 m)

BIBLIOGRAPHY

[1] Greenhouse gas. <https://en.wikipedia.org/wiki/Greenhouse_gas>.
[2] P2. Emissions from Coal-Fired Power Generation by Osamu Ito. Workshop on IEA High Efficiency, Low Emissions Technology Roadmap. New Delhi; November 29, 2011.
[3] International Finance Corporation and World Bank (IFC-WB). Environmental, Health and Safety Guidelines for Thermal Power Plants; 2008.
[4] Bozzuto C, editor. Clean combustion technologies. 5th ed. Alstom; 2009.
[5] Stultz SC, Kitto JB, editors. STEAM its generation and use. The Babcock and Wilcox Company; 2005.
[6] Gordon G, Peishakov I. Dust collection and gas cleaning. Mir Publishers; 1972.
[7] El-Wakil MM. Power plant technology. McGraw-Hill Book Company; 1984.
[8] White HJ. Industrial electrostatic precipitation. Addison-Wesley Publishing Company Ltd.; 1963.
[9] Feldman PL, Kumar KS. Research-Cottrell. Effects of Wide Plate Spacing in Electrostatic Precipitators. Paper presented at the 83rd Annual Meeting and Exhibition, Pittsburg, Pennsylvania. June 24–29, 1990.
[10] Singer JG, (ed.). Combustion-fossil power. Combustion Engineering Inc.; 1991.
[11] Assessment and Requirement of Bag Filter vis-à-vis ESP in Thermal Power Plants. Central Pollution Control Board – Ministry of Environment And Forests, India; March 2007.
[12] World Energy Council, 18th Congress, Buenos Aires; October 2001.
[13] Makansi J. Special report controlling SO_2 emissions. Power; March 1993.
[14] Makansi J. Special report reducing NOx emissions. Power; May 1993.
[15] Lehigh Energy Update. Controlling Mercury Emissions from Coal-fired Power Plants; February 2010;28(1).

CHAPTER 15

Codes and Standards for Power Plant Design and Operation

15.1 INTRODUCTION

In the most basic sense, codes are laws [1,2]. Any violation of codes, if not followed wherever applicable, can result in legal trouble. Codes are passed by legislature and/or governmental bodies, and under their authority of jurisdictions the right to use the codes is controlled. Codes lay down guidelines to ensure minimum safeguards to the public by regulating specific objectives of a particular subject. They are written by experts and professionals with knowledge and expertise in a particular field.

Let us consider building codes; most countries have their jurisdictions in these codes. Even within a country local authorities, such as municipalities, have their own building codes. They are meant to protect the health, safety, and welfare of future occupants before giving permission to build the residential and commercial buildings. Building codes are laws that tell people what types of buildings can be built in a particular area and what procedures need to be followed to build them in that particular area. These codes also specify even the smallest details for all types of buildings. Since determining such building codes is an arduous task that requires many laborious hours, once they are finalized they are considered as templates for adoption called "model building codes" for the purpose of adoption by various local authorities in a country.

The same is true of codes applicable to various fields of engineering, including thermal power plants. When designers, manufacturers, erectors, operators, and users of equipment follow these codes they can confidently and comfortably ensure the safety, security, healthiness, welfare, etc., of plant personnel along with people at large as well as equipment of thermal power plants.

Standards are technical documents, produced by recognized authorities, and they are not enforceable by law [1,2]. Standards lay down the guidelines necessary to ensure that products, procedures, or materials are of acceptable quality and fit for their purpose. They provide technical instructions for promoting the safety, reliability, productivity, and efficiency of equipment or system falling in the scope of various fields of engineering. Standards apply to everything from chalk to cheese, pin to process, tube to turbine, motor to medical equipment, and so on.

Thermal Power Plant
DOI: http://dx.doi.org/10.1016/B978-0-12-801575-9.00015-9

The objective of a standard is to educate the public regarding the standardization and unification of industrial standards, thereby contributing to the improvement of technology and the enhancement of production and efficiency. Standards can be voluntarily followed to achieve the minimum levels of quality that are respected in a particular industry. Following a standard procedure the cost of production may be curtailed or down time reduced. Standardization of various products also facilitates harnessing the benefit of interchangeability. A standardardized product is readily accepted in the market.

Codes and standards can be used repeatedly. They are dynamic documents that are continuously updated to accommodate new ideas, evolutions, inventions, and developments.

15.2 BENEFITS

The benefits of adopting a code or a standard include:

Benefit 1. If the population of a country uses efficient appliances, that country can enhance the overall efficiency of their national economy. A study a few years back revealed that if improvements in energy efficiency averted 20% of Pakistan's projected energy demand over the next 25 years, Pakistan would need 10 billion US dollars less for capital investments in power plants, transmission lines, and fuel [3].

Benefit 2. Standards also enhance consumer welfare by boosting energy efficient equipment. Following standard guidelines, the number of consumer appliance models and features available to consumers has increased worldwide with lower than expected impact on prices. The average cost of operating a new consumer appliance has dropped, in spite of enhanced features and increased sizes.

Benefit 3. Local businesses may flourish and become more profitable by making efficient equipment and improved products available to the public. Even in the global marketplace local appliances, consumer goods, heavy industrial equipment, etc., of a country would be more competitive and attractive.

Benefit 4. By adopting standard products and following standard procedures a country will be in a position to meet climate change goals and reduce urban/regional pollution. Reducing energy consumption decreases both carbon and other emissions from fossil-fueled power plants.

15.3 ISSUING ORGANIZATIONS

Standards are issued by a vast multitude of organizations. Most countries have a single standards issuing agency, e.g., the Bureau of Indian Standards in India and the British Standards Institution in the UK, but in the United States voluntary standards are largely the responsibility of industrial associations such as the ASTM (American Society for Testing and Materials).

Worldwide there are two major non-treaty standards organizations – the International Standards Organization (ISO) and the International Electrotechnical Commission (IEC). In more recent years, international standards agencies have become more important. For example, CEN (Comité Européen de Normalisation) – European Committee for Standardization – promotes voluntary technical harmonization in Europe, while the IEC promotes international standardization in electrical technology.

Practically all developed countries and many developing countries have their own national standards (Table A15.1). Over the years some of these national standards have also been accepted as international standards and are being used in many countries outside the country of origin. Country-specific national standards may have to be used at times while dealing with that specific country.

Some of the issuing organizations of commonly used codes and standards are as follows –

- **IS/BIS – Indian Standards/Bureau of Indian Standards**. BIS is engaged in formulation of Indian standards for 14 sectors, i.e., basic and production engineering, chemicals, civil engineering, electronics and information technology, electro-technical, food and agriculture, mechanical engineering, management and systems, medical equipment and hospital planning, metallurgical engineering, petroleum, coal and related products, transport engineering, textile and water resources.
- **ISO – International Organization for Standardization**. ISO is a non-governmental international standard-setting body made up of representatives from national standards bodies and was established in 1947. The mission of ISO is to promote the development of standardization and related activities in the world with a view to facilitating the international exchange of goods and services. For the world's largest developer of standards the principal activity is development of intellectual, scientific, technological, and economical standards and the ISO standards cover a wide variety of items ranging from medical equipment to shipbuilding.
- **IEC – International Electrotechnical Commission**. The IEC is the leading global organization that prepares international standards for all electrical, electronic, and related technologies. These often serve as the basis for national standardization. In order to promote international unification, all national committees express an international consensus of opinion on a particular subject and adopt the text of the IEC recommendation as their national standard.
- **ANSI – American National Standards Institute**. ANSI serves as administrator and coordinator of the United States private sector voluntary standardization system. ANSI provides a forum for the development of American national standards. ANSI supports the development and approval of national voluntary standards, develops

accreditation programs, and serves as U.S. representative to the International Standards Organization (ISO). Members include trade associations, labor unions, professional societies, standards developing organizations, private industry, academe, and government.

- **ASME – American Society of Mechanical Engineers**. ASME is one of the leading organizations in the world developing codes and standards. ASME publishes codes and standards for the engineering profession, the public, industry, and government.
- **ASTM – American Society for Testing and Materials**. ASTM is a private sector organization which provides standards that are used internationally in research and development, product testing, and quality systems. ASTM International is an international standards organization that develops and publishes voluntary technical standards for a wide range of materials, products, systems, and services. Standards are developed within committees, and new committees are formed as needed, upon request of interested members.
- **API – American Petroleum Institute**. API is a leader in the development of petroleum and petrochemical equipment and operating standards covering topics that range from drill bits to environmental protection. These embrace proven, sound engineering and operating practices and safe, interchangeable equipment and materials. Many have been incorporated into state and federal regulations and adopted by ISO for worldwide acceptance.
- **AWWA – American Water Works Association**. AWWA is an international nonprofit scientific and educational society dedicated to the improvement of water quality and supply. It is the world's largest organization of water supply professionals, representing the entire industry spectrum of treatment plant operators and managers, scientists, environmentalists, manufacturers, academicians, regulators, utilities, and others who hold genuine interest in water supply and public health. AWWA is the authoritative resource for knowledge, information, and advocacy to improve the quality and supply of water in North America and beyond.
- **NFPA – National Fire Protection Association**, United States. NFPA develops and maintains a variety of codes and standards concerning all areas of fire safety. The mission of the international non-profit NFPA is to reduce the worldwide burden of fire and other hazards on the quality of life by providing and advocating scientifically-based consensus codes and standards, research, training, and education.
- **IEEE – Institute of Electric and Electronic Engineers**. The IEEE standards cover the fields of electrical engineering, electronics, radio, and the allied branches of engineering, and the related arts and sciences.
- **EPA – Environmental Protection Agency**, United States. EPA consolidates a variety of federal research, monitoring, standard-setting, and enforcement activities to ensure environmental protection. EPA's mission is to protect human health and

to safeguard the natural environment – air, water, and land – upon which life depends. From regulating auto emissions to banning the use of DDT; from cleaning up toxic waste to protecting the ozone layer; from increasing recycling to revitalizing inner-city brownfields, EPA's achievements have resulted in cleaner air, purer water, and better protected land.

- **EPRI –Electric Power Research Institute**, United States. EPRI is an independent, non-profit center for public interest energy and environmental research. EPRI brings together members, participants, the Institute's scientists and engineers, and other leading experts to work collaboratively on solutions to the challenges of electric power. These solutions span nearly every area of electricity generation, delivery, and use, including health, safety, and environment.
- **BSI – British Standards Institution**. British Standards is the National Standards Body of the UK responsible for facilitating, drafting, publishing, and marketing standards and related information products.
- **CSA – Canadian Standards Association**. CSA is a non-profit membership-based association serving business, industry, government, and consumers in Canada and the global marketplace. As a solutions-oriented organization, CSA works in Canada and around the world to develop standards that address real needs, such as enhancing public safety and health, advancing the quality of life, helping to preserve the environment, and facilitating trade.
- **DIN – Deutsches Institut für Normung**. The German Institute for Standardization is the German national organization for standardization and is that country's ISO member body. Since 1975 it has been recognized by the German government as the national standards body and represents German interests at international and European level. One of the earliest, and surely the most well-known, is DIN 476, the standard that introduced the A4, etc., paper sizes in 1922. This was later adopted as international standard ISO 216 in 1975.
- **VDI – Verlag des Vereins Deutscher Ingenieure**. VDI the Association of German Engineers is a financially independent and politically unaffiliated, non-profit organization of engineers and natural scientists. VDI is today the largest engineering association in Western Europe. In Germany, it is recognized as the representative of engineers both within the profession and in the public arena. As the leading institution for training and technology transfer among experts, it is also a partner at the preliminary stages of the decision-making process in matters of technological policy and for all questions that engineers face in their professional or public lives.
- **GOST – Gosudartsvennye Standarty**. This is Russian National Standards. GOST refers to a set of technical standards maintained by the Euro-Asian Council for Standardization, Metrology and Certification (EASC), a regional standards organization operating under the auspices of the Commonwealth of Independent States (CIS).

GOST standards, which were the cornerstone of the standardization reform, played a crucial role in standardizing and optimizing every facet of the design, production, and distribution of goods produced in the USSR. Virtually everything--from needles to shoes and from bicycles to intercontinental ballistic missiles--was mass-produced by government-owned enterprises in compliance with applicable - GOST standard.

- **GB (Guojia Biaozhun)** – China National Standard. GB sets mandatory and recommended standards regulated by SAC (Standardization Administration of the People's Republic of China). GB standards cover many areas, like other national standards such as ANSI in the United States. Mandatory standards are prefixed "GB." Recommended standards are prefixed "GB/T" (T from Chinese language tuījiàn meaning "recommended"). A standard number follows "GB" or "GB/T."
- **SAC – Standardization Administration of the People's Republic of China.** SAC is authorized by the State Council and under the control of AQSIQ to exercise the administrative functions and carry out centralized administration for standardization in China.

15.4 COMMONLY USED CODES AND STANDARDS

Some of the commonly used codes and standards for design and operation of power plants by mechanical and power engineers are presented below.

15.4.1 ASME boiler and pressure vessel code

ASME BPVC is an internationally recognized code for the latest rules of safety governing the design, fabrication and inspection of boilers, pressure vessels, and nuclear power plant components during construction. The objective of the rules is to assure reasonably certain protection of life and property and to provide a margin for deterioration in service.

ISO 16528-1 – Boilers and Pressure Vessels—Part 1 – Performance Requirements. This specifies performance requirements for the construction of boilers and pressure vessels, to ensure the integrity of the pressure boundary. Guidance is given on safety modes together with the criteria for satisfying these.

ISO 16528-2 – Boilers and Pressure Vessels—Part 2 – Procedure Fulfilling Requirements of ISO 16528-1. This provides a mechanism and a standard format for national/regional standard issuing bodies to demonstrate that their standards fulfill performance requirements of ISO 16528-1.

NFPA 85 – Boiler and Combustion Systems Hazards Code. This code covers design, installation, operation, maintenance, and training of "single burner boilers," "multiple burner boilers," "stokers," and "atmospheric fluidized bed boilers." This code is also applicable to "pulverised fuel systems" and "fired or unfired HRSGs." It deals with strength of the structure, operation and maintenance procedures, combustion and draft control equipment, safety interlocks, alarms, trips, and other related controls that are essential to safe equipment operation.

IBR – Indian Boiler Regulations. These regulations shall apply to all Indian boilers including those working on the principles of natural circulation, forced circulation and forced flow with no fixed steam and water line.

IS 15685 – 2006 – Purchaser's Data Sheet for Power Boiler. Scope covers the minimum technical data to be supplied by a purchaser, while placing an enquiry or order to a supplier for the purchase of a utility boiler. The information given by a purchaser according to this data sheet will enable a manufacturer or supplier to assess the exact requirement of the purchaser and recommend to the most suitable type of equipment.

IS 15696 – 2006 – Supplier's Data Sheet for Power Boiler. This standard covers the drawings, documents and minimum technical parameters to be provided by the manufacturers/suppliers against enquiry or quotation of a purchaser. The information provided by the manufacturer/supplier according to this data sheet would enable the purchaser to assess his exact requirement in selecting a most suitable type of equipment.

BS EN 60045-1 – 1993 – Guide for Steam Turbines Procurement. The purpose of this standard is to enable an intending purchaser to state his requirement clearly to potential suppliers, and to make him aware of options and alternatives he may wish to consider.

IEC 60045-1 – 1991 {IS 14205 (Part 1)} – Steam Turbine Part 1 Specification. The purpose of this part is to make an intending purchaser aware of options and alternatives, which he may wish to consider, and to enable him to state his technical requirements clearly to potential suppliers.

ASME TDP–1 – Recommended Practices for the Prevention of Water Damage to Steam Turbines Used for Electric Power Generation. The recommended practices are concerned primarily with the prevention of water damage to steam turbines used for fossil fuel fired electric power generation. The practices cover design, operation, inspection, testing and maintenance, of all systems that have a potential for allowing water to enter the turbine, to prevent any unusual accumulations.

HEI – Heat Exchange Institute, Inc., United States. The HEI is acknowledged worldwide as the leading standards development organization for heat exchange and vacuum apparatus. The HEI maintains its commitment to the technical advancement of the industry it serves. Standards for Closed Feedwater Heaters, Standards and Technical Specifications for Deaerators, Standards for Power Plant Heat Exchangers, Standards for Direct Contact Barometric and Low Level Condensers, Method and Procedure for the Determination of Dissolved Oxygen. 2005 – Performance Standard for Liquid Ring Vacuum Pumps; 2006 – 10th Edition of Standards for Steam Surface Condensers; 2007 – Standards for Steam Jet Vacuum Systems

IS-1520 – Design Standard of Pumps.

BHRA (British Hydraulic Research Association) – Design of Sump in Pump House.

HI – Hydraulic Institute, Inc., United States. The objective of a standard is to educate the public regarding the standardisation and unification of industrial standards, and thereby to contribute to the improvement of technology and the enhancement of production of efficiency.

BS 806 – Ferrous Piping Systems for and in Connection with Land Boilers.

ANSI B 16.5 – Steel Pipe Flanges and Flange Fittings.

ANSI B 16.9 – Wrought Steel Butt Welding Fittings.

ANSI B 16.11 – Socket Welding Ends.

ANSI B 16.25 – Butt Welding Ends.

ANSI B 16.34 – Steel Valves – Flanged and Butt Welding Ends.

ANSI B31.1 – Standards of Power Piping. This code prescribes minimum requirements for the design, materials, fabrication, erection, test, and inspection of power and auxiliary service piping systems for electric generation stations, industrial institutional plants, central and district heating plants. The code covers boiler external piping for power boilers and high temperature, high pressure water boilers.

ANSI B31.2 – Standards of Fuel Gas Piping. This is a good reference for the design of gas piping systems (from the meter to the appliance).

ANSI B 36.10 – Welding and Seamless Wrought Steel Pipe.

API 600 – Steel Gate Valves (Flanged and Butt Welding Ends).

MSS-SP-58 – Materials, Design and Fabrication of Hangers and Supports.

MSS-SP-89 – Fabrication and Installation of Steel Hangers.

AWWA-C-504 – Standard for Rubber Seated Butterfly Valves.

IS-780 and IS-2906 – Specification for Sluice Valves for Water Works.

IS-8154 / IS-8183 / IS-9842 – Standard for Thermal Insulation.

TAC – Tariff Advisory Committee (for approval of Fire Fighting System), India. TAC controls and regulates the rates, advantages, terms and conditions that may be offered by insurers in respect of General Insurance Business relating to Fire, Marine (Hull), Motor, Engineering and Workmen Compensation.

NFPA 850 – Fire Protection for Electric Generating Plants and High Voltage Direct Current Converter Stations. This document provides recommendations (not requirements) for fire prevention and fire protection for electric generating plants and high voltage direct current converter stations.

NFPA 30 – Flammable and Combustible Liquids Code.

NFPA 54 – National Fuel Gas Code. This code is a safety code that applies to the installation of fuel gas piping systems, fuel gas utilization equipment, and related accessories. Coverage of piping systems shall extend from the point of delivery to the connections with each gas utilization device. For other than undiluted liquefied petroleum gas systems, the point of delivery shall be considered the outlet of the service meter assembly or the outlet of the service regulator or service shutoff valve where no meter is provided.

For undiluted liquefied petroleum gas systems, the point of delivery shall be considered the outlet of the final pressure regulator, exclusive of line gas regulators, in the system.

NFPA 70 – National Electrical Code. The National Electrical Code (NEC), or NFPA 70, is a standard for the safe installation of electrical wiring and equipment. It is part of the National Fire Codes series published by the National Fire Protection Association (NFPA). The NEC codifies the requirements for safe electrical installations into a single, standardized source.

NFPA 37 – Standard for the Installation and Use of Stationery Combustion Engines and Gas Turbines. This standard establishes criteria to minimise the hazards of fire and provides minimum fire safety requirements during the installation and operation of stationery combustion engines and gas turbines.

NFPA 11 – Standard for Low Expansion Foam/Combined Agent Systems.

NFPA 11A – Standard for Medium and High Expansion Foam Systems.

NFPA 11C – Standard for Mobile Foam Apparatus.

NFPA 13 – Standard for the Installation of Sprinkler Systems.

NFPA 16 – Standard on Deluge Foam-Water Sprinkler and Foam-Water Spray Systems.

NFPA 16A – Recommended Practice for the Installation of Closed-Head Foam-Water Sprinkler Systems.

NFPA 18 – Standard on Wetting Agents.

NFPA 2001 – Standard on Clean Air Fire Extinguishing Systems. This standard provides the use and guidance for purchasing, designing, installing, testing, inspecting, approving, listing, operating, and maintaining engineered or pre-engineered total flooding clean agent fire extinguishing systems, so that such equipment will function as intended throughout its life. It does not cover fire extinguishing systems that use carbon dioxide or water as the primary extinguishing media. There are many agents included in this standard. However, the most common agents used for total flooding systems in normally occupied areas are HFC-227ea (FM-200), IG-541 (Inergen), and HFC-23 (FE-13).

NFPA 20 – Standard for the Installation of Centrifugal Fire Pumps.

NFPA 22 – Standard for Water Tanks for Private Fire Protection.

NFPA 24 – Standard for the Installation of Private Fire Service Mains and Their Appurtenances.

NFPA 72E – Standard on Automatic Fire Detectors.

NFPA 214 – Standard on Water-Cooling Towers. This standard applies to fire protection for field-erected and factory-assembled water-cooling towers of combustible construction or those in which the fill is of combustible material.

BS 4485-PART 2 – Water cooling towers. Methods for performance testing.

BS 4485-PART 3 – Water cooling towers. Code of practice for thermal and functional design.

BS 4485-PART 4 – Water cooling towers. Code of Practice for Structural Design and Construction.

CTI Code ATC 105 – Acceptance Test Code for Water Cooling Towers. The purpose of this code is to describe instrumentations and procedures for the testing and performance evaluation including thermal capability of water-cooling towers.

ASME PTC 23 – Water Cooling Tower – Performance Test Codes.

BS EN 12952 – Acceptance Tests on Stationery Steam Generators of the Power Station Type.

DIN 1942 – Acceptance Testing of Steam Generators.

15.4.2 ASME PTC 4 – Fired Steam Generators – Performance Test Codes

This code establishes procedures for conducting performance tests of fuel fired steam generators. It contains rules and procedures for the conduct and reporting of steam generator testing, including mandatory requirements for pretest arrangements, instruments to be employed, their application and methods of measurement, testing techniques, and methods of calculation of test results.

ASME PTC 4.2 – Coal Pulverizers – Performance Test Codes.

ASME PTC 4.3 – Air Heaters – Performance Test Codes.

ASME PTC 4.4 – Gas Turbine Heat Recovery Steam Generators – Performance Test Codes.

IS – 10391 – Code of Practice for Chemical Cleaning of Boilers. The primary reasons for chemical cleaning of boilers are to prevent tube failures and improve unit availability. Chemical cleaning removes mill scale and serves to remove atmospheric rust, which inevitably accumulates to some degree during erection.

IS – 10392 – Specification for Feed Water and Boiler Water for Low and Medium Pressure Land Boilers. Boilers require good quality water for their safe and efficient operation. Natural water accumulates impurities rendering it unfit for use in boilers without treatment. Treated water also increases safety of boilers during operation.

ANSI/ASME PTC 38 – Determining the Concentration of Particulate Matter in a Gas Stream. This code prescribes methods for measuring the average weight concentration of particulate matter carried in a gas stream confined in a duct or stack. Data thus obtained may be utilized for such purposes as the performance evaluation of gas cleaning apparatus/systems, compliance testing in conjunction with emission control and performance regulations promulgated by regulatory agencies.

COINDS 17 – Emissions Regulations Part One–Standards for Stack Emissions (Central Pollution Control Board (CPCB), India).

COINDS 20 – Emissions Regulations Part Three–Standard methods for Sampling and Analysis of Air Pollutants (CPCB, India).

BS 752 – Test Code for Acceptance of Steam Turbines.

BS 5968 – Methods of Acceptance Testing of Industrial Type Steam Turbines.

DIN 1943 – Acceptance Testing of Steam Turbines.

IEC 953 – 1 {IS 14198 (Part 1)} – Rules for Steam Turbine Thermal Acceptance Tests – Part 1 Method A – High Accuracy for large condensing Steam Turbines.

IEC 953 – 2 {IS 14198 (Part 2)} – Rules for Steam Turbine Thermal Acceptance Tests – Part 2 Method B – Wide Range of Accuracy for Various Types and Sizes of Turbines.

ASME PTC 6 – Steam Turbines – Performance Test Codes. This code provides procedures for the accurate testing of steam turbines. It contains rules and procedures for the conduct and reporting of steam turbine testing, including mandatory requirements for pretest arrangements, instruments to be employed, their application and methods of measurement, testing techniques, and methods of calculation of test results. It also contains procedures and techniques required to determine enthalpy values within the moisture region and modifications necessary to permit testing within the restrictions of radiological safety requirements in nuclear plants.

ASME PTC 6A – Appendix A to Test Codes for Steam Turbines.

ASME PTC 6 Report – Guidance for Evaluation of Measurement Uncertainty in Performance Tests of Steam Turbines.

ASME PTC 6S Report – Procedures for Routine Performance Tests of Steam Turbines.

ASME PTC 6.2 – Steam Turbines in Combined Cycles – Performance Tests Codes.

IEC 1064 – Acceptance Tests for Steam Turbine Speed Control System.

ASME PTC 12.1 – Closed Feedwater Heaters – Performance Test Codes.

ASME PTC 12.2 – Steam-Condensing Apparatus – Performance Test Codes.

ASME PTC 12.3 – Deaerators – Performance Test Codes.

API - 616 – Combustion Gas Turbines for General Refinery Services.

ASME PTC 22 – Gas Turbine Power Plants – Performance Test Codes.

ASME PTC 30 – Air Cooled Heat Exchangers – Performance Test Codes.

ASME PTC PM – Performance Monitoring Guidelines for Steam Power Plants.

These guidelines establish procedures for monitoring steam cycle performance parameters in a routine, ongoing and practical manner and constitute a set of non-mandatory guidelines to promote performance monitoring activities. They provide procedures for validation and interpretation of data, determination of performance characteristics and trends, determination of sources of performance problems, analysis of the performance in relation to the process, determination of losses due to degradation, possible corrective actions, and performance optimization.

ISO 7919 – 1 – Mechanical Vibration of Non-Reciprocating Machines – Measurements on Rotating Shafts and Evaluation- Part 1 – General Guidelines.

IEC 2372 (IS 11724) – Basis for Specifying the Rules in Evaluating the Mechanical Vibration of Machines in the Operation Range of 10 to 200Rev/S. Basis for Specifying Evaluation Standards.

BS 4675 – Mechanical Vibration in Rotating and Reciprocating Machinery Part 1 – Basis for Specifying Evaluation Standards for Rotating Machines with Operating Speeds from 10 to 200 Revolutions Per Second.

VDI 2056 – Vibration of Rotor.

VDI 2060 – Static and Dynamic Balancing of Rotor.

IEC 651 (IS 9779) – Sound Level Meters.

BS 5969 – Specification for Sound Level Meters.

ISO 562 – Hard Coal and Coke – Determination of Volatile Matter.

ISO 1988 – Hard Coal – Sampling.

IS 5456 – 2006 – Testing Positive Displacement Type Air Compressors and Exhausters – Code of Practice

IS-3177 and IS-807 – Standard for Cranes, Hoist, etc.

SIS-055900/ SSPC-VIS-1.67 / DIN 55928 (Part 4) / BS 4232 / IS-1477 (Part I) – Surface Preparation for Painting.

SSPC-PA-I / DIN 55928 (Part 4) / IS-1477 (Part II) – Application of Painting.

Govt. of India's Extra Ordinary Gazette Notification (dated 31.12.1993) and Schedule-I – General Standards for Discharge of Environmental Pollutants Part-A – Liquid Effluents and Standards for Discharge of Liquid Effluents from Thermal Power Plants.

TEMA – Tubular Exchanger Manufacturer Association.

NEMA – Motors and Generators Standard.

IS 803 – Code of Practice for Design, Fabrication and Erection of Vertical Mild Steel Cylindrical Welded Oil Storage Tank

Regulations of Ministry of Environment and Forests (MOEF), India, with Amendments to date:

- Water (Prevention and Control of Pollution) Act 1974
- Water (Prevention and Control of Pollution) Cess Act 1977
- Air (Prevention and Control of Pollution) Act 1981
- Environment (Protection) Act 1986
- Environment Protection Rules 1986
- Public Liability Insurance Act 1991
- Hazardous Wastes (Management and Handling) Rules 1989
- Manufacture Storage and Import of Hazardous Chemical Rules 1989
- Noise Pollution (Control and Regulation) Rules 1999
- Notification on Use of Ash Generated in Thermal Power Plants 1999

BIBLIOGRAPHY

[1] Introduction to ASME Codes and Standards.
[2] Geren Ronald L. Codes vs. Standards, RLGA Technical Services No. 3, October 2004.
[3] Asian Development Bank (ADB) Pakistan: Sustainable Energy Efficiency Development Program, August 2009.

Table A15.1 Standards organizations of nations

Name of country	Short title	Name of issuing organization
Algeria	IANOR	Institut Algerien de Normalisation
Argentina	IRAM	Instituto Argentino de Normalizacion
Armenia	SARM	National Institute of Standards and Quality
Australia	SAI	Standards Australia International
Austria	ONDRM	Austrian Standards Institute
Bangladesh	BSTI	Bangladesh Standards and Testing Institution
Belarus	BELST	Committee for Standardization, Metrology and Certification of Belarus
Belgium	IBN	Belgian Institution for Standardization
Bolivia	IBNORCA	Instituto Boliviano de Normalizacien y Calidad
Bosnia and Herzegovina	BASMP	Institute for Standards, Metrology and Intellectual Property of Bosnia and Herzegovina
Brazil	ABNT	Associao Brasileira de Normas Tecnicas
Brunei Darussalam	CPRU	Construction Planning and Research Unit, Ministry of Development
Bulgaria	BDS	Bulgarian Institute for Standardization
Canada	SCC	Standards Council of Canada
Chile	INN	Instituto Nacional de Normalizacion
China	SAC	Standardization Administration of China
China	CSSN	China Standards Information Center
Colombia	ICONTEC	Instituto Colombiano de Normas Tecnicas y Certificacion
Costa Rica	INTECO	Instituto de Normas Tecnicas de Costa Rica
Croatia	DZNM	State Office for Standardization and Metrology
Cuba	NC	Oficina Nacional de Normalizacion
Czech Republic	CSNI	Czech Standards Institute
Denmark	DS	Dansk Standard
Ecuador	INEN	Instituto Ecuatoriano de Normalizacion
Egypt	EO	Egyptian Organization for Standardization and Quality Control

(*Continued*)

Table A15.1 (Continued)

Name of country	Short title	Name of issuing organization
El Salvador	CONACYT	Consejo Nacional de Ciencia y Tecnologia
Estonia	EVS	Eesti Standardikeskus
Ethiopia	QSAE	Quality and Standards Authority of Ethiopia
Finland	SFS	Finnish Standards Association
France	AFNOR	Association Francaise de Normalisation – The French National Standards Agency
Germany	DIN	Deutsches Institut für Normung
Greece	ELOT	Hellenic Organization for Standardization
Grenada	GDBS	Grenada Bureau of Standards
Guatemala	COGUANOR	Comision Guatemalteca de Normas
Guyana	GNBS	Guyana National Bureau of Standards
Hong Kong	ITCHKSAR	Innovation and Technology Commission
Hungary	MSZT	Magyar Szabvnygyi Testlet
Iceland	IST	Icelandic Council for Standardization
India	BIS	Bureau of Indian Standards
Indonesia	BSN	Badan Standardisasi Nasional
Iran	ISIRI	Institute of Standards and Industrial Research of Iran
Ireland	NSAI	National Standards Authority of Ireland
Israel	SII	The Standards Institution of Israel
Italy	UNI	Ente Nazionale Italiano di Unificazione
Jamaica	JBS	Bureau of Standards, Jamaica
Japan	JISC	Japan Industrial Standards Committee
Jordan	JISM	Jordan Institution for Standards and Metrology
Kazakhstan	KAZMEMST	Committee for Standardization, Metrology and Certification
Kenya	KEBS	Kenya Bureau of Standards
Republic of Korea	KATS	Korean Agency for Technology and Standards
Kuwait	KOWSMD	Public Authority for Industry, Standards and Industrial Services Affairs
Kyrgyzstan	KYRGYZST	State Inspection for Standardization and Metrology
Latvia	LVS	Latvian Standard
Lebanon	LIBNOR	Lebanese Standards Institution
Lithuania	LST	Lithuanian Standards Board
Luxembourg	SEE	Service de l'Energie de l'Etat, Organisme Luxembourgeois de Normalisation
Malaysia	DSM	Department of Standards Malaysia
Malta	MSA	Malta Standards Authority

(*Continued*)

Table A15.1 (Continued)

Name of country	Short title	Name of issuing organization
Mauritius	MSB	Mauritius Standards Bureau
Mexico	DGN	Direccion General de Normas
Moldova	MOLDST	Department of Standardization and Metrology
Morocco	SNIMA	Service de Normalisation Industrielle Marocaine
Netherlands	NEN	Nederlandse Norm, maintained by the Nederlands Normalisatie Instituut (NNI)
New Zealand	SNZ	Standards New Zealand
Nicaragua	DTNM	Direccion de Tecnologia, Normalizacion y Metrologia
Nigeria	SON	Standards Organisation of Nigeria
Norway	NSF	Norges Standardiseringsforbund
Oman	DGSM	Directorate General for Specifications and Measurements
Pakistan	PSQCA	Pakistan Standards and Quality Control Authority
Palestine	PSI	Palestine Standards Institution
Panama	COPANIT	Comision Panameoa de Normas Industriales y Tecnicas
Papua New Guinea	NISIT	National Institute of Standards and Industrial Technology
Peru	INDECOPI	Instituto Nacional de Defensa de la Competencia y de la Proteccion de la Propiedad Intelectual
Philippines	BPS	Bureau of Product Standards
Poland	PKN	Polish Committee for Standardization
Portugal	IPQ	Instituto Portuguis da Qualidade
Romania	ASRO	Asociatia de Standardizare din Romania
Russian Federation	GOST-R	State Committee of the Russian Federation for Standardization, Metrology and Certification
Saudi Arabia	SASO	Saudi Arabian Standards Organization
Seychelles	SBS	Seychelles Bureau of Standards
Serbia and Montenegro	ISSM	Institution for Standardization of Serbia and Montenegro
Singapore	SPRING SG	Standards, Productivity and Innovation Board
Slovakia	SUTN	Slovak Standards Institute
Slovenia	SIST	Slovenian Institute for Standardization
South Africa	SABS	South African Bureau of Standards
Spain	AENOR	Asociacion Espanola de Normalizacion y Certificacion

(Continued)

Table A15.1 (Continued)

Name of country	Short title	Name of issuing organization
Sri Lanka	SLSI	Sri Lanka Standards Institution
Sweden	SIS/SEN/SMS	Swedish Standards Institute
Switzerland	SNV	(Schweizerische Normenvereinigung) – Swiss Standards.
Syrian Arab Republic	SASMO	Syrian Arab Organization for Standardization and Metrology
Taiwan	BSMI	Bureau of Standards, Metrology and Inspection
Tanzania	TBS	Tanzania Bureau of Standards
Thailand	TISI	Thai Industrial Standards Institute
Turkey	TSE	Trk Standardlari Enstits
Uganda	UNBS	Uganda National Bureau of Standards
Ukraine	DSSU	State Committee on Technical Regulation and Consumer Policy of Ukraine
United Kingdom	BSI	British Standards Institute
Uruguay	UNIT	Instituto Uruguayo de Normas Tecnicas
USA	ANSI	American National Standards Institute
USA	NIST	National Institute of Standards and Technology
Venezuela	FONDONORMA	Fondo para la Normalizacion y Certificacion de la Calidad
Vietnam	TCVN	Directorate for Standards and Quality

APPENDIX A

Power from Renewable Energy

A.1 INTRODUCTION

Renewable energy resources are those that come from natural energy flows of the earth such as solar, wind, geothermal, ocean thermal, tides, waves, hydraulic, agricultural residue, firewood, plant growth, animal dung, municipal wastes, and even domestic garbage. They naturally occur, recur, sometimes periodically, do not deplete, are almost inexhaustible, replenished constantly, and are free for the taking [1]. They are also called alternate sources of energy or non-conventional sources of energy.

By and large, renewable energy sources, with the exception of biomass, are clean, almost never causing environmental pollution be it thermal, chemical, dust bearing, toxic wastes, or obnoxious gaseous constituents, and sometimes are even noise free. Thus, they are environmentally attractive for replacing conventional fuels, i.e., coal, fuel oil, natural gas, etc., in the area of power generation along with other areas such as fuel in motor vehicles. One particular advantage of utilizing renewable energy in the power sector is that it can substantially reduce greenhouse gas emissions.

As of 2010 about 17% of total worldwide energy consumption is sourced from renewable energy (Figure A.1) [2]. In the power sector, however, renewable energy plays a much bigger role. In 2013 generation of about 22% of worldwide electricity was provided by renewable energy [3]. This trend is encouraging against the backdrop of already depleting fossil fuels, along with ever-increasing rise in fossil fuel prices. The growth of power produced from renewable energy sources up to 2011 is shown in Figure A.2 [2].

While conventional energy sources are found in only a limited number of countries worldwide, renewable energy resources exist in a wide range of areas spreading across many countries globally. If the full potential of renewable energy available in any country could be harnessed it help reduce dependence on costly imported conventional fuels. Thus, renewable energy sources could easily be used in rural and remote areas for the generation of electricity, instead of using energy from conventional fuels. In 2011 Mr. Ban Ki-moon, the Secretary General of the United Nations, said "renewable energy has the ability to lift the poorest nations to new levels of prosperity [4]."

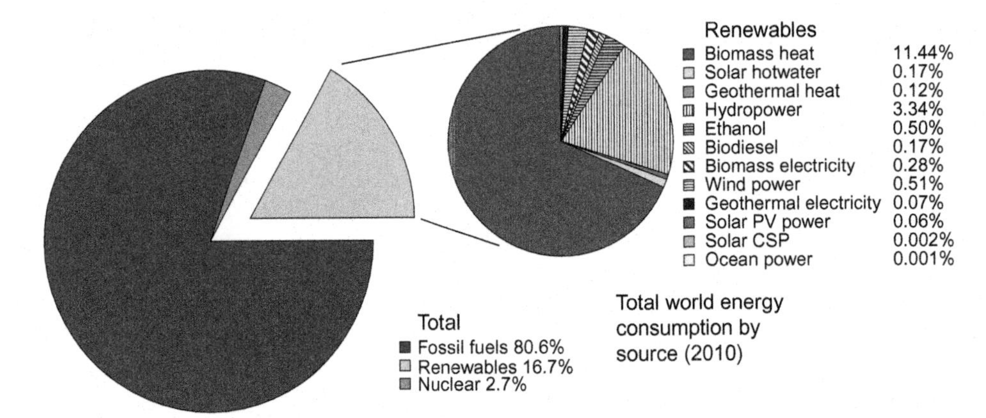

Figure A.1 Sources of worldwide energy consumption [2].

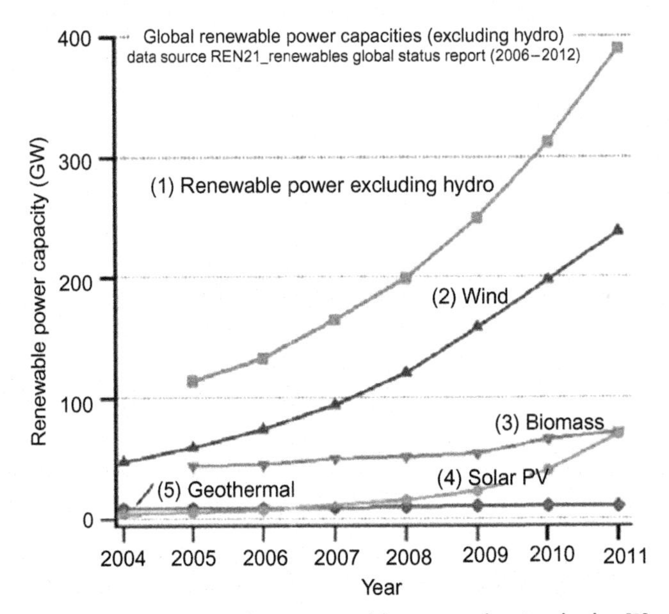

Figure A.2 Growth of a power capacity from renewable energy, barring hydro [2].

A.2 SOLAR ENERGY

Every moment the total quantity of solar energy incident upon the earth is immense. If the full potential of this incident energy could be harnessed effectively it would satisfy worldwide energy needs. Notwithstanding such potential, the main drawback of solar energy is it is very diffusing, cyclic, at times undependable, and it doesn't work at night without a battery-type storage device. It also suffers from atmospheric interference from clouds, particulate matter, gases, etc. To serve a community using

Figure A.3 A view of solar collectors [2].

solar energy a vast area of land will be needed to generate enough energy. As a result solar technologies are very expensive [1].

Solar radiation may either be converted into thermal or photovoltaic energy to generate electricity. In the thermal type, solar thermal collectors (Figure A.3) [2], which are mirrors, gather the sun's energy (which spreads over a large area) and focus this distributed energy into a small beam. This high-energy beam is then directed onto receivers containing the working fluid of a thermodynamic cycle to produce mechanical work and electricity.

Photovoltaic systems consist of direct-conversion devices in the form of cells made of semiconductor materials. As sunlight hits these cells they convert the solar radiant-energy light photons into electric current using the photoelectric effect and generate electricity.

In 2011, the International Energy Agency (IEA) stated that "the development of affordable, inexhaustible and clean solar energy technologies will have huge longer-term benefits. It will increase countries' energy security through reliance on an indigenous, inexhaustible and mostly import-independent resource, enhance sustainability, reduce pollution, lower the costs of mitigating climate change, and keep fossil fuel prices lower than otherwise [5]."

A.3 WIND ENERGY

Wind power may be considered as a modified solar power, since wind is the effect of solar heat [1]. As the air temperature increases, hot air starts rising from the earth's surface causing a drop in the atmospheric pressure near the surface. This drop in

Figure A.4 A view of wind mills [2].

pressure pulls cold air to the evacuated area. Since the earth is warmer in the daytime and cooler at night the air above the earth's surface warms and cools at different rates, causing ripples in air above the earth's surface, which is called "wind."

Although solar radiation is received across the total surface of the earth, it gets absorbed in or reflected from surfaces bearing sand, stone, water, etc. Thus, winds are stronger and more consistent in offshore and high altitude sites than those at land. Experience reveals that the average wind speed offshore is about 90% greater than on land. This movement of air or airflow may be used to run wind mills or wind turbines (Figure A.4) [2]. The power output of a wind mill varies with the wind speed in a proportion cube of the wind speed [1]. Thus, a small increase in wind speed causes a sharp rise in power output.

Wind power is pollution-free, i.e., it doesn't cause air, soil, or water pollution. Its source of energy is free and it is virtually inexhaustible in producing electricity. However, wind power is only suitable in areas with generally steady winds. The first windmill to drive an electric generator was built in the 19th century.

A.4 GEOTHERMAL ENERGY

Geothermal energy is the natural heat from the core of the earth. This energy may be sourced from shallow ground or the upper 3 m deep of the earth's surface as hot water at a temperature ranging from 283 K to 289 K to heat and cool buildings. Geothermal energy may be in the form of steam from hot rock found thousands of meters beneath the earth's surface. The heat has manifested itself for thousands of years. To access this vast source of heat, cold water is injected down one well, circulating it through hot fractured rock, and then hot water is drawn from another well. Thus, the energy is clean and sustainable.

When geothermal energy is spouted as natural steam from the earth it is used to generate electricity in similar way as the energy used in conventional steam power plants having turbo-generators. Electricity is also generated by using the heat from the hot water to boil water in a separate heat exchanger generating steam. Geothermal power is cost effective, reliable, sustainable, and environmentally friendly.

The most powerful geothermal energy may also be sourced from an even deeper area into the core of the earth's surface having extremely high temperatures of molten rock called magma. Unfortunately, the available technology is unable to recover this high-temperature energy directly from magma.

As of 2015 the United States is the highest producer of geothermal power in the world with an installed capacity of 3548 MW, which is 28% of worldwide geothermal power capacity [6].

There are two types of geothermal power plants: dry steam (Figure A.5) and flash steam types (Figure A.6) [6].

In dry steam type hot water on reaching the surface is vaporized into almost dry steam at a pressure of about 0.8 MPa and temperature of about 423 K–488 K. Flash steam stations tap high-pressure hot water from the interior of the earth. When hot water reaches earth's surface pressure drops to about 0.8 MPa resulting in flashed steam at a temperature ranging from 447 K–588 K.

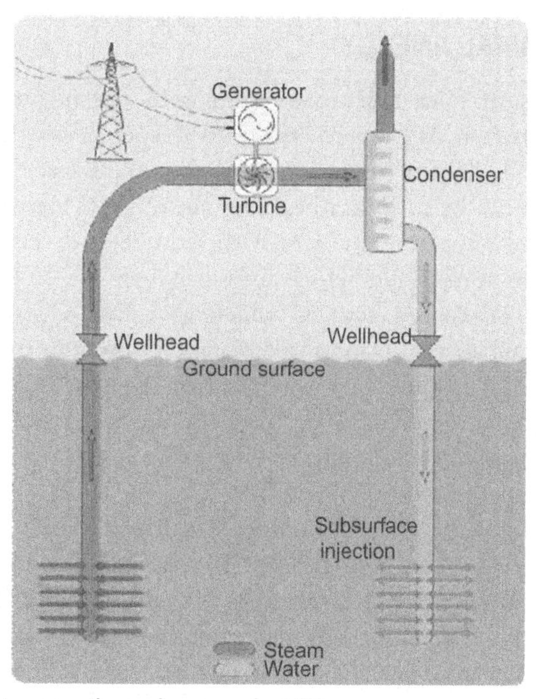

Figure A.5 Dry steam type geothermal power plant [6].

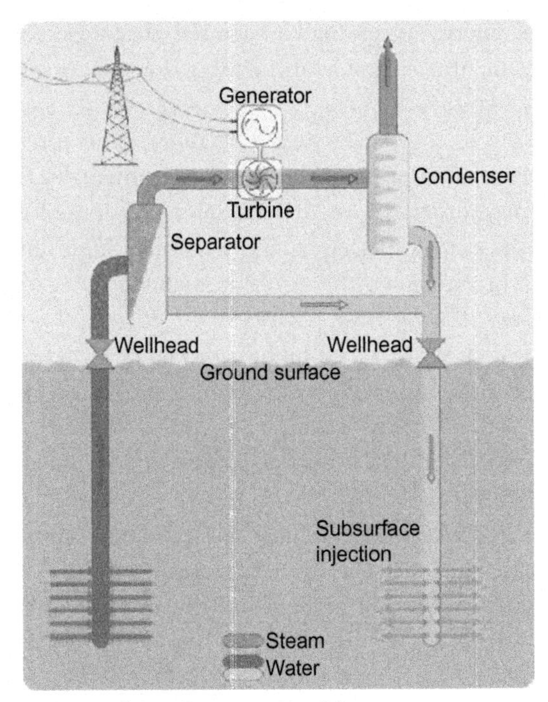

Figure A.6 Flash steam type geothermal power plant [6].

A.5 OCEAN THERMAL ENERGY

Seas and oceans absorb solar radiation, resulting in ocean currents and moderate temperature gradients from the water surface to deeper below. Since the surface of water receives direct sunlight it is warmer, but below the surface, it is very cold. This temperature gradient can be used in a heat engine to make energy to generate power [7]. Since the surface temperature varies with latitude and season the temperature difference between the surface and below varies between 15–22 K [1] (Figure A.7).

Using this type of energy source, which is called Ocean Thermal Energy Conversion, or OTEC, a large quantum of electrical power in the range of gigawatts could be produced. But to build a power plant of this size large diameter of intake pipe submerged to ocean depth of a kilometer or more would be required. Cost of building such plant would be economically prohibitive.

There are three types of OTEC plants:

Closed Cycle: In a closed cycle OTEC plant pressurized warm surface water is passed through a heat exchanger where heat transfer takes place to vaporize a low boiling point fluid, such as ammonia (Table A.1). This vapor is then expanded through a turbine to generate electricity. Thereafter, the vapor is condensed into liquid in a separate heat exchanger by cold water from deep below the surface for recycling (Figure A.8).

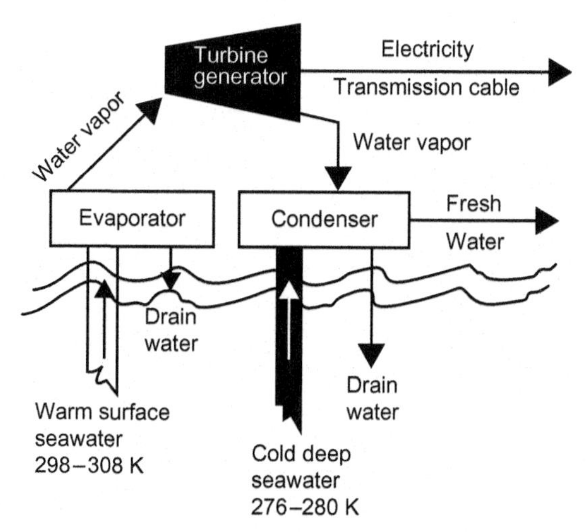

Figure A.7 Typical ocean thermal energy power plant.

Table A.1 Properties of ammonia

Boiling Point at 101.3 kPa:	239.5 K
Critical Point:	11.28 MPa & 405.4 K
Latent Heat of Vaporization at 101.3 kPa:	1371.2 kJ/kg
Vapor Pressure at 294 K:	888 kPa
Liquid Density at Boiling Point:	682 kg/m^3
Gas Density at Boiling Point:	0.86 kg/m^3
Gas Density at 101.3 kPa and 288 K:	0.73 kg/m^3
Gas Specific Gravity at 101.3 kPa and 294 K:	0.597
Gas Specific Volume at 101.3 kPa and 294 K:	1.411 m^3/kg
Gas Auto-ignition Temperature:	903 K

Open Cycle: In an open cycle OTEC plant pressurized warm surface water is supplied to a low-pressure closed container, where water flashes into steam. This steam then operates a low-pressure turbine to generate electricity. The exhaust steam is condensed by supplying low-temperature water from deep below the surface.

Hybrid: A hybrid OTEC plant is a combination of the above two plants.

A.6 TIDAL ENERGY

Ocean tides, caused by lunar and solar gravitational attractions, result in the rise and fall of waters with ranges that vary daily and seasonally [7]. During high tide water is stored in reservoirs behind dams. During low tide this water is released and the potential energy of such tidal waves is utilized generate power in the same way as in a

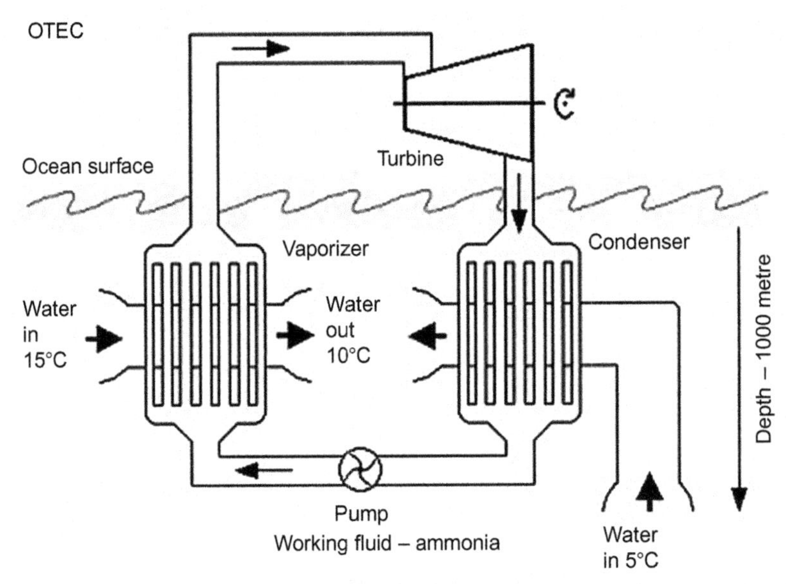

Figure A.8 Closed cycle OTEC power plant. *Source: http://www.orionsarm.com/eg-article/48571957d8a5b.*

conventional hydel power plant. The main drawback of this type of plant is that it essentially requires large underwater turbines to generate power (Figure A.9). As a result, the cost of this plant is prohibitively high compared to a river dam hydel power plant.

A.7 WAVE ENERGY

When wind blows over the surface of oceans large waves with high kinetic energy are generated. This energy is used in turbines to generate power [7]. The working principle of operating such a plant is explained as follows (Figure A.10):

i. The incoming wave as it rises inside a chamber forces the air out the chamber and
ii. As the air is driven out it rotates a turbine to generate electricity.

This type of plant can only be located where wave power has high potential.

A.8 HYDRAULIC TURBINE

A hydraulic turbine is a machine that converts the potential energy of supplied water into mechanical energy of a rotating shaft, which in turn drives a generator to produce electricity. This type of turbine operates on either the impulse or reaction principle [8]. Hydraulic turbines can be used for either vertical shaft or horizontal shaft.

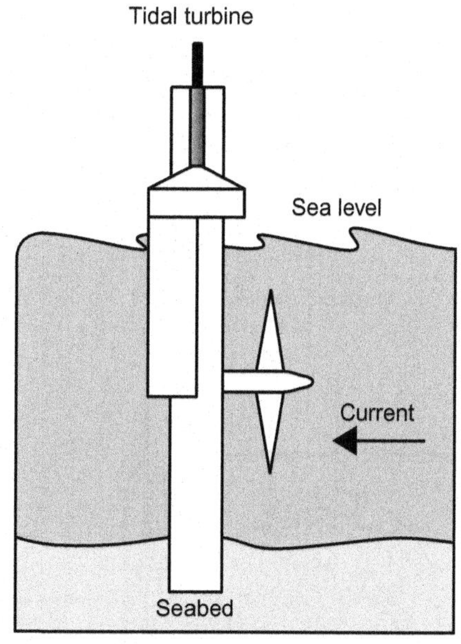

Figure A.9 Typical tidal turbine.

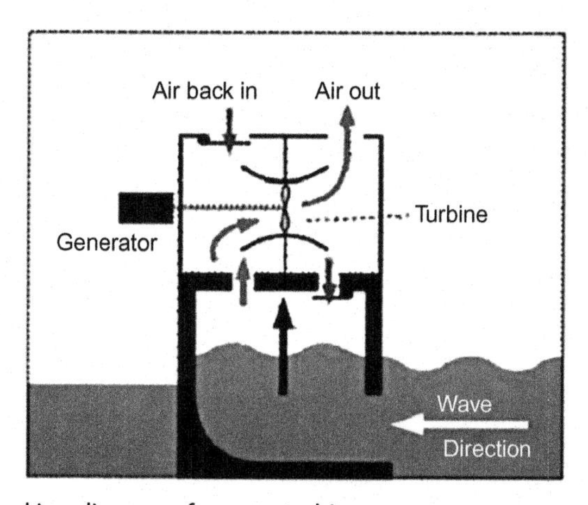

Figure A.10 Typical working diagram of a wave turbine.

Based on the operational requirements along with head of water available this turbine may be classified into three categories [8]:

Pelton Wheel: Pelton Wheel, invented by Lester Allen Pelton in 1880, is an example of an axial-flow impulse turbine where a fall in pressure of liquid takes place only in the nozzles of the machine, but remains unchanged while flowing through the rotor. It is of relatively low (specific) speed (between 11 and 27 rpm) [8], suitable for high heads (100−1700 m), low flow applications, and receives water supply directly from the

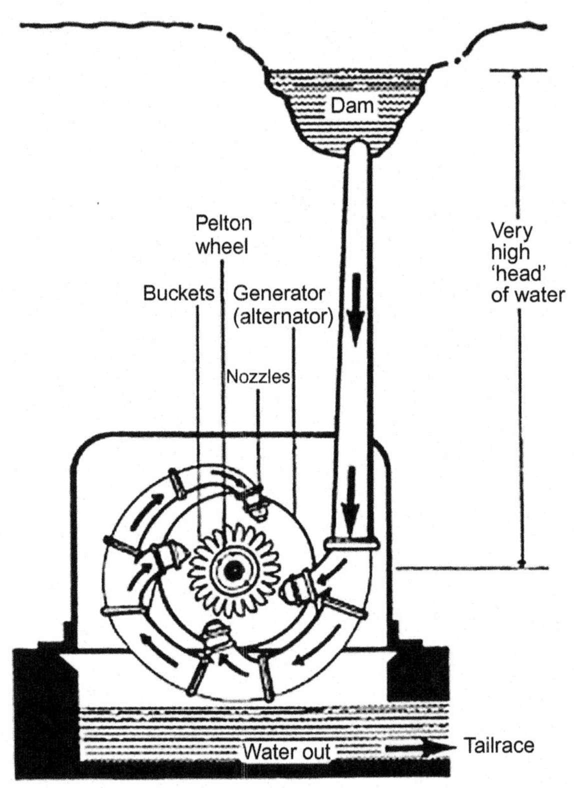

Figure A.11 Layout of a pelton wheel turbine. *Source: http://www.see.murdoch.edu.au/resources/info/Tech/hydro/large.html.*

pipeline. Pelton wheels generally receive water from reservoirs with downward gradients up to 2,000 meters (Figure A.11).

The wheel is built in such a way that on the periphery of its runner spoon-shaped buckets or cups are fitted. After passing through nozzles the water strikes these buckets, which in turn results in rotating the shaft to produce mechanical energy. This energy is then converted to electric power.

Francis Turbine: This is an example of an inward radial-flow reaction turbine and is of relatively medium (specific) speed (between 38 and 380 rpm), suitable for medium heads (between 9 and 185 m) [8]. In Francis Turbine high pressure water enters the turbine with radial inflow and leaves the turbine axially through the draft tube (Figure A.12). The high-pressure water while passing through guide vanes, located at the periphery of the turbine shaft, rotate the shaft for producing power. The speed of the shaft is maintained constant at various loads by changing the vane angle.

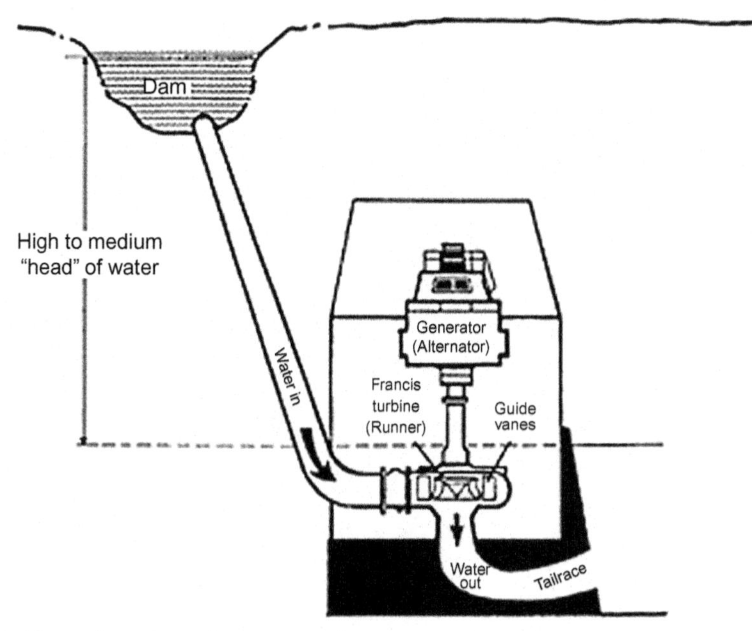

Figure A.12 Water flow through Francis turbine. *Source: http://www.see.murdoch.edu.au/resources/info/Tech/hydro/large.html.*

Kaplan Turbine: This type of turbine is an axial-flow reaction turbine also called as a propeller type. It is of relatively high (specific) speed (between 305 and 535 rpm) and suitable for low heads (varying from 3 to 15 m), and thus it is essential to pass large flow rates of water through a Kaplan turbine to produce power [8]. At the inlet of the turbine a ring of fixed guide vanes is fitted, where a passage is provided between the guide vanes and the rotor. The rotor is fitted with variable pitch blades, similar to a propeller, the angle of which can be changed in accordance with the load demand and maintaining constant speed. Incoming water enters the passage in the radial direction and is forced to exit in the axial direction, which in turn rotates the shaft to produce power (Figure A.13).

[NOTE: The specific speed of a turbine is defined as the speed of a geometrically similar turbine that would develop 1kW under 1m head.

$$\text{Specific speed, rpm } N_S = (NP^{\frac{1}{2}})/H^{\frac{5}{4}}$$

where

N = the normal working speed, rpm
P = power output of the turbine, kW and
H = the net or effective head in meters]

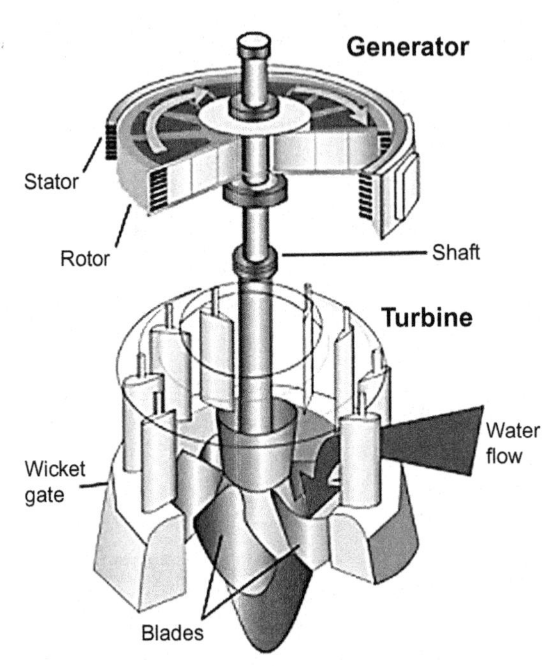

Figure A.13 Sectional view of a Kaplan turbine. *Source: en.wikipedia.org/wiki/Water_turbine.*

A.9 ENERGY FROM BIOMASS

Biomass is a source of renewable energy and is abundantly available. It is an organic matter, which has been used as long as humans have been burning wood to make fire. Biomass is produced from solar energy by photosynthesis. As such, it absorbs the same amount of carbon when growing as it releases when consumed as a fuel. Therefore, it does not add to carbon dioxide in the atmosphere and eventually mitigates further increase of greenhouse gas emission. In contrast, combustion of fossil fuels only enhances greenhouse gas emission without any recovery.

Biomass is derived from numerous sources, including plant material, byproducts from timber industry, bagasse, agricultural crops, and urban garbage (Figure A.14). Animal and human waste and decomposed garbage waste may be used to produce methane for combustion.

Biomass can regrow over a relatively short period of time. It is an important source of energy and the most important fuel worldwide after coal, oil, and natural gas. It can be used directly through combustion in a furnace to boil water, and steam thus produced is used in a conventional turbo-generator to produce electricity. Methane produced from biomass can be used directly in combustion turbines to generate electricity. Biomass can also be used indirectly by converting it to ethanol, which is a

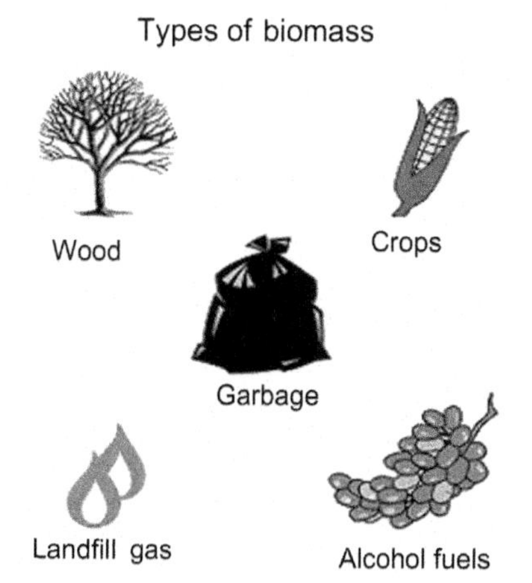

Figure A.14 Some example of biomass.

biofuel, and using it for special cars and motor vehicles, which may eventually reduce our dependence on gasoline for driving cars.

Some of the biomass used to generate bio-power comprises the following:

- **i.** Agricultural wastes, i.e., wheat straw, rice husk, jute stick, etc.
- **ii.** Energy crops, i.e., bagasse, bamboo, special type of grass, e.g. switchgrass, etc.
- **iii.** Wood and forest residues, i.e., dry leaves, twigs, etc.
- **iv.** Wood wastes, i.e., sawdust, wood shavings
- **v.** Clean industrial and municipal wastes
- **vi.** Cattle dung, poultry litter, etc.

BIBLIOGRAPHY

[1] El-Wakil MM, Power Plant Technology. McGraw-Hill Book Company; 1984.
[2] Renewable Energy, <Wikipedia.en.wikipedia.org/> wiki Renewable_energy; 2014.
[3] International Energy Agency. Renewable energy, <http://www.iea.org/aboutus/faqs/renewable-energy/>; 2014.
[4] Steve L. U.N. Secretary-General: Renewables Can End Energy Poverty. Renewable Energy World; 2011.
[5] Solar Energy Perspectives: The last paragraph of Executive Summary. <https://www.iea.org/Textbase/npsum/solar2011SUM.pdf>.
[6] Geothermal Electricity, <en.wikipedia.org/wiki/Geothermal_electricity>; 2015.
[7] ELECTRONIC BUS. <electronicsbus.com/ocean-energy-tidal-wavepower-generation-osmotic-power-deep-ocean-thermal-energy-harvesting/>.
[8] Lewitt EH. Hydraulics and Fluid Mechanics. English Language Book Society. 10th Edition; 1963.

APPENDIX B

Power from Nuclear Energy

B.1 INTRODUCTION

Compared with the energy-liberating potential of fossil fuels, energy liberated from nuclear fuel is astronomically high. While only 1 kg of U^{235} is capable of liberating about 85×10^6 MJ of heat by going through nuclear fission, a quantity of about 5×10^6 kg of coal with a high heating value of 17 MJ/kg has to be burnt to get an equivalent amount of energy.

There are two different methods of energy generation by a nuclear fuel, i.e., either by nuclear fission or by nuclear fusion. When heavy fissile elements are subjected to nuclear fission in a nuclear reactor, chain reactions of these elements take place, releasing a huge quantum of energy. In nuclear fusion, simple atomic nuclei are fused together to form complex nuclei (as in case of fusion of hydrogen isotopes to form helium) and in the process liberates a large quantum of energy. The process of nuclear fusion, also known as thermonuclear reaction, is very difficult to control even today. Therefore, at present, the main source of nuclear energy is only available from nuclear fission.

The most common fissile radioactive heavy metals are naturally occurring isotope of uranium, U^{235}, artificial isotope of uranium, U^{233}, and artificial element plutonium, P^{239}. In a nuclear reactor a controlled chain reaction of nuclear fission of these heavy elements takes place. The nuclear energy thus liberated is converted into heat that is removed from the reactor by a coolant, e.g., liquid sodium, pressurized water, etc. Hot coolant is then passed through another heat exchanger, an unfired steam generator, where water is circulated as a coolant, which absorbs heat and results in generation of steam. Steam thus generated is used in a conventional turbo-generator to produce electricity. On one hand, this steam generator is eco-friendly, since it does not emit any carbon dioxide, sulfur, or mercury. But on the other hand, a major concern of a nuclear power plant is that the area surrounding the nuclear reactor is potentially radioactive. Further, nuclear wastes, if not disposed of carefully, could have a devastating effect on living beings and inanimate objects including the environment.

There are various types of nuclear reactors used to generate electricity. Some of these reactors, commonly used globally, are discussed in the following.

B.2 PRESSURIZED-WATER REACTOR (PWR)

A PWR power plant is composed of two loops in series, i.e., the coolant loop, called the primary loop, and the water-steam or working fluid loop, also known as the secondary loop [1].

The pressurized water in the primary loop passes over the reactor core to act as moderator and coolant. The coolant does not flow to the turbine; it picks up reactor heat and transfers it to the working fluid in the secondary loop, which is an unfired steam generator that produces steam. The steam is then used in a Rankine cycle to drive a turbine and generate electricity (Figure B.1 [2] and Figure B.2 [3]). Coolant pressure and temperature are usually maintained at about 15.5 MPa and 618 K [3], which is lower than the saturation temperature, and thus boiling of the coolant inside the reactor is avoided.

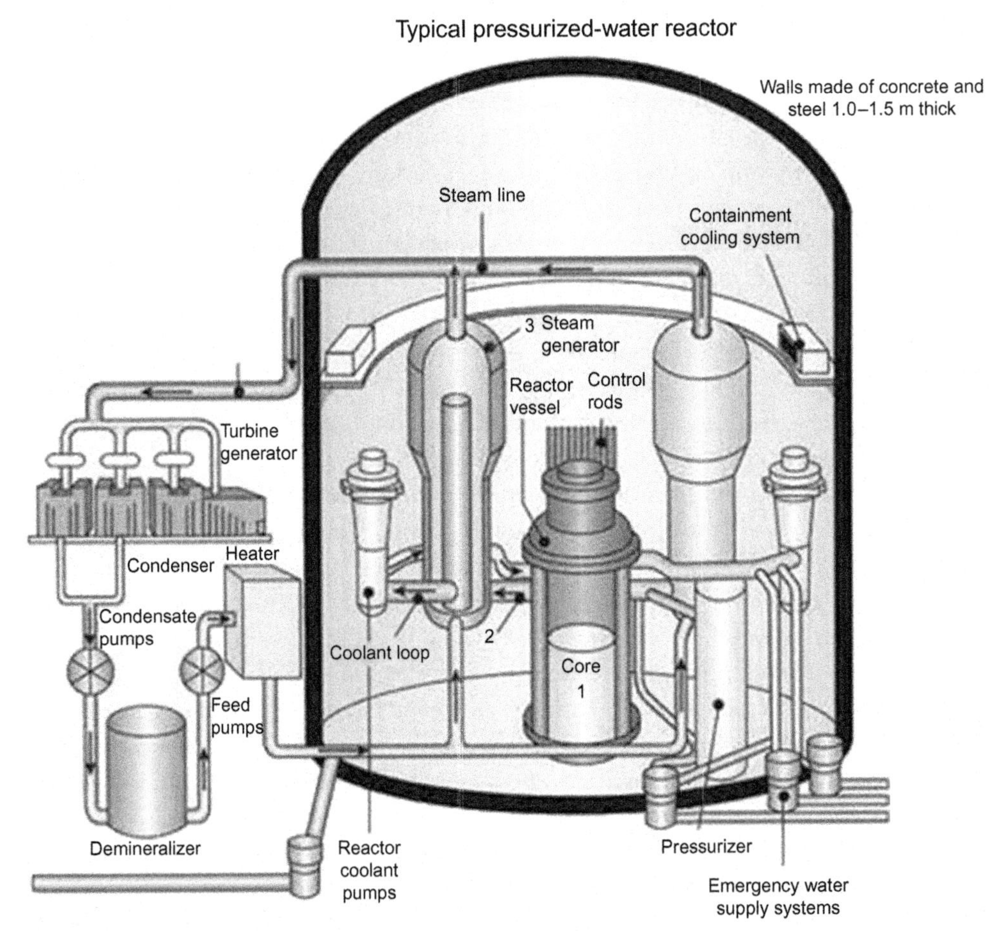

Figure B.1 Sectional view of a typical pressurized water reactor [2].

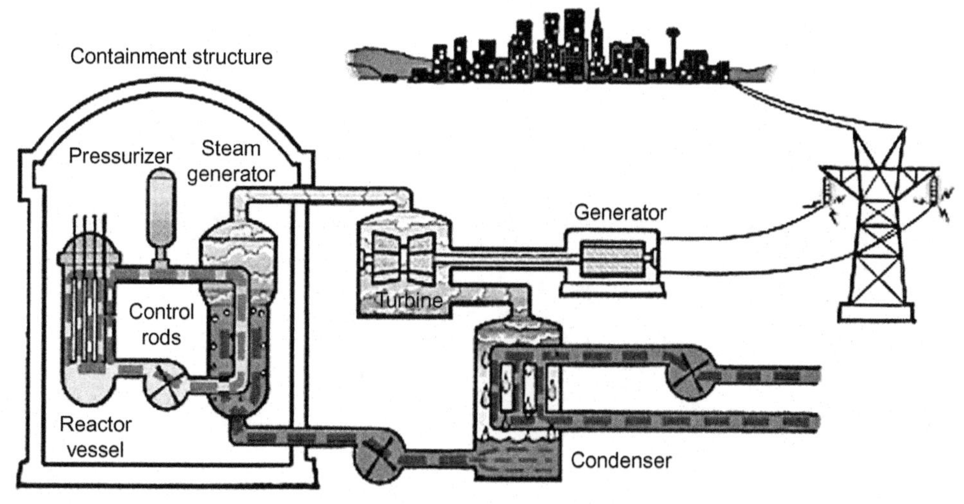

Figure B.2 Flow diagram of a pressurized water reactor [3].

In a PWR enriched uranium is used as fuel. The reactor's core generally contains 150–200 fuel assemblies [2]. During the design stage all precautions are taken to ensure that boiling of pressurized water does not take place within the reactor. The most attractive advantage of a PWR is that in the event of a fuel leak in the core radioactive contaminants would not pass to the turbine and condenser. Another advantage is that the pressure and temperature (about 10.5 MPa and 588 K, respectively) of the steam coming from the PWR is higher than the pressure and temperature available from a boiling water reactor. Disadvantage of PWR is the reactor is more complicated and more costly to construct [4].

B.3 BOILING-WATER REACTOR (BWR)

In a BWR the same water loop serves as moderator, coolant for the core, and steam source for the turbine. In a BWR, the coolant is in direct contact with the heat-producing nuclear fuel and boils in the same compartment in which the fuel is located (Figure B.3) [5]. Liquid enters the reactor core at the bottom, where it receives sensible heat to saturation plus some latent heat of vaporization. When it reaches at the top of the core, it is converted into a very wet mixture of liquid and vapor. The vapor is then separated from the liquid in a steam separator, not shown in Figure B.3, and flows through a turbine to generate power. The condensate is pumped back to the bottom of the reactor.

The main concern of adopting a BWR is any fuel leak, which might take place within the reactor, would make the water radioactive. The radioactivity is then propagated to the turbine and the rest of the loop. A typical operating pressure for a BWR is about 7 MPa, about half the coolant pressure of PWR. The temperature of steam coming out of a BWR corresponds to a saturation temperature at 7 MPa pressure, i.e., about 558 K [4].

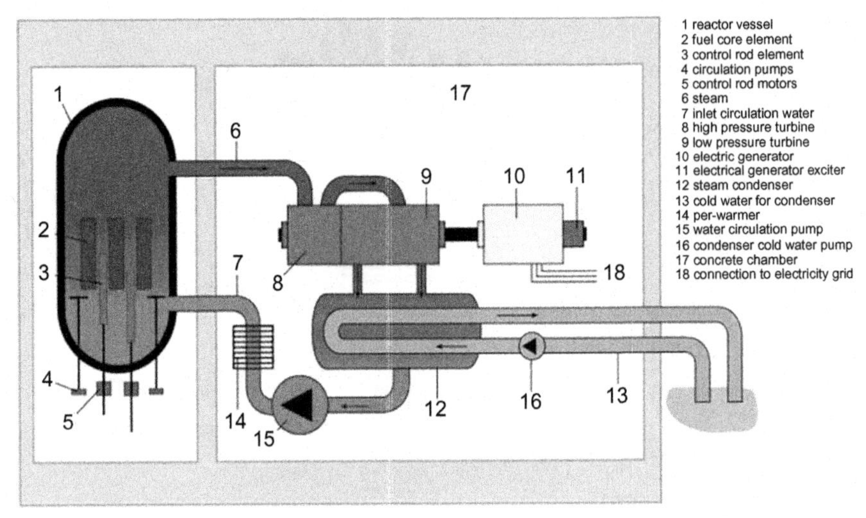

Figure B.3 Flow diagram of a boiling water reactor [5].

B.4 GAS-COOLED REACTOR (GCR)

A GCR is cooled by a gas (helium or carbon dioxide) [4], unlike the liquid coolant, which is used in a PWR and BWR. During the process of cooling the reactor, the gas absorbs heat; this hot coolant then can be used either directly as the working fluid of a combustion turbine to generate electricity or indirectly to generate steam, which in turn is used in a turbo-generator to produce electricity (Figure B.4) [6,7].

There are two types of GCR. In one type both natural and enriched uranium fuels are used with CO_2 as the coolant and graphite as the moderator. In the other type only enriched fuels are used where the coolant is helium and the moderator is heavy water.

The major advantages of adopting this technology are:

i. Higher steam temperature, typically about 838 K, could be availed with GCR. As a result, higher plant efficiency could be obtained compared to the water cooled design.

ii. GCR is less prone to land hazards that could be experienced with water cooled/ moderated reactors.

B.5 HEAVY-WATER REACTOR (HWR)

In a HWR, or more precisely a pressurized heavy-water reactor (PHWR), the fuel used is natural uranium with pressurized heavy water (deuterium oxide: D_2O) as coolant-moderator (Figure B.5) [8]. Since heavy water is pressurized it can be heated to higher temperatures without boiling, as is the case with pressurized water in a PWR.

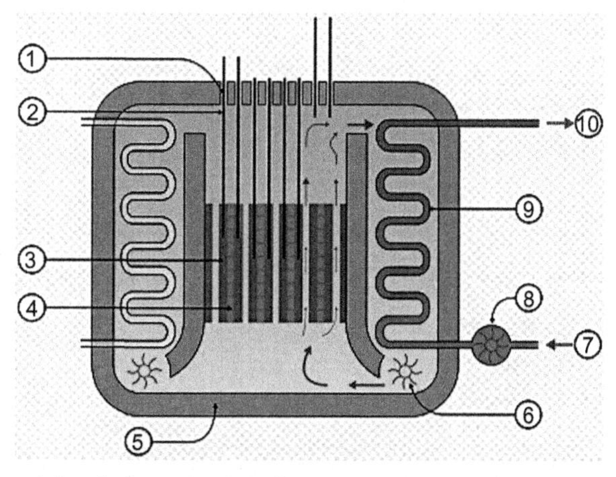

1. Charge tubes 2. Control rods 3. Graphite moderator 4. Fuel assemblies
5. Concrete pressure vessel and radiation shielding 6. Gas circulator 7. Water
8. Water circulator 9. Heat exchanger 10. Steam

Figure B.4 Sectional view of a typical gas cooled reactor [6].

A HWR requires large moderator-to-fuel ratios, thus requiring large diameter pressure-vessels. The production costs of D_2O are very high compared with the cost of H_2O. But this higher cost is traded-off to a certain extent by using natural uranium, since in the case of PWR additional cost is incurred for producing enriched uranium.

One "proprietary" version of HWR is CANDU, which stands for "Canada Deuterium Uranium" (Figure B.6) [9]. The reactor was designed by Atomic Energy of Canada Limited and is currently marketed by Mississauga, Ontario-based Candu Energy Inc. The pressure and temperature of the steam at the outlet of CANDU are typically 4 MPa and 524 K, respectively.

B.6 FAST-BREEDER REACTOR (FBR)

The fuel used in a FBR is generally P^{239}, but U^{235} is also used in certain reactors. Fast breeder reactors are so named because of their design to breed fuel by producing more fissionable fuel than they could consume. In a FBR neutrons are not slowed down to thermal energy by a moderator. Coolant and other reactor materials moderate the neutrons. Reactors are cooled by liquid sodium, which also acts as the heat-transfer medium. In lieu of sodium, lithium may also be used as the coolant, but sodium being the most abundant is used commonly [4].

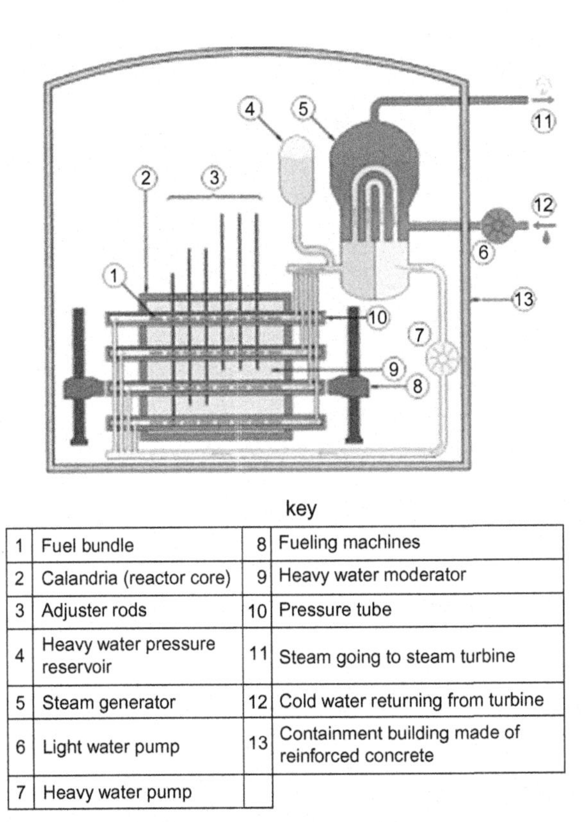

key

1	Fuel bundle	8	Fueling machines
2	Calandria (reactor core)	9	Heavy water moderator
3	Adjuster rods	10	Pressure tube
4	Heavy water pressure reservoir	11	Steam going to steam turbine
5	Steam generator	12	Cold water returning from turbine
6	Light water pump	13	Containment building made of reinforced concrete
7	Heavy water pump		

Figure B.5 Sectional view of a typical heavy water reactor [8].

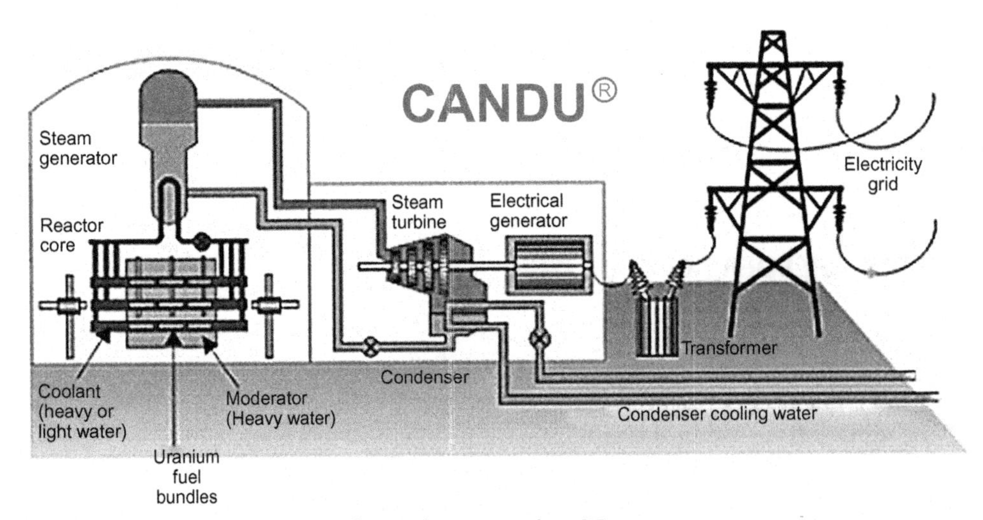

Figure B.6 General arrangement of a candu reactor plant [9].

One major disadvantage of a FBR is it requires higher degree of enrichment of the fuel than do the water moderated reactors. The advantages of using sodium compared to using pressurized water are:

i. At atmospheric pressure sodium can be liquefied at a moderate temperature of 371 K and it boils at 1156 K. Therefore, the operating temperature range of sodium is large at atmospheric pressure and it does not require further pressurization.
ii. Thermal conductivity of sodium is very large (84 $W.m^{-1}.K^{-1}$ at 298 K), as compared to that of water (0.016 $W.m^{-1}.K^{-1}$ at 398 K). Hence, it is an efficient heat transfer medium.

The disadvantages of using sodium are:

i. It is opaque, which hinders visual maintenance.
ii. Sodium is an extremely reactive chemical; when comes in contact with air or water it starts burning. On contact with water sodium may even cause explosion, thus safety is at stake.
iii. During operation, radioactive sodium-24 may be formed by neutron activation. Therefore, there is also a slight radiation hazard.

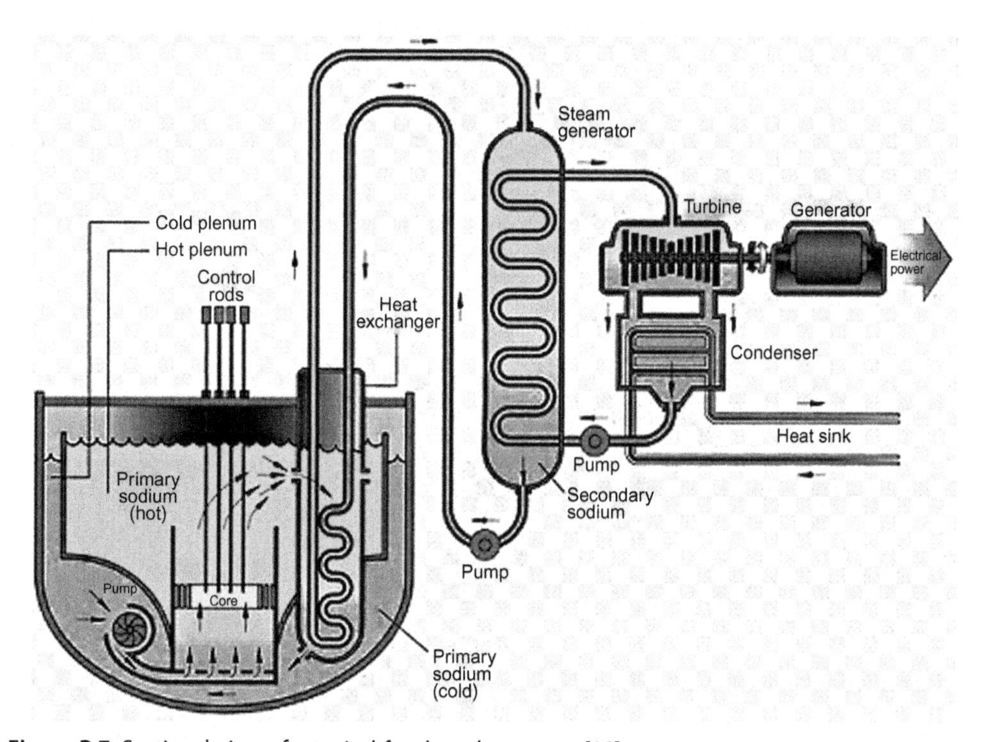

Figure B.7 Sectional view of a typical fast-breeder reactor [10].

In order to get rid of radioactive risks, a separate design has been developed, where an intermediate loop is introduced between the primary and secondary loop. This new reactor is called a liquid-metal fast-breeder reactor, or LMFBR [4]. In this reactor liquid sodium is used as a coolant in the primary loop of the reactor. Heat contained in the primary coolant is exchanged with the intermediate coolant, which is also a liquid metal, either sodium or a combination of sodium and potassium, NaK, in a primary heat exchanger. Hot intermediate coolant is then introduced into a secondary heat exchanger, i.e., the steam generator, as a source of heat to boil water and generate steam. A sectional view of a typical LMFBR is shown in Figure B.7 [10].

In a typical design steam conditions at the superheater outlet are about 10 MPa and 759 K.

BIBLIOGRAPHY

[1] Pressurized Water Reactor, <www.nucleartourist.com/type/pwr.htm>; 2009.
[2] Pressurized Water Reactors, <www.nrc.gov/reactors/pwrs.html>; 2013.
[3] Pressurized Water Reactor, <en.wikipedia.org/wiki/Pressurized_water_reactor>.
[4] hyperphysics.phy-astr.gsu.edu/hbase/nucene/reactor.html; 2014.
[5] Boiling Water Reactor, Wikipedia, the free encyclopedia. <http://en.wikipedia.org/wiki/Boiling_water_reactor>; 2014.
[6] Gas Cooled Reactors, <upload.wikimedia.org/wikipedia/commons/thumb/e/e1/AGR_reactor_schematic.svg/2000px-AGR_reactor_schematic.svg.png>.
[7] Ricky, In: Stonecypher, L. (Ed.), Gas Cooled Reactors in Nuclear Plants. <www.brighthubengineering.com/power-plants/2498-gas-cooled-reactors-in-nuclear-plants/>; 2008.
[8] Pressurized Heavy-Water Reactor (PHWR), <nuclearstreet.com/nuclearpower-plants/w/nuclear_power_plants/320.pressurized-heavy-water-reactor-phwr.aspx>.
[9] CANDU Heavy-Water Reactor, <www.nuclearfaq.ca>; 2013.
[10] Fast-Breeder Reactor, <upload.wikimedia.org/wikipedia/commons/thumb/d/d8/SodiumCooled_Fast_Reactor_Schemata.svg/1280px-Sodium-Cooled_Fast_Reactor_Schemata.svg.png>.

Appendix C

Table A.1 Conversion factors

Parameter	SI units	Metric system of units	Imperial and US system of units
Length	1 m	100 cm	3.281 ft
		1000 mm	39.37 in
			1.094 yd
	0.01 m	1 cm	0.3937 in
	0.0254 m	2.54 cm	1 in
	0.3048 m	30.48 cm	12 in
			1 ft
	0.9144 m	91.44 cm	3 ft
			1 yd
	1.828 m	1.828 m	6 ft
			1 fathom
	1000 m	1 km	0.621 miles
			0.539 nautical miles
	1609 m	1.609 km	1760 yd
			1 mile
	1853 m	1.853 km	1 nautical mile
Area	1 m^2	10000 cm^2	1550 in^2
		10^6 mm^2	10.764 ft^2
			1.196 yd^2
	0.0929 m^2	929.03 cm^2	144 in^2
			1 ft^2
	0.836 m^2	8361 cm^2	9 ft^2
			1 yd^2
	10^{-4} m^2	1 cm^2	0.155 in^2
	10^4 m^2	1 hectare	2.47 acres
	4049 m^2	0.405 hectare	1 acre
			4840 yd^2
	1 mm^2	10^{-2} cm^2	0.00155 in^2
	645.2 mm^2	6.45 cm^2	1 in^2
	10^6 m^2	1 km^2	1196836 yd^2
	2.59×10^6 m^2	259 hectares	1 $mile^2$
			640 acres

(*Continued*)

Table A.1 (Continued)

Parameter	SI units	Metric system of units	Imperial and US system of units
Volume	1×10^{-3} m^3	1000 ml	0.2200 Imp. gallon
		1000 cc	0.2642 US gallon
	3.785×10^{-3} m^3	3.785 l	0.833 Imp. gallon
			1 US gallon
	4.55×10^{-3} m^3	4.55 l	1 Imp. gallon
			1.2 US gallon
	1 m^3	1000 l	35.31 ft^3
			1.308 yd^3
	10^{-6} m^3	1 cc	0.0610 in^3
	0.02832 m^3	28.32 l	1728 in^3
			1 ft^3
	0.7646 m^3	764.6 l	27 ft^3
			1 yd^3
	0.02685 Nm^3	26850 Ncc	1 scf
	1 Nm^3	10^6 Ncc	37.244 scf
Mass	1 kg	1000 g	2.205 lb
	0.4536 kg	453.6 g	1 lb
			16 oz
			7000 gr
	64.8×10^{-6} kg	64.8 mg	1 gr
	6.350 kg	6.350 kg	14 lb
			1 stone
	1×10^{-3} kg	1 g	0.03528 oz
	0.0283 kg	28.34 g	1 oz
	100 kg	1 quintal	220.5 lb
	1000 kg	1 ton	2205 lb
			0.984 ton
	907 kg	907 kg	2000 lb
			1 short ton
	1016 kg	1016 kg	2240 lb
			1 ton
			20 cwt (hundredweight)
Density	1 m^3 kg^{-1}	1000 cc/g	16.02 ft^3/lb
	0.06243 m^3 kg^{-1}	62.43 cc/g	1 ft^3/lb
Specific weight	1 kg.m^{-3}	0.001 g/cc	0.0624 lb/ft^3
	1000 kg.m^{-3}	1 g/cc	6.24 lb/ft^3
	160.26 kg.m^{-3}	0.1602 g/cc	1 lb/ft^3
	0.022886 kg.m^{-3}	2.2886×10^{-5} g/cc	1 gr/ft^3
Pressure & stress	1 Pa	0.102 mmwg	1.45×10^{-4} psi
	1 Nm^{-2}	0.102 kg/m^2	
		7.5×10^{-3} mmHg	
	100 kPa	1 Bar (b)	14.5 psi
		1.02 kg/cm^2	
		750 mmHg	

(*Continued*)

Table A.1 (Continued)

Parameter	SI units	Metric system of units	Imperial and US system of units
	101.3 kPa	1013 mb	14.7 psi
		1.033 kg/cm^2	29.92 inHg
		760 mmHg	
	98.1 kPa	1 kg/cm^2	14.22 psi
		10 mwg	28.94 inHg
		736 mmHg	
	1 kPa	102 kg/m^2	0.145 psi
		102 mmwg	0.295 inHg
		7.503 mmHg	
	0.1333 kPa	1 mmHg	0.193 psi
	6.895 kPa	0.0703 kg/cm^2	1 psi
	9.81 Pa	1 mmwg	0.001422 psi
	249 Pa	0.00254 kg/cm^2	1 inwg
		25.4 mmwg	
	1 MPa	10 b	145 psi
	1 Nmm^{-2}	10.2 kg/cm^2	
	15.44 Nmm^{-2}	1.575 kg/mm^2	1 Ton/in^2
	9.80 Nmm^{-2}	1 kg/mm^2	0.635 Ton/in^2
Temperature	273 K	0°C	32°F/492 R
	0 K	−273°C	−460°F/0 R
	255.22 K	−17.78°C	0°F / 460 R
Heat, Power, Work & CV	1 W	0.2389 cal/s	9.478×10^{-4} Btu/s
	1 $J.s^{-1}$		
	1 kW.s	0.2389 kcal	0.948 Btu
	1 kJ	102 m.kgf	738 ft.lbf
	1 kW.hr	860 kcal	3413 Btu
	3600 kJ		
	4.186 kJ	1 kcal	3.9686 Btu
		427 m.kgf	3088 ft.lbf
	1.055 kJ	0.252 kcal	1 Btu
		107.6 m.kgf	778 ft.lbf
	9.80 J	2.342×10^{-3} kcal	9.29×10^{-3} Btu
		1 m.kgf	7.23 ft.lbf
	1 $kJ.kg^{-1}$	0.2389 kcal/kg	0.430 Btu/lb
	4.186 $kJ.kg^{-1}$	1 kcal/kg	1.8 Btu/lb
	2.326 $kJ.kg^{-1}$	0.556 kcal/kg	1 Btu/lb
	1 $kJ.m^{-3}$	0.2389 cal/l	0.02684 Btu/ft^3
	4.186 $kJ.m^{-3}$	1 cal/l	0.11235 Btu/ft^3
	37.26 $kJ.m^{-3}$	8.90 cal/l	1 Btu/ft^3
	39.302 $KJ.Nm^{-3}$	9.389 $kcal/Nm^3$	1 Btu/scf
	1 $kJ.m^{-3}.K^{-1}$	0.2389 $kcal/m^3.°C$	0.0149 $Btu/ft^3.°F$
	4.186 $kJ.m^{-3}.K^{-1}$	1 $kcal/m^3.°C$	0.06237 $Btu/ft^3.°F$
	67.116 $kJ.m^{-3}.K^{-1}$	16.03 $kcal/m^3.°C$	1 $Btu/ft^3.°F$
	1 $W.m^{-2}$	23.89×10^{-6} $cal/cm^2.s$	0.316 $Btu/ft^2.h$
	41.876 $kW.m^{-2}$	1 $cal/cm^2.s$	13233 $Btu/ft^2.h$

(Continued)

Table A.1 (Continued)

Parameter	SI units	Metric system of units	Imperial and US system of units
	3.165 W.m^{-2}	75.57 × 10^{-6} cal/cm^2.s	1 Btu/ft^2.h
	1 kJ.m^{-2}	0.2389 kcal/m^2	0.0881 Btu/ft^2
	4.186 kJ.m^{-2}	1 kcal/m^2	0.3687 Btu/ft^2
	11.35 kJ.m^{-2}	2.712 kcal/m^2	1 Btu/ft^2
	1 kJ.m^{-2}.K^{-1}	0.2389 kcal/m^2.°C	0.0489 Btu/ft^2.°F
	4.186 kJ.m^{-2}.K^{-1}	1 kcal/m^2.°C	0.2048 Btu/ft^2.°F
	20.44 kJ.m^{-2}.K^{-1}	4.883 kcal/m^2.°C	1 Btu/ft^2.°F
Specific Heat	1 kJ.kg^{-1}.K^{-1}	0.2389 kcal/kg.°C	0.2389 Btu/lb.°F
	4.186 kJ.kg^{-1}.K^{-1}	1 kcal/kg.°C	1 Btu/lb.°F
Thermal Conductivity	1 W.m^{-1}.K^{-1}	2.389 × 10^{-3} cal/cm.s.°C	0.578 Btu/ft.h.°F
	418.41 W.m^{-1}.K^{-1}	1 cal/cm.s.°C	241.84 Btu/ft.h.°F
	1.73 W.m^{-1}.K^{-1}	4.135 × 10^{-3} cal/cm.s.°C	1 Btu/ft.h.°F
Heat Transfer Coefficient	1 W.m^{-2}.K^{-1}	2.389 × 10^{-5} cal/cm^2.s.°C	0.176 Btu/ft^2.h.°F
	41858.5 W.m^{-2}.K^{-1}	1 cal/cm^2.s.°C	7367 Btu/ft^2.h.°F
	5.682 W.m^{-2}.K^{-1}	1.357 × 10^{-3} cal/cm^2.s.°C	1 Btu/ft^2.h.°F
Velocity	1 m.s^{-1}	3.6 km/h	196.86 fpm
	0.278 m.s^{-1}	1 km/h	54.68 fpm
	5.08 × 10^{-3} m.s^{-1}	0.0183 km/h	1 fpm
	1 m.s^{-1}	3.6 km/h	2.236 m/h
	0.278 m.s^{-1}	1 km/h	0.621 m/h
	0.447 m.s^{-1}	1.609 km/h	1 m/h
Flow	1 m^3.s^{-1}	3600 m^3/h	2118.6 cfm
	2.78 × 10^{-4} m^3.s^{-1}	1 m^3/h	0.589 cfm
	4.72 × 10^{-4} m^3.s^{-1}	1.698 m^3/h	1 cfm
	1 Nm^3.s^{-1}	3600 Nm^3/h	2234.64 scfm
	2.778 × 10^{-4} Nm^3.s^{-1}	1 Nm^3/h	0.6207 scfm
	4.475 × 10^{-4} Nm^3.s^{-1}	1.611 Nm^3/h	1 scfm
	6.31 × 10^{-5} m^3.s^{-1}	0.2271 m^3/h	1 US gpm
	7.59 × 10^{-5} m^3.s^{-1}	0.273 m^3/h	1 Imperial gpm
	1 kg.s^{-1}	3.6 tph	7938 lb/h
	0.278 kg.s^{-1}	1 tph	2205 lb/h
	1.26 × 10^{-4} kg.s^{-1}	4.535 × 10^{-4}tph	1 lb/h
	1 m^3.s^{-1}	3600 m^3/h	18.975 Imperial MGD
	2.78 × 10^{-4} m^3.s^{-1}	1 m^3/h	5.275 × 10^{-3} Imperial MGD
	0.0527 m^3.s^{-1}	189.58 m^3/h	1 Imperial MGD
	1 m^3.s^{-1}	3600 m^3/h	22.827 US MGD
	2.78 × 10^{-4} m^3.s^{-1}	1 m^3/h	6.341 × 10^{-3} US MGD
	0.0438 m^3.s^{-1}	157.71 m^3/h	1 US MGD
Miscellaneous	0.43 kg.GJ^{-1}	1.8 × 10^{-3} g/kcal	1 lbm/MBtu

Table A.2 Basic SI units (systeme international d'unites)

Sl. No.	Physical quantity	Unit	Symbol	Definition of unit
1	Length	meter	m	—
2	Mass	kilogramme	kg	—
3	Time	second	s	—
4	Temperature	kelvin	K	—
5	Electric Current	ampere	A	—
6	Luminous Intensity	candela	cd	—
7	Amount of Substance	mole	mol	—
8	Plane Angle	radian	rad	—
9	Solid Angle	steradian	sr	—
10	Frequency	hertz	Hz	s^{-1}
11	Energy	joule	J	$kg.m^2.s^{-2}$
12	Force	Newton	N	$J.m^{-1} = kg.m.s^{-2}$
13	Power	watt	W	$J.s^{-1} = kg.m^2.s^{-3}$
14	Pressure	pascal	Pa	$N.m^{-2} = J.m^{-3} = kg.m^{-1}.s^{-2}$
15	Electric Charge	coulomb	C	A.s
16	Electric Potential Difference	volt	V	$W.A^{-1} = J.s^{-1}.A^{-1} = kg.m^2.s^{-3}.A^{-1}$
17	Electric Resistance	ohm	—	$V.A^{-1} = W.A^{-2} = J.s^{-1}.A^{-2} = kg.m^2s^{-3}.A^{-2}$
18	Electric Capacitance	farad	F	$C.V^{-1} = kg^{-1}.m^{-2}.s^4.A^2$
19	Magnetic Flux	Weber	Wb	$V.s = W.s.A^{-1} = J.A^{-1} = kg.m^2.s^{-2}.A^{-1}$
20	Magnetic Flux Density	tesla	T	$Wb.m^{-2} = V.s.m^{-2} = W.s.A^{-1}.m^{-2} = J.A^{-1}.m^{-2} = kg.s^{-2}.A^{-1}$
21	Inductance	henry	H	$V.s.A^{-1} = W.s.A^{-2} = J.A^{-2} = kg.m^2.s^{-2}.A^{-2}$
22	Conductance	Siemens	S	$V^{-1}.A = W^{-1}.A^2 = J^{-1}.s.A^2 = kg^{-1}.m^{-2}.s^3.A^2$
23	Dynamic Viscosity	poiseuille	Pl	$N.s.m^{-2} = J.s.m^{-3} = kg.m^{-1}.s^{-1}$
24	Luminous Flux	lumen	lm	cd.sr
25	Illumination	lux	lx	$cd.sr.m^{-2}$

Table A.3 Prefixes used in SI units

Sl. No.	Prefix	Symbol	Factor
1	tera	T	10^{12}
2	giga	G	10^{9}
3	mega	M	10^{6}
4	kilo	k	10^{3}
5	hecto*	h	10^{2}
6	deca*	da	10^{1}
7	deci*	d	10^{-1}
8	centi*	c	10^{-2}
9	milli	m	10^{-3}
10	micro	μ	10^{-6}
11	nano	n	10^{-9}
12	pico	p	10^{-12}
13	femto	f	10^{-15}
14	atto	a	10^{-18}

(NOTES: 1. Prefixes marked with * should normally be avoided;
2. It is preferable to write $J.s^{-1}$, instead of J/s.)

INDEX

Note: Page numbers followed by "*f*" and "*t*" refer to figures and tables, respectively.

B

D

E

G

H

O

P

R

S

T

U

V

W

Y

Z